CRC HANDBOOK SERIES IN CLINICAL LABORATORY SCIENCE

David Seligson , Sc.D., M.D.
Editor-in-Chief

SECTIONS AND SECTION EDITORS

Each section contains one or more volumes.

CRC Handbook Series in Clinical Laboratory Science

David Seligson, Editor-In-Chief
Chairman and Professor
Department of Laboratory Medicine
School of Medicine
Yale University

Section H: Virology and Rickettsiology

Volume I

Part 1

Section Editors

Gueh-Djen Hsiung
Professor of Laboratory Medicine
Yale University
School of Medicine
Chief, Virology Laboratory
Veterans Administration Hospital
West Haven, Connecticut

Robert H. Green
Professor of Medicine
University of Connecticut School of Medicine
Clinical Professor of Medicine
Yale University School of Medicine
Chief of Medicine
Middlesex Memorial Hospital
Middletown, Connecticut

CRC PRESS, Inc.
2255 Palm Beach Lakes Blvd. · West Palm Beach, Fla. 33409

Library of Congress Cataloging in Publication Data (Revised)

Main entry under title:

CRC handbook series in clinical laboratory science.

Spine title: Handbook series in clinical laboratory science.
Includes bibliographies and index.
CONTENTS: Section A. Nuclear medicine. —Section E. Clinical microbiology.—[etc.]— Section H. Virology and rickettsiology.
1. Medical laboratories. 2. Diagnosis, Laboratory. I. Seligson, David. II. Chemical Rubber Company , Cleveland. III. Title: Handbook series in clinical laboratory science. [DNLM: 1. Virology—Period. 2. Rickettsiaceae infections—Period. 3. Rickettsial infections—Period. W1 C56SD]
RB37.C18 616.07'56 76-27688
ISBN O-8493-7000-0 (set)

This book presents data obtained from authentic and highly regarded sources. Reprinted material is quoted with permission, and sources are indicated. A wide variety of references are listed. Every reasonable effort has been made to give reliable data and information but the editor and the publisher cannot assume responsibility for the validity of all materials or for the consequences of their use.

International Standard Book Number
Complete Set 0-8493-7000-0
Virology and Rickettsiology Section, Volume I, Part 1 0-8493-7061-2

Library of Congress Card No. 76-27688

Printed in the United States

PUBLISHERS PREFACE

In 1913, when the First Edition of the *Handbook of Chemistry and Physics* appeared, scientific progress, particularly in chemistry and physics, had produced an extensive literature but its utility was seriously handicapped because it was fragmented and unorganized. The simple but invaluable contribution of the *Handbook of Chemistry and Physics* was to provide a systematic compilation of the most useful and reliable scientific data within the covers of a single volume. Referred to as the ''bible'', the Handbook soon became a universal and essential reference source for the scientific community. The latest edition published represents more than 65 years of continuous service to millions of professional scientists and students throughout the world.

In the years following World War II, scientific information expanded at an explosive rate due to the tremendous growth of research facilities and sophisticated analytical instrumentation. The single-volume Handbook concept, although providing a high level of convenience, was not adequate for the reference requirements of many of the newer scientific disciplines. Due to the sheer quantity of useful and reliable data being generated, it was no longer feasible or desirable to select only that information which could be contained in a single volume and arbitrarily to reject the remainder. **Comprehensiveness** had become as essential as **convenience.**

By the late 1960's, it was apparent that the solution to the problem was the development of the multi-volume Handbook. This answer arose out of necessity during the editorial processing of the *Handbook of Environmental Control.* A hybrid discipline or, to be more precise, an interdisciplinary field such as Environmental Science could be logically structured into major subject areas. This permitted individual volumes to be developed for each major subject. The individual volumes, published either simultaneously or by some predetermined sequence, collectively became a multi-volume Handbook series.

The logic of this new approach was irrefutable and the concept was promptly accepted by both the scientist and science librarian. It became the format of a growing number of CRC Handbook Series in fields such as Materials Science, Laboratory Animal Science, and Marine Science.

Within a few years, however, it was clear that even the multi-volume Handbook concept was not sufficient. It was necessary to create an information structure more compatible with the dynamic character of scientific information, and flexible enough to accommodate continuous but unpredictable growth, regardless of quantity or direction. This became the objective of a ''third generation'' Handbook concept.

This latest concept utilizes each major subject within an information field as a ''Section'' rather than the equivalent of a single volume. Each Section, therefore, may include as many volumes as the quantity and quality of available information will justify. The structure achieves permanent flexibility because it can, in effect, expand ''vertically'' and ''horizontally''. Any section can continue to grow (vertically) in number of volumes, and new sections can be added (horizontally) as and when required by the information field itself. A key innovation which makes this massive and complex information base almost as convenient to use as a single-volume Handbook is the utilization of computer technology to produce up-dated, cumulative index volumes.

The *Handbook Series in Clinical Laboratory Science* has the distinction of being the prototype for the "sectionalized, multi-volume Handbook series". Currently underway are additional information programs based on the same organizational design. These include information fields such as Nutrition, Energy, and Agricultural Science which are of critical importance not only to scientific progress but to the advancement of the total quality of life.

We are confident that the "third generation" CRC Handbook comprises a worthy contribution to both information science and the scientific community. We are equally certain that it does not represent the ultimate reference source. We predict that the most dramatic progress in the management of scientific information remains to be achieved.

B. J. Starkoff
President
CRC Press, Inc.

PREFACE
CRC HANDBOOK SERIES IN CLINICAL LABORATORY SCIENCE

According to the *Webster — Merriam Dictionary,* 3rd ed., a handbook is

1a. a book capable of being conveniently carried as a ready reference,
1b. a concise reference book covering a particular field of knowledge,
2a. a bookmaker's book of bets.

In order to accommodate the information in the field of clinical laboratory science we shall need more than 20 volumes. A series of volumes such as these, that will weigh in the order of 24 kg (53 pounds), will not fit the meaning of 1a: nor will 2a fit, although we are gambling that we can get the appropriate information together in a reasonable time. Definition 1b is more appropriate to our goal which is to accumulate, arrange, and present the qualitative and quantitative clinical laboratory data related to men and women (of all ages), so that anyone who needs to extract some information from the vast amount available will have it at his or her fingertips, provided he or she is within arm's length of these volumes.

To understand the biology of man, to know man, sick and well, requires knowledge of the body constituents and their interrelationships; the dynamic relationships of these constituents; and information about tools, chemicals and techniques used to get these data. Notation of the source of the data is important to scholars. It is our purpose to present this information in tabular and graphic form in a systematic way in order to help readers retrieve it.

If we are successful, we will make it possible for a user to find any information related to constituents of blood, body fluids, body tissues, and so on, whether they be organic or inorganic molecules, viruses, bacteria, or parasites. Users of these volumes should be able to find easily the composition of a buffer, the half-life of an isotope, or the life cycle of a parasite if that information has been used to study man or if the chemicals or particles are found in man.

We shall consider this *Handbook Series in Clinical Laboratory Science* successful if it meets most of our goals and if it provides the nidus for continuing growth. The editor of the series, the executive editor of each volume, the advisory boards, contributors, and publisher will make every effort to improve subsequent editions of this *Handbook.* The user will undoubtedly find errors of omission and commission or will have new information (all of which we hope will be called to our attention.) All of us should strive to improve these volumes so that users and contributors the world over will find them useful in knowing more about man.

David Seligson
Editor-in-Chief

DAVID SELIGSON, EDITOR-IN-CHIEF
CRC HANDBOOK SERIES IN CLINICAL LABORATORY SCIENCE

Dr. David Seligson is Professor and Chairman of the Department of Laboratory Medicine (Clinical Pathology) and Professor of Pathology in the School of Medicine at Yale University. He is the Director of the Department of Clinical Laboratories of the Yale-New Haven Hospital. He has an Sc.D. in Biochemistry from Johns Hopkins University (1942) and an M.D. from the University of Utah (1946). He has been a President of the American Association of Clinical Chemists and the Academy of Clinical Laboratory Physicians and Scientists. He is a Fellow of the American Society of Clinical Pathologists, College of American Pathologists, American College of Physicians, and American Board of Clinical Chemistry.

He has been editor of *Standard Methods of the American Association of Clinical Chemists* (Volumes 2,3, and 4) and he is a member of other editorial boards.

Dr. Seligson has an honorary Masters degree from Yale University and has received the following awards:

John G. Reinhold Award, Philadelphia Section of the American Association of Clinical Chemists, 1966

Fellowship of Medici Publici, University of Utah College of Medicine and Medical Alumni Association, 1966

Donald D. Van Slyke Award, American Association of Clinical Chemists 1971

Ames Award, American Association of Clinical Chemists, National Meeting, Seattle, Washington, August 1971

Gerald T. Evans Award, Academy of Clinical Laboratory Physicians and Scientists, June 1974

Honorary Member Association of Clinical Biochemists, London, April 1976

PREFACE
SECTION H: VIROLOGY AND RICKETTSIOLOGY
VOLUME I

The main purpose of this volume is to provide both the laboratory investigator and the physician with a clear, concise account of the current status of clinical virology. Descriptions of specimens to be collected and diagnostic techniques to be applied in relation to the suspected etiologic agent are provided. The bulk of the text deals with descriptions of the morphological, biochemical, and biological characteristics of the various virus families, as well as methods used for the laboratory diagnosis of viral infections. These descriptions include many standard and newly developed procedures for recognition and characterization of viruses that will be useful to laboratory workers, as well as to clinicians who are interested in the diagnosis and management of patients suspected of having viral illnesses.

Through the use of the tables, charts, and illustrations in this handbook, the reader can readily find the characteristics of each major virus group, as well as descriptions of the pathogenesis of each viral disease. The text is further broken down into the commonly recognized anatomical systems with clinical descriptions of the viral infections that predominantly affect each system.

The authors of this volume are well known in their respective fields; their descriptions of viruses and viral diseases, as well as the numerous illustrations, provide a large body of information that brings new dimensions to the clinical and diagnostic aspects of viral diseases.

The editors gratefully acknowledge the contributions of the many authors whose manuscripts comprise this volume, and who so patiently and tolerantly have deal with editorial suggestions while awaiting its publication.

Particular thanks are extended to Mrs. Dorothy Wong and Mrs. Kari Hastings for their editorial assistance, Mrs. Barbara Meek for illustrations and other art work, and Mrs. V. Mercadante for the secretarial assistance that made the publication of this book possible.

G.D.H.
R.H.G.

G. D. HSIUNG AND ROBERT H. GREEN
SECTION EDITORS
SECTION H: VIROLOGY AND RICKETTSIOLOGY
VOLUME I

G. D. Hsiung, Ph.D. is Professor of Laboratory Medicine at Yale University School of Medicine and Chief of the Virology Laboratory at the Veterans Administration Hospital in West Haven, Connecticut. Dr. Hsiung received her B.S. degree in Biology at Ginling College, China in 1942 and her doctoral degree in Microbiology and Public Health at Michigan State University in 1951. She came to Yale University School of Medicine as a research fellow in 1953 and remained as a faculty member.

Dr. Hsiung is a fellow of the American Academy of Microbiology, a member of the American Society for Microbiology, the American Association of Immunologists, the Society for Experimental Biology and Medicine, and the Tissue Culture Association. She has served as consultant to the Pan American Health Organization of the World Health Organization, the Viral and Rickettsiology Registry Committee of the American Type Culture Collection, and several other organizations. Dr. Hsiung is also active as an editor of the *Journal of Clinical Virology* and as a reviewer for the National Science Foundation and the Career Development Program of the Veterans Administration. She is the author of *Diagnostic Virology*, published by Yale University Press, and over one hundred scientific papers dealing with clinical and experimental virology. Her particular research interests are concerned with latent viruses, the oncogenic potential of viruses, and the pathogenesis of viral diseases.

Robert H. Green, M.D. is chairman of the Department of Medicine at Middlesex Memorial Hospital, Professor of Medicine at the University of Connecticut School of Medicine, and Clinical Professor of Medicine at the Yale University School of Medicine. He received his M.D. degree from the Johns Hopkins University School of Medicine, and, after serving residencies in internal medicine, obtained his early training and experience in virology under the supervision of Dr. Joseph E. Smadel in the laboratories of Dr. Thomas M. Rivers at the Rockefeller Institute. He subsequently held appointments as Professor of Medicine at New York University School of Medicine, Professor of Pathology and Medicine at the Yale University School of Medicine, Professor of Medicine and Dean of the School of Medicine at the Medical University of South Carolina, as well as various posts in the Veterans Administration. Dr. Green's research has been devoted largely to clinical and laboratory studies of viral infections, especially those caused by vaccinia, rubella, influenza, and poliovirus.

FOREWORD
SECTION H: VIROLOGY AND RICKETTSIOLOGY
VOLUME I

"Demand not that events should happen as you wish; but wish them to happen as they do happen, and you will go on well." This is from the *Enchiridion* of Epictetus, a handbook or manual of the first century which summarizes a larger work on his philosophical principles. The composition of handbooks was old in his day and has gone on steadily ever since. In scientific activity at present, the writing of handbooks appears to increase more or less in proportion to the rate of development of the new material that is rapidly being added. In the case of virology, its range has extended greatly during recent years and, consequently, we find several guidebooks available.

Even so, it is often difficult, when one seeks a significant bit of information bearing upon laboratory or clinical data, to come upon it readily and in sufficient detail to enable one to discover how the events actually happen so that work can proceed as quickly as possible. I believe that this volume of the *CRC Handbook Series in Clinical Laboratory Science* will provide for these requirements in a highly satisfactory way.

The authors of the various chapters are well versed in their subjects through vast clinical and/or laboratory experience. They are people to whom one would immediately turn when in need of a particular piece of advice or direction. Their presentations are direct and easily understood. The authors are concerned with their specialties and so are able to present the fundamentals with the latest words on procedural developments. In respect to all the writers in general, I cannot help but refer specifically to one of the Section Editors, Dr. Hsiung, whose earlier book on diagnostic virology has been a constant source of aid to us and to many others in the field, and whose contributions here will be equally or even more useful.

The over-all arrangement seems excellent from the point of clarity, with the viruses listed according to their nucleic acid content and with a final section of the diseases of various anatomical systems. This order should make the transfer from the laboratory data to the clinical attributes a straightforward procedure. In reading various sections of the book I have also been impressed with the ease that the latest information about diagnostic methods can be found as well as the lacunae where further work is demanded.

Although, in a sense, this book is first, and most notably a technical treatise, it is much more. Anyone who wishes to become acquainted directly with the present state of microbial virology and rickettsiology would make a fine beginning if he were to read the book straight through. Therefore, it will become I think, not only a true *Enchiridion*, a manual that is frequently in the hand, but a direct introduction to students interested in this subject from a broader point of view.

John F. Enders
University Professor Emeritus
Harvard University

ADVISORY BOARD
SECTION H: VIROLOGY AND RICKETTSIOLOGY
VOLUME I

CONTRIBUTORS
SECTION H: VIROLOGY AND RICKETTSIOLOGY
VOLUME I
PART 1

Jordi Casals
Department of Epidemiology
and Public Health
Yale University School of Medicine
New Haven, Connecticut

Richard W. Compans
Department of Microbiology
University of Alabama Medical Center
Birmingham, Alabama

Richard L. Crowell
Department of Microbiology
and Immunology
The Langbord Virus Laboratory
Hahnemann Medical College
Philadelphia, Pennsylvania

John E. Dahlberg
Laboratory of RNA Tumor Viruses
National Cancer Institute
Bethesda, Maryland

Caroline K. Y. Fong
Virology Laboratory
Veterans Administration Hospital
West Haven, Connecticut

Charles J. Gauntt
Department of Microbiology
The University of Texas
Health Science Center
San Antonio, Texas

Markus Gasser
Institute of Microbiology and Hygiene
University Basel
Basel, Switzerland

Robert H. Green
Department of Medicine
Middlesex Memorial Hospital
Middletown, Connecticut

I. H. Holmes
Department of Microbiology
University of Melbourne
Parkville, Victoria
Australia

Hope E. Hopps
Bureau of Biologics
Food and Drug Administration
Bethesda, Maryland

G. D. Hsiung
Virology Laboratory
Veterans Administration Hospital
West Haven, Connecticut

Burton J. Landau
The Langbord Virus Laboratory
Hahnemann Medical College
Philadelphia, Pennsylvania

Hans Loffler
Institute of Microbiology and Hygiene
University Basel
Basel, Switzerland

Kenneth McIntosh
Department of Pediatrics
Medical Center
University of Colorado
Denver, Colorado

Joseph L. Melnick
Department of Virology and Epidemiology
Baylor College of Medicine
Houston, Texas

Harry M. Meyer
Bureau of Biologics
Food and Drug Administration
Bethesda, Maryland

Fredrick A. Murphy
College of Veterinary Medicine and
Biomedical Sciences
Colorado State University
Fort Collins, Colorado

Kiyoto Nakamura
Department of Microbiology
University of Alabama Medical Center
Birmingham, Alabama

Erling Norrby
Department of Virology
Karolinska Institutet
School of Medicine
Stockholm, Sweden

Peter Palese
Department of Microbiology
Mount Sinai School of Medicine
New York, New York

Paul D. Parkman
Bureau of Biologics
Food and Drug Administration
Bethesda, Maryland

C. A. Phillips
Department of Medicine
College of Medicine
University of Vermont
Burlington, Vermont

Mary B. Ritchey
Department of Microbiology
Mount Sinai School of Medicine
New York, New York

Robert E. Shope
Department of Epidemiology
Yale University School of Medicine
New Haven, Connecticut

Rex S. Spendlove
Department of Biology
Utah State University
Logan, Utah

TABLE OF CONTENTS
SECTION H: VIROLOGY AND RICKETTSIOLOGY
VOLUME I
PART 1

TABLE OF CONTENTS
SECTION H: VIROLOGY AND RICKETTSIOLOGY
VOLUME I
PART 2

General Aspects

CLINICAL VIROLOGY AND VIRAL DIAGNOSIS

G. D. Hsiung and R. H. Green

Introduction

While extensive studies have been made in recent years in molecular virology, the laboratory diagnosis of virus diseases is still not widely practiced. Viral diagnostic facilities generally are available only in connection with research laboratories, and are rarely operated within a clinical microbiology setting. To a large extent, this situation is a reflection of the stringent procedures and specialized knowledge necessary for conducting virological tests in a clinical laboratory. Furthermore, the clinical virologists are also faced with past prejudices, expressed in the attitude that by the time virus diagnosis was achieved, the patient had either recovered or was beyond recovery. Fortunately, this adage no longer holds true since rapid diagnosis of certain viral diseases has become a reality. Today some viral diseases are treatable and in others, prognostic implications may be crucial. The operation of a clinical virology section similar to that of a clinical bacteriology laboratory may soon become an important function of clinical laboratory science.[1-4]

Specimen Collection for Viral Studies*

Specimen collection

Success in isolating viruses from clinical materials is largely dependent upon the proper collection and handling of specimens. Ideally, materials for viral studies should be collected and inoculated at the bedside since many viruses deteriorate upon standing. Specimens for virus study should be taken as early as possible in the course of the illness (Figure 1) and, if the patient is hospitalized, on the date of admission. All specimens collected as swabs should be placed in sterile vials containing 1 to 2 ml of carrying fluids, for example, 0.2% gelatin or bovine albumin in Hanks' balanced salt solution (BSS) with antibiotics. Other specimens such as body fluids and tissues should be collected aseptically and placed in sterile containers. Since many viruses are extremely labile, the specimens should be kept moist and delivered to the laboratory promptly.

Choice of specimens

The choice of specimens is largely dependent upon the clinical syndrome and/or the virus suspected. Due to the wide range of viruses responsible for similar clinical syndromes, it is common practice to include a set of specimens. Table 1 summarizes the preferred types of specimens to be collected from patients in accordance with the clinical syndromes which are commonly associated with different viruses. Blood samples are ordinarily used for serodiagnosis and, if possible, should be obtained from all patients in addition to the specimens indicated. Paired sera, one collected at time of acute illness and one at convalescence, are very helpful, particularly when virus isolation procedures are unsuccessful. Whole blood, especially the leukocytes, may be used for virus isolation in selected cases.

Storage, transport, and testing of specimens

As mentioned above, all freshly collected specimens should be kept moist and cool (not frozen) and delivered to a virology laboratory as soon as possible. In the laboratory, all specimens received should be processed immediately for virus detection since

* Parts of this and some of the following sections are modified from G. D. Hsiung, *Diagnostic Virology*, Yale University Press, New Haven, 1973.

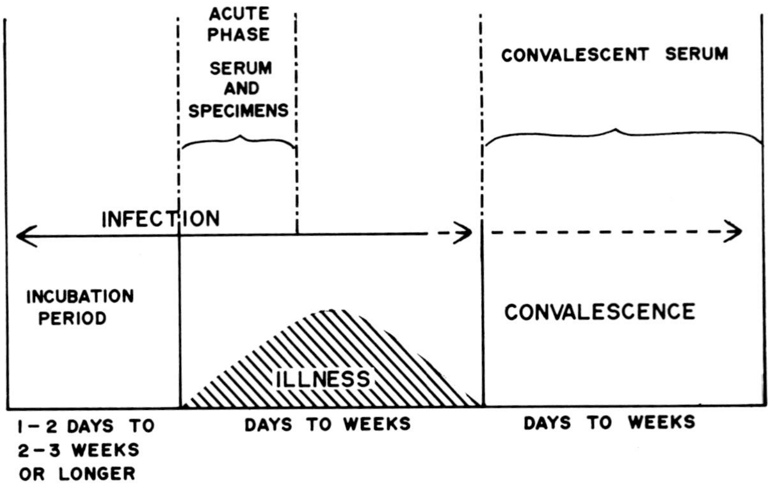

FIGURE 1. Time schedule for collecting specimens for virus studies.

many viruses are extremely sensitive to freezing-thawing or prolonged standing in the absence of living susceptible cells. If storage is necessary, specimens should be held at 4°C and tested within 24 hr. If they must be kept for a longer time, they should be frozen at −70°C or lower (not at −20°C). In certain instances the successful recovery of infectious virus is dependent upon the viability of the tissue cells in which the viruses are present. Therefore, tissues obtained at biopsy or surgery should be trypsinized to retain intact cells and used for explant cultures or cocultivation with susceptible cell monolayers.

It should be noted that the laboratory cannot be expected to carry out appropriate studies without having available a clinical history and/or reference to viruses suspected. This information assists greatly in the selection of the proper testing system, since each virus type has its preferred host spectrum. (For details, see separate chapter for different virus groups). Unfortunately, there is not a single cell system available for cultivation of all human viruses.

In the laboratory, viral diagnosis can be made in several different ways. These include (1) virus isolation procedures, (2) serologic tests, and (3) direct microscopy depending upon specimens available. In selected instances, special procedures such as cultivation of tissue cells, cocultivation with susceptible cells, and/or organ cultures for "slow" viruses may also be applied.

Methods Commonly used for Virus Isolation

Although the basic principles involved in the isolation and identification of viral agents are essentially the same as those for the isolation of bacteria, the procedures necessary for culturing viruses pose a problem since there is not a single cell culture system in which most or all viruses will grow. Therefore, the methods used for virus detection may vary from laboratory to laboratory. These variations depend upon the availability of cell culture systems and small animals, as well as upon the knowledge and experience of the individual in charge. Figure 2 shows a flow chart of procedures which are commonly used for the isolation and identification of viruses from clinical specimens in a viral diagnostic laboratory. The selection of test systems, whether cell culture or animal inoculation, is based upon the type of virus suspected as well as clinical diagnosis and history. However, it is generally recognized that the more techniques used, the greater the chance of recovering an infectious agent.

Table 1
SPECIMENS TO BE COLLECTED FOR VIRUS STUDIES

Clinical syndrome	Viruses commonly associated	Specimens to be collected for virus isolation[f]								
		Nasal washing	Throat swab	Rectal swab	Stool	CSF	Urine	Blood[a]	Others	Autopsy specimens
Respiratory Tract										
Pneumonia										
Child	RS	X	X					X	Sputum[b]	
	Parainfluenza 1,2,3	X	X					X	Sputum	
	Influenza A	X	X					X	Sputum	Lung
Adult	Influenza A,B	X	X					X	Sputum	
Immunosuppressed	Cytomegalovins	X	X				X	+X		Lung
	Varicella-Zoster					X		X	Skin lesion	Lung and other organs
	HSV	X	X					X	Brain, skin lesion	Lung and other organs
Bronchiolitis	RS	X	X					X	Brain, skin lesion	
	Parainfluenza 1,2,3	X	X					X	Brain, skin lesion	
	Adenoviruses[c]		X	X	X			X	Brain, skin lesion	
Croup	Parainfluenza 1,2,3	X	X					X	Brain, skin lesion	
	RS	X	X					X		
URI	Rhinoviruses	X	X	X	X			X		
	RS	X	X							
	Adenoviruses[e]	X	X	X	X			X		
	Parainfluenza viruses[e]	X	X					X		
	Enteroviruses[e]	X	X	X	X			X		
Pharyngitis	Herpes simplex		X					X		
	Influenza A and B	X	X					X		
	Parainfluenza viruses	X	X		X			X		
	Rhinoviruses	X	X	X	X			X		
	EB Virus	X	X					X		
	Adenoviruses[c]	X	X	X	X		X	X		Conjunctival secretion

Table 1 (continued) SPECIMENS TO BE COLLECTED FOR VIRUS STUDIES

Clinical syndrome	Viruses commonly associated	Specimens to be collected for virus isolation[f]								
		Nasal washing	Throat swab	Rectal swab	Stool	CSF	Urine	Blood[a]	Others	Autopsy specimens
Digestive Tract										
Gastroenteritis	Rotavirus				X[d]			X		
	Norwalk agent				X[d]			X		
Hepatitis	Hepatitis A				X[d]			X	Liver	Liver
	Hepatitis B							X	Liver	Liver
	Cytomegalovirus		X				X	X		
Cutaneous and mucous membrane										
Vesicular	Small pox							X	Skin lesions	
	Varicella-Zoster							X	X	
	Herpes simplex							X	X	
Exanthematous	Measles		X					X		
	Rubella		X					X	Tissue	
	Coxsackie A9,16		X					X		
	Molluscum contagiosum								Skin lesions	
Central nervous system (acute)										
Encephalitis	Herpes simplex		X			X		X	Brain biopsy	Brain tissue
	Mumps		X				X	X		Brain
	Measles	post infection							X	
	Influenza									
	EEE, WEE and other Togaviruses							X		Brain
Aseptic Meningitis	Coxsackie, Echo, Poliovirus	X	X	X				X		CNS spinal cord
Rabies	Rabies virus								Saliva, corneal secretion, skin biopsy	Brain

Table 1 (continued)
SPECIMENS TO BE COLLECTED FOR VIRUS STUDIES

Clinical syndrome	Viruses commonly associated	Specimens to be collected for virus isolation[f]								
		Nasal washing	Throat swab	Rectal swab	Stool	CSF	Urine	Blood[a]	Others	Autopsy specimens
Central nervous system (chronic)										
SSPE	Measles									
PML	Papova								Brain biopsy	Brain tissue
Cardiovascular										
Myocarditis	Coxsackie B virus	X	X	X				X	Pericardial fluid	Heart and pericardial
	Echovirus[e]								Pericardial and/or myocardial biopsy	
Muscular										
Pleurodynia	Same		X	X	X			X		
Congenital										
Rubella syndrome	Rubella virus		X				X	X	Placental tissue	Various fetal tissues
Cytomegalovirus	Cytomegalovirus						X	X		
Neonatal herpes	Herpes simplex virus	X								Fetal tissue
Lymphoreticular										
Infectious mononucleosis	EB virus		X					+X		
Regional lymphadenitis	Adenovirus		X	X	X			X		
	Herpes simplex virus	X						X		

[a] + Whole blood or buffy coat for virus isolation; × acute and convalescent sera for antibody study.
[b] If available.
[c] Less common.
[d] Electron microscopy.
[e] Rare.
[f] Best specimens.

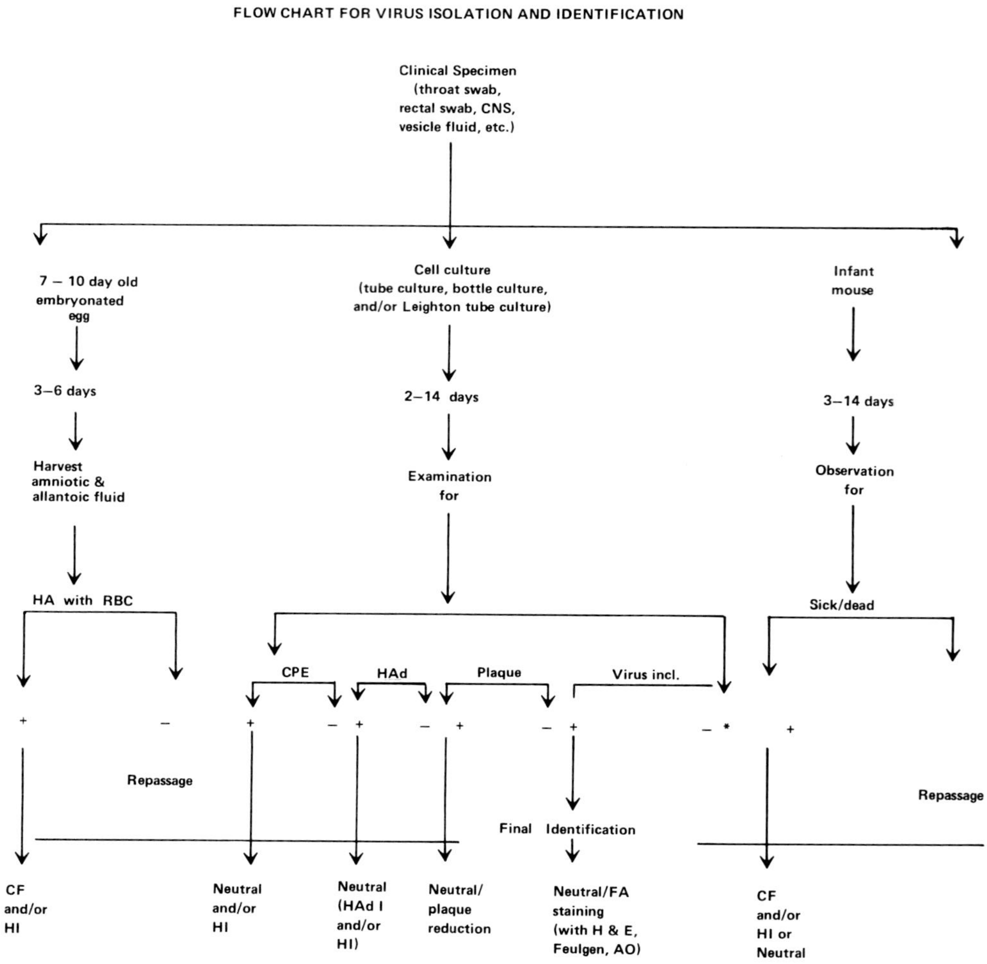

FIGURE 2. Flow chart for virus isolation and identification.* When a new agent is suspected, physical-chemical methods including ultrafiltration, ether sensitivity and chemical inhibitors can be used; serologic methods of choice for final identification vary according to the virus isolate. (From Hsiung, G. D., *Diagnostic Virology, An Illustrated Handbook*; Yale University Press, New Haven, 1973. With permission.)

Cell Culture System

Various comprehensive studies on the susceptibility of different cell culture systems to viral infection have shown that there is no single cell type which can serve as the cell type of choice for all virus isolation procedures (Table 2). For routine purposes, however, the most commonly occurring human viruses can be isolated in one or more of the following types of cell culture: primary human embryonic kidney cells (HEK); primary monkey kidney (MK) cell cultures such as Rhesus or African green monkey; a human diploid cell strain such as Wl-38; and a continuous human heteroploid cell line such as Hep-2. Various combinations of any of the above should provide the necessary degree of susceptibility to infection to ensure the isolation of most viruses. An additional cell type, rabbit kidney (RK) cells, is useful because of its specific susceptibility to herpes simplex virus, and it thus provides a means of preliminary identification of this herpesvirus when used in conjunction with other cell cultures.

Cellular Changes in Monolayer Cultures

Once the cell cultures are inoculated with clinical specimens, they are incubated at

Table 2
COMMON PROCEDURES FOR VIRUS ISOLATION

Virus group	Virus type	Cell Culture System[a] HEK	MK	WI-38	Hep-2	RK	Method of recognition in cell culture	Newborn mouse	Embryonated eggs
RNA									
Picornaviridae	Poliovirus	+ +	+ +	+ +	+ +	−	Rapid CPE	−	−
	Coxsackie A	±	±	±	−	−	CPE by certain types	+ +	−
	Coxsackie B	+ +	+ +	+ +	+ +	−	Rapid CPE	+ +	−
	Echovirus	+ +	+ +	+ +	±	−	CPE	−	−
	Rhinovirus	+	+	+ +	±	−	CPE[b]	−	−
Togaviridae	Alphavirus	±	±	±	−	−	CPE and/or plaque formation under agar overlay	+ +	−
	Flavivirus								
Bunyaviridae	Bunyavirus							+ +	−
Rubiviridae	Rubella	−	GMK + +	−	−	−	Interference	−	−
Reoviridae	Reovirus	−	+ +	−	−	−	CPE (Cytoplasmic inclusions)	−	−
	Rotavirus	−	−	−	−	−	No CPE in cell culture Electron microscopy of stool extract	−	−
Coronaviridae	Coronavirus	−	−	−	−	−	CPE[b]	−	−
Myxoviridae	Influenza	+	− to + +	−	−	−	Hemadsorption	−	+ +
	Parainfluenza	+	+ +	−	−	−	Hemadsorption	−	−
	Measles	+	+	−	+ +	−	Syncytial cell formation (nuclear and cytoplasmic inclusions)	−	−
	Respiratory syncytial	−	−	−	+ +	−	Syncytial cell formation	−	−

Table 2 (continued)
COMMON PROCEDURES FOR VIRUS ISOLATION

Virus group	Virus type	Cell Culture System[a] HEK	MK	WI-38	Hep-2	RK	Method of recognition in cell culture	Newborn mouse	Embryonated eggs
DNA									
Parvoviridae	Parvovirus	−	−	−	−	−	No CPE[c]	−	−
Papovaviridae	Papovavirus	−	+[d]	−	−	−	CPE	−	−
Adenoviridae	Adenovirus	+ +	±	+	+ +	−	CPE (nuclear inclusion)	−	−
Herpesviridae	Herpes simplex	+ +	±	+	+	+ +	Rapid CPE (nuclear inclusion)	+	−
	Varicella-Zoster	−	−	+ +	−	−	Slow CPE (nuclear inclusion)	−	−
	Cytomegalovirus	−	−	+ +	−	−	Slow CPE (nuclear inclusion)	−	−
	Epstein-Barr virus	−	−	−	−	−	Cultivation of leukocytes for FA	−	+
Poxviridae	Poxvirus (variola)	−	−	−	−	−	No CPE in cell culture	−	+
	Vaccinia	+ +	+ +	+ +	+ +	+ +	Rapid CPE (cytoplasmic inclusion)	+	+

[a] Primary cell cultures: HEK, human embryonic kidney; MK, monkey kidney; RK, rabbit kidney; continuous cell lines: WI-38 a human diploid cell strain; Hep-2, a human heterploid cell line. − = No reaction, + peq light reaction, + + = good reaction, ± = questionable reaction.

[b] Organ culture.

[c] See parvovirus chapter.

[d] Monkey papovavirus induces CPE in African green monkey kidney cells.

33 to 35°C and examined daily or every other day for 2 weeks or longer for evidence of virus induced cellular changes, primarily cytopathic effect (CPE). The appearance of CPE is dependent to some extent on both the concentration of virus present in the original specimen and the growth rate of the virus in question. In certain cases cytopathology is characteristic for a particular virus group, and the pattern can be used to make a tentative decision as to the major virus group to which the isolate probably belongs. As illustrated in Figure 3, CPE induced by an enterovirus in MK cell cultures (Figure 3C) and Hep-2 cells (Figure 3D) is distinguishably different from that induced by the adenovirus (Figures 3E and 3F) or herpes simplex virus (Figures 3G and 3H). For more details, see Chapter 4. Some agents which fail to induce recognizable CPE may be detected by indirect techniques, for example, recognition of myxovirus infected MK cell cultures by hemadsorption (Figure 3I) or detection of rubella virus by the interference phenomena. Many viruses, including respiratory syncytial virus and the parainfluenza viruses induce syncytial cell formation in Hep-2 cells (Figure 3J). Occasionally, blind passages may be necessary to reach high enough concentrations of virus to produce evidence of its presence.

Plaque Formation Under Agar Overlay

The formation of virus plaques under agar-containing overlay medium is somewhat analagous to bacterial colonies on solid media. This is done in monolayer cultures either in bottles or petri dishes. After the fluid is removed from the monolayer cultures and the specimen is inoculated, the cell sheet is covered with agar overlay medium containing neutral red, a vital dye. As virus replication occurs, cell degeneration or death is demonstrated by the presence of decolorized areas or plaques while the surrounding live cells remain stained by the dye (Figure 4). Plaque size and shape are often characteristic for different viruses, and plaque morphology is a useful tool, especially for the recognition and characterization of the enteroviruses (Figure 4A). Furthermore, the plaque method appears to be more sensitive than the CPE method for the isolation of enteroviruses from clinical specimens in which the number of virus particles is too small to be detected by other means. Certain virus strains do not produce CPE but, nevertheless, do form plaques, as in the case of certain arboviruses. In cases of mixed infections the plaque method is far more efficient than CPE for the recognition of the presence of two types of viruses (Figure 4B).

Host Cell Spectrum and Cellular Reaction as a Means of Presumptive Identification

In view of the fact that different cell systems show differing degrees of susceptibility to different viruses, it is possible to make a presumptive identification as soon as certain types of cell culture show various reactions: i.e., type of CPE, pattern of plaque morphology under agar overlay medium, and presence or absence of hemadsorption. Once a presumptive identification of a new isolate has been made, one can then go on to determine the serotype within the particular virus group, by specific serologic procedures (see Chapter 2). Thus the laboratory frequently can make a tentative identification and report it to the patient's physician prior to the final serologic identification.

Problems of Endogenous Virus Contaminants

From time to time primary cell cultures have been found to harbor one or more types of endogenous viruses. (See Chapter 41). These are particularly frequent in primary kidney cells derived from monkey tissues[6,7] as well as from some nonprimate tissues.[8,9] Many of these endogenous agents induce CPE or hemadsorption or they may even interfere with replication of the virus in the specimens being tested. Since the presence of mycoplasmas has been reported in a number of cell lines,[10,11] it is good

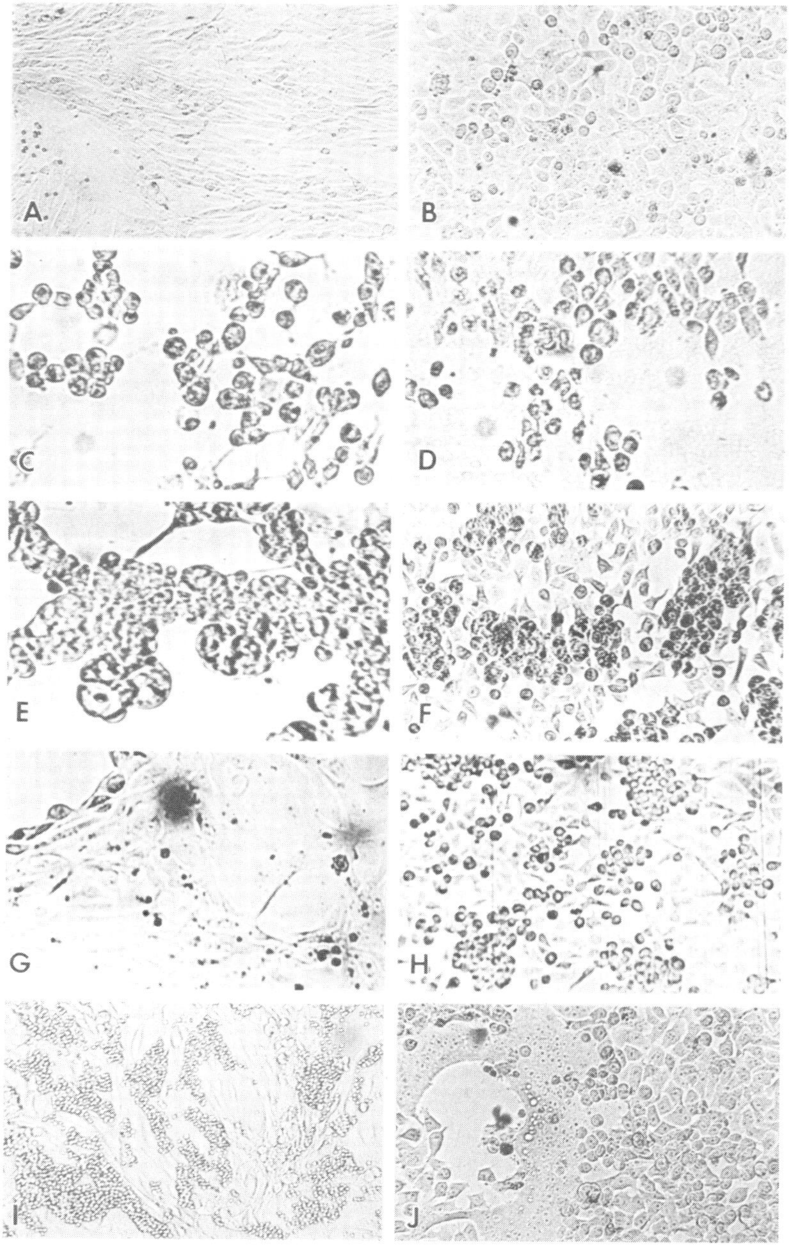

FIGURE 3. Virus induced cellular changes in primary monkey kidney cell cultures (panel at left) and Hep-2 cell cultures (panel at right). (Magnification × 100.) (A and B) Noninfected control cultures; (C and D) CPE induced by poliovirus type 1; (E and F) CPE induced by adenovirus type 3; (G and H) CPE induced by herpes simplex virus type 1 and (I and J) Hemadsorption and syncytial cell formation induced by parainfluenza virus types 1 and 3, respectively. (From Hsiung, G. D., *Mt. Sinai J. Med. N.Y.*, 44, 1, 1977. With permission.)

practice to examine periodically for the presence of these organisms in the cell strains, or cell lines being used. (See Chapter 42). The diagnostic virologist should always be aware of the possibility of endogenous contaminants and must be able to recognize their presence and not confuse them with virus isolates from clinical specimens.

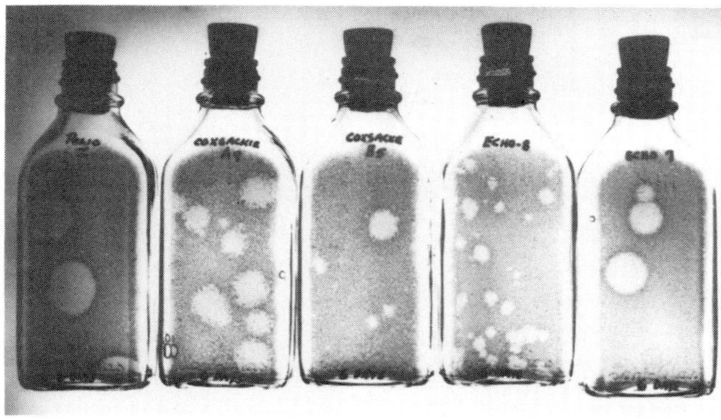

A

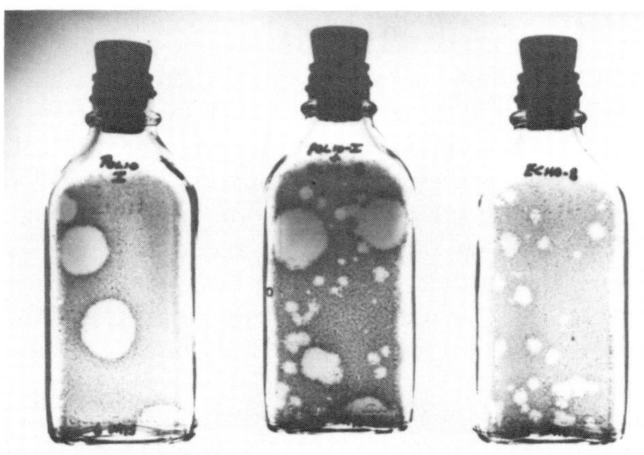

B

FIGURE 4. Plaque morphology of enteroviruses in rhesus monkey kidney cell monolayers under agar overlay medium 6 days after inoculation. (A) Left to right — Polio-1, Coxsackie A-9, Coxsackie B5, Echo-8, Echo-7. (B) Bottle at left Poliovirus type 1 alone; bottle ar right, Echovirus type 8 alone; center bottle shows two large plaques of Poliovirus and small plaques of Echovirus.

Newborn Mice

Although animal inoculation has been largely replaced by tissue culture systems for isolating viruses, newborn mice are still the host system of choice for isolating certain types of Coxsackie A viruses and most arboviruses. It is costly and certainly impractical to inoculate all specimens into newborn mice. Only in selected instances, for example, patients with herpangina or with suspected arbovirus encephalitis, should specimens be inoculated into newborn mice.

Embryonated Chicken Eggs

Embryonated eggs are still preferable for the isolation of certain strains of influenza A virus[12] although influenza B viruses are readily detected in MK cell cultures. If smallpox virus is suspected, the inoculation of the chorioallantoic membrane is a useful technique for differentiating this virus from vaccinia virus and from herpes simplex virus by pocks on the egg membrane.

Serodiagnosis

A large number of tests and procedures have been developed in recent years to measure a variety of immunologic parameters in patients' sera, especially with the development of microtiter technique.[13] The chapter following this not only describes these tests, but it also explains the purpose for which the tests are used and the rationale behind the actual procedures. In certain instances, where virus is not or cannot be recovered, serologic procedures may provide an important means for establishing a viral etiology for a particular clinical syndrome. Table 3 illustrates examples of diagnoses which were made on the basis of serologic tests. Furthermore, procedures such as neutralization and hemagglutination-inhibition tests play an important role in the definitive identification of virus isolates.

Immunofluorescence techniques have long been applied for the diagnosis of human influenza virus in nasal smears[14] and have been used also for the diagnosis of other respiratory viruses.[15] The methods are still not widely used for diagnosis of virus infection of the respiratory tract. Part of the problem may be due to the small quantity of virus particles or antigen available as well as the difficulties inherent in selecting a suitable reagent for the hundreds of possible virus types that may be present in a clinical specimen. The techniques described by Sommerville for rapid diagnosis of virus infection of the central nervous system by examination of cell pellets from cerebral spinal fluid[16] or for determining the presence of viral antigen in leukocytes and macrophages[17] appear to be promising, but in each instance preliminary cultivation procedures are necessary as has been described for the detection of Epstein-Barr virus in cultured leukocytes using fluorescent antibody techniques.[18] Certainly the method offers a great advantage in the detection of rabies virus infection in brain tissue when only one virus type is involved.

The Use of Electron Microscopy for Detection and Recognition of Viruses

The size and shape of a virus can be determined by direct electron microscopic (EM) examination. The techniques most commonly used include negative staining of virus particles with electron-dense salts of phosphotungstic acid or the preparation of ultrathin sections of cells or tissues suspected of harboring a virus. (For detailed illustrations, see Chapter 4). EM has been used for the diagnosis of pox virus and herpesvirus infection for almost 30 years,[19] but this technique has not been widely used until recently. Recognition of more than one kind of virus-like particle in HBs Ag positive sera of patients with hepatitis was made possible by EM.[20] Moreover, the development of immune EM, a new method for the recognition of virus particles in clinical specimens as well as for serodiagnosis, has ushered in a whole new era in diagnostic virology.

Rapid Diagnosis and Recognition of New Viral Agents

Visualization of poxvirus particles in extracts of crusts or in vesicle fluids can be made within a few hours when EM examination is undertaken. Where smallpox is suspected the electron microscope can be helpful in providing an early presumptive diagnosis.[21] Other investigators have reported the advantages of using EM for the differentiation of herpesviruses and poxvirus in skin lesions.[22] In fact, the recognition of a new herpesvirus, the Epstein-Barr virus, in cultured lymphoblastic cells derived from a Burkitt's lymphoma patient, was originally made by electron microscopy.[23] The first human papovavirus particle was observed by EM in a human skin papilloma as early as 1953,[24] but methods for cultivation of this virus are still not available. The recogni-

Table 3
EXAMPLES FOR ILLUSTRATING INTERPRETATION OF LABORATORY TESTS FOR VIRAL ANTIBODY STUDIES

Case number	Clinical history diagnosis	Age (Years)	Serum samples[a]	Complement-fixation test: Influenza A	Influenza B	Para 1	Para 2	Para 3	Herpes	Adeno	Other tests: NT[b] Adeno-2	Other tests	Laboratory interpretation
1	Pneumonia	3	A	< 8	< 8	< 8	< 8	< 8	8	128	< 20		No diagnostic rise by CF[b] but N.T. showed fourfold rise to the virus isolated as Adeno-2
			C	< 8	< 8	< 8	< 8	< 8	< 8	256	80		
											CF Mumps	RS	
2	"Flu"	5	A	< 8	< 8	< 8	8	16	16	64	< 8	8	Diagnostic rise to both mumps and para-2. These two viruses have common antigens, infection probably due to mumps
			C	< 8	< 8	< 8	32	16	32	64	32	8	
3	"Flu"	9 mo	A	8	< 8	32	< 8	32	8	64	8	16	Diagnostic rise to both para-2 and 3, since para-3 antibody was high infection due to para-2 with heterotypic rise to para-3
			C	8	8	64	32	128	8	128	8	8	
4	"Flu"	4	A	< 8	< 8	< 8	< 8	8	64	8	< 8	< 8	Possible mixed infection with para-2 and RS virus
			C	< 8	< 8	< 8	32	8	64	16	< 8	32	

[a] A = acute; C = convalescent.
[b] CF = complement-fixation test; N. T. = neutralization test.

(Courtesy of Dr. F. Michalski; St. Michael's Hospital, Newark, N.J.)

tion of another human papovavirus in the brain cells of patients with a rare demyelinating disease, progressive multifocal leukoencephalopathy (PML), was also made by EM[25] several years before the virus was isolated.[26,27] Thus, EM can be very helpful in the diagnosis of viral diseases.

Immune Electron Microscopy

The technique of immune EM has advanced rapidly and provides a useful, more sensitive refinement than simple EM examination. For example, EM examination of stool filtrates mixed with convalescent serum revealed 22 to 27 nm virus particles, not previously detected, in stool samples of patients with nonbacterial gastroenteritis[28] (see Chapter 33). Similarly, the discovery of reovirus-like particles, 67 to 73 nm in diameter, in stool samples from patients with infantile diarrhea has also been reported.[29-31] Thus, the field of diagnostic virology has been broadened through the use of immune EM, particularly in dealing with fastidious agents for which cultivation procedures have not been developed. In those instances where a particular virus infection is suspected and attempts at virus isolation have been unsuccessful, serodiagnosis, using immune EM, may be employed.

Detection of "Slow" Viruses by Cultivation of Tissue Cells

It has become apparent in recent years that in order to recover "slow" viruses from infected tissues, especially those in a latent state, the inoculum must contain viable cells, i.e., cultivation of the infected cells must be carried out in order to recover infectious virus. Although the original isolation of a human adenovirus was obtained from a spontaneously degenerated cell culture derived from human adenoid tissue,[32] cell culture propagation for isolating human viruses has not been a common practice. On the other hand, animal viruses, particularly those of simian origin, have been isolated to a large degree as the result of spontaneous degeneration of kidney cell cultures that were being used for virus research or for virus vaccines.[6,7]

Cultivation and Cocultivation

The terms cultivation and cocultivation refer to those techniques whereby fresh tissues are dispersed with trypsin and the resultant cells are grown out in monolayer cultures for long-term cultivation and observation for virus induced cellular changes, or the trypsin-dispersed cells are cocultivated with susceptible cell monolayer cultures and the latter are observed for the development of cytopathic effect. Although it has been known for many years that varicella-zoster virus could be passaged best using infected viable cells,[33] the superiority of viable cells over tissue homogenates (a method commonly used for virus isolation) for recovery of infectious virus from infected tissues has only recently been recognized, especially when dealing with slow virus infections of the central nervous system.[26,27,34,35] These cultivation and cocultivation techniques will probably receive wide use in the near future for the recovery of viruses in cases of other chronic diseases of suspected viral etiology.

Organ culture

The use of organ culture techniques in virology is a relatively recent development which has facilitated the isolation and recognition of many respiratory viruses. Essentially the method entails the cultivation of small tissue explants in special medium and under conditions in which growth of undifferentiated cells is restricted while the normal architecture and function of the tissue is maintained. Many rhinovirus strains were originally isolated by the use of nasal or tracheal epithelial organ cultures derived from human embryos.[36] Coronaviruses also have been isolated by the use of ciliated tracheal organ culture.[37] Undoubtedly, by cultivating explants of other tissues, additional viruses will be isolated and associated with human diseases of presently unknown etiology.

Characterization of "New" Viral Agents

As described above, procedures used for the isolation and identification of viruses are now fairly well standardized. With highly trained personnel, a virology laboratory can perform most of the routine diagnostic tests and can identify agents commonly encountered in a clinical laboratory. However, the identification of a "new" virus or a new variant strain of a known agent may present problems. It may be necessary to characterize the physical and chemical properties of the new agents, according to nucleic acid type (RNA or DNA), size and shape, and ether sensitivity, in order to facilitate identification.

Nucleic Acid Determination

The nucleic acid type can be determined by a number of methods. However, since these methods are indirect, interpretation of results should be considered tentative until more refined methods of nucleic acid extraction are employed. In general, the multiplication of DNA viruses is inhibited by 5-fluoro-2′-deoxyuridine (FrdU) or 5-bromo-2′-deoxyuridine (BrdU), whereas multiplication of RNA viruses is not affected by these compounds. In selected instances, nucleic acid type can be determined by the use of the acridine orange staining technique and/or the Feulgen reaction.[38] The appearance of a yellowish-green inclusion in the nucleus after acridine orange staining or a brilliant red inclusion in a cell stained by the Feulgen reaction is an indication that the nucleic acid of the virus is probably DNA.

Ether Sensitivity Test

Determination of ether sensitivity is readily accomplished by exposing the virus suspension to ethyl ether for 1 to 2 hr followed by testing for virus infectivity. Loss of infectivity indicates sensitivity due to the destruction of the lipid-containing envelope surrounding the virus particle whereas retention of infectivity indicates a lack of such a lipid-containing structure and consequently a resistance to ether treatment.

Ultrafiltration for Virus Size Determination

Although the size and shape of a virus can be determined directly by EM, other, more accessible means for estimating virus particle size have been developed. One of the simplest and oldest methods for determining virus particle size is ultrafiltration. This is usually accomplished by passing a virus suspension through a series of collodion membranes of graded porosity. The size of the virus is measured grossly by its ability to pass through or be retained by the filter. Despite some limitations and complications that have been encountered, the availability of millipore membrane filters and suitable equipment has brought ultrafiltration procedures within the scope of a conventional clinical laboratory.[39] The equipment used for filtering small amounts of virus-containing fluid includes a 2-ml syringe attached to Swinny adapters connected to a hypodermic needle. After clarification through a membrane with a 450- or 300-nm pore diameter, the us fluid is further filtered through a test membrane having average pore sizes of 100 or 50 nm. A 15 or 20 ml Vacutainer tube is used to collect the filtrate. Virus infectivity of the suspension is determined before and after filtration, and virus size is estimated according to the virus infectivity retained in the filtrate. Large size viruses lose their infectivity when 100 nm and 50 nm average pore diameter membranes are used. Medium size viruses pass through 100-nm, but not through 50-nm membranes; and small viruses pass through both the 100- and 50-nm membranes. Thus, by utilizing a few simple tests to determine certain physical-chemical properties as well as estimating the size of a virus it is possible to place a newly isolated agent within one of the major virus groups (see Chapter 4).

Summary

As with other tests in the clinical laboratory, it is necessary to interpret the results of virus studies in relation to the clinical history of the patient. It must be kept in mind that isolation of any agent, even though accompanied by demonstration of a significant rise in specific antibody, does not necessarily mean that the isolate is the etiologic agent for the current disease. Thus, while much can be contributed to the diagnosis of an infection by laboratory studies, this information must be evaluated together with the patient's illness before a specific virus can be designated as the etiological agent of the disease.

Sufficient laboratory testing procedures are now available to provide rapid and accurate diagnosis of most of the common viral infections. With the commercial availability of various cell cultures and standard reagents, isolation and identification of the more common viral agents can be easily accomplished in a conventional clinical virology laboratory in a manner similar to the services provided by a bacteriology laboratory. Through the application of these procedures, a number of other services to the patient and medical staff can be attained. For example, one can prevent the unnecessary use of antibacterial drugs, furnish a more accurate prognosis and thus enhance the physician's clinical acumen, and provide evidence of a potential viral epidemic and thus permit the timely application of appropriate public health measures. Furthermore, some antiviral drugs are now available and there is every possibility that more will exist in the future. For the treatment of certain viral diseases the establishment of a laboratory diagnosis is essential in order to select the proper chemotherapy as illustrated in the recent report on herpes simplex encephalitis.[40]

The recent application of immune EM procedures has led to recognition of a hitherto unknown association between many viruses and particular clinical syndromes. The isolation of "slow" viruses from the central nervous system through the use of cultivation and cocultivation techniques has reinforced the importance of clinical virology in modern medicine. With the present increased interest in the development of various virological/immunological techniques in clinical laboratories, it is anticipated that the services of a diagnostic virology laboratory will undoubtedly be increased and further utilized in the very near future.

GENERAL REFERENCES

Swain, R. H. A. and Dodds, T. C., *Clinical Virology,* Livingstone, Edinburgh, 1967.

Lennette, E. H. and Schmidt, N. J., *Diagnostic Procedures for Viral and Rickettsial Infections,* 4th ed., American Public Health Association, New York, 1969.

Hsiung, G. D., *Diagnostic Virology. An Illustrated Handbook,* Yale University Press, New Haven, 1973.

Kurstak, E. and Morisset, R., *Viral Immunodiagnosis,* Academic Press, New York, 1974.

Grist, N., Ross, C. A., and Bell, E. J., *Diagnostic Methods in Clinical Virology,* 2nd ed., Lippincott, New York, 1974.

Fenner, F. and White, D. O., *Medical Virology,* 2nd ed., Academic Press, New York, 1976.

REFERENCES

1. **ASCP Research Symposium, 6th,** Laboratory diagnosis of viral infections: recent advances and their clinical application, *Am. J. Clin. Pathol.,* 57, 731, 1972.
2. **Herrmann, E. C., Jr.,** New concepts and developments in applied diagnostic virology, *Prog. Med. Virol.,* 17, 221, 1974.
3. **Hsiung, G. D.,** Laboratory diagnosis of viral infections: general principles and recent developments, *Mt. Sinai J. Med. N.Y.,* 44, 1, 1977.
4. **Douglas, R. G., Jr. and Hall, C. B.,** Significance of virus isolation and identification and virus serology in the care of patients, in *Significance of Medical Microbiology in the Care of Patients,* Lorian, V., Ed., Williams & Wilkins, Baltimore, 1977, 159.
5. **Hsiung, G. D.,** *Diagnostic Virology. An Illustrated Handbook,* Yale University Press, New Haven, 1973.
6. **Hsiung, G. D.,** Latent virus infections in primate tissues with special reference to simian viruses. *Bacteriol. Rev.,* 32, 185, 1968.
7. **Hull, R. N., The simian viruses,** in *Virol. Monogr.,* 2,1,1968.
8. **Smith, K. O.,** Adventitious viruses in cell cultures, *Prog. Med. Virol,* 12, 302, 1970.
9. **Swack, N. S. and Hsiung, G. D.,** Endogenous agents in primary cell cultures with special reference to latent viruses, *In Vitro,* 10, 260, 1974.
10. **Stanbridge, E.,** Mycoplasmas and cell cultures, *Bacteriol. Rev.,* 35, 206, 1971.
11. **Barile, M. F.,** Mycoplasmal contamination of cell cultures: Mycoplasma-virus-cell culture interactions, in *Contamination in Tissue Culture,* Fogh, J., Ed., Academic Press, New York, 1973, 131.
12. **Coleman, M. T. and Dowdle, W. R.,** Properties of the Hong Kong influenza virus. I. General characteristics of the Hong Kong virus, *Bull. W.H.O.,* 41, 415, 1969.
13. **Sever, J. L.,** Application of microtechnique to viral serological investigations, *J. Immuno.,* 88, 320, 1962.
14. **Liu, C.,** Rapid diagnosis of human influenza infection from nasal smears by means of fluorescein-labeled antibody, *Proc. Soc. Exp. Biol. Med.,* 92, 883, 1956.
15. **Gardner, P. S.,** Immunofluorescence and immunoperoxidase methods for the rapid viral diagnosis of respiratory viral infections. Workshop on New and Useful Methods in Viral Diagnosis, National Institutes of Health, 1978,
16. **Sommerville, R. G.,** Rapid identification of neurotropic viruses by immunofluorescent technique applied to cerebrospinal fluid cellular deposits, *Arch. Ges. Virusforsch,* 19, 63, 1966.
17. **Sommerville, R. G.,** Rapid diagnosis of viral infections by immunofluorescent staining of viral antigens in leucocytes and macrophages, *Prog. Med. Virol.,* 10, 398, 1968.
18. **Henle, G. and Henle, W.,** Immunofluorescence in cells derived from Burkitt's lymphoma, *J. Bacteriol.,* 91, 1248, 1966.
19. **Nagler, F. P. O. and Rake, G.,** The use of the electron microscope in diagnosis of variola, vaccinia, and varicella, *J. Bacteriol.,* 55, 45, 1948.
20. **Dane, D. S., Cameron, C. H., and Briggs, M.,** Virus-like particles in serum of patients with Australia-antigen-associated hepatitis, *Lancet,* 1, 695, 1970.
21. **Cruickshank, J. G., Bedson, H. S., and Watson, D. H.,** Electron microscopy in the rapid diagnosis of smallpox, *Lancet,* 2, 527, 1966.
22. **Williams, M. G., Almeida, J. D., and Howatson, A. F.,** Electron microscope studies on viral skin lesions. A simple and rapid method of identifying virus particles, *Arch. Dermatol.,* 86, 290, 1962.
23. **Epstein, M. A., Achong, B. G., and Barr, Y. M.,** Virus particles in cultured lymphoblasts from Burkitt's lymphoma, *Lancet,* 1, 702, 1964.
24. **Bunting, H.,** Close-packed array of virus-like particles within cells of a human skin papilloma, *Proc. Soc. Exp. Biol. Med.,* 84, 327, 1953.
25. **Zu Rhein, G. M. and Chou, S. M.,** Particles resembling papova viruses in human cerebral demyelinating disease, *Science,* 148, 1477, 1965.
26. **Padgett, B. L., Walker, D. L., Zu Rhein, G. M., Eckroade, R. J., and Dessel, B. H.,** Cultivation of papova-like virus from human brain with progressive multifocal leucoencephalopathy, *Lancet,* 1, 1257, 1971.
27. **Weiner, L. P., Herndon, R. M., Narayan, O., Johnson, R. T., Shah, K., Rubinstein, L. J., Preziosi, T. J., and Conley, F. K.,** Isolation of virus related to SV40 from patients with progressive multifocal leukoencephalopathy, *N. Engl. J. Med.* 286, 385, 1972.
28. **Kapikian, A. Z., Wyatt, R. G., Dolin, R., Thornhill, T. S., Kalica, A. R., and Chanock, R. M.,** Visualization by immune electron microscopy of a 27-nm particle associated with acute infectious nonbacterial gastroenteritis, *J. Virol.,* 10, 1075, 1972.
29. **Bishop, R. F., Davidson, G. P., Holmes, I. H., and Ruck, B. J.,** Letter: Evidence for viral gastroenteritis, *N. Engl. J. Med.,* 289, 1096, 1973.

30. **Flewett, T. H., Bryden, A. S., and Davies, H.,** Virus particles in gastroenteritis, *Lancet,* 2, 1497, 1973.
31. **Kapikian, A. Z., Kim, H. W., Wyatt, R. G., Rodriguez, W. J., Cline, W. L., Parrott, R. H., and Chanock, R. M.,** Reoviruslike agent in stools: association with infantile diarrhea and development of serologic tests, *Science,* 185, 1049, 1974.
32. **Rowe, W. P., Huebner, R. J., Gilmore, L. K., Parrott, R. H., and Ward, T. G.,** Isolation of a cytopathogenic agent from human adenoids undergoing spontaneous degeneration in tissue culture, *Proc. Soc. Exp. Biol. Med.,* 84, 570, 1953.
33. **Weller, T. H.,** Serial propagation in vitro of agents producing inclusion bodies derived from varicella and herpes zoster, *Proc. Soc. Exp. Biol. Med.,* 83, 340, 1953.
34. **Horta-Barboasa, L., Fuccillo, D. A., Sever, J. L., and Zeman, W.,** Subacute sclerosing panencephalitis: isolation of measles virus from a brain biopsy, *Nature (London),* 221, 974, 1969.
35. **Payne, F. E., Baublis, J. V., and Itabashi, H. H.,** Isolation of measles virus from cell cultures of brain from a patient with subacute sclerosing panencephalitis, *N. Engl. J. Med.,* 281, 585, 1969.
36. **Tyrrell, D. A. J.,** Rhinoviruses, *Virol. Monogr.,* 2, 67, 1968.
37. **McIntosh, K., Dees, J. H., Becker, W. B., Kapikian, A. Z., and Chanock, R. M.,** Recovery in tracheal organ cultures of novel viruses from patients with respiratory disease, *Proc. Natl. Acad. Sci. U.S.A.,* 57, 933, 1967.
38. **Mayor, H. D. and Melnick, J. L.,** Intracellular and extracellular reactions of viruses with vital dyes, *Yale J. Biol. Med.,* 34, 340, 1961—62.
39. **Hsiung, G. D.,** Use of ultrafiltration for animal virus grouping, *Bacteriol. Rev.,* 29, 477, 1965.
40. **Whitley, R. J., Soong, S., Dolin, R., Galasso, G. J., Chien, L. T., Alford, C. A., and the Collaborative Study Group,** Adenine arabinoside therapy of biopsy-proved herpes simplex encephalitis, *N. Engl. J. Med.,* 297, 289, 1977.

IMMUNODIAGNOSIS

Markus Gasser and Hans Löffler

Introduction

In clinical laboratory diagnostic work, immunity to viruses and rickettsiae is explored mainly for two purposes, (1) diagnosis of current infection and, (2) assessment of the state of immunity both in individuals and in populations. While there is a great variety of serological methods available for detecting humoral immunity, tests for assessing cell-mediated immunity are also at hand. The choice of the particular method to be used may largely depend on practical considerations, but regardless of the method used, it should conform to the aims of the investigation.

Aims

The clinical diagnosis of viral and rickettsial disease is established, with the aid of the laboratory, primarily by isolation and identification of the pathogenic microorganism. Immunodiagnosis, or serodiagnosis, may be used merely to supplement the direct diagnostic approach as in the tracing of contacts during surveillance of epidemics; it may be necessary for diagnosis where cultivation of the pathogenic organisms is difficult, fastidious, or time consuming, as in rubella, herpes zoster, infectious mononucleosis, and in some arbovirus and rickettsial infections. Serodiagnosis is also needed once the disease is in progress, when further shedding of virus cannot be expected. This is usually after the first few days of illness in the adult patient, whereas in children, virus shedding may continue for some time.

For current infection to be proven, a fourfold or greater rise in antibody titer observed in paired sera, collected in the acute phase of illness and 10 to 21 days thereafter (convalescent phase), is required. The rise emerges more slowly in children than in adults. A single high titer, especially of complement-fixing antibodies which usually last only for several weeks, may also indicate recent infection. Low titers, on the other hand, with no rise or a moderate rise, are assumed to be anamnestic responses resulting from previous infection. Distinguishing between current and previous infection is most desired in cases of rubella and cytomegalovirus infection during pregnancy, and can be accomplished by determining the class of immunoglobulins (Ig) to which specific antibodies belong. In the course of the immune response antibodies, of the IgM and IgG class appear successively, the predominance of IgM antibodies suggesting current infection. The level of IVGM antibodies may be maintained, however, in persistent infections. In the neonate, antibodies of the IgM class are formed actively in response to infection, while IgG antibodies are acquired passively by transplacental passage.

Assessment of antiviral immunity is useful in judging the need for vaccination individually, e. g., rubella in women of child bearing age, but it serves also, as is recommended in rabies preexposure vaccination, to evaluate the response to a given vaccine. Moreover, the efficacy of a particular type or product, whole vaccine vs. split vaccine for instance, or of a particular immunization scheme may be evaluated. In studies relating to immunodeficiency or immunosuppressive therapy, antiviral immunity, one of the parameters of immune reactivity, may be assessed by neutralization tests. It should be noticed, however, that locally secreted IgA antibodies or cell-mediated immune mechanisms might be more relevant to protection than is serum IgG antibody.

The aims of serological surveys are either to assess immunity of groups or populations to epidemic diseases, or to control the actual spread, as mentioned. By means of serology, latent infections, of course, are also detected. Finally, in diseases of hitherto unknown origin, such as subacute sclerosing panencephalitis, it is felt that the demonstration of a close relationship between disease and a certain microorganism, as suggested by immunodiagnosis and isolation procedures, might prove of value as a marker and possibly enhance the understanding of pathogenesis and etiology.

Methods

Immunological tests suitable for immunodiagnosis require specificity, sensitivity, and convenience. These characteristics relate to both the antigen used in the test and to the method of testing. Specificity relies basically on the quality of the antigen used, the preparation of specific antigens from viral and rickettsial agents, therefore, being most important. From the same microorganism various antigens may be prepared, representing antigenically different sites, for example, the inner part or core of virus particles and surface structures of the envelope or capsid. For screening of rickettsial infections (except Q fever and rickettsialpox) and of infectious mononucleosis, heterogeneous antigens are satisfactory, i.e., Proteus O-X19, O-X2, and O-XK, and red blood cells of certain species, respectively. Sensitivity depends mainly on the test method and the way it is performed. Labeling techniques such as radioimmunoassay are most sensitive, followed by neutralization and passive agglutination, while complement-fixation and precipitation methods are somewhat less sensitive. However, consistency of results is as important as the level of sensitivity. Since standardization is not easily achieved, and day to day variations may occur with most methods, paired sera should be tested in parallel. Convenience includes the ease of performance, the relative cost, and the time required for completion. For purposes of convenience, microtechniques are preferred. However, a method used conveniently in one laboratory, may be found inconvenient in another, depending upon the facilities available.

For serologic examination, serum is the usual specimen, but sero-mucous secretions and cerobrospinal fluid (CSF) are tested on occasion, because secretory antibodies found in saliva, nasal fluid, or tears, are components of local immunity. Occurrence of antibodies in CSF is explained either by damage of blood-brain barrier or by production taking place locally within the central nervous system.

Though most immunological tests do not involve infectious microorganisms, there is always the risk of hepatitis B virus being present in the sera of patients or controls, and appropriate precautions should be taken. In the following paragraphs the principles of the methods presently in use are outlined, references being made to detailed descriptions.

Determination of Immunoglobulin Class

A number of methods have been used to determine the immunoglobulin class of specific antibodies. In tests employing immunofluorescence techniques, the use of labeled-antihuman IgM and IgG is the obvious procedure.[1,2] Treatment of the serum sample with reducing agents like 2-mercaptoethanol results in degradation of IgM.[3] Adsorption on staphylococcal protein A removes IgG activity from the serum.[4,5] Density gradient ultracentrifugation[6,7] and gel filtration[8,9] are methods which separate proteins on the basis of their size. The fractions obtained by separation techniques may be checked for purity by single radial diffusion, and each fraction is tested separately for antibody content.

Precipitation Methods

Counterimmunoelectrophoresis is a double diffusion method, the sensitivity and speed of which is increased by the application of a constant current. Two wells have to be in line in an electrophoresis gel. During electrophoresis, α-globulins move towards the cathode, the technique being therefore applicable to antigens with motility towards the anode. Counterimmunoelectrophoresis currently is being used in large scale sample testing for hepatitis B antigen, and the corresponding antibodies.[10]

Single radial diffusion is based on the diffusion of the test sample from a well into a gel in which the corresponding antibody is incorporated. A precipitate is formed around the well, the diameter of which is proportional to the antigen concentration. By single radial diffusion the immunoglobulin content of the serum or serum fractions can be estimated. [11] For immunological studies with influenza virus, a single radial diffusion test has been developed in which the antibody diffuses, while the virus is incorporated in the gel and does not diffuse. A zone of opalescence, presumably caused by primary antigen-antibody interaction, rather than a precipitate is formed.[12]

Passive Agglutination and Hemolysis

Viruses can be coated onto larger particles which are then agglutinable by viral antibodies. Polystyrene, latex, bentonite, and other inert particles, or red blood cells, are carriers for this type of test. Coating occurs spontaneously, merely by incubation at appropriate pH, or requires chemicals such as tannic acid or chromic chloride. Conversely, antibodies may be attached to carrier particles and compete for antigen with antibodies in the test serum (passive agglutination inhibition). Passive agglutination and agglutination inhibition methods have been used for the demonstration of antibodies against arboviruses, enteroviruses, and hepatitis B virus.[13-15]

The use of red blood cells as carriers offers an advantage, in that by addition of complement, lysis may be induced. Single radial hemolysis combines the methods of single radial diffusion and passive hemolysis. Antigen coated red blood cells and complement are components of the gel. The amount of complement, as long as it is in excess, is not critical. The reaction of diffusing antibodies with antigen results in the formation of a zone of lysis, the diameter of which is linearly related to antibody concentration. Tests have been developed for measurement of antibodies against influenza, [16,17] rubella, [18,19] mumps.[20] and coronaviruses.[21] This method may well be more widely used in the future, especially in the screening of large numbers of serum samples.

Complement Fixation Test

Complement fixation is the most widely used and extensively studied serologic test in diagnostic virology.[22-24] The method is based on the binding of complement by antigen-antibody complexes. Two systems, each consisting of antigen and antibody, are competing for complement. Test serum, antigen, and complement (contained in guinea pig serum) are allowed to react first, followed by addition of the indicator system, i.e., sheep red blood cells and rabbit serum containing sheep red blood cell antibodies. In the absence of antibodies in the test system, available complement induces lysis of the red blood cells in the indicator system, whereas, when antibodies are present lysis does not occur. Since the amount of complement participating in the test is critical, two hemolysing units usually being optimal, the preliminary procedure is to titrate the guinea pig serum for complement content, and to inactivate complement activity of test serum by heat (56°C for 1/2 hr). Involving five biologically variable reagents, the test requires appropriate controls, including a serum control for anticomplementary activity of the test serum.

Complement fixation is applicable to those viruses and rickettsiae from which satisfactory antigens can be prepared. Much effort has been spent on the production of improved antigen preparations, and antigens from more recently recognized pathogens like rotavirus have been introduced. [25,26] Most of the antigens used represent internal nucleoproteins, and the corresponding antibodies therefore do not necessarily correlate with immunity. The specificity of the antigens is commonly rather broad, often relating to virus groups, such as influenza viruses, and adenoviruses, which provides an advantageous method of screening.

Neutralization

Biological activities of microorganisms can be impaired or inhibited by specific antibodies. Neutralization tests imitate in vitro the effect the humoral immune response is supposed to exert in vivo. Neutralizing antibodies are directed against surface structures of virus particles, and when properly performed, neutralization tests are very sensitive. Moreover, because of their narrow specificity, antibodies to subtypes of certain viruses are distinguishable.

Neutralization of infectivity requires the use of infectious viruses as antigens. For a given host system, the 50% infectious dose (ID_{50}) of virus has to be determined. Neutralization is performed by incubating 100 ID_{50} of the virus with the serum to be tested for a set time, after which, the mixture is inoculated into the susceptible host system. Neutralization in tissue culture is manifested by lack of cytopathic effect (CPE). Constant virus-varying serum relations allow the antibody level to be expressed as a titer. Many variations of the method are possible, utilizing different host and indicator systems. Instead of CPE, metabolic changes may be determined by changes in pH of the tissue culture fluid [27, 28] Neutralizing antibodies are presumed to prevent viruses from adsorbing to the cells, and participation of complement in the reaction may result in the actual destruction of virus particles,[29] increasing slightly the sensitivity of the method. [30,31] Less attention is now paid to the relation of infectious to noninfectious particles within a viral suspension, though empty virus particles might bind antibodies.

Neutralization of hemagglutination, or hemagglutination inhibition, is applicable to viruses which agglutinate certain red blood cells, i.e., myxoviruses and rubella virus, among others. Viruses used in hemagglutination inhibition need not be infectious. The hemagglutination inhibiting serum titer is determined against a constant virus concentration (4 hemagglutinating units), and serum samples should be heat-inactivated (56°C for 1/2 hr) and treated with receptor-destroying enzyme, kaolin, or dextran sulfate prior to testing, to remove nonspecific inhibitors. [32,33]

Labeling Techniques

By labeling of antigen or antibody with fluorescent dye, radioactive isotopes, or enzymes, antigen-antibody reactions can be made detectable, and trace amounts of either antigen or antibody may be revealed. Though the main purpose of the immunofluorescence (IF) technique is to identify virus isolates rapidly, IF also has been used to demonstrate the presence of antibodies which are otherwise difficult to detect. For example, antibodies to Epstein-Barr virus and cytomegalovirus usually are measured by the indirect IF technique. [1, 34] A particularly useful refinement of this method is the determination of the immunoglobulin class of antibodies. However, IF is not easy to quantify objectively.

Solid-phase radioimmunoassay (RIA) employs a solid phase of sepharose beads, cellulose particles, or staphylococcal protein A for attachment to either antigen or antibody, and I-125 as antigen or antibody marker. A variety of procedures have been designed. In RIA tests for antibodies to hepatitis viruses A and B, antigen is fixed on polystyrene beads, and antibody in test serum either competes for the antigen with known labeled antibody or is detected by the addition of free labelled antigen in a sandwich effect.[35,36]

Enzyme-immunoassay (EIA) replaces the radioactive label by stable enzyme markers such as horseradish peroxidase or alkaline phosphatase. Antigen is attached to a polystyrene carrier surface, and serum is added. After washing, enzyme- labeled antiglobulin is allowed to react, and washed in turn. Enzyme substrate is then added, which is hydrolysed by captured enzyme and gives a color, the optical density of which is directly proportional to the amount of antibody in the test serum. EIA techniques have been used so far in the study of antibodies to a variety of viruses,[37,38] and may become a valuable tool in diagnostic virology.

Tests for Cell-mediated Immunity

In mumps, measles, herpes zoster, and other infections, cell-mediated immune mechanisms seem to be of considerable importance. Cell-mediated immunity may be assessed in vivo or in vitro. Skin tests which have been performed in several infections, have been compared with other immune responses. [39,40] Lymphocyte activation, the most pertinent in vitro test for cell-mediated immunity, depends on the biochemical changes that immune lymphocytes undergo upon encounter with the antigen. The increase in deoxyribonucleic acid (DNA) synthesis is usually measured by the uptake of tritiated thymidine into the newly synthesized DNA.[41-43] Testing of lymphocyte activation requires the use of lymphocyte separation and culture techniques.[44,45] Cell-mediated immunity against virus-infected target cells which is of relevance to the understanding of pathogenesis, has so far not been used for diagnostic purposes.

GENERAL REFERENCES

Lennette, E. H. and Schmidt, N. J., Ed., *Diagnostic procedures for viral and rickettsial diseases,* American Public Health Association, New York, 19]9.

Russell, W. O., Laboratory diagnosis of viral infections: recent advances and their clinical application, *Am. J. Clin. Pathol.,* 57, 731, 1972.

Hsiung, G. D., *Diagnostic Virology. An Illustrated Handbook,* Yale University Press, New Haven, 1973.

Tyrrell, D. A. J., Immunological methods in virology, in *Handbook of Experimental Immunology,* Vol. 3, Weir, D. M., Ed., Blackwell Scientific Publications, Oxford, 1973, Chap. 37.

Herrmann, E. C., Jr., New concepts and developments in applied diagnostic virology, *Progr. Med. Virol.,* 17, 221, 1974.

Grist, N. R., Ross, C. A., and Bell, E. J., *Diagnostic Methods in Clinical Virology,* Blackwell Scientific Publications, Oxford, 1974.

Rose, N. R. and Friedman, H., Eds., *Manual of Clinical Immunology,* Sect. E, American Society for Microbiology, Washington, D. C., 1976.

REFERENCES

1. **Haire, M. and Hadden, D. S. M.,** Rapid diagnosis of rubella by demonstrating rubella-specific IgM antibodies in the serum by indirect immunofluorescence, *J. Med. Microbiol.,* 5, 237, 1972.
2. **Schmitz, H. and Scherer, M.,** IgM antibodies to Epstein-Barr virus in infectious mononucleosis, *Arch. Gesamte Virusforsch,* 37, 332, 1972.
3. **Banatvala, J. E., Best, J. M., Kennedy, E. A., Smith, E. E., and Spence, M. E.,** A serologic method for demonstrating recent infection by rubella virus, *Br. Med. J.,* 3, 285, 1967.
4. **Ankerst, J., Christensen, P., Kjellén, L., and Kronvall, G.,** A routine diagnostic test for IgA and IgM antibodies to rubella virus: absorption of IgG with *Staphylococcus aureus, J. Inf. Dis.,* 130, 268, 1974.
5. **Handsher, R. and Fogel, A.,** Modified Staphylococcal absorption method used for detecting rubella-specific immunoglobulin M antibodies during a rubella epidemic, *J. Clin. Microbiol.,* 5, 588, 1977.
6. **Vesikari, T. and Vahera, A.,** Rubella: a method for rapid diagnosis of a recent infection by demonstration of the IgM antibodies, *Br. Med. J.,* 1, 221, 1968.
7. **Desmyter, J., South, M. A., and Rawls, W. E.,** The IgM antibody response in rubella during pregnancy, *J. Med. Microbiol.,* 4, 107, 1971.
8. **Burgin-Wolff, A., Hernandez, R., and Just, M.,** Separation of Rubella IgM, IgA and IgG antibodies by gel filtration on agarose, *Lancet* 2, 1278, 1971.
9. **Gupta, J. D., Peterson, C., Stout, M. and Murphy, A. M.,** Single-sample diagnosis of recent rubella by fractionation of antibody on Sephadex G-200 column. *J. Clin. Pathol.,* 24, 547, 1971.

10. **Gocke, D. J. and Howe, C.**, Rapid detection of Australia antigen by counterimmunoelectrophoresis, *J. Immunol.*, 104, 1031, 1970.
11. **Mancini, G., Carbonara, A. O., and Heremans, J. F.**, Immunochemical quantitation of antigens by single radial immunodiffusion, *Immunochemistry*, 2, 235, 1965.
12. **Schild, G. C., Henry-Aymard, M., and Pereira, H. G.**, A quantitative single-radial-diffusion test for immunological studies with influenza virus, *J. Gen. Virol.*, 16, 231, 1972.
13. **Likar, M.**, Latex agglutination test for the detection of arbovirus antibodies, *Pathol., Microbiol.*, 37, 157, 1971.
14. **Schmidt, W. A. K. and Klein, M.**, Ein Latex-Agglutinations-Hemmtest (LAH Test) zur Bestimmung von Antikörpern gegen Enteroviren, *Arch. Gesamte Virusforsch.*, 44, 255, 1974.
15. **Vyas, G. N. and Schulman, N. R.**, Hemagglutination assay for antigen and antibody associated with viral hepatitis, *Science*, 170, 332, 1970.
16. **Russell, S. M., McCahon, D., and Beare, A. S.**, A single radial hemolysis technique for the measurement of influenza antibody, *J. Gen. Virol.*, 27, 1, 1975.
17. **Probert, M. and Russell, S. M.**, Measurement of parainfluenza-3 virus antibody by the single radial hemolysis technique, *J. Clin. Microbiol.*, 2, 157, 1975.
18. **Skaug, K., Orstavik, I., and Ulstrup, J. C.**, Application of the passive hemolysis test for the determination of rubella virus antibodies, *Acta Pathol. Microbiol. Scand. Sect. B*, 83, 367, 1975.
19. **Grillner, L. and Strannegard, O.**, Evaluation of the hemolysis-in-gel test for the screening of rubella immunity and the demonstration of recent infection, *J. Clin. Microbiol.*, 3, 86, 1976.
20. **Grillner, L. and Blomberg, J.**, Hemolysis-in-gel and neutralization tests for determination of antibodies to mumps virus, *J. Clin. Microbiol.*, 4, 11, 1976.
21. **Hierholzer, J. and Tannock, G. A.**, Quantitation of antibody to nonhemagglutinating viruses by single radial hemolysis: serological test for human coronaviruses, *J. Clin. Microbiol.*, 5, 613, 1977.
22. **Rice, C. E.**, The use of the complement-fixation test in the study and diagnosis of viral diseases in man and animals. A review. Part I. Introduction, *Can. J. Comp. Med.*, 24, 126, 1960.
23. **Schmidt, N. J., and Lennette, E. H.**, Recent advances in the sero- diagnosis of virus infections, *Progr. Med. Virol.*, 3, 1, 1961.
24. Standardized diagnostic complement fixation method and adaptation to micro test. Public Health Service Monograph 74, U. S. Printing Office, Washington, 1965.
25. **Blacklow, N. R.**, Echeverria, P., and Smith, D. H., Serological studies with reovirus-like enteritis agent, *Inf. Immun.*, 13, 1563, 1976.
26. **Gust, I. D., Pringle, R. C., Barnes, G. L., Davidson, G. P., and Bishop, R. F.**, Complement-fixing antibody response to rotavirus infection, *J. Clin. Microbiol.*, 5, 125, 1977.
27. **Löffler, H. and Vogt, W.**, Ein Mikrofarbtest zur Bestimmung neutralisierender Antikörpen gegen Poliomyelitisvirus, *Schweiz. Med. Wochenschr.*, 87, 337, 1957.
28. **Kyriazopoulou, V. G. and Bell, E. J.**, A micrometabolic inhibition test for the estimation of poliovirus neutralizing antibodies, *Bull. W.H.O.*, 47, 171, 1972.
29. **Welsh, R. M., Jr., Lampert, P. W., Burner, P. A., and Oldstone, M. B. A.**, Antibody-complement interactions with purified lymphocytic choriomeningitis virus, *Virology*, 73, 59, 1976.
30. **Yoshino, K. and Taniguchi, S.**, Studies on the neutralizing of herpes simplex virus, 1. Appearance of neutralization antibodies having different grades of complement requirement, *Virology*, 26, 44, 1965.
31. **Takabayashi, K. and McIntosh, K.**, Effect of heat-labile factors on the neutralization of vaccinia virus by human sera, *Inf. Immun.* 8, 582, 1973.
32. **Isaacs, A. and Bozzo, A.**, The use of v. cholerae filtrates in destruction of non-specific inhibitor in ferret sera. *Br. J. Exp. Pathol.*, 32, 325, 1951.
33. **Spence, L.**, Kaolin treatment of sera for removal of non-specific inhibitors to Asian strains of influenza virus, *Proc. Soc. Exp. Biol. Med.*, 103, 425, 1960.
34. **Chiang, W., Wentworth, B. B., and Alexander E. R.**, The use of an immunofluorescent technique for the determination of antibodies to cytomegalovirus strains in human sera, *J. Immunol.*, 104, 992, 1970.
35. **Ginsberg, A. L., Conrad, M. E., Bancroft, W. H., Ling, C. M., and Overby, L. R.**, Antibody to Australia antigen: detection with a simple radioimmunoassay, incidence in military populations and role in the prevention of hepatitis B with gamma globulin, *J. Lab. Clin. Med.*, 82, 317, 1973.
36. **Purcell, R. H., Wong, D. C., Moritsugu, Y., Dienstag, J. L., Routenberg, J. A., and Boggs, J. D.**, A microtiter solid-phase radioimmunoassay for hepatitis A antigen and antibody, *J. Immunol.*, 116, 349, 1976.
37. **Engvall, E. and Perlmann, P.**, Enzyme-linked immunosorbent assay, ELISA. III. Quantitation of specific antibodies by enzyme-labeled anti-immuno-globulin in antigen-coated tubes, *J. Immunol.*, 109, 129, 1972.
38. **Bidwell, D. E., Bartlett, A., and voller, A.**, enzyme immunoassays for viral diseases, *J. Inf. Dis.*, 136, Suppl., 274, 1977.

39. Thong, Y. H., Steele, R. W., Vincent, M. M., Hensen, S. A., and Bellanti, J. A., Impaired in vitro cell-mediated immunity to rubella virus during pregnancy, *N. Engl. J. Med.*, 289, 604, 1973.
40. Geme, J. W. St., Jr., Yamauchi, T., Eisenklam, E. J., Noren, G. R., Aase, J. M., Jurmain, R. B., Henn, R. M., Gabel, M. C., Hollister, A. W., and Paumier, R. Immunologic significance of the mumps virus skin test in infants, children and adults, *Am. J. Epidemiol.*, 101, 253, 1975.
41. Gerber, P. and Lucas, S. J., In vitro stimulation of human lymphocytes by Epstein-Barr virus, *Cell. Immunol.*, 5, 318, 1972.
42. Smith, K. A., Chess, L., and Mardiney, M. R., Jr., The characteristics of lymphocyte tritiated thymidine incorporation in response to mumps virus, *Cell. Immunol.*, 5, 597, 1972.
43. Jordon, G. W. and Merigan, T. C., Cell-mediated immunity to varicella-zoster virus: in vitro lymphocyte responses, *J. Infect. Dis.* 130, 495, 1974.
44. Bŏyum, A., Separation of leukocytes from blood and bone marrow, *Scand. J. Clin. Lab. Invest. Suppl.*, 21, 97, 1968.
45. Schellekens, Th. A. and Eijsvoogel, v. P., Lymphocyte transformation in vitro: I. Tissue culture conditions and quantitative measurements, *Clin. Exp. Immunol.*, 3, 571, 1968.

CLASSIFICATION OF VIRUSES

Joseph L. Melnick

Introduction

Until about 1950, little was known about viruses other than their pathogenic effect in causing diseases; thus, any efforts at classification tended to focus on host response rather than on properties of the virus particle. At present, we are approaching the end of an important phase of discovery and characterization of animal viruses. The knowledge thus gained has made it possible to establish and broadly define groupings for these agents, and it is now becoming apparent that most of the major groups of viruses of vertebrates — at least of man and the animals important to man — have been recognized and described. Many of these virus groupings, initially established on tentative and provisional bases, now appear to form "real" families and genera in which the members are indeed related in fundamental ways. For example, the validity of the original grouping of the enteroviruses based on an enteric habitat and small size is being borne out by current studies utilizing sophisticated techniques of modern molecular virology to compare the genetic makeup of different members of the group and their mode of replication.

The shift in emphasis from sketching the broad outlines of the virus kingdom to filling in essential details about viruses and their interrelationships has been recognized by a change in the name of the International Committee on Nomenclature of Viruses (ICNV) to the International Committee on Taxonomy of Viruses (ICTV). The ICNV was established in 1966 by the International Association of Microbiological Societies (IAMS) and published its first report in 1971.[1] Work of the Study Groups and Subcommittees of the ICTV is proceeding, and other active investigators also are examining and reporting data of their special areas of virology from a taxonomic viewpoint. Reports of these groups appear regularly in *Intervirology,* the new journal of the Virology Section of the IAMS. Official decisions of the ICTV made at its meetings held during the Third International Congress for Virology, Madrid, September 10—17, 1975 have been summarized by ICTV President Frank Fenner,[2] and a full report of the results of the work of the ICTV and its committees from 1970 - 1975 has been published as a special issue of *Intervirology*[3]

In addition to the ICTV and its Study Groups, general resources for the data presented here include the author's own reviews on taxonomy of animal viruses, published regularly in an annual series,[4] and a current review by F. A. Murphy,[5] the current Chairman of the Vertebrate Virus Subcommittee of the ICTV. Tables 1, 2, and 3 are schematic diagrams showing separation of viruses of vertebrates into 16 families.[4] Viruses having an RNA genome and cubic symmetry of the nucleocapsid are shown in Table 1, those with an RNA genome and noncubic or unknown capsid symmetry are shown in Table 2, and those with a DNA genome are shown in Table 3.

RNA Viruses (Tables 1 and 2)

Picornaviridae[6]

The smallest of the viruses with RNA genomes exist in at least four genera and several hundred species. At least 70 members of the *Enterovirus* genus are known to infect man; these include polioviruses; coxsackieviruses; echoviruses; and (classified in recent years) new enterovirus serotypes that are assigned sequential numbers, e.g., enterovirus-68, etc., rather than being placed in poorly defined subgroups. Well over 100 viruses infecting human beings belong to the genus *Rhinovirus.* Large numbers of agents from both of these genera also exist for other species.

Table 1
CURRENT CLASSIFICATION OF ANIMAL VIRUSES WITH AN RNA GENOME AND CUBIC CAPSID SYMMETRY

Nucleic acid core	RNA	RNA	RNA	RNA	RNA	RNA	RNA	RNA	RNA	RNA
Capsid symmetry	Cubic	Cubic	Cubic	Cubic	Cubic	Cubic	Cubic	Cubic	Cubic	Cubic
Virion: naked or enveloped	Naked	Naked	Naked	Naked	Naked	Naked	Enveloped	Enveloped	Enveloped	Enveloped
Site of capsid assembly	Cytoplasm	Cytoplasm	Cytoplasm	Cytoplasm	Cytoplasm	Cytoplasm	Cytoplasm	Cytoplasm	Cytoplasm	Cytoplasm
Site of nucleocapsid envelopment							Surface membrane	Surface membrane	Intracytoplasmic membranes	Intracytoplasmic membranes
Reaction to ether (or other lipid solvents)	Resistant (or relatively resistant)[a]	Resistant (or relatively resistant)[a]	Resistant (or relatively resistant)[a]	Resistant (or relatively resistant)[a]	Resistant (or relatively resistant)[a]	Resistant (or relatively resistant)[a]	Sensitive	Sensitive	Sensitive	Sensitive
Number of capsomeres	32	32	32	32	92	32	32 or 42	32 or 42	?	?
Diameter of virion (nm)[b]	20 - 30	20 - 30	40	60–80	75	70	60	60	40	40
Molecular weight of nucleic acid in virion ($\times 10^6$)	2.5	2.8	2–3	12	15	12	4	4	3–4	3–4
Genus names	*Enterovirus*[c]	*Rhinovirus*	*Calicivirus*	*Orbivirus*	*Reovirus*[c]	*Rotavirus*: a probable genus[c]	*Alphavirus*	*Rubivirus*	*Pestivirus*	*Flavivirus*
Family names	Picornaviridae	Picornaviridae		Reoviridae	Reoviridae	Reoviridae	Togaviridae	Togaviridae	Togaviridae	Togaviridae
Type species (or typical member)	Poliovirus type 1	Human rhinovirus type 1A	Vesicular exanthema virus of swine	Bluetongue virus	Human reovirus type 1	Acute diarrhea viruses of young children and of newborn calves and mice	Sindbis virus	Rubella virus	Mucosal disease virus of cattle	Yellow fever virus

Note: Among the RNA-containing viruses of vertebrates, all have single-stranded RNA except members of Reoviridae, the RNA of which is double stranded

[a] For most genera of Picornaviridae and Reoviridae, infectivity is resistant to treatment with lipid solvents; however, the orbiviruses are only partly resistant.

[b] Diameter or diameter × length.

[c] Most of the RNA-containing viruses are sensitive to pH 3 treatment; exceptions are the enteroviruses, reoviruses, and rotaviruses.

Modified from Melnick, J. L., *Prog. Med. Virol.*, 22, 211, 1976. With permission.

Table 2
CURRENT CLASSIFICATION OF ANIMAL VIRUSES WITH AN RNA GENOME AND CAPSID SYMMETRY OTHER THAN CUBIC

Nucleic acid core	RNA							
Capsid symmetry	Helical, probably helical, unsymmetric, or unknown							
Virion: naked or enveloped	Enveloped							
Site of capsid assembly	Cytoplasm							
Site of nucleocapsid envelopment	Surface membrane						Intracytoplasmic membranes	
Reaction to ether (or other lipid solvents)	Sensitive						Sensitive	
Diameter of helix (nm)	6–9	12–15	18	18				2–3
Diameter of virion (nm)[a]	80–120	150–300	150–300	60x180	~100	50–300	80–130	100
Molecular weight of nucleic acid in virion (x 10^6)	4	6–8	6–8	3–4	10–12	3–5	9?	6–7
Genus names	*Influenzavirus*	*Pneumovirus*	*Paramyxovirus*; *Morbillivirus*	*Vesiculovirus*; *Lyssavirus*		*Arenavirus*	*Coronavirus*	*Bunyavirus*; (Other viruses)
Subfamily names					Oncovirinae; Spumavirinae; Lentivirinae			
Family names	Orthomyxoviridae	Paramyxoviridae	Paramyxoviridae	Rhabdoviridae	Retroviridae	Arenaviridae	Coronaviridae	Bunyaviridae
Type species (or typical member)	Influenza virus type A: strain AO/WS/33 (HON1)	Respiratory syncytial virus	Newcastle disease virus; Measles virus	Vesicular stomatitis virus; Rabies virus	Rous sarcoma virus; Foamy viruses; Visna virus	Lymphocytic chorio-meningitis virus	Avian infectious bronchitis virus (IBV)	Bunyamwera virus

Note: Among the RNA-containing viruses of vertebrates, all have single-stranded RNA except members of Reoviridae, the RNA of which is double stranded.

[a] Diameter or diameter × length.

Modified from Melnick, J. L., *Prog. Med. Virol.*, 22, 211, 1976. With permission.

The picornavirus genome is one piece of linear, single-stranded RNA of low molecular weight (approximately 2.5 × 10^6). The RNA is infectious and serves as its own messenger for protein translation. The enteroviruses are acid stable and have a buoyant density in CsCl of approximately 1.34 gm/cm³. In contrast, the rhinoviruses are acid labile and have a higher buoyant density (approximately 1.4 gm/cm³).

Another group of viruses, the caliciviruses, seem to belong in a separate family despite some similarities to the picornaviruses. This group includes viruses which cause swine vesicular exanthema and a respiratory disease of cats and several viruses that infect sea lions.

The diseases caused by picornaviruses range from severe paralysis (paralytic poliomyelitis) to aseptic meningitis, pleurodynia, myocarditis, skin rashes, and common colds. Inapparent infection is very common. Different viruses may produce the same syndrome; on the other hand, the same picornavirus may cause more than a single syndrome. After decades of investigation, it now appears that hepatitis A virus is an enterovirus.

Reoviridae[7]

Members of this virus family share a property unique among the RNA-containing viruses of vertebrates, i.e., they have a double-stranded rather than single-stranded RNA genome. The genome consists of several segments. The capsid has a double shell and the structure of the outer capsid layer is indistinct, but icosahedral symmetry has been demonstrated in the inner capsid layers of all three recognized groups within the family that infect vertebrates: the genera *Reovirus* and *Orbivirus* and the probable genus *Rotavirus*. Members of the genus *Reovirus* have been considered to have 92 capsomeres (however, this number is presently being restudied), while the other two groups have 32 capsomeres. The capsomeres of the orbiviruses are unusually large (10 to 15 nm wide) and appear to be ring shaped. The human reoviruses are found in the enteric tract, but their association with disease is not clear. Members of this genus recovered from lower animals are similar to those of man. The *Orbivirus* genus includes viruses that infect vertebrates but also infect invertebrates; some have been considered to be arboviruses (see below). Several have been recovered only from insects. The diseases caused by orbiviruses include bluetongue, African horsesicknesss, Colorado tick fever of man, and epizootic hemorrhagic disease of deer. The members of the probable genus *Rotavirus* that infect human beings are being increasingly recognized as major pathogens of nonbacterial infantile diarrhea; the gastroenteritis syndrome is clinically more severe and of longer duration than the illness caused by the 27-nm "Norwalk agent" and occurs in sporadic rather than epidemic form. It is one of the commonest childhood illnesses throughout the world, and is a leading cause of death in underdeveloped countries. Other members of this antigenically interrelated rotavirus group include calf diarrhea virus, the virus of epizootic diarrhea of infant mice, SA 11 virus of monkeys, and similar viruses from swine and other species. Much of the initial study of the rotaviruses was accomplished by use of electron microscopy and immune electron microscopy; thus far, isolation of the human members of the group has not been achieved in cell culture. The virus replicates in fetal intestinal organ cultures, and antibody to the virus can be assayed by indirect immunofluorescence.[8]

In addition to the members of Reoviridae that infect vertebrates, there are at least two other groups within the family that are considered genera although they do not yet have official generic names: the cytoplasmic polyhedrosis group of insect viruses and the plant reovirus group.

Togaviridae[5]

Members of this family include most arboviruses* of antigenic groups A and B now classed in the genera *Alphavirus* (group A) and *Flavivirus* (group B) and in newly designated genera which include nonarbo togaviruses, rubella (*Rubivirus*), and the mucosal disease virus group (*Pestivirus*). The virions are spherical, 40 to 70 nm in diame-

* One important and well-known virus group name that does not appear in the diagrams is an ecologic category, the arbovirus group.[9] The more than 350 arthropod-borne viruses survive through a complex cycle involving vertebrate hosts and arthropods which serve as vectors, transmitting the viruses by their bites. This grouping, based on transmission, remains a useful one despite the wide diversity of its members in regard to properties of the virion. The vast majority of arboviruses have now been sufficiently well characterized to permit their taxonomic placement. Their classic serologic interrelationships previously delineated by arbovirologists have been found to be paralleled by morphologic similarities, and these serologic relationships have tremendously speeded the taxonomic placement process by providing clues as to where to look. Once some of the members of a serologic group have been characterized in terms of biophysical and biochemical properties, attention of taxonomists can be focused on their antigenic relatives. Arboviruses are now included in a number of families, primarily Togaviridae, Bunyaviridae, Rhabdoviridae, and Reoviridae.

ter, and have a lipoprotein envelope with lipid and virus-specified glycopeptide tightly applied to an icosahedral nucleocapsid. The genome is a single molecule of single-stranded RNA. The alphaviruses and flaviviruses include many of the major human arboviral pathogens. The viruses of Venezuelan, eastern, and western equine encephalitis are alphaviruses and the viruses of yellow fever, dengue, Japanese encephalitis, St. Louis encephalitis, Omsk hemorrhagic fever, and Russian spring-summer encephalitis are flaviviruses. Thus far, rubella virus is the only member assigned to the genus *Rubivirus*, but equine arteritis virus may also belong to this genus. Members of the *Pestivirus* genus include bovine virus diarrhea (mucosal disease complex) virus, hog cholera (European swine fever) virus, border disease virus, and probably also lactic dehydrogenase virus of mice and simian hemorrhagic fever virus.

Orthomyxoviridae[10]

Those recognized to date are the influenza viruses, which may be spherical, elongated, or filamentous. Most members of the family have "spikes" projecting from the surface of the envelope; these are glycosylated protein peplomers 10 to 14 nm long and 4 nm in diameter consisting of two types, the hemagglutinin and the neuraminidase. The helically symmetric ribonucleoprotein capsid has RNA strands 6 to 9 nm in diameter (6 nm for influenza types A and B and 9 nm for type C). The RNA is single stranded and made up of eight segments. An RNA-dependent RNA polymerase is associated with purified virions. During replication the helical nucleocapsid is first detected in the nucleus, whereas the hemagglutinin and neuraminidase are formed in the cytoplasm. The virus matures by budding at the cell surface membrane. The orthomyxoviruses are sensitive to dactinomycin.

The genus *Influenzavirus* has been established and includes viruses of types A and B; type C is considered only a "probable genus." Antigenic variation is common, especially among type A members. Recombination occurs with high frequency within species, but not between types. Type A influenza viruses include agents of human, equine, and swine influenza and fowl plague. For types B and C, only human strains are known.

Paramyxoviridae[3]

Usually spherical, these virions may also be pleomorphic. They are 150 nm or larger in diameter, and filamentous forms may be several thousand nanometers long. On the lipid bilayer envelope are surface projections. The RNA is single stranded and in unsegmented, linear form; the helical nucleocapsid is 12 to 18 nm in diameter. Virions are formed in the cytoplasm by budding from the plasma membrane. Infectivity is sensitive to ether, acid, and heat, but paramyxoviruses are resistant to dactinomycin. The genera include *Paramyxovirus* (parainfluenza viruses, mumps virus, Newcastle disease virus, Yucaipa, and other avian paramyxoviruses), *Morbillivirus* (the viruses of measles, canine distemper, rinderpest, and peste de petite ruminant), and *Pneumovirus* (respiratory syncytial viruses of man and cattle and pneumonia virus of mice). Members of the genus *Paramyxovirus* contain both hemagglutinin and neuraminidase in the virion. *Morbillivirus* members contain hemagglutinin but not neuraminidase in the viral envelope, while members of the genus *Pneumovirus* contain neither hemagglutinin nor neuraminidase in the virions. Members of the Paramyxoviridae are genetically stable and genetic recombination does not occur.

Rhabdoviridae[3]

Members of this family have enveloped virions that are rod shaped, resembling a bullet (with one end rounded and the other flattened), or bacilliform. Enclosed within the lipoprotein envelope and membrane protein is the long, tubular nucleocapsid with helical symmetry. The genome of single-stranded RNA is in unsegmented, linear form.

Members of some genera multiply in arthropods as well as in vertebrates or higher plants; others multiply only in insects. Infectivity is sensitive to ether, acid , and heat. The genera that infect vertebrates are *Lyssavirus* (including rabies virus, Duvenhage virus, and Mokola virus, all infecting man; Lagos bat virus; and several agents thus far isolated only from insects) and *Vesiculovirus* (including vesicular stomatitis virus and a number of antigenically related viruses). Among the vesiculoviruses are Chandipura virus of man), Flanders-Hart Park virus (of mosquitoes and birds), Kern Canyon virus (of bats), Piry virus (of opossums), and probably also bovine ephemeral fever virus, Mt. Elgon virus (of bats), Egtved virus (viral hemorrhagic septicemia virus, of trout), and several other fish viruses. In addition, a genus *Sigmavirus* has been established for the rhabdoviruses of Drosophila, and there are also numerous rhabdoviruses of plants. Marburg virus, a simian virus highly pathogenic for man, is rhabdovirus-like in most characteristics, but has very elongated forms.

Retroviridae[11,12]

Much remains to be firmly established concerning this family, but much progress has been made. The members include not only the RNA tumor viruses ("oncornaviruses" and "leukoviruses"), which are now assigned to the subfamily Oncovirinae, but also the slow viruses of the Maedi/Visna group, now subfamily Lentivirinae, and the foamy virus group (agents which form syncytia in cell cultures), now assigned to the subfamily Spumavirinae.

Members of the family characteristically have a reverse transcriptase (RNA-dependent DNA polymerase) within the virion. In the most thoroughly studied members, the lipoprotein envelope encloses an inner shell with icosahedral symmetry and a central core or nucleocapsid with helical symmetry. The genome is a single molecule of single-stranded RNA which dissociates readily into 2 or 3 pieces. Infectivity is sensitive to ether, acid, and heat. Replication of the viral RNA involves a DNA provirus which is integrated into host DNA.

Arenaviridae[13]

Arenaviridae have spherical or pleomorphic virions with a dense lipid bilayer membrane bearing surface projections. The virion core contains electron-dense RNA-containing granules approximately 20 to 30 nm in diameter which resemble ribosomes in size, shape, and density. The RNA is single stranded and consists of four large and one to three small segments. Virions are formed by budding from the surface membrane. Viral RNA is probably transcribed by virion polymerase into complementary RNA which probably acts as mRNA. Infectivity is sensitive to ether, acid, and heat. Most member viruses have a single restricted rodent host in which persistent infection occurs accompanied by viremia, viruria, or both; spread to other mammals and to man occurs, but is rare. Members include lymphocytic choriomeningitis virus which infects mice, but may spread to man; Lassa virus; and members of the Tacaribe complex — Junin and Machupo viruses of South American hemorrhagic fevers, Pichinde virus, and several other viral agents which have been isolated from arthropods but for which natural transmission cycles involving arthropod vectors have not been demonstrated.

Coronaviridae[14]

This family is named for unique petal-shaped or club-shaped peplomers which project from the envelope and, in negatively stained electron micrographs, form a fringe resembling the solar corona. The interior structure of the virion is not fully resolved, but is probably a loosely wound, helically symmetric nucleocapsid. The genome consists of one large molecule of single-stranded RNA. Infectivity is sensitive to ether, acids, and heat. Nucleocapsids develop in the cytoplasm and mature by budding through intracytoplasmic membranes. Several serotypes of human coronaviruses have been isolated from patients with acute upper respiratory tract illnesses, primarily through the use of human embryonic tracheal and nasal organ cultures. There are several coronaviruses indigenous to lower animals. They include avian infectious bronchitis virus, turkey bluecomb disease virus, mouse hepatitis virus, porcine transmissible gastroenteritis virus and hemagglutinating encephalitis virus, calf neonatal diarrhea coronavirus, and at least two rat coronaviruses.

Bunyaviridae[15]

This family is the largest and most recently recognized taxonomic grouping assigned to an antigenically interrelated set of arboviruses. There are at least 150 members; more than 85 belong to the Bunyamwera supergroup of arboviruses, which consists of 10 serologically cross-related groups and several ungrouped arboviruses. The virions are spherical and on their unit-membrane envelope have surface projections which may be randomly placed or clustered in arrays with icosahedral symmetry; the envelope contains at least one virus-specified glycopeptide. The internal ribonucleoprotein is helically wound and symmetric, with long strands 2 to 2.5 nm broad. The single-stranded RNA probably exists in three circular segments. The virions develop in the cytoplasm and mature by budding through intracytoplasmic membranes into smooth-surfaced vesicles in the Golgi region or nearby. Infectivity is sensitive to ether, acid, and heat. Virus particles hemagglutinate. Members of the family produce a number of important diseases of man and domestic animals, e. g., California encephalitis, Crimean hemorrhagic fever, sandfly fever, Rift Valley Fever, and Nairobi sheep disease. Most Bunyaviridae members are transmitted by mosquitoes, but some are tick-borne. In addition to the viruses already assigned to the genus *Bunyavirus* (confined at present to members of the serologic Bunyamwera supergroup), a number of other arboviruses are not serologically related, but are generally very similar to bunyaviruses in most properties. These agents are considered to be possible members of the family, perhaps in different genera yet to be designated. The largest antigenic grouping among these is the Uukuniemi group, antigenically unrelated to *Bunyavirus* and transmitted by ticks.

DNA Viruses (Table 3)

Parvoviridae[16]

Originally named picodnaviruses to reflect their small size and DNA-containing genome,[17] the family of Parvoviridae now includes two named genera, *Parvovirus* and *Densovirus,* and a "probable genus," the adeno-associated virus group. A typical member is adeno-associated satellite virus, several serotypes of which are indigenous to man. As shown beginning at the top of Table 3, these are DNA-containing viruses that have cubic symmetry and a naked (unenveloped) nucleocapsid. During replication, capsid assembly takes place in the nucleus of the host cell. Infectivity of these viruses is resistant to ether and other lipid solvents, the capsid has 32 capsomeres, the diameter of the virus particle is 18 to 26 nm, and the molecular weight of the nucleic is 1.5 to 2.2×10^6. The capsomeres which form the outer layer of the nucleocapsid are each 2 to 4 nm in diameter.

The genus *Parvovirus* includes autonomously replicating members of the group that infect vertebrates: hamster osteolytic H viruses; latent rat viruses (Kilham rat virus, X14 virus); minute virus of mice (MVM); and parvoviruses of swine, cattle, cats, and other species. The members of the *Densovirus* genus are viruses of insects, but also are capable of producing cytopathic effects in L cells of vertebrates; they replicate autonomously. However, the adeno-satellite viruses are defective and cannot multiply

Table 3
CURRENT CLASSIFICATION OF ANIMAL VIRUSES WITH A DNA GENOME

Nucleic acid core	DNA	DNA	DNA	DNA	DNA	DNA
Capsid symmetry	Cubic	Cubic	Cubic	Cubic	Cubic	Complex
Virion: naked or enveloped	Naked	Naked	Naked	Enveloped	Enveloped	Complex coats
Site of capsid assembly[a]	Nucleus	Nucleus	Nucleus	Cytoplasm	Nucleus	Cytoplasm
Site of nucleocapsid envelopment				Cytoplasmic membrane	Nuclear membrane	
Reaction to ether (or other lipid solvents)	Resistant	Resistant	Resistant	Sensitive[b]	Sensitive	Resistant[c]
Number of capsomeres	32	72	252	1500	162	
Diameter of virion (nm)[d]	18–26	45–55	70–90	130	100[e]	230 x 300
Molecular weight of nucleic acid (x 10⁶)	1.5–2.2	2.4–5.0	20-30	130	54–92	160
Genus names	*Parvovirus*; (*Adeno-associated virus group*); *Densovirus*	*Papillomavirus*; *Polyomavirus*	*Mastadenovirus*; *Aviadenovirus*	*Iridovirus*	*Herpesvirus*; (*many other genera not yet defined*)	*Orthopoxvirus*[c]; *Avipoxvirus*[c]; *Capripoxvirus*; *Leporipoxvirus*; *Parapoxvirus*; *Entomopoxvirus*
Family names	Parvoviridae	Papovaviridae	Adenoviridae	Iridoviridae[b]	Herpesviridae	Poxviridae
Type species (or typical member)	Latent rat virus (Kilham); Adeno-associated satellite virus (AAV) type 1; Densonucleosis virus of *Galleria*	Rabbit papilloma (Shope papilloma) virus; Polyoma virus	Human adenovirus type 1; CELO virus	(African swine fever virus: iridescent virus of *Tipula*)	Herpes simplex virus	Vaccinia virus

Note: Among the DNA-containing viruses of vertebrates, all have double-stranded DNA except members of the Parvoviridae, the DNA of which is single stranded within the virion.

[a] For the DNA viruses whose capsid assembly takes place in the nucleus, a phase of replication occurs in the cytoplasm, as evidenced by the detection of viral messenger RNA associated with polyribosomes.
[b] Some members of Iridoviridae are enveloped (particularly the African swine fever virus and other similar viruses of vertebrates); the insect iridescent virus group members have no envelope, but a lipid fraction is present in these viruses also.
[c] Members of the genera *Orthopoxvirus* and *Avipoxvirus* are ether-resistant, although other members of Poxviridae are sensitive to another lipid solvent (chloroform).
[d] Diameter or diameter × length.
[e] The naked virus is 100 nm in diameter; however, enveloped virions range up to 200 nm.

Modified from Melnick, J. L., *Prog. Med. Virol.*, 22, 211, 1976. With permission.

in the absence of a replicating adenovirus which serves as a "helper virus." Herpesvirus can act as a partial helper; in cells coinfected with herpesvirus, infectious satellite DNA and capsid proteins are made, but they are not assembled into satellite virions.

Members of the Parvoviridae are the only DNA-containing viruses of vertebrates in which the DNA genome is single stranded within the virion; all the others shown in Table 3 have double-stranded DNA. In the case of adeno-satellite viruses and densoviruses, separate virions contain single strands of positive or negative DNA; these strands are complementary and come together outside the cell to form a double strand. In the members of the genus *Parvovirus,* however, the DNA in the virion is a positive strand only. Members of this genus show marked preference for actively dividing cells, have been shown to be transmissible transplacentally, and are receiving attention for their special disease potential in fetuses and neonates.[18,19]

Papovaviridae[20,21]

These relatively small, ether-resistant viruses contain double-stranded DNA in circular form. Many are unusually heat stable, surviving temperatures that inactivate most viruses. The representatives that infect human beings are the papilloma or wart virus and SV40-like viruses such as JC virus, which has been isolated from the brain tissue of patients with progressive multifocal leukoencephalopathy (PML), or BK virus, which has been isolated from the urine of immunosuppressed recipients of renal transplants.[22] Other members include papilloma viruses of several vertebrate species, polyoma and K viruses of mice, and vacuolating viruses of monkeys (SV40) and rabbits. These viruses have relatively slow growth cycles characterized by replication within the nucleus. Papovaviruses produce latent and chronic infections in their natural hosts, and all are tumorigenic in at least some animal host species. The genome, a single cyclic molecule of double-stranded DNA, integrates into cellular chromosomes of transformed cells. The capsid antigens of JC and BK viruses are unique, but the tumor antigen induced by each cross-reacts antigenically with that induced by SV40.

When SV40 and adenoviruses replicate together within the same cell, they may interact to form various kinds of SV40/adenovirus "hybrid" virus particles in which a portion of the SV40 genome is covalently linked to incomplete or complete adenovirus DNA and is carried within an adenovirus capsid.

Adenoviridae[23]

Among these medium-sized viruses at least 33 serotypes infect man, and there are distinct serotypes for a number of other species. The virion is a nonenveloped, isometric particle with 252 capsomeres, each 7 to 9 nm in diameter. Vertex capsomeres are antigenically distinct from the others and carry one or two filamentous projections. The adenovirus genome is a single linear molecule of double-stranded DNA.

Adenoviruses have a predilection for mucous membranes and may persist for years in lymphoid tissue. Some of the adenoviruses cause acute respiratory diseases, febrile catarrhs, pharyngitis, and conjunctivitis. Human adenoviruses rarely cause disease in laboratory animals, but certain serotypes produce tumors in newborn hamsters. Common antigens are shared by all mammalian adenoviruses, which now have been classed as members of the *Mastadenovirus* genus; these antigens differ from the corresponding antigens of the members of the genus *Aviadenovirus.*

Iridoviridae[24,25]

The most-studied members of this family are the members of the insect iridescent virus group (e.g., *Tipula* iridescent virus), now placed in the genus *Iridovirus.* However, other important viruses that are considered to be probable members of this family include African swine fever virus and a large number of viruses of frogs and of fish. No human iridovirus is known. The vertebrate iridoviruses are enveloped, whereas the iridoviruses that infect insects have a lipid fraction present in the virion, but do not have an envelope as such. The genome is a single very large molecule of double-stranded DNA.

Herpesviridae[26]

The herpesvirus virion has a lipid-containing envelope surrounding an icosadeltahedral capsid with 162 partially hollow capsomeres. Within the capsid is a core consisting of linear, double-stranded DNA wrapped around a protein spool.

Herpesviruses are noteworthy for their ability to establish latent and/or persistent infections. Latent infections may last for the lifetime of the host, even in the presence of circulating antibodies. Special interest has been generated also by the association of Epstein-Barr (EB) herpesvirus with human Burkitt's lymphomas and nasopharyngeal carcinomas and by the possible role of the genital herpesvirus, herpes simplex type 2, in cancer of the uterine cervix. Several simian herpesviruses have been shown to be oncogenic in experimentally infected animals. Infections of heterologous species are in many cases very serious; examples are infection of man by one of the simian herpesviruses and the infection of cattle by pseudorabies virus (a virus indigenous to swine). Human diseases include oral and genital herpes, chickenpox, and shingles due to varicella/zoster virus, cytomegalic inclusion disease, and infectious mononucleosis.

Clearly, there are numerous members of the herpesvirus family; however, because of the nature of the persons working in this area, classification has not been agreed upon and to date only one genus has been formally recognized: *Herpesvirus,* the herpes simplex virus group. A classification of herpesviruses into subgroups A and B based on whether the virus was cell free or cell associated proved useful before the days of molecular virology.[27]

Poxviridae[28]

These large viruses are brick shaped or ovoid, with a complex virion structure. An external coat contains lipid and tubular or globular protein structures and encloses one or two lateral bodies and an internal body (core) which contains the genome. The virion contains more than 30 structural proteins and several enzymes, including a DNA-dependent RNA polymerase. The genome consists of a single molecule of double-stranded DNA. Genetic recombination occurs within genera. Nongenetic reactivation occurs both within and between genera of the poxviruses that infect vertebrates. Most poxviruses of vertebrates share at least one antigen and members of each genus of vertebrate poxviruses have other antigens in common. This is the major DNA-containing virus family the members of which replicate entirely within the cytoplasm. A number of them produce intracytoplasmic inclusion bodies. The family has been divided into six genera. The genus *Orthopoxvirus* includes the poxviruses of man. This genus produces a hemagglutinin separate from the virion; it is serologically specific and is a lipid-rich pleomorphic particle 50 to 65 nm in diameter.

Table 4 lists the family names of the major virus groups that infect vertebrates together with the cryptogram approved by the ICTV for each family.[3] The cryptograms are the current forms developed by the ICTV to serve as a code of symbols providing a compact and uniform summation of certain fundamental properties of viruses of animals, plants, insects, and bacteria. Each cryptogram consists of four sets of symbols, as described in Table 4.

Table 4
CRYPTOGRAMS OF VERTEBRATE VIRUS FAMILIES AS APPROVED BY THE ICTV[3]

Family	Symbols of cryptogram								
	1st pair[a]		2nd Pair[b]		3rd pair[c]		4th set[d]		
	NA /	NA strand(s)	mol wt NA (millions)	/ %NA	Particle outline (plus envelope or occluding protein) /	Nucleocapsid outline	Host(s) /	Transmission /	Vector(s)
Parvoviridae	[D /	1 :	1.5—2.2	/ 19—32 :	S /	S :	I, V /	C, I, O, R]	
Papovaviridae	[D /	2 :	3—5	/ 12 :	S /	S :	V /	O, Ve	/ Ac, Si]
Adenoviridae	[D /	2 :	20—30	/ 12—17 :	S /	S :	V /	I, O, R]	
Iridoviridae	[D /	2 :	130—160	/ 12—16 :	S or Se /	S :	I, V /	C, I, O, Ve	/ Ac]
Herpesviridae	[D /	2 :	92—102	/ 8.5 :	Se /	S :	(F),(I),V /	C, O, R]	
Poxviridae	[D /	2 :	130—240	/ 5—7.5 :	X /	* :	I,V /	O, R, Ve	/ Ac, Di, Si]
Picornaviridae	[R /	1 :	2.3—2.8	/ 30 :	S /	S :	I, V /	I, O, R]	
Reoviridae	[R /	2 :	Σ12—Σ19	/ 15—30 :	S or So /	S :	I, S, V /	I, O, Ve	/ Ac,Au,Di]
Togaviridae	[R /	1 :	3.5— 4	/ 5—8 :	Se /	S :	I, V /	C, I, O, R, Ve	/ Ac, Di]
Orthomyxo-viridae	[R /	1 :	Σ4	/ 1 :	Se /	E :	V /	R]	
Paramyxoviridae	[R /	1 :	5—8	/ 1 :	Se /	E :	V /	O, R]	
Rhabdoviridae	[R /	1 :	3.5—4.6	/ 2—3 :	Ue /	E :	I,S,V /	C, O, Ve	/ Ap,Au,Di]
Retroviridae	[R /	1 :	7—10	/ 2 :	Se /	* :	V /	C, I, O, R]	
Arenaviridae	[R /	1 :	Σ3.2—Σ5.6	/ * :	Se /	* :	V /	C, O]	
Coronaviridae	[R /	1 :	9	/ * :	Se /	E :	V /	I, R]	
Bunyaviridae	[R /	1 :	Σ6—Σ7	/ * :	Se /	E :	I, V /	C, Ve	/ Ac, Di]

Note: *, property is not known or not fully known; ( ), enclosed information is doubtful or unconfirmed.

[a] 1st pair: Type of nucleic acid / strandedness of nucleic acid (NA/NA strand[s]).

Symbols for type of nucleic acid: R = RNA, D = DNA.
Symbols for strandedness: 1 = single stranded, 2 = double stranded.

[b] 2nd pair: Molecular weight of nucleic acid (in millions) / percentage of nucleic acid in infective particles (mol wt NA/% NA).
This set describes the genome of infective particles. The genome of some viruses is divided. When different pieces of the genome occur together in one particle, the symbol Σ indicates the total molecular weight of the pieces in the particle (e.g., Reoviridae: R/2 : Σ 12 to Σ 19), but when the pieces occur in different particles, the composition of each particle type is listed separately (e.g., tobacco rattle virus: R/1 : $\frac{2\cdot 3}{5} + \frac{0\cdot 6 - 1\cdot 3}{5}$).

Table 4 (continued)
CRYPTOGRAMS OF VERTEBRATE VIRUS FAMILIES AS APPROVED BY THE ICTV[3]

[c] 3rd pair: Outline of particle, presence of envelope or occluding protein/outline of nucleocapsid (the nucleic acid plus the protein most closely in contact with it).

Symbols for both properties:

S = essentially spherical
E = elongated with parallel sides, ends not rounded
U = elongated with parallel sides, end(s) rounded
X = complex or none of above
e in first term = presence of viral envelope
o in first term = virion occluded in viral protein matrix

[d] 4th set (two or three terms): Kind(s) of host infected/mode(s) of transmission/ kind(s) of vector.

Third term is omitted if no vector is known.

If an invertebrate animal which transmits the virus to a plant or animal supports multiplication of the virus, the letter "I" appears in the first term (as well as "S" and/or "V"), "Ve" in the second term, and the appropriate abbreviation(s) for the specific vector(s) in the third term. Many viruses of both plants and vertebrates are transmitted mechanically by invertebrate vectors; "I" does not then appear in the first term, but "Ve" occurs in the second term and the appropriate abbreviation(s) for the vector(s) in the third term.

Symbols for kinds of host:

A = alga
B = bacterium
F = fungus
I = invertebrate
S = seed plant
V = vertebrate

Symbols for modes of transmission:

C = congenital
I = intestinal tract (ingestion)
O = contact (contaminated environment)
R = respiratory tract (inhalation)
Ve = invertebrate vector

Symbols for kinds of vector:

Arachnida
Ac = Acarina (mite and tick)
Insecta
Ap = Aphididae, Hemiptera (aphid)
Au = Auchenorrhyncha, Hemiptera (leaf-, plant-, or tree-hopper)
Cl = Coleoptera (beetle)
Di = Diptera (fly and mosquito)
Si = Siphonaptera (flea)
Th = Thysanoptera (thrips)

Other vectors:

Fu = Fungi, Chytridiales and Plasmodiophales (fungus)
Ne = Nematoda (nematode)

For example, Table 4 indicates the following properties for the family Parvoviridae:

- The genome is DNA / single stranded.
- The nucleic acid has a molecular weight of 1.5 to 2.2×10^6 / constituting approximately 12% of the infective particle.
- The particle is essentially spherical in shape / the nucleocapsid is also spherical.
- Both invertebrate and vertebrate hosts are known / transmission may be congenital (C), by ingestion (I), by contact with contaminated environment (O), or through the respiratory tract by inhalation (R).

To take another example, Reoviridae members are shown to have

- An RNA genome / double stranded.
- Nucleic acid in segments with molecular weight totaling 12 to 19×10^6 / nucleic acid constituting 15 to 30% of the infective particle.
- Particle shape either spherical or spherical with the virion occluded in a viral protein matrix / spherical nucleocapsid shape.
- Hosts include invertebrates, seed plants, and vertebrates / transmission by ingestion, contact, or by means of an invertebrate vector / vectors include Acarina (mites and ticks), Auchenorrhyncha (leaf-, plant-, or tree-hopper), and Diptera (flies or mosquitoes).

GENERAL REFERENCES

Wildy, P., Classification and nomenclature of viruses: first report of the International Committee on Nomenclature of Viruses, *Monogr. Virol.*, Vol. 5, 1971.

Fenner, F., The classification and nomenclature of viruses. Summary of results of meetings of the International Committee on Taxonomy of Viruses in Madrid, September 1975, *Intervirology*, 6, 1, 1975/1976.

Fenner, F., Classification and nomenclature of viruses: second report of the International Committee on Taxonomy of Viruses, *Intervirology*, 7, 1, 1976.

Melnick, J. L., Taxonomy of viruses, 1976, *Prog. Med. Virol.*, 22, 211, 1976; also taxonomy articles in preceding volumes of this series beginning with Vol. 9, 1967.

Murphy, F. A., Taxonomy of vertebrate viruses, in *CRC Handbook of Microbiology*, Vol. 2, 2nd ed., Laskin, A. I. and Lechevalier, H., Eds., CRC Press, Cleveland, 1978, in press .

REFERENCES

1. **Wildy, P.**, Classification and nomenclature of viruses: first report of the International Committee on Nomenclature of Viruses, *Monogr. Virol.*, Vol. 5, 1971.
2. **Fenner, F.**, The classification and nomenclature of viruses. Summary of results of meetings of the International Committee on Taxonomy of Viruses in Madrid, September 1975, *Intervirology*, 6, 1, 1975/1976.
3. **Fenner, F.**, Classification and nomenclature of viruses: second report of the International Committee on Taxonomy of Viruses, *Intervirology*, 7, 1, 1976.
4. **Melnick, J. L.**, Taxonomy of viruses, 1976, *Prog. Med. Virol.*, 22, 211, 1976; also taxonomy articles in preceding volumes of this series beginning with Vol. 9, 1967.
5. **Murphy, F. A.**, Taxonomy of vertebrate viruses, in *CRC Handbook of Microbiology*, Vol. 2, 2nd ed., Laskin, A. I. and Lechevalier, H., Eds., CRC Press, Cleveland, 1978, in press.
6. **Melnick, J. L., Agol, V. I., Bachrach, H. L., Brown, F., Cooper, P. D., Fiers, W., Gard, S., Gear, J. H. S., Ghendon, Y., Kasza, L., LaPlaca, M., Mandel, B., McGregor, S., Mohanty, S. B., Plummer, G., Rueckert, R. R., Schaffer, F. L., Tagaya, I., Tyrrell, D. A. J., Voroshilova, M., and Wenner, H. A.**, Picornaviridae, *Intervirology*, 4, 303, 1974.
7. **Joklik, W. K. et al.**, Reoviridae, *Intervirology*, in press.
8. **Holmes, I. H.**, Viral gastroenteritis, *Prog. Med. Virol.*, 25, in press, 1979.
9. **Berge, T. O.**, International Catalogue of Arboviruses Including Certain Other Viruses of Vertebrates, 2nd ed ., DHEW Publ. No. (CDC) 75-8301, U. S. Department of Health, Education, and Welfare, Washington, D. C., 1975.
10. **Dowdle, W R., Davenport, F. M., Fukumi, H., Schild, G. C., Tumova, B., Webster, R. G. and Zakstelskaja, L. Ya.**, Orthomyxoviridae, *Intervirology*, 5, 245, 1975.
11. **Dalton, A. J., Melnick, J. L., Bauer, H., Beaudreau, G., Bentvelzen, P., Bolognesi, D., Gallo, R., Graffi, A., Haguenau, F., Heston, W., Huebner, R., Todaro, G., and Heine, U. I.**, The case for a family of reverse transcriptase viruses: Retraviridae, *Intervirology*, 4, 201, 1974.
12. **Vogt, P. K.**, The oncovirinae, in *Proceedings of the Official Opening, Consultation and International Workshop on Advances in Comparative Virology*, World Health Organization Collaborating Centre for Collection and Evaluation of Data on Comparative Virology, Munich, 1976, 327.
13. **Pfau, C. J., Bergold, G. H., Casals, J., Johnson, K. M., Murphy, F. A., Pedersen, I. R., Rawls, W. E., Rowe, W. P., Webb, P. A., and Weissenbacher, M. C.**, Arenaviruses, *Intervirology*, 4, 207, 1974.
14. **Tyrrell, D. A. J., Almeida, J. D., Cunningham, C. H., Dowdle, W. R., Hofstad, M. S., McIntosh, K., Tajima, M., Zakstelskaya, L. Ya., Easterday, B. C., Kapikian, A., and Bingham, R. W.**, Coronaviridae, *Intervirology*, 5, 76, 1975.
15. **Porterfield, J. S., Casals, J., Chumakov, M. P., Gaidamovich, S. Ya., Hannoun, C., Holmes, I. H., Horzinek, M. C., Mussgay, M., Oker-Blom, N., and Russell, P. K.**, Bunyaviruses and Bunyaviridae, *Intervirology*, 6, 13, 1975/76.
16. **Bachmann, P. A., Hoggan, M. D., Melnick, J. L., Pereira, H. G., and Vago, C.**, Parvoviridae, *Intervirology*, 5, 83, 1975.
17. **Mayor, H. D. and Melnick, J. L.**, Small deoxyribonucleic acid-containing viruses (picodnavirus group), *Nature*, 210, 331, 1966.
18. **Kilham, L. and Margolis, G.**, Problems of human concern arising from animal models of intrauterine and neonatal infections due to viruses: a review. I. Introduction and virologic studies, *Prog. Med. Virol.*, 20, 113, 1975.

19. **Margolis, G. and Kilham, L.**, Problems of human concern arising from animal models of intrauterine and neonatal infections due to viruses: a review. II. Pathologic studies, *Prog. Med. Virol.*, 20, 144, 1975.
20. **Melnick, J. L.**, Papova virus group , *Science*, 135, 1128, 1962.
21. **Melnick, J. L., Allison, A. C., Butel, J. S., Eckhart, W., Eddy, B. E., Kit, S., Levine, A. J., Miles, J. A. R., Pagano, J. S., Sachs, L., and Vonka, V.**, Papovaviridae, *Intervirology*, 3, 106, 1974.
22. **Padgett, B. L. and Walker, D. L.**, New human papovaviruses, *Prog. Med. Virol.*, 22, 1, 1976.
23. **Norrby, E., Bartha, A., Boulanger, P., Dreizin, R. S., Ginsberg, H. S., Kalter, S. S., Kawamura, H., Rowe, W. P., Russell, W. C., Schlesinger, R. W., and Wigand, R.**, Adenoviridae, *Intervirology*, 7, 117, 1976.
24. **Bellett, A. J. D.**, The iridescent virus group, *Adv. Virus Res.*, 13, 225, 1968.
25. **Kelly, D. C. and Robertson, J. S.**, Icosahedral cytoplasmic deoxyviruses, *J. Gen. Virol.*, 20(Suppl.), 17, 1973.
26. **O'Callaghan, D. J. and Randall, C. C.**, Molecular anatomy of herpesviruses: recent studies, *Prog. Med. Virol.*, 22, 152, 1976.
27. **Melnick, J. L., Midulla, M., Wimberly, I., Barrera-Oro, J. G., and Levy, B. M.**, A new member of the herpesvirus group isolated from South American marmosets, *J. Immunol.*, 92, 596, 1964.
28. **Fenner, F., Pereira, H. G., Porterfield, J. S., Joklik, W. K., and Downie, A. W.**, Family and generic names for viruses approved by the International Committee on Taxonomy of Viruses, June 1974, *Intervirology*, 3, 193, 1974.

ADDENDUM

As virus taxonomy is in a phase of active advancement, certain of the details given here have been superseded by new developments and decisions since the time of writing. However, the fundamental principles of virus taxonomy, the general approach to taxonomic problems, and most of the major groupings of vertebrate viruses, are well established. For new information about virus properties, as well as new recommendations and decisions on taxonomy, the reader is referred to current and future volumes of *Intervirology* and *Progress in Medical Virology*.

LIGHT AND ELECTRON MICROSCOPY OF VIRUS-INFECTED CELLS

C. K. Y. Fong and G. D. Hsiung

Introduction

Light and electron microscopy, either alone or in combination, provide excellent tools for the laboratory diagnosis of virus infection. Although viral cytopathology remains one of the most commonly employed methods for recognizing the presence of virus in an infected cell or cell culture, it is only through the use of electron microscopy that the virus particle itself, its size and shape, can be seen directly.

Electron microscopy has been used for the diagnosis of poxvirus infections for almost 30 years. However, it is only recently that its application has been extended to encompass the recognition and identification of a variety of other viral diseases. The techniques most commonly used include negative staining with electron dense salts such as phosphotungstic acid or the preparation of ultrathin sections of cells or tissues suspected of harboring virus.

In the absence of facilities for the electron microscopic examination of specimens, the most reliable method for detecting evidence of viruses is to promote virus growth in cell cultures which can be examined by means of light microscopy of either live, unfixed cell cultures, or fixed, stained preparations. Each virus group often exhibits a distinctive characteristic cytopathic effect (CPE) in a specific cell culture system(s), thereby providing clues through which a presumptive identification of a clinical isolate can be made. In some instances, where typical CPE is not produced in cell cultures, evidence of virus growth can be obtained by fixing and staining infected cells followed by examination for the presence of virus induced intracytoplasmic and/or intranuclear inclusions that are also generally typical of the virus in question.

The major virus groups included in this volume are summarized schematically in Figure 1.* Viruses representative of each major virus group will be described briefly and illustrated by both light and/or electron micrographs. The major objective is to facilitate and expedite the recognition of viruses or virus-induced cellular changes by illustrating the various ways in which each virus group can be identified.

The textual information included in this chapter is only a base outline to supplement the illustrative material and should not be used as a sole source of information. The reader is referred to the individual chapters in this volume dealing with each of the major virus groups or other general references in order to obtain the more detailed information provided therein.

RNA Viruses

Picornaviridae

The picornaviruses, smallest of the single-stranded RNA viruses, are comprised of two ether-resistant groups: the enteroviruses and the rhinoviruses; the former are subdivided into polioviruses, Coxsackieviruses and echoviruses. The picornaviruses are 25 to 30 nm in diameter and possess icosahedral symmetry. Primate cells are the only cell type susceptible to infection with this group of viruses.[1] Virus replication and maturation take place in the cytoplasm where aggregates of mature virus particles in crystalline arrays can be found occasionally in the infected cells.[2,3]

Among the enteroviruses, polioviruses provide the best example for illustrating virus-induced cytopathic effect (CPE) in cells of primate origin. Primary monkey kidney cell monolayers and passaged cell lines including Hep-2 and HeLa cells are highly susceptible to poliovirus infection (Figures 2B, 2E, 3B, and 3E). On the other hand, cer-

* All figures for this chapter follow text.

tain members of the enterovirus family, for example certain types of echoviruses, induce CPE in rhesus monkey kidney cells (Figures 2C and 3C) but not in Hep-2 (Figure 2F) or Patas monkey kidney cells (Figure 3F).[4]

Under electron microscopic examination, poliovirus infected cells, which have usually degenerated, reveal aggregates of poliovirus particles in the cytoplasm (Figure 4). A large aggregate of poliovirus particles in crystalline arrays is illustrated in Figure 5.

Togaviridae

Togaviruses, which include those viruses formerly known as group A and group B arboviruses, are 40 to 60 nm in diameter and are ether-sensitive, RNA-containing viruses consisting of a small icosahedral capsid and a lipid-containing envelope.[5] Replication of the togaviruses occurs in the cytoplasm of infected cells; maturation and envelopment take place by budding at the plasma membrane or into cytoplasmic vacuoles.[6,7] Togaviruses generally do not induce distinct CPE or inclusions in cell cultures, although cellular granulation has been observed in many togavirus infected cells. The representative cytopathology seen in chick embryo fibroblast cells and Vero cells infected with Sindbis virus is illustrated in Figure 6.

Electron microscopic examination of Sindbis virus-infected, cultured cells has revealed the presence of many intracytoplasmic spherical nucleocapsids, approximately 30 nm in diameter (Figure 7). Enveloped virus particles 55 nm in diameter are formed by budding at the plasma membrane (Figure 8) or into cytoplasmic vacuoles.

Rubiviridae

Rubella virus represents an RNA virus group that resembles togavirus in both morphology and morphogenesis[8-11] The virus possesses a lipid-containing, ether-sensitive envelope, is spherical in shape, and is 60 nm in diameter. The virus replicates in the cytoplasm and matures by budding at the plasma membrane or into cytoplasmic vacuoles.

Under light microscopic examination, rubella virus infected cells generally do not exhibit cytopathology, therefore virus interference techniques are usually employed in order to demonstrate virus growth (see Rubella Virus Chapter). In a Vero cell monolayer infected with rubella virus, no apparent evidence of virus-induced CPE or inclusions in stained preparations are observed. However, inhibition of cell mitosis and some vacuolation and granulation were observed in the rubella virus-infected Vero cells (Figure 9B), as compared with the uninfected cells (Figure 9A). Under electron microscopic examination in a suitable cell system such as Vero cells or BHK-21 cells, virus particles are seen in the cytoplasmic vacuoles (Figure 10). Budding at the plasma membrane (Figure 11) as well as in the extra-cellular space (Figure 12) is commonly found.

Reoviridae: (Diplornaviridae)

The reoviridae family is composed of several genera: reovirus, rotavirus, orbivirus, cytoplasmic polyhedrosis virus, and wound tumor virus. The name "reovirus" standing for "respiratory enteric orphan" virus was coined because these viruses are found in both the respiratory and enteric tracts of man but are not commonly associated with any disease.[12] However, in 1973, a reovirus-like agent was found to be frequently associated with infantile enteritis[13-15] Reoviruses were the first viruses shown to possess double-stranded RNA, [16,17] They are medium sized (70 nm in diameter), ether-resistant viruses that exhibit cubic symmetry.

In cell culture, reoviruses show a broad host spectrum and can infect cells of both primate and nonprimate origin.[4] Figure 13B shows a guinea pig embryonic cell monolayer infected with reovirus type 1. The granulation and degeneration of cells usually associated with reovirus infection are difficult to distinguish from the nonspecific degeneration that accompanies cell aging. However, hemotoxylin-eosin staining of the infected cells reveals intracytoplasmic eosinophilic inclusions (Figure 13D and Plate A*, Figure 3) typical of reovirus infection. By acridine orange staining, the intracytoplasmic inclusions of reovirus show greenish yellow fluorescence, indicating the presence of double-stranded RNA (Plate A, Figure 6). With fluorescent antibody technique, viral antigens are detected in the intracytoplasmic inclusions (Plate A, Figure 5).

Under electron microscopic examination, the cytoplasmic inclusions can be shown to contain virus aggregates (Figure 14) and sometimes, microtubule structures (Figure 15). Individual mature virus particles negatively stained with sodium phosphotungstate show surface structures characteristic of reovirus (Figures 16 and 17).

Coronaviridae

Coronavirus, a RNA-containing virus, is characterized by its distinctive morphology. The virions possess club-shaped projections on their surface giving the appearance of a crown or corona around the virus particle. This group includes avian infectious bronchitis virus, rat coronavirus, mouse hepatitis virus, and human coronavirus. All of them are capable of causing respiratory diseases in their homologous animal host species.[18-22]

Electron microscopic examination shows that morphologically, the coronaviruses are spherical, with a wide range in size varying from 80 to 160 nm in diameter. The cores are somewhat dense, but many are translucent and appear ring-like in structure. The club-shaped surface projections are usually seen on negatively stained particles, and are not commonly seen on the virions prepared by thin sectioning. Figure 18 and 19 illustrate cornavirus particles seen in the liver of an infected mouse.

Myxoviridae

The name myxovirus was derived from the fact that these viruses have an affinity for mucoprotein receptors which can be destroyed following treatment with receptor-destroying enzyme (RDE). Subsequently, it became apparent that these viruses fall into two distinct groups, the orthomyxoviruses and the paramyxoviruses.[23]

Orthomyxoviruses

The orthomyxovirus group includes the influenza virus groups A, B, and C, all of which agglutinate erythrocytes from a variety of avian and mammalian species, and then elute, following destruction of the receptors by viral neuraminidase. Orthomyxoviruses are ether sensitive RNA-containing viruses, including both spherical and filamentous forms, 90 to 120 nm in cross section, that are enveloped with a lipid-containing membrane which is covered with surface projections or spikes composed of hemagglutinin and neuraminidase.[24] Newly isolated influenza A viruses generally replicate well in the cells of the amniotic-allantoic membrane of embryonated chicken eggs, whereas laboratory adapted strains of influenza virus grow well in cell cultures.

Orthomyxoviruses can replicate in cell cultures without inducing distinct CPE. However, following the addition of guinea pig erythrocytes to the infected cell monolayers, hemadsorption occurs indicating the presence of virus in the infected cells (Figure 20B). Occasionally, varying degrees of cytopathic effect are observed in influenza B virus infected cells (Figures 20C and 20D).

Electron microscopic examination of influenza virus-infected chick amniotic membrane cells reveals many filamentous forms of virions (Figure 21). A laboratory-adapted influenza A_0 virus, strain WSN, grows well in many cell cultures and produces numerous virus particles, both spherical and filamentous forms (Figure 22). Influenza

* Plates A and B follow page 48. Color figures are not sequential.

B virus, which exhibits a similar morphology (Figure 23), replicates in monkey kidney cells and when guinea pig erythrocytes are added to culture fluid containing influenza virus, hemagglutination occurs [25] as the virus particles attach to the red blood cells (Figure 24).

Paramyxoviruses

The paramyxoviruses closely resemble the orthomyxoviruses morphologically and biologically in that they are also ether-sensitive RNA viruses. However, they are larger, 150 to 300 nm in size. This group includes not only the parainfluenza viruses, but also mumps and Newcastle Disease Virus (NDV). Primary monkey kidney cell cultures are highly sensitive to infection with parainfluenza virus types 1, 2, 3, 4, and 5, although no obvious CPE can be detected in virus infected cells.[26] Demonstration of the presence of virus is dependent upon the hemadsorption of guinea pig erythrocytes onto virus infected cell monolayers as illustrated in Figure 25B and 25D. Paramyxoviruses can cause multinucleated syncytial cell formations in certain cell systems (Plate A, Figure 2). The electron micrographs illustrate viral nucleocapsids 18 nm in cross section, exhibiting a helical structure, that is commonly seen in the cytoplasm of parainfluenza virus infected monkey kidney cells (Figure 26). Occasionally, virus particles budding at the plasma membrane (Figure 27) or enveloped virus particles covered with spikes can be seen (Figure 28).

Pseudomyxovirus

Although measles and respiratory syncytial (RS) viruses resemble the myxoviruses morphologically, in that they are large, ether-sensitive RNA viruses containing helical ribonucleoprotein, they are distinctly different both biologically and serologically. [27-31] For example, measles virus is capable of agglutinating monkey erythrocytes, but RDE treatment does not abolish this phenomenon, as is observed with the myxoviruses.[34] Furthermore, RS virus does not produce hemagglutinin.[32] Thus the name pseudomyxovirus or metamyxovirus has been suggested for this group of viruses.

Measles virus induces easily recognizable multinucleated syncytial cells in primary human or monkey kidney cells (Figure 29A) which resemble those induced by monkey foamy virus which belongs to a different virus group. However, the hemadsorption of monkey erythrocytes onto measles virus infected monolayers (Figure 29B) and the intranuclear and intracytoplasmic eosinophilic inclusions produced in measles virus infected cells (Figure 29C, 29D and Plate A, Figure 1) are not seen in cells infected with foamy virus.

Electron microscopic observation of measles virus-infected cells reveals the presence of viral nucleocapsids in both the nucleus (Figure 30) and the cytoplasm (Figure 31). Intact virions can be seen at the extracellular surface (Figure 31, insert).

Rhabdoviridae

The rhabdoviruses are large, bullet-shaped, enveloped virions, 65 × 190 nm in size that are flat on one end and rounded at the other.[33, 34] Rabies virus and Marburg agents, both of which belong to this group, occur naturally in wild animals and are lethal to man.[35] Vesicular stomatitis virus (VSV) is frequently found in rodents and produces CPE in many primary and passaged cell culture types (Figure 32).

Electron microscopic studies of VSV in chick embryo cells have shown that virus particles acquire envelopes by budding into cytoplasmic vacuoles or at plasma membrane (Figure 33). Rabies virus particles obtained directly from the tissues of an infected fox can be seen in Figures 34 and 35. Virions are present in the intracytoplasmic inclusions and can also be seen budding from the cytoplasmic membrane to form complete virus particles similar to those found in VSV-infected cells.

DNA Viruses
Parvoviridae

The parvoviruses are small size, ether-resistant viruses containing single-stranded DNA. Members of this group include the H viruses (hamster osteolytic viruses), Kilham rat virus, X-14 rat virus, Kirk rat virus, minute virus of mice, and adeno-associated viruses.[36-38] Morphologically, parvoviruses are spherical, 18 to 25 nm in diameter, and possess cubic symmetry. Parvoviruses replicate in the nucleus of susceptible cells and, in some instances, produce intranuclear basophilic inclusions (Figure 36B). Some members of the parvovirus group are defective and can only replicate in the presence of a helper virus. The best known example of this phenomenon is adeno-associated virus (AAV). AAV requires the active replication of an adenovirus, which serves as a helper virus, in order to initiate its own replicative cycle. Figure 37 illustrates an AAV aggregate adjacent to an adenovirus aggregate in an infected nucleus, and Figure 38 shows purified H-1 virions negatively stained with uranyl acetate.

Papovaviridae

The papovavirus family includes the papilloma viruses of man and animals, polyoma virus of mice, and vacuolating virus of monkeys (SV40).[39] Papovavirus, isolated from man, showed the antigenic properties of SV40.[40] Most viruses in this group are capable of inducing tumors in experimental animals. They are ether resistant, relatively small and contain double-stranded DNA. The papilloma viruses are slightly larger (50 to 55 nm in diameter) than the polyoma and SV40 viruses, which are 45 to 50 nm. All papovaviruses replicate and assemble in the nucleus of susceptible cells.[41-45]

In African green monkey kidney (AGMK) cell cultures, SV40 induces characteristic cytoplasmic vacuolation which can be easily recognized in unfixed preparations (Figures 39B and 39C). Following virus replication in the nucleus, basophilic intranuclear inclusions are seen distinctly (Figures 39E, 39F, and Plate B, Figure 6). SV40 is a common endogenous virus in Rhesus monkey kidney (RhMK) cells; however, the vacuolation seen in AGMK cells is not evident in the RhMK cells. Therefore, evidence of SV40 replication in RhMK cells can only be found by examining hematoxylin and eosin stained preparations for the presence of basophilic intranuclear inclusions.[41] Under electron microscopic examination, SV40 virions are easily recognized in the nucleus of infected AGMK cells (Figure 40). Virus particles, released from the nucleus to the cytoplasm and extracellular space, are often associated with membranes (Figure 41).

Human papovavirus particles were first observed in warts taken directly from patients. Figure 42 illustrates a human wart virus aggregate arranged in crystalline arrays which was observed in a human skin tumor cell. Papovavirus particles can also be seen in an enlarged epithelial cell from the urothelium (Figure 43) and in a negatively stained preparation of a urine sample (Figure 44) taken from a renal transplant patient undergoing immunosuppressive drug therapy.

Adenoviridae

Adenoviruses are of medium size (70 to 80 nm) exhibit cubic symmetry, and contain double-stranded DNA. They are not enveloped and are ether resistant. Virus replication takes place in the nucleus. Some adenoviruses can induce tumors in rodents.[46-49]

Human adenoviruses replicate best in cell cultures derived from human tissues. Grape-like clusters of large rounded cells, typical of adenovirus-infected cells, can be easily recognized (Figure 45B). In hematoxylin and eosin stained preparations, adenoviruses induce distinctive intranuclear inclusions that are eosinophilic during early infection and basophilic during the late stages of infection. In accordance with typical growth characteristics and the physical appearance of the inclusions, human adenoviruses can be further subdivided into different groups. Adenovirus types 1, 2, 5, and 6 produce similar nuclear changes (Figure 45D and Plate B, Figure 3) whereas the nuclear inclusions of types 3, 4 and 7 exhibit different characteristics (Figure 45E and Plate B, Figure 4).[47]

Under electron microscopic examination, virions of some adenovirus types appear to be scattered throughout the nucleus (Figure 46). In other adenovirus types, the particles are arranged in crystalline arrays in the infected nucleus[48-49] (Figures 47 and 48). Negatively stained adenovirus preparations show the distinct surface structure of the virions (Figure 49), which consists of 252 capsomeres.

Herpesviridae

Members of the herpesvirus group that have been isolated from man include herpes simplex virus (HSV) types 1 and 2, cytomegalovirus (CMV), varicella-zoster virus (VZV), and Epstein-Barr virus (EBV). Herpesviruses are classified as large viruses, 150 to 160 nm in diameter, containing double-stranded DNA. The icosahedral nucleocapsid, 100 nm diameter, is surrounded by an ether-sensitive, lipid-containing envelope. Herpesviruses replicate and assemble in the nucleus to produce nucleocapsids that are enveloped by budding at the inner nuclear membrane or into the cytoplasmic membrane.[50-53]

Herpes simplex virus has a wide host range both in animals and in cell culture. Characteristic virus-induced CPE and intranuclear inclusions appear in both human and nonhuman cells, especially rabbit kidney cells (Figures 50B, 50E and Plate B, Figure 2). In contrast, CMV and VZV possess a narrow host cell spectrum, inducing CPE only in human cells (Figure 50C) and not in nonprimate cells (Figure 50F).

Electron micrographs of rabbit kidney cells infected with HSV types 1 and 2 illustrate typical herpesvirus nucleocapsids (Figures 51 and 52). Enveloped mature virions can be seen in nuclear vacuoles and in the extracellular space. Characteristic membrane duplication often occurs in herpesvirus infected cells (Figure 52A). The membranes are utilized for envelopment of nucleocapsids. In addition, HSV type 2 induces characteristic filamentous structures (Figure 52B) which are not apparent in HSV type 1 infected cells.

Direct electron microscopic examination of clinical specimens taken from patients in whom HSV infection is suspected is very useful for rapid virus diagnosis. Figure 53A and 53B illustrate a case where HSV type 1 particles were observed in thin sections of a brain biopsy.

CMV[54-58] and VZV[59-61] both induce typical focal areas of CPE in human fibroblast cell cultures but they replicate more slowly than HSV. An example of CMV induced CPE in WI-38 cells is illustrated in Figure 54. Intranuclear eosinophilic inclusions surrounded by a halo are characteristic for CMV and VZV infected cells (Figure 54B and Plate B, Figure 1). Electron micrographs of CMV infected WI-38 cells are shown in Figure 55. Intranuclear inclusions containing many nucleocapsids of typical herpesvirus morphology, with electron dense precursor, have accumulated in the nucleus. Enveloped virions can be seen in nuclear vacuoles (Figure 56A), in cytoplasmic vacuoles (Figure 56B), and in the extracellular space (Figure 56C). In addition to enveloped virions, electron-dense bodies are scattered throughout the cytoplasm of CMV infected cells in culture (Figure 55).

In VZV infected cells, most extracellular virions appear to be distorted or partially distorted, a characteristic of VZV in infected cell cultures[61] (Figure 57).

Poxviridae

The poxvirus group includes the largest and most complex viruses of all the animal viruses known to date.[62-64] They are ether-resistant DNA viruses with brick- or oval-

Plate A

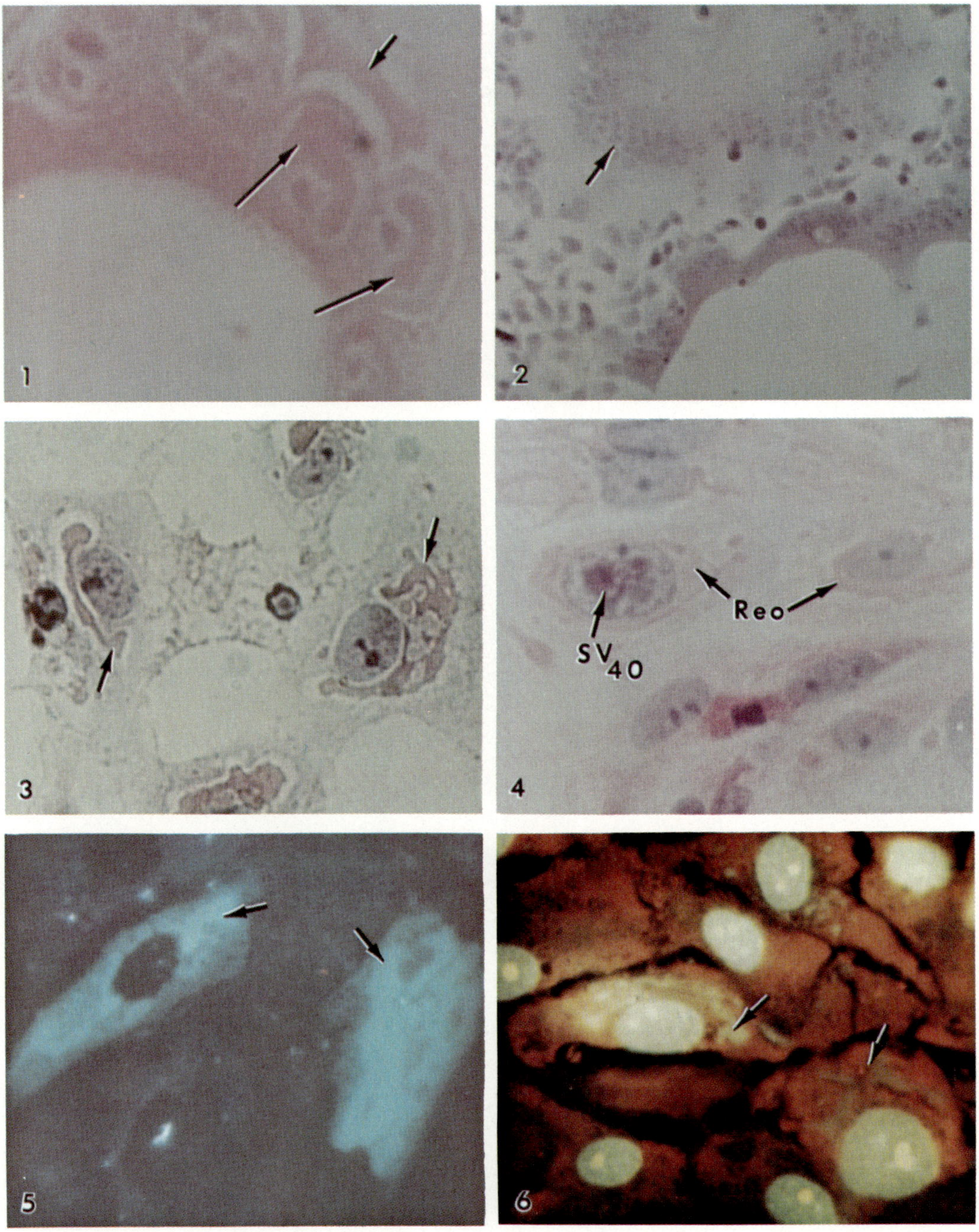

FIGURE 1. Measles virus infected human embryonic kidney cells (HEK); note: eosinophilic intranuclear inclusions (long arrows) and intracytoplasmic inclusions (short arrow). (Hematoxylin-eosin; magnification X 970. **FIGURE 2.** Parainfluenza virus type 3 infected Hep-2 cells; note: a large multinucleated syncytial cell (arrow). (Hematoxylin-eosin; magnification X 100.) **FIGURE 3.** Reovirus type 1 infected African green monkey kidney (AGMK) cells; note: eosinophilic intracytoplasmic inclusions (arrows). (Hematoxylin-eosin; magnification X 400.) **FIGURE 4.** AGMK cells infected with both SV40 and reovirus showing basophilic intranuclear inclusions of SV40 and eosinophilic intracytoplasmic inclusions of reovirus. (Hematoxylin-eosin; magnification X 400.) **FIGURE 5.** Reovirus type 1 infected AGMK cells treated with fluorescein-conjugated antiserum to reovirus type 1; the intracytoplasmic fluorescence (arrows) indicates the presence of viral antigen. (Magnification X 400.) **FIGURE 6.** Reovirus infected AGMK cells; the greenish yellow fluorescent intractyoplasmic inclusions (arrows) indicate the presence of double-stranded RNA. (Acridineorange; magnification X 400.)

Plate B

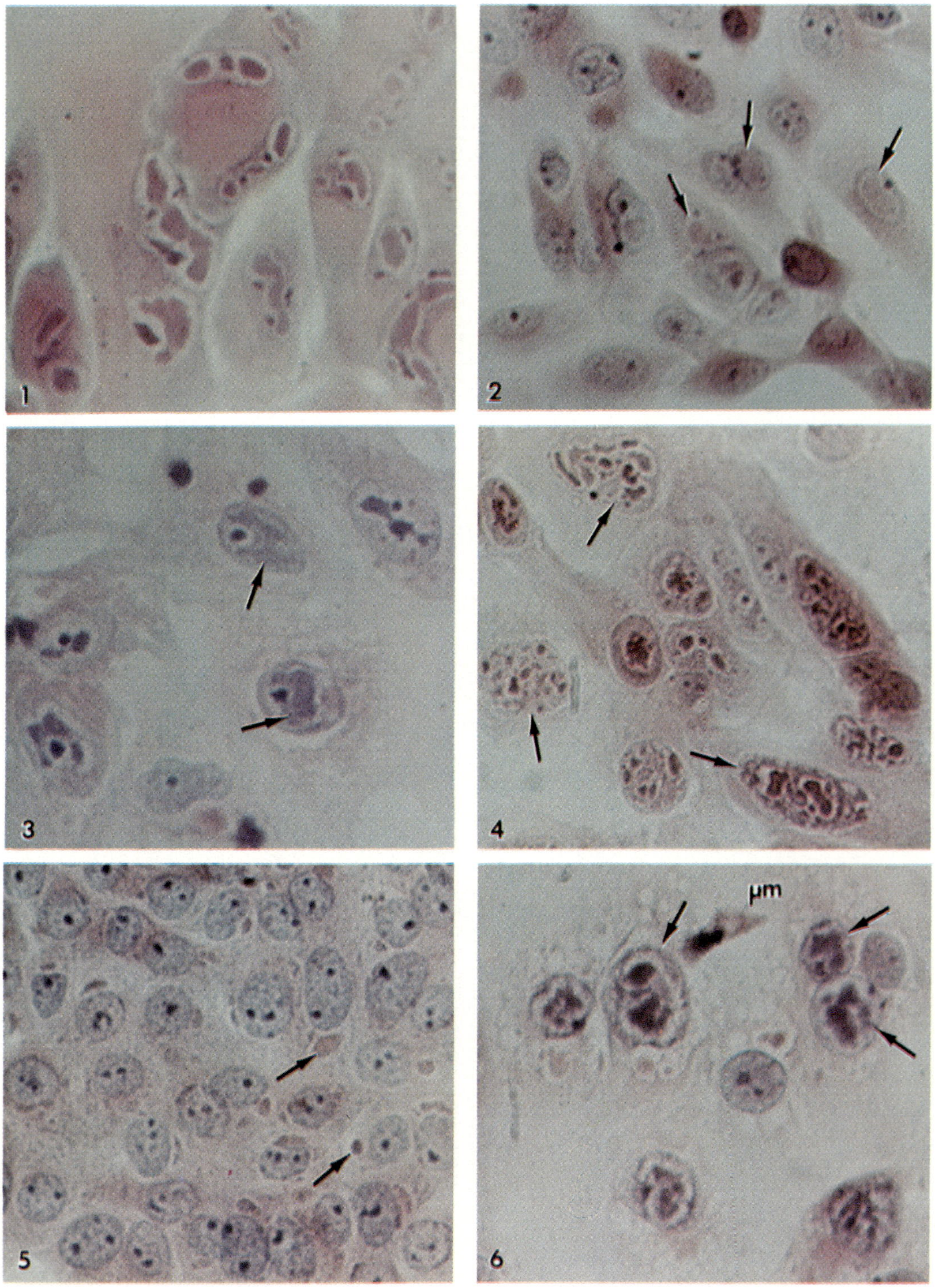

FIGURE 1. Human cytomegalovirus infected WI-38 cells; note: eosinophilic intranuclear inclusions. (Hematoxylin-eosin; magnification X 400.) **FIGURE 2.** Herpes simplex virus infected rabbit kidney cells; note: eosinophilic intranuclear inclusions (arrows). (Hematoxylin-eosin; magnification X 400.) **FIGURE 3.** Adenovirus type 2 infected HEK cells; note: basophilic intranuclear inclusions (arrows). (Hematoxylin-eosin; magnification X 400.) **FIGURE 4.** Adenovirus type 7 infected HEK cells; note: multiple crystal-like basophilic intranuclear inclusions (arrows). (Hematoxylin-eosin; magnification X 400.) **FIGURE 5.** Vaccinia virus infected Rhesus monkey kidney (RhMK) cells; note: eosinophilic intracytoplasmic inclusions (arrows). (Hematoxylin-eosin; magnification X 400.) **FIGURE 6.** SV40 infected AGMK cells; note: typical basophilic intranuclear inclusions (arrows). (Hematoxylin-eosin; magnification X 400.)

shaped structures 300 × 240 × 100 nm in dimension. Smallpox (variola major), alastrim (variola minor), and molluscum contagiosum viruses are members of the human poxvirus group.

Vaccinia virus, an animal representative of the poxvirus group, replicates and assembles in the cytoplasm of susceptible cells. Virus growth can be inferred by the observation of CPE in fluid cultures (Figure 58B) and can be recognized by the development of typical multinucleated giant cells. In hematoxylin and eosin stained preparations, virus induced eosinophilic intracytoplasmic inclusions are also indications of the presence of the virus (Figure 58C and Plate B, Figure 5). Vaccinia virus has a wide host cell spectrum and can be propagated in vitro in a variety of cell types; variola on the other hand, has a more restricted host spectrum.

Electron microscopic examination of poxvirus infected cells reveals that each virion consists of a dumbbell shaped core and two lateral bodies enclosed by an outer coat composed of complex tubular lipoprotein subunits (Figures 59 and 60). Direct electron microscopic examination of specimens suspected of harboring a member of this group has been used for diagnosis of poxvirus infection for the last 2 decades.[65, 66]

Mixed Viral Infections

The simultaneous infection of a cell culture by two different viruses is defined as a mixed infection.[67-73] The occurrence of mixed infections is not uncommon since a large number of animal species, especially simian primates whose tissues are used in the preparation of cell cultures, harbor a variety of endogenous viruses.[72] Occasionally clinical specimens may contain two viruses. In some instances, for example, two enteroviruses in the same stool samples as has been observed following administration of live poliovirus vaccine, differences in virus plaque morphology are very useful for recognizing the presence of two viruses (See Clinical Virology Chapter, Figure 4B). However, when this is not possible, the presence of two viruses may be detected by light and/or electron microscopy.

Figure 61 illustrates the presence of an adenovirus and a parainfluenza virus in an infected Hep-2 cell culture. This mixed infection was recognized following examination of a hematoxylin and eosin stained preparation. The presence of an adenovirus, which induced the intranuclear inclusions observed among the multinucleated syncytial cells induced by a paramyxovirus, would have gone unnoticed if hematoxylin and eosin staining had not been applied to the infected culture cells. Mixed viral infections are more commonly seen in primary monkey kidney cells, a species which exhibits a high incidence of endogenous virus infections, especially SV40 and monkey foamy agents.[72]

Mixed infection with SV40 and measles virus has also been seen (Figure 62). The basophilic intranuclear inclusions induced by SV40 and the eosinophilic intranuclear and intracytoplasmic inclusion induced by measles virus can both be observed (Figures 62A and 62B). Occasionally a single cell can be doubly infected with both viruses, SV40 virus inclusions in the nucleus and measles virus inclusions in the cytoplasm. Furthermore, both SV40 virions and measles virus nucleocapsids were found in an infected nucleus during electron microscopic examination (Figure 63).

Mixed infections with SV40 and reovirus (Figure 64 and Plate A, Figure 4) and SV40 and foamy virus (Figure 65) have also been observed. In the latter, typical basophilic intranuclear inclusions induced by SV40 were observed in a syncytial cell surrounded by cytoplasmic vacuoles induced by foamy virus.

Figure 66 illustrates a mixed infection with both SV40 and SV5 (paramyxovirus) in the same cell. SV40 virions were seen in the nucleus while nucleocapsids of SV5 were present in the cytoplasm of the same cell.

Concluding Remarks

The illustrative material and textual outlines within this chapter are intended to provide a supplement to the various other chapters in order to facilitate the recognition of the major groups of viruses by the methods available to most diagnostic virology laboratories. This has been carried out by bringing together in a visual presentation typical examples of each of the different virus groups, as seen by both light and electron microscopy, that are illustrative of those aspects of each group that a clinical virologist could anticipate encountering in the course of attempting to identify a virus isolated from clinical material.

By using both light microscopy to examine cultures for either virus induced cytopathology in unfixed cell cultures or for viral inclusions in fixed, stained preparations, and electron microscopy for direct visualization of virus particles, the investigators can expect to make a preliminary, presumptive identification in fairly short order. This chapter was designed to facilitate this process and to help in choosing the specific characteristics that would be most helpful in making a definitive diagnosis of viral infection.

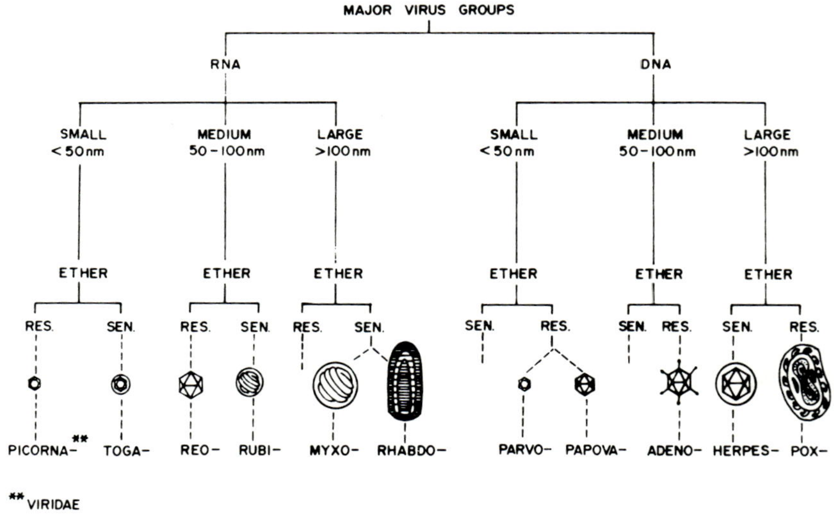

FIGURE 1. A schematic diagram of the major groups of viruses isolated from man and animals. (Modified from Hsiung, G. D., *Diagnostic Virology,* Yale University Press, New Haven, Conn., 1973, 52. With permission.)

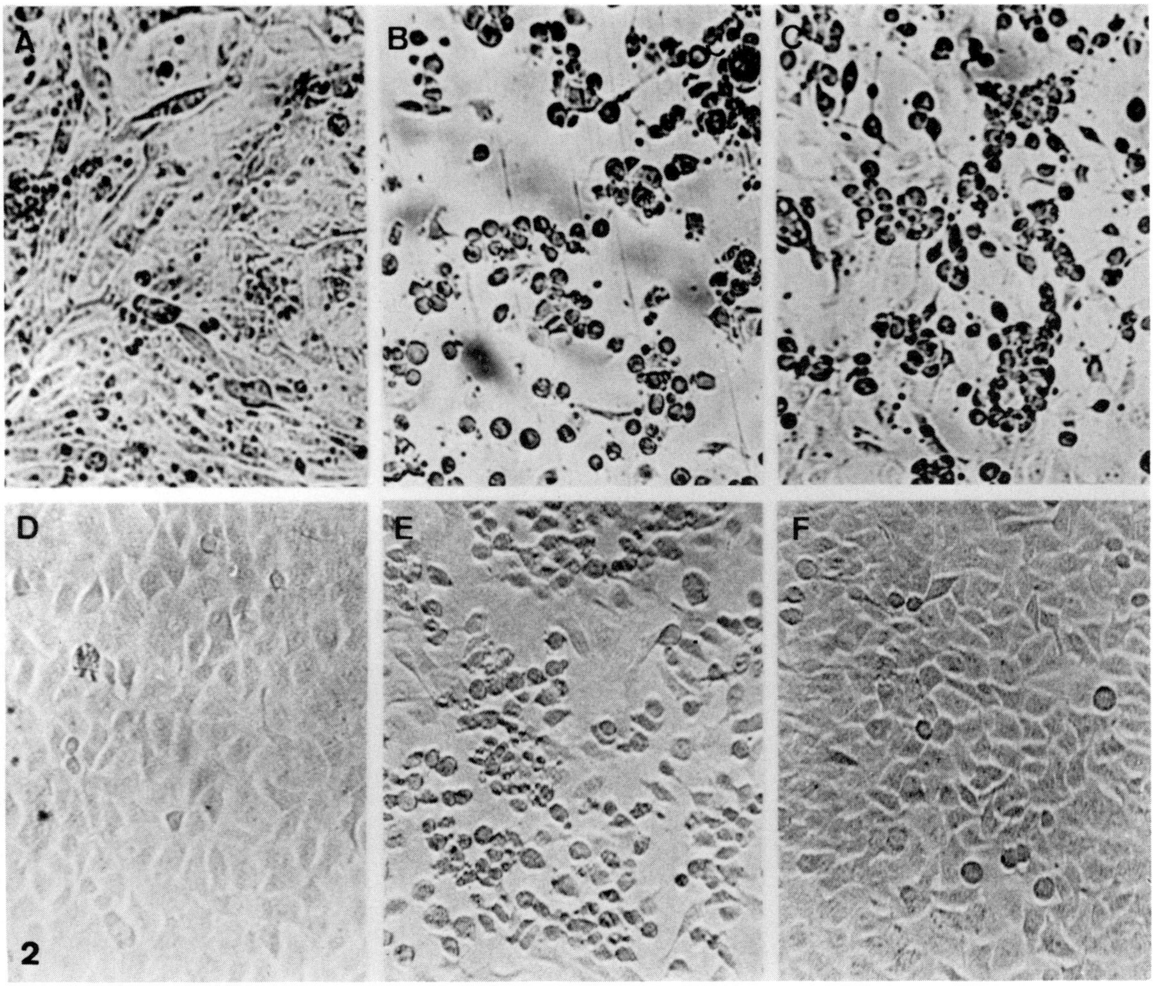

FIGURE 2. Cytopathic effect induced by poliovirus and echovirus in monolayer cultures of rhesus monkey kidney (RhMK) and Hep-2 cells. (Unfixed and unstained preparations; magnification × 100.) (A) Uninfected RhMK cells. (B) RhMK cells infected with poliovirus type 1 (LSC strain). (C) RhMK cells infected with echovirus type 11. (D) Uninfected Hep-2 cells. (E) Hep-2 cells infected with poliovirus type 1 (LSC). (F) Hep-2 cells infected with echovirus type 11. (Note: absence of CPE.)

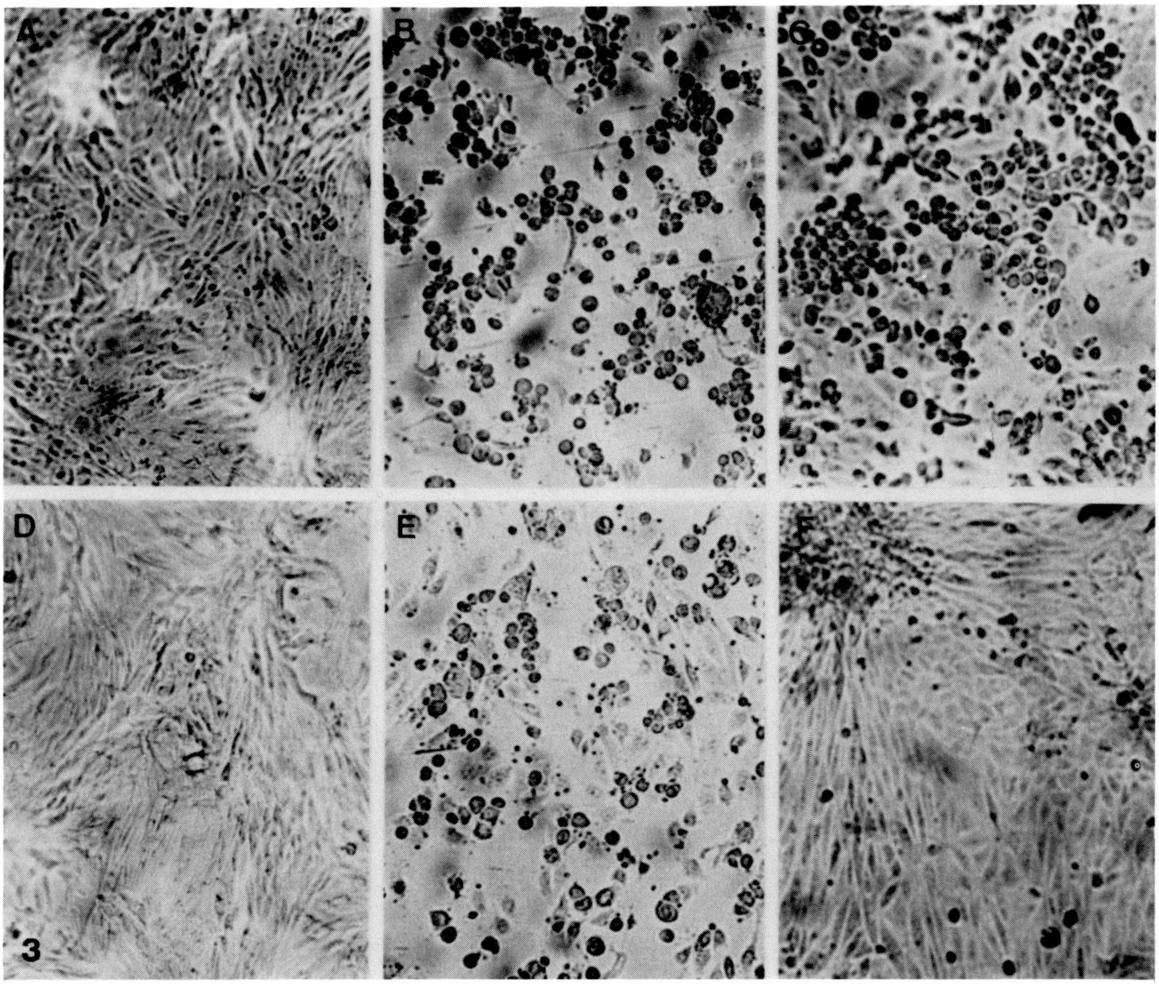

FIGURE 3. Cytopathic effect induced by poliovirus and echovirus in monolayer cell cultures of RhMK and patas monkey kidney. (Unfixed and unstained preparations; magnification × 100.) (A) Uninfected RhMK cells. (B) RhMK cells infected with poliovirus type 3. (C) RhMK cells infected with echovirus type 11. (D) Uninfected patas monkey kidney cells. (E) Patas monkey kidney cells infected with poliovirus type 3. (F) Patas monkey kidney cells infected with echovirus type 11. (Note: absence of CPE.)

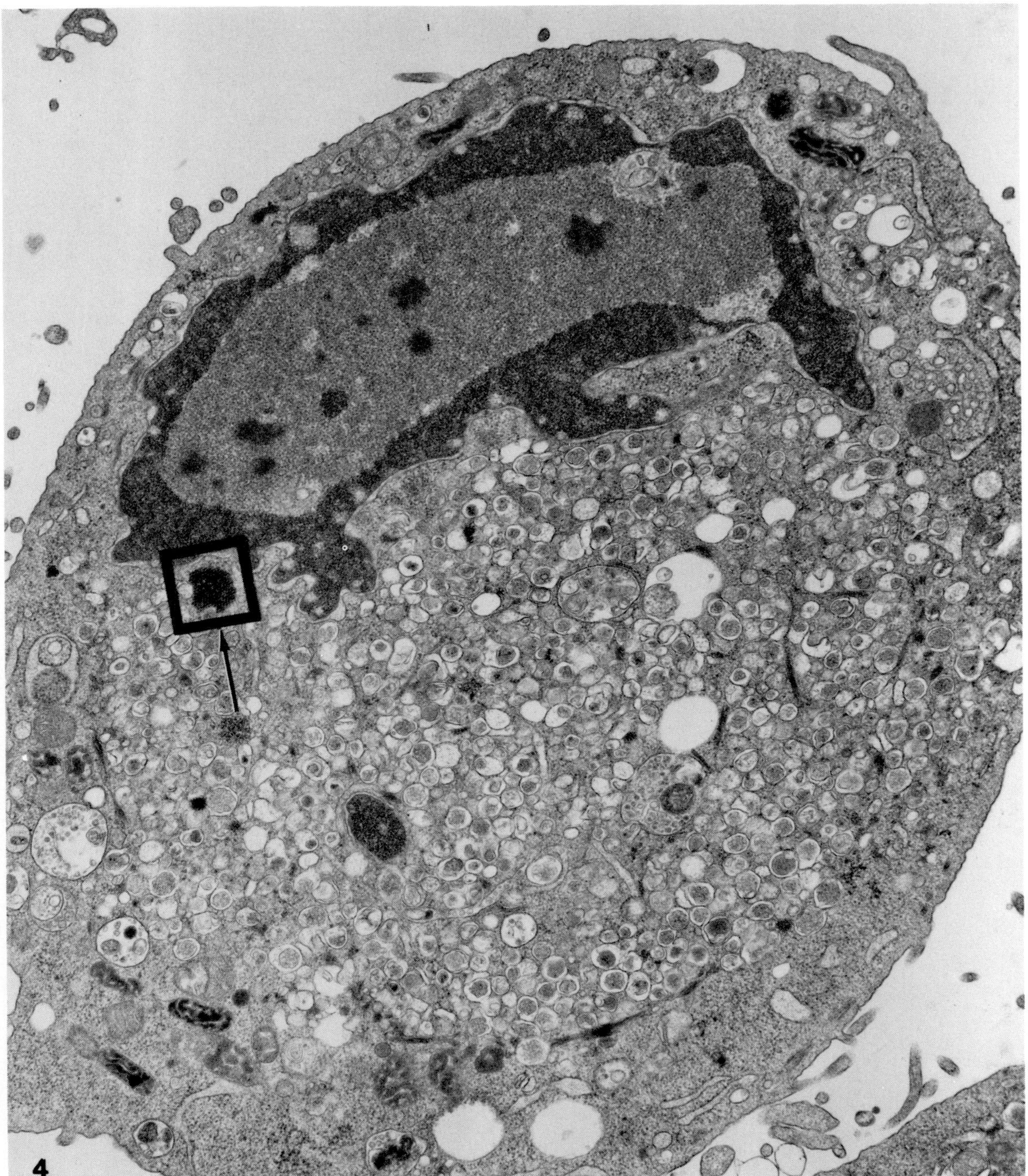

FIGURE 4. Electron micrograph of poliovirus type 1 infected Hep-2 cell. Note: the nucleus is pushed to one side of the cell, the chromatin is highly condensed and the cytoplasm is vacuolated. One viral aggregate is seen in the cytoplasm (arrow). (Magnification × 16,320.)

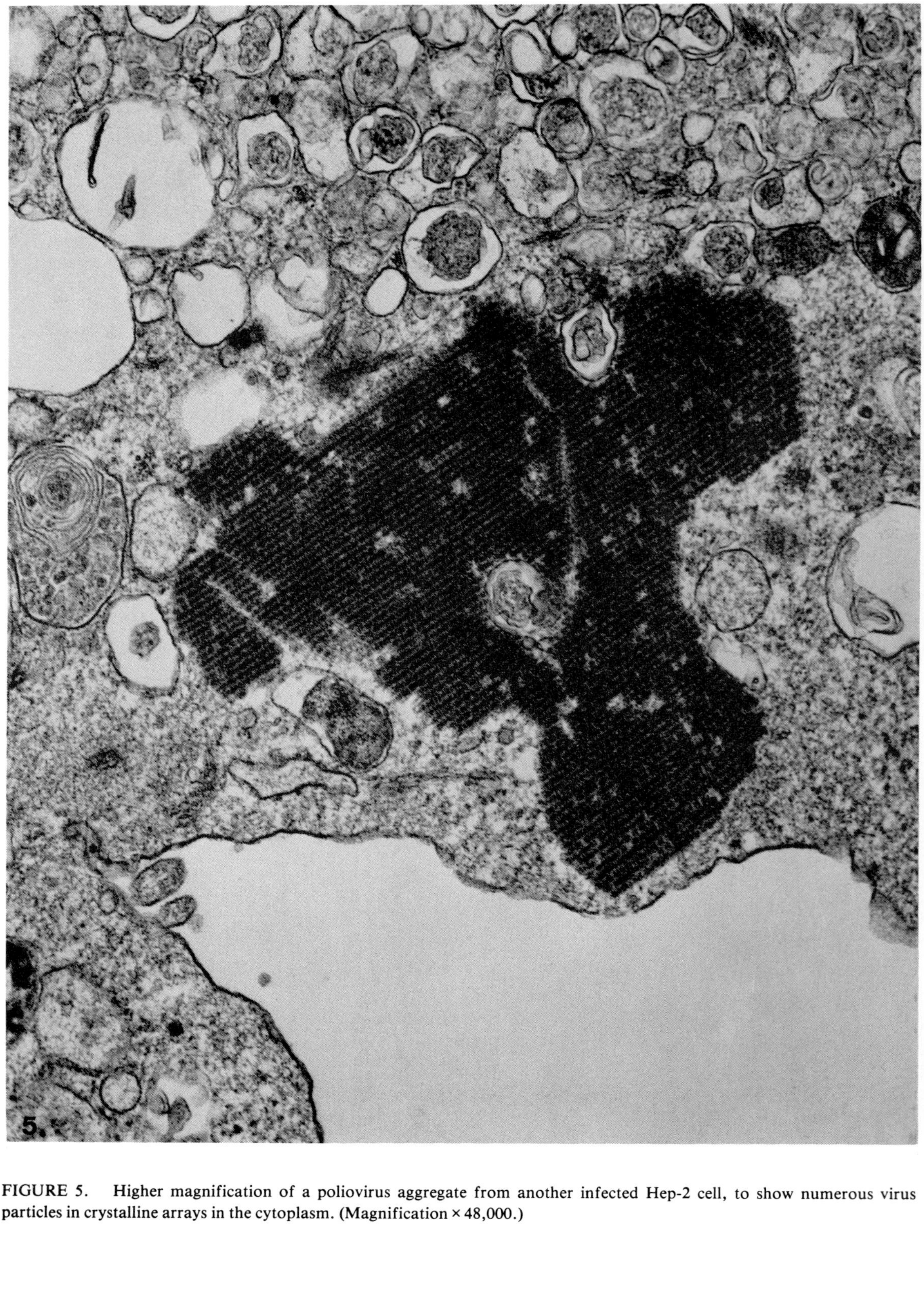

FIGURE 5. Higher magnification of a poliovirus aggregate from another infected Hep-2 cell, to show numerous virus particles in crystalline arrays in the cytoplasm. (Magnification × 48,000.)

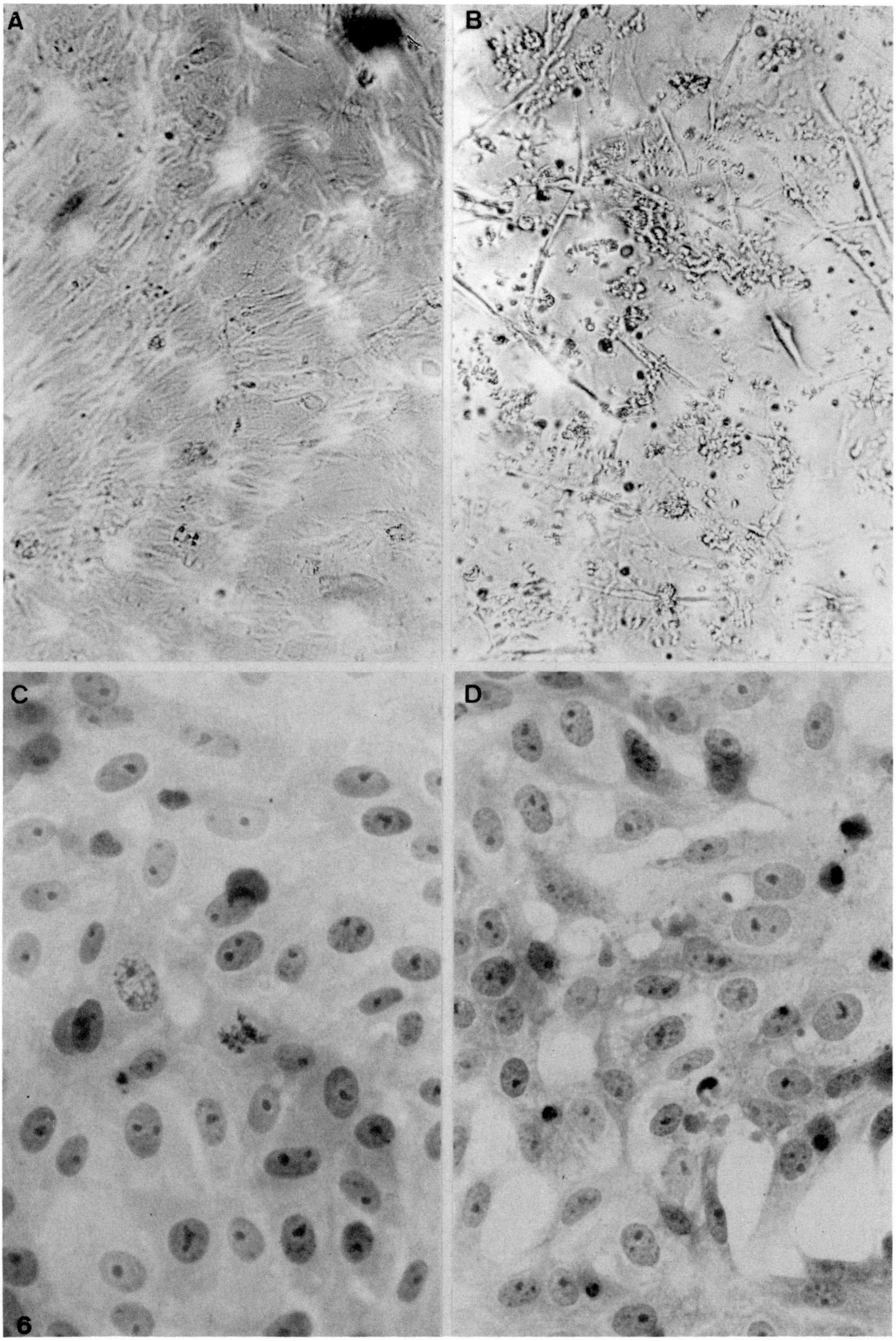

FIGURE 6. Cytopathic effect induced by Sindbis virus in chick embryo (CE) and Vero cells. [(A and B) Unfixed and unstained cells; magnification × 100. (C and D) Hematoxylin and eosin stained; Magnification × 400.] (A) Uninfected CE cells. (B) Sindbis virus infected CE cells. (Note: rounding and granulation.) (C) Uninfected Vero cells. (D) Sindbis virus infected Vero cells. (Note: some granulation and vacuolation in the cytoplasm of infected cells.)

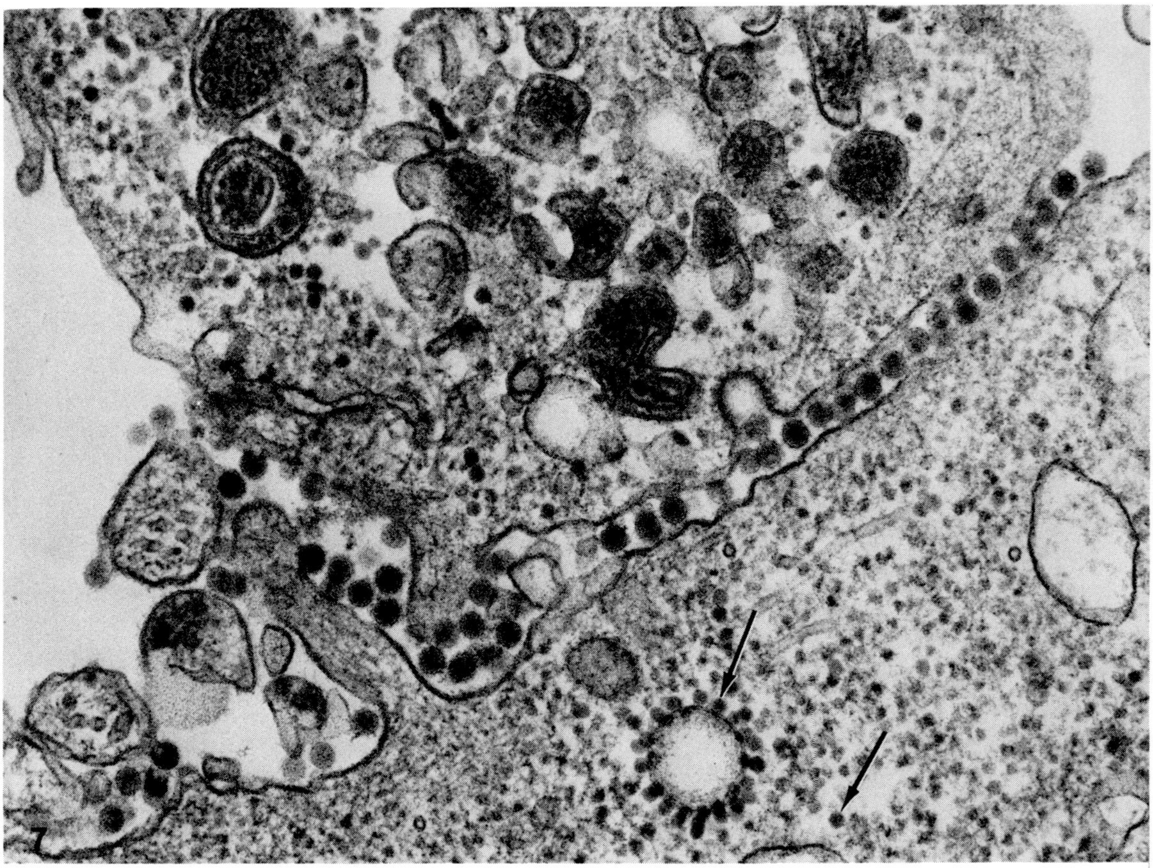

FIGURE 7. Electron micrograph of a portion of a Sindbis virus-infected Vero cell showing intracellular nucleocapsids (arrows), 30 nm, and extracellular enveloped virus particles, 55 nm. (Magnification × 64,800.)

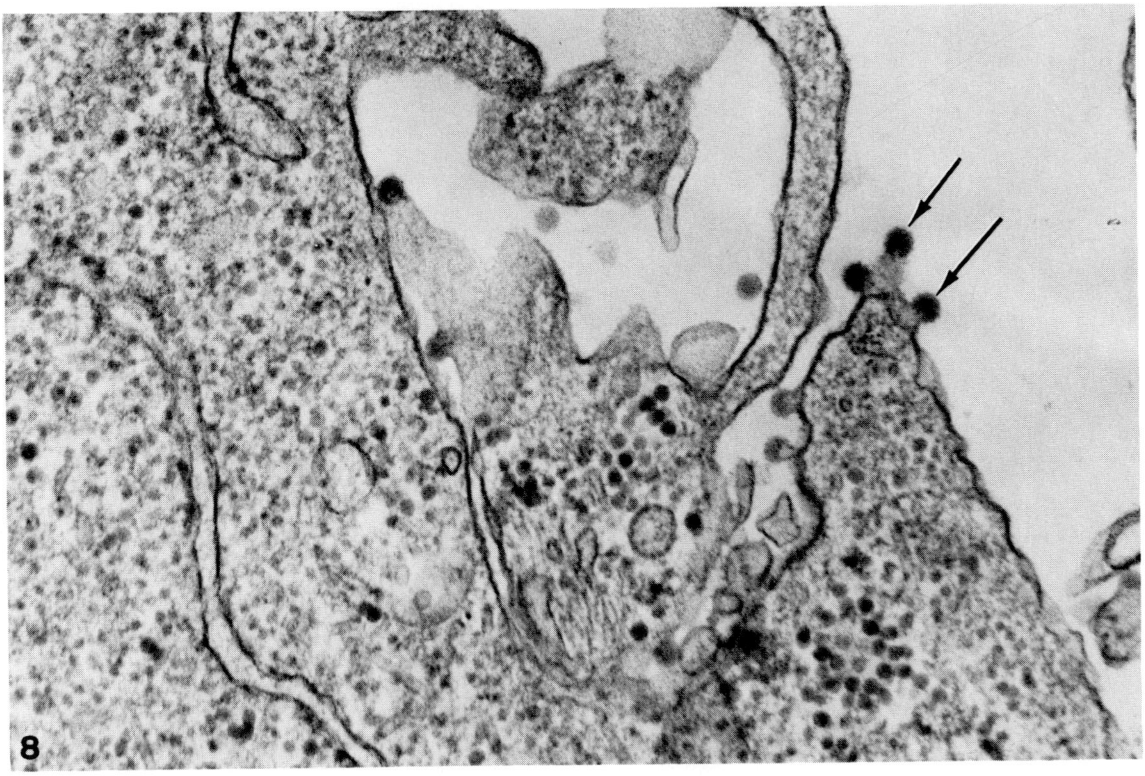

FIGURE 8. Several Sindbis virus nucleocapsids in the process of budding at the Vero cell plasma membrane to form enveloped virions (arrows). (Magnification × 64,800.)

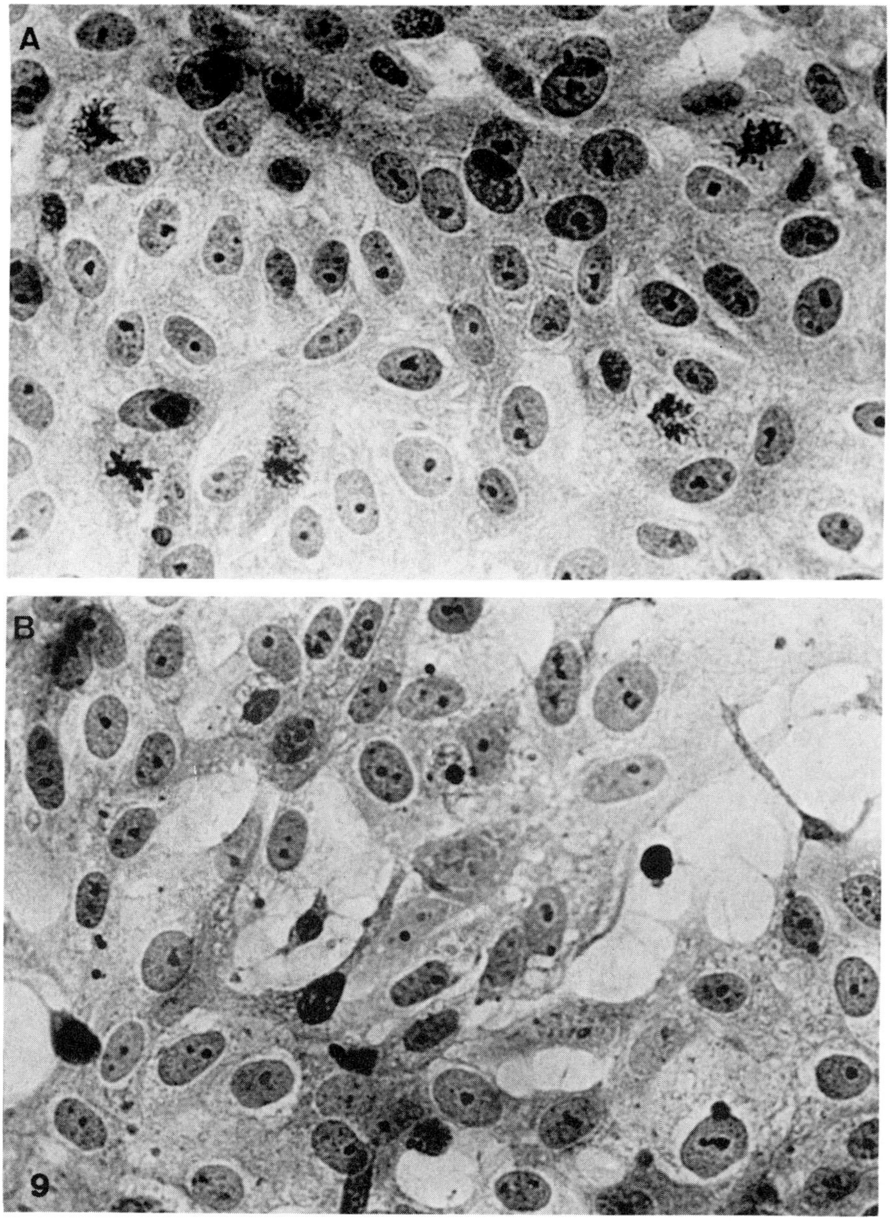

FIGURE 9. Vero cells infected with rubella virus, 5 days after infection. (Hematoxylin and eosin stained preparations; magnification × 400.) (A) Uninfected Vero cells. (B) Rubella virus infected Vero cells; note: granulation and vacuolation in the cytoplasm of infected cells.

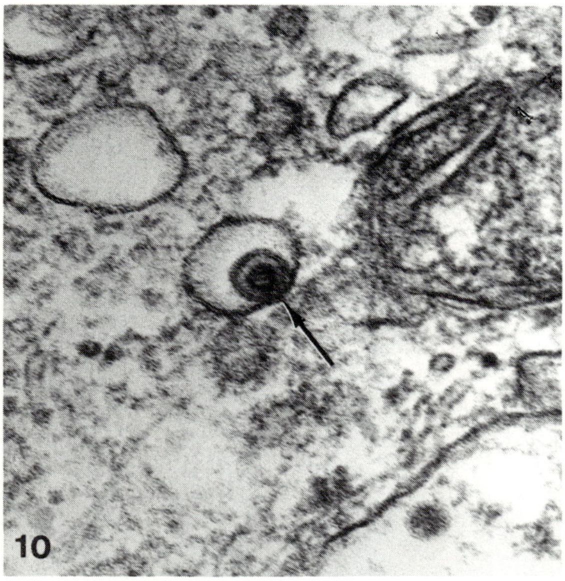

FIGURE 10. Electron micrograph of a portion of Vero cell infected with rubella virus to show a single enveloped virus particle, 60 nm, in a cytoplasmic vacuole (arrow). (Magnification × 100,800.)

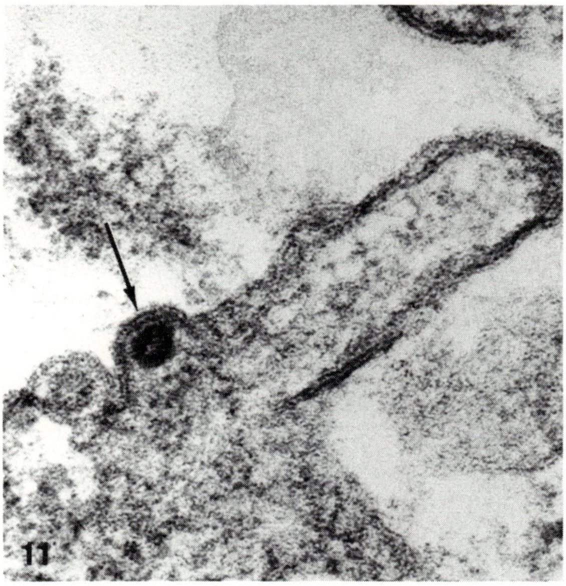

FIGURE 11. A rubella virus capsid in the process of budding at the Vero cell plasma membrane (arrow). (Magnification × 168,000.)

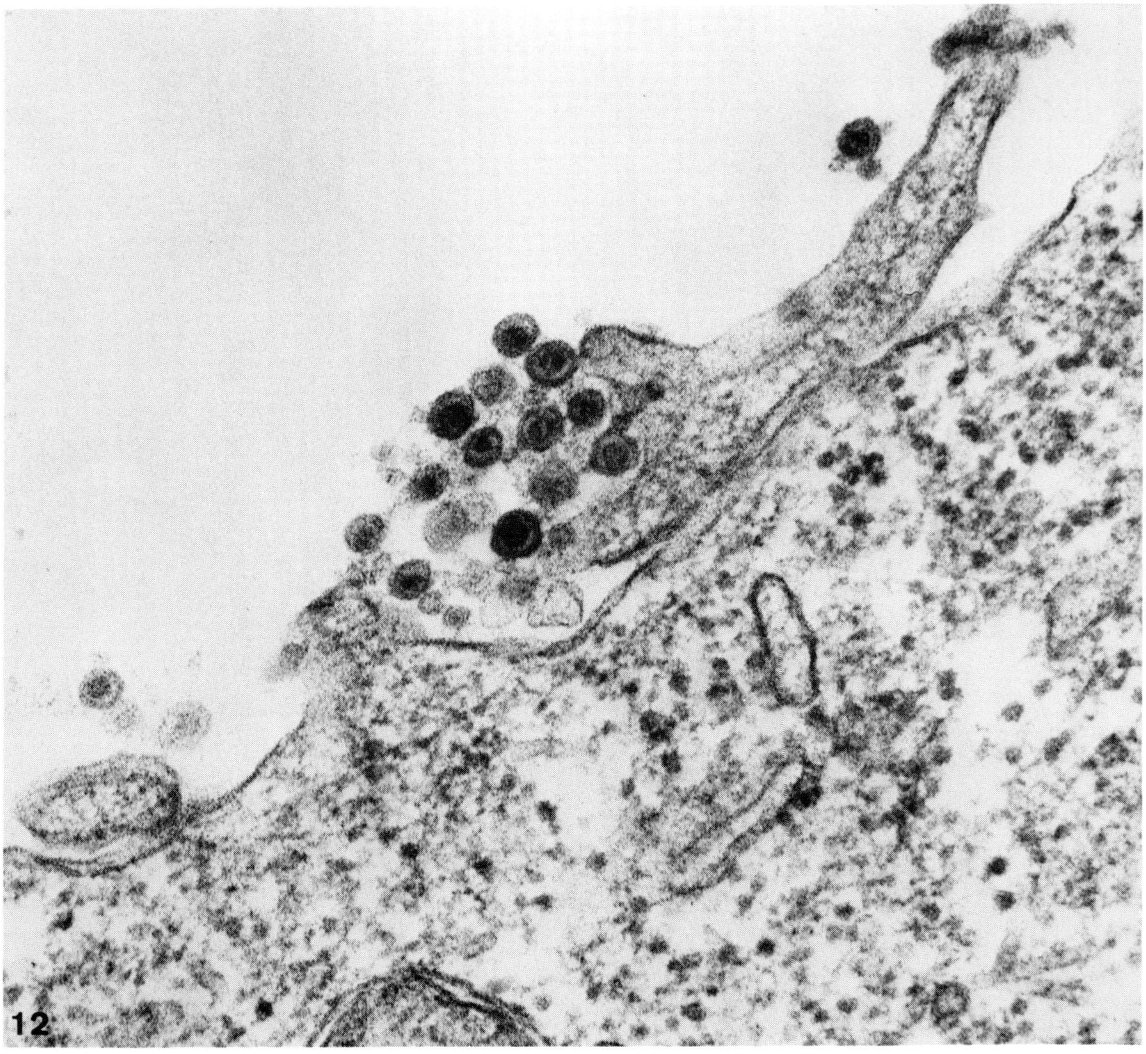

FIGURE 12. Rubella virus particles at extracellular space of an infected Vero cell. (Magnification × 100,800.)

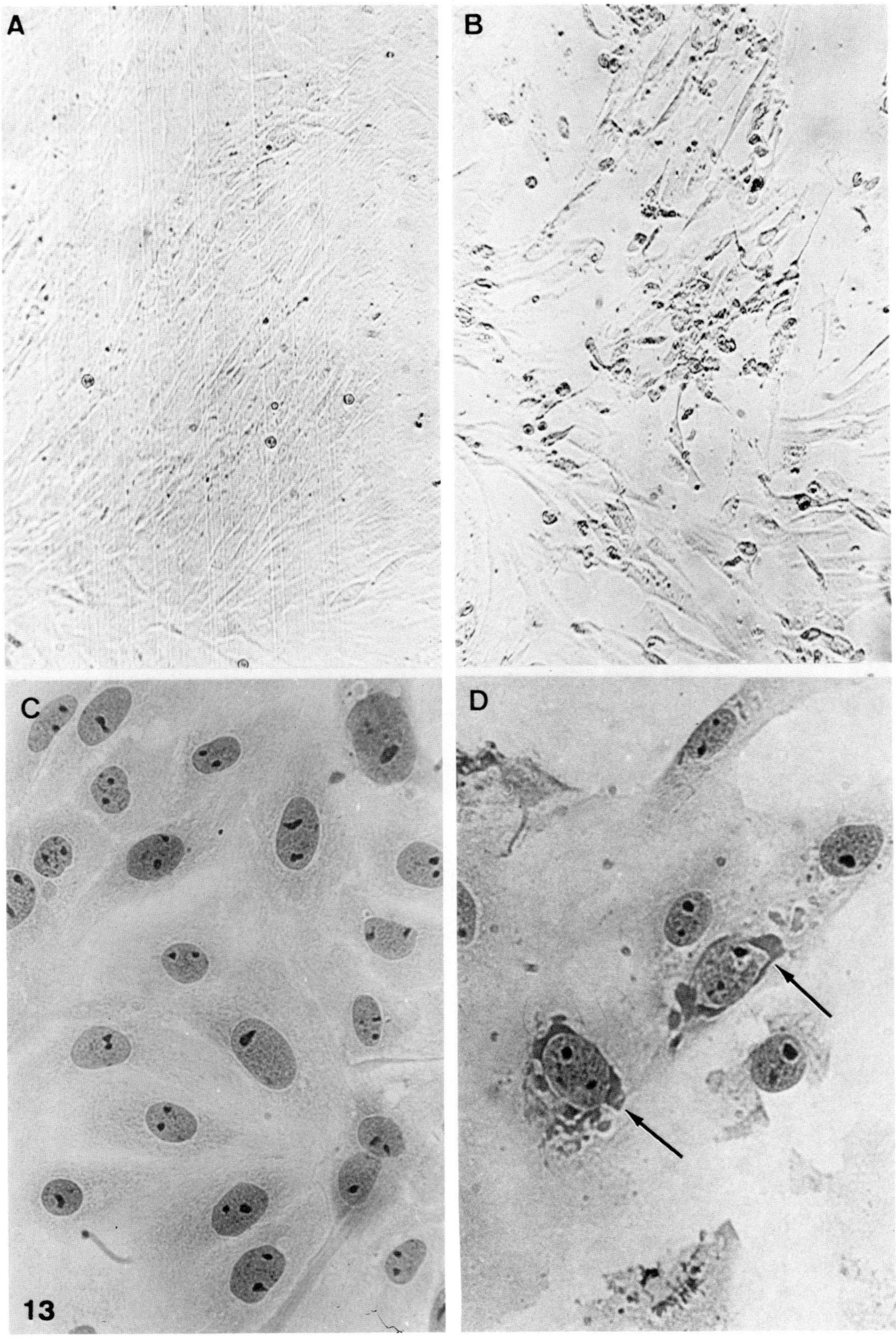

FIGURE 13. Reovirus infected nonprimate and primate cell cultures. [(A and B) unfixed, unstained; magnification × 100. (C and D) Hematoxylin and eosin stained; magnification × 400.]
(A) Uninfected guinea pig embryo fibroblast (GPEF) cells. (B) GPEF cells infected with reovirus type 1. (Note: granulation and degeneration of infected cells.) (C) Uninfected RhMK cells. (D) RhMK cells infected with reovirus type 1 showing intracytoplasmic eosinophilic inclusions (arrows).

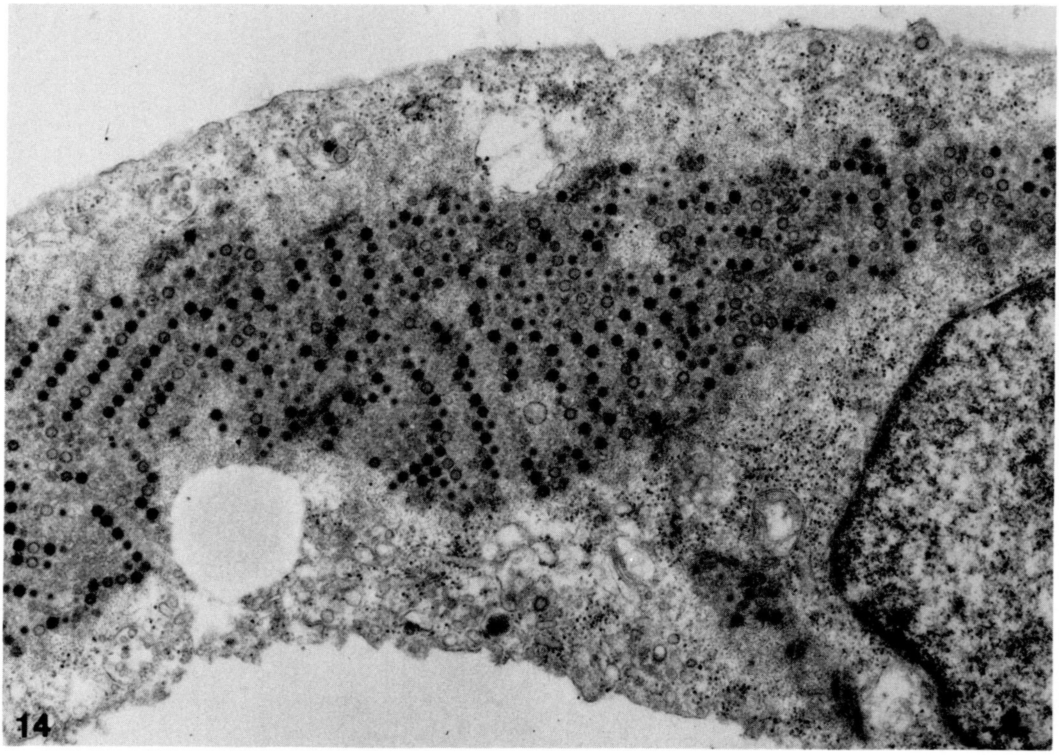

FIGURE 14. Electron micrograph of RhMK cell infected with reovirus type 1 showing intracytoplasmic inclusion containing virus particles. (Magnification × 19,680.)

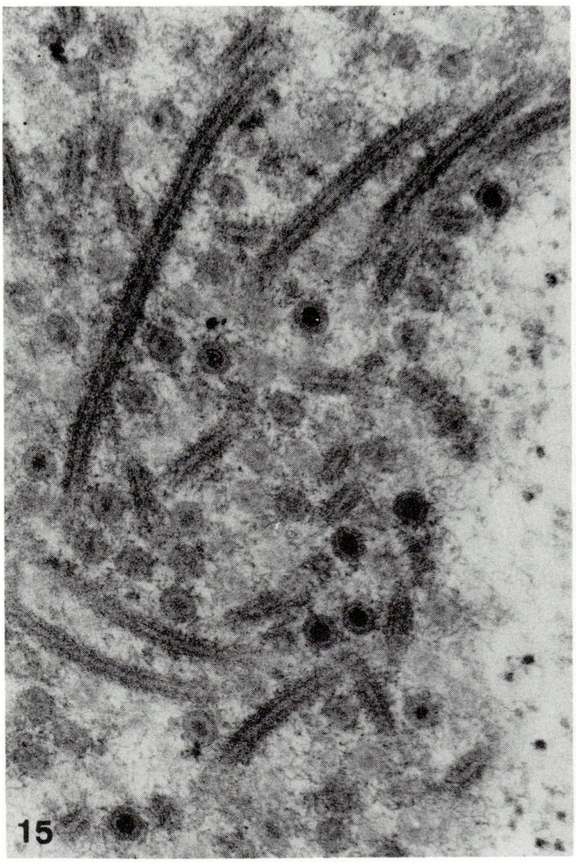

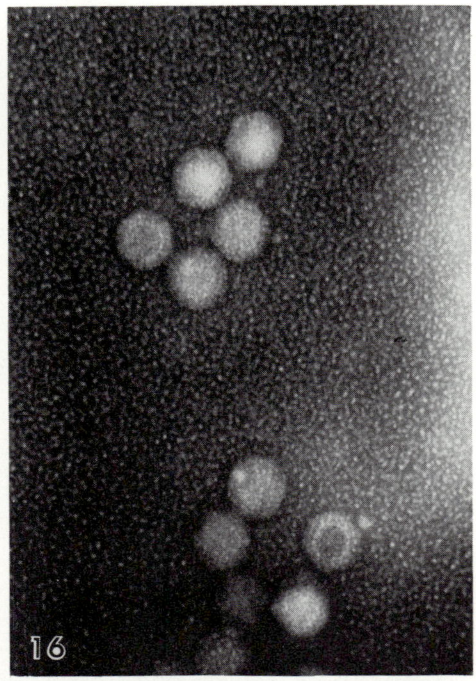

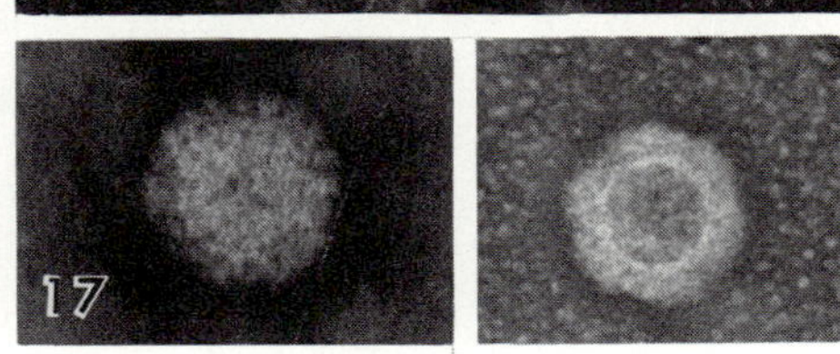

FIGURE 15. A portion of a reovirus type 3 infected RhMK cell showing microtubules associated with a viral inclusion. (Magnification × 64,800.) FIGURE 16. Reovirus particles negatively stained with phosphotungstic acid (PTA). (Magnification × 100,800.) FIGURE 17. Higher magnification of PTA stained reovirus particles showing distinctive capsomeres (left), and a virus with empty core (right). (Magnification × 168,000.)

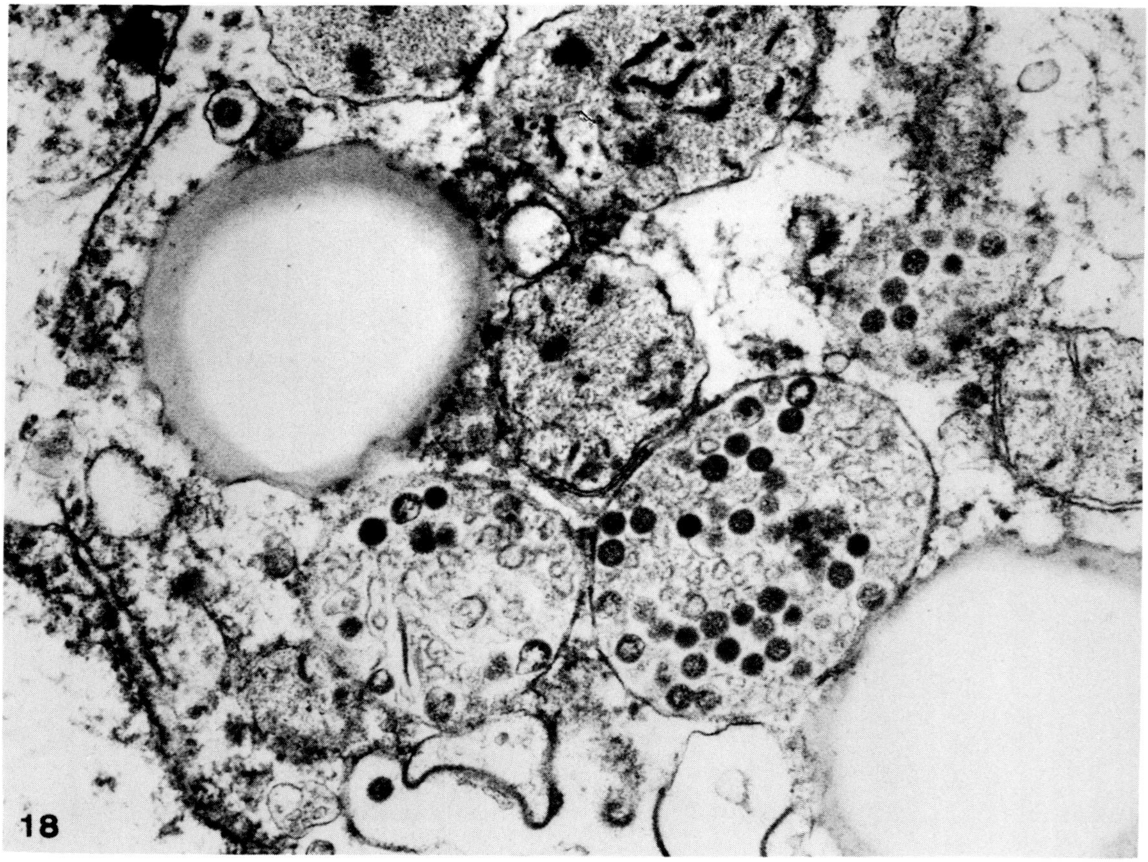

FIGURE 18. Electron micrograph of coronavirus particles in a liver cell from an infected mouse. Virus particles are seen in cytoplasmic vacuoles. (Magnification × 48,000.) (Courtesy of F. Bia.)

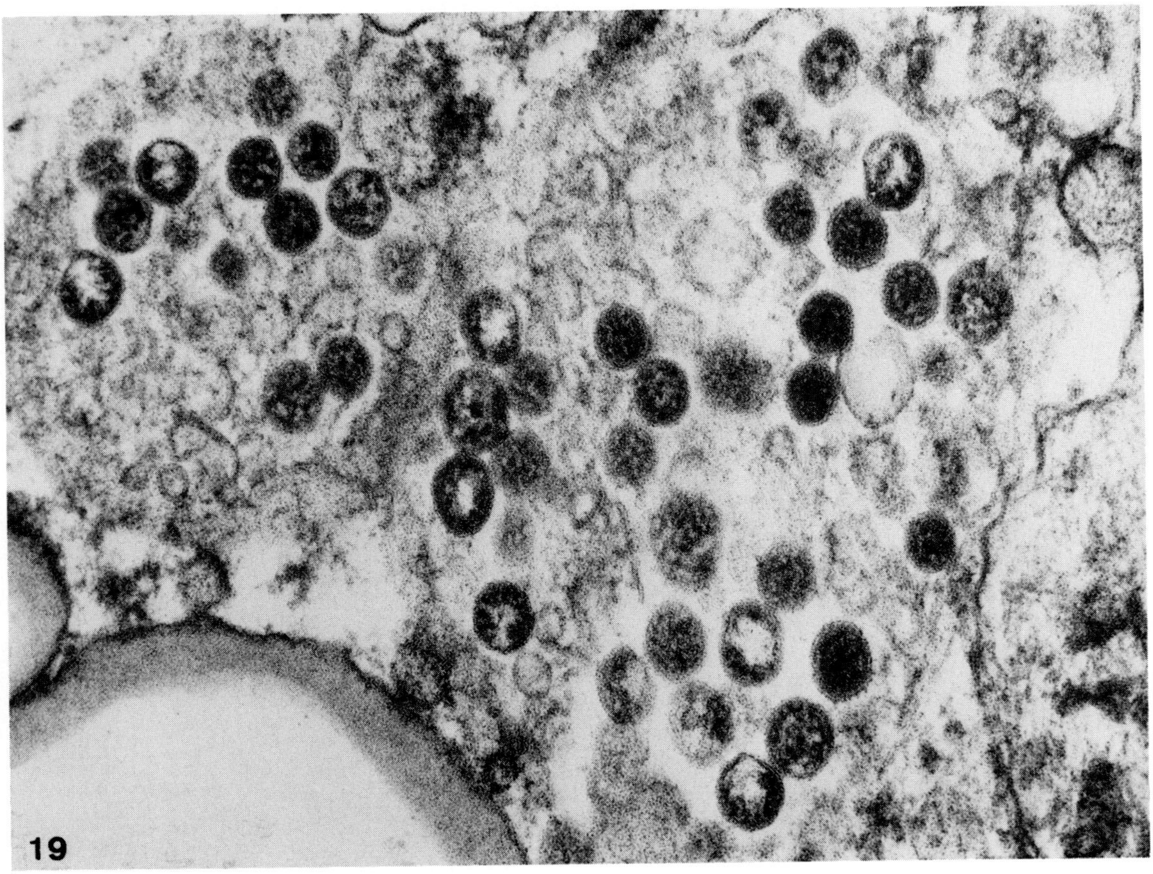

FIGURE 19. Same sample as in Figure 18, at a higher magnification showing distinctive virus structure. (Magnification × 100,800.) (Courtesy of F. Bia.)

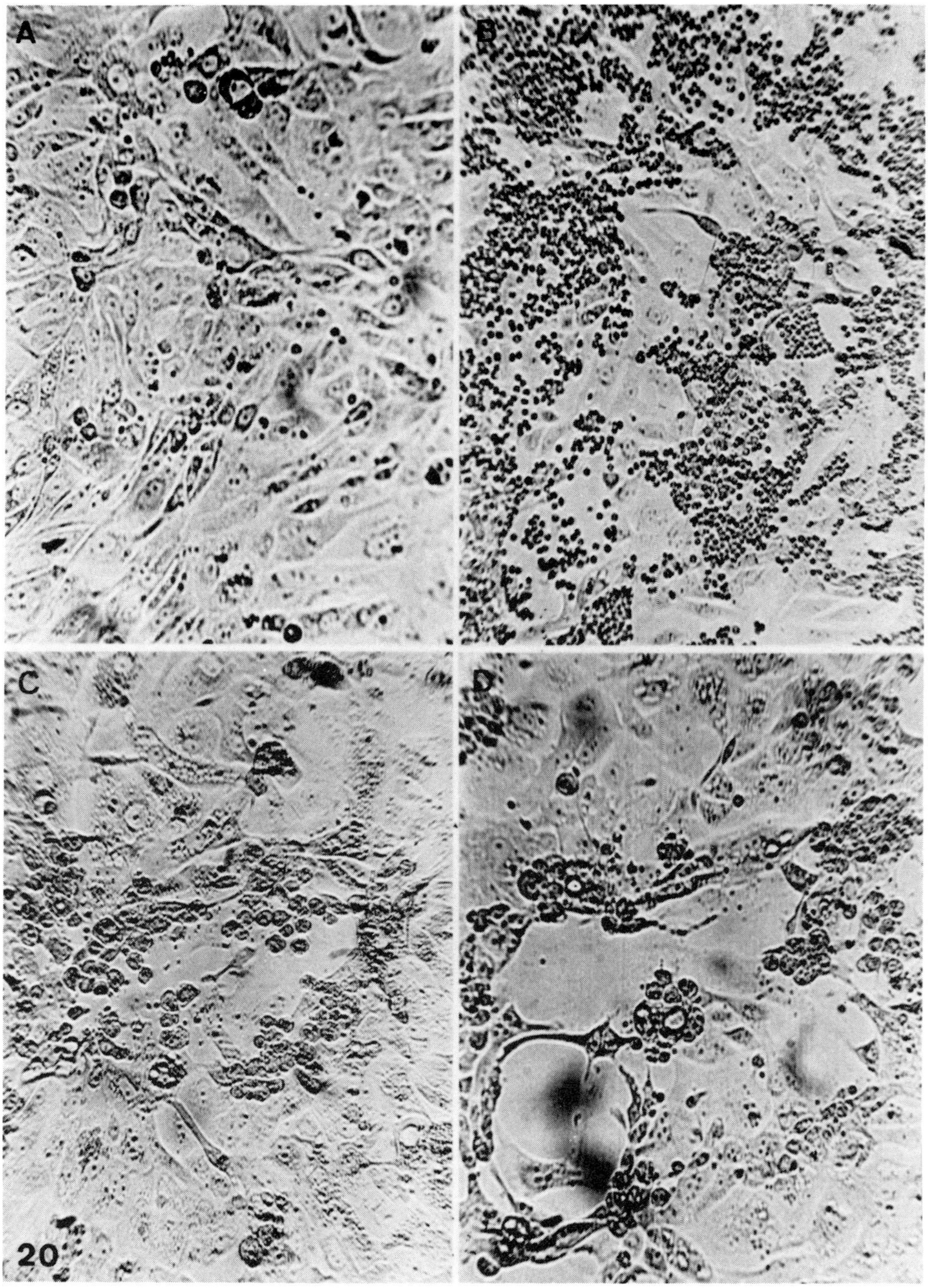

FIGURE 20. African green monkey kidney (AGMK) cells infected with influenza virus, showing hemadsorption and cytopathic effect. (Unfixed, unstained preparations; magnification × 100.) (A) Uninfected AGMK cells, (B) AGMK cells infected with influenza virus B/Hong Kong, showing positive hemadsorption of guinea pig erythrocytes, (C) AGMK cells infected with influenza B virus showing early CPE, and (D) Same as in C, showing late advanced CPE.

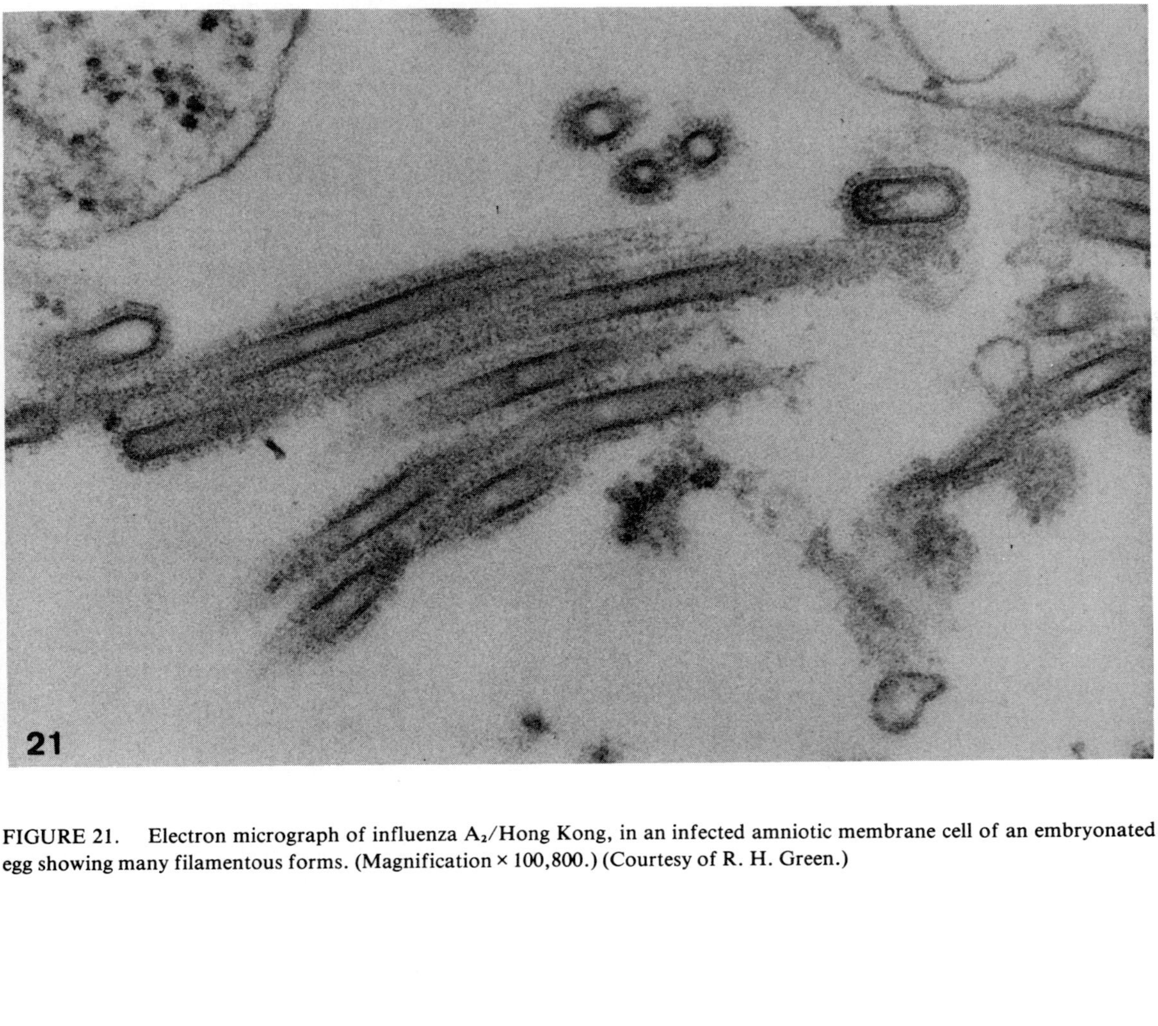

FIGURE 21. Electron micrograph of influenza A_2/Hong Kong, in an infected amniotic membrane cell of an embryonated egg showing many filamentous forms. (Magnification × 100,800.) (Courtesy of R. H. Green.)

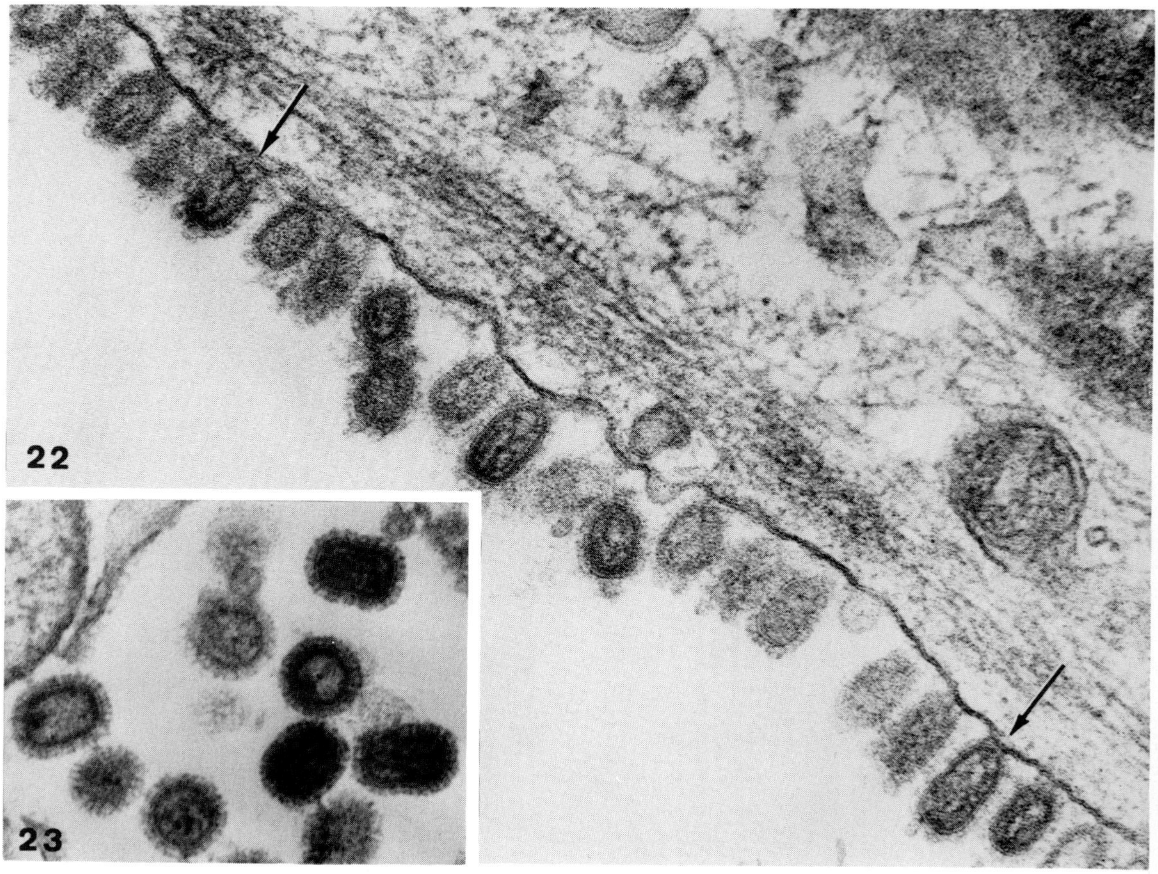

FIGURE 22. African green monkey kidney cell (AGMK) infected with influenza virus A_o/WSN. Many enveloped virus particles, with spikes, spherical and oval shaped, are seen at the cell surface, or in the process of budding at the plasma membrane (arrows). (Magnification × 100,800.) FIGURE 23. AGMK infected with influenza virus B/Hong Kong, showing similar morphology as shown in Figure 22. (Magnification × 100,800.)

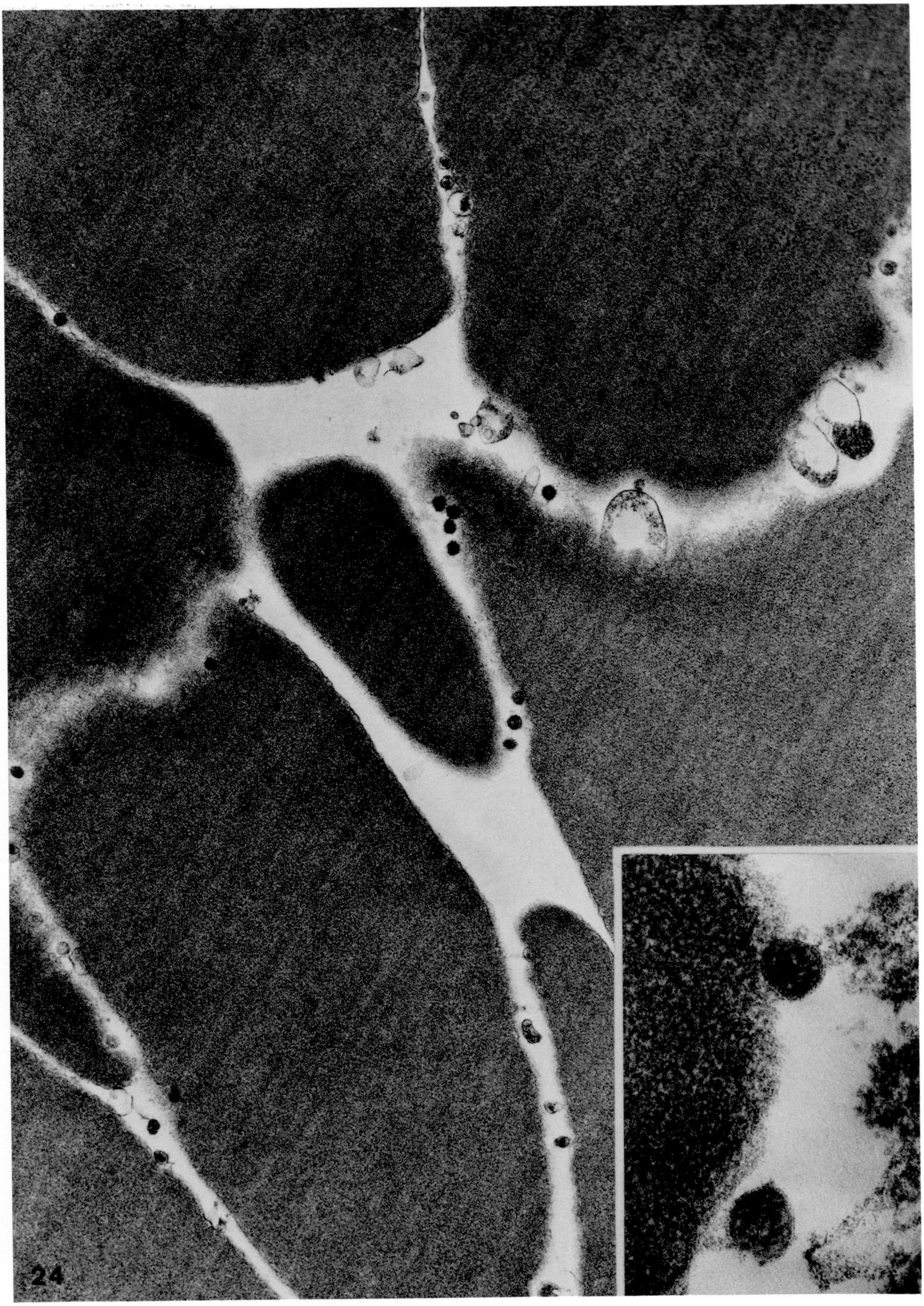

FIGURE 24. Electron micrograph showing interaction of influenza virus with guinea pig erythrocytes. Note: virus particles attached to and between the red blood cells. (Magnification × 24,000., Insert: × 100,800.)

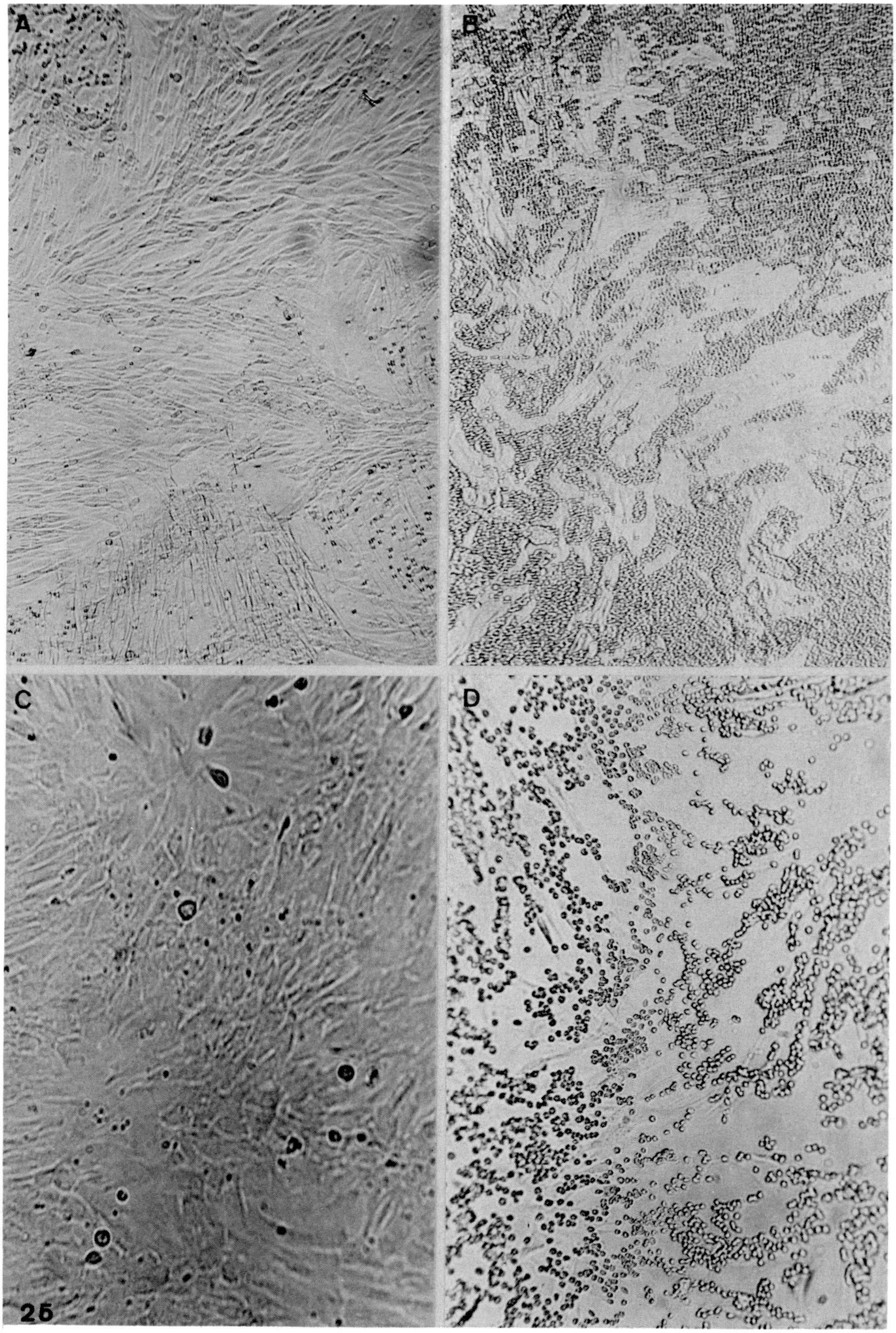

FIGURE 25. RhMK and guinea pig embryo fibroblast (GPEF) cells infected with parainfluenza viruses showing hemadsorption with guinea pig red blood cells. (Unfixed, unstained; magnification × 100.) (A) Uninfected RhMK cells, (B) RhMK cells infected with parainfluenza virus type 1 showing positive hemadsorption with guinea pig blood cells, (C) Uninfected GPEF cells, and (D) GPEF cells infected with parainfluenza virus type 3, showing extensive hemadsorption.

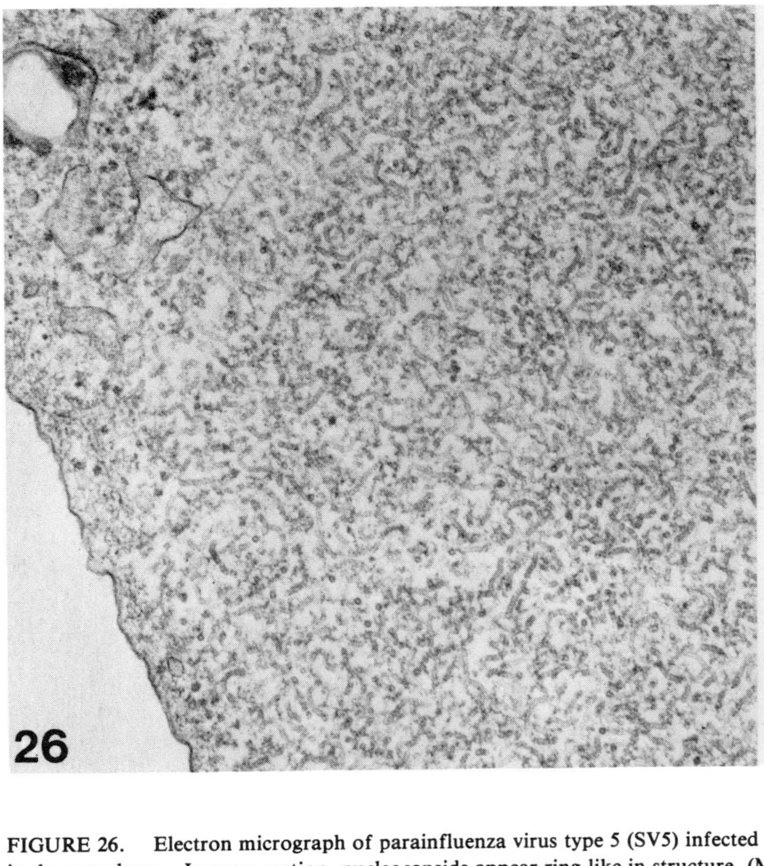

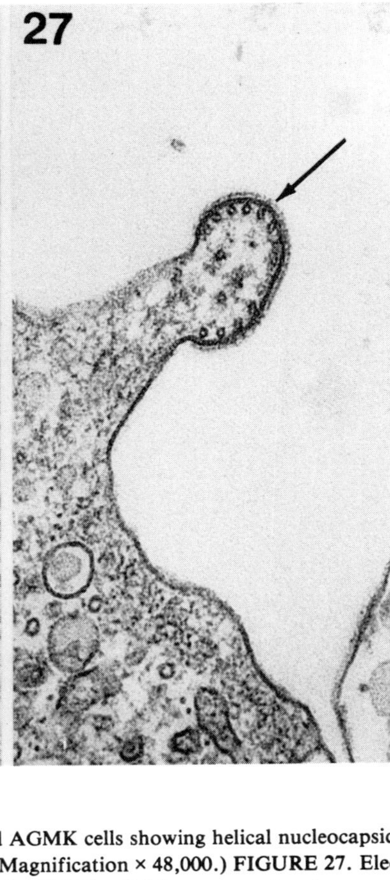

FIGURE 26. Electron micrograph of parainfluenza virus type 5 (SV5) infected AGMK cells showing helical nucleocapsids in the cytoplasm. In cross section, nucleocapsids appear ring-like in structure. (Magnification × 48,000.) FIGURE 27. Electron micrograph of parainfluenza virus type 3 budding at the plasma membrane of a RhMK cell. The outer surface is covered with 10 nm spikes (arrow). (Magnification × 70,000.)

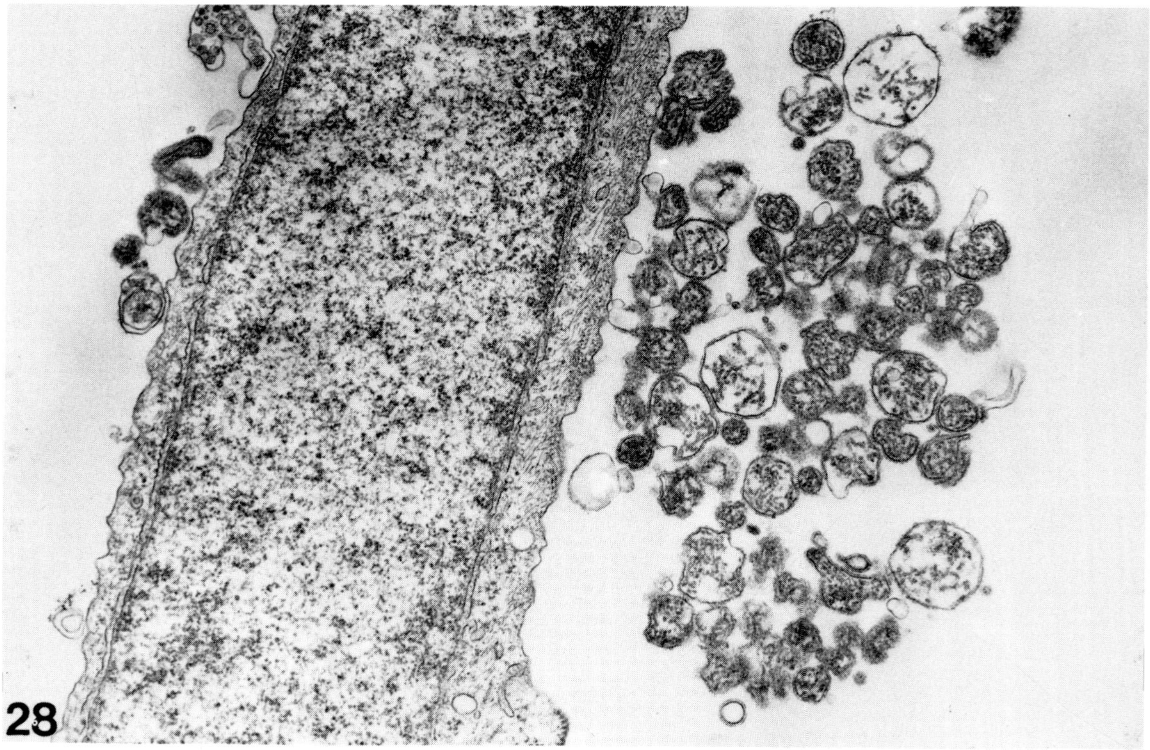

FIGURE 28. Parainfluenza virus type 3 infected RhMK cell showing variable sizes of extracellular virus particles. Some virus particles appear to have disintegrated. (Magnification × 24,000.)

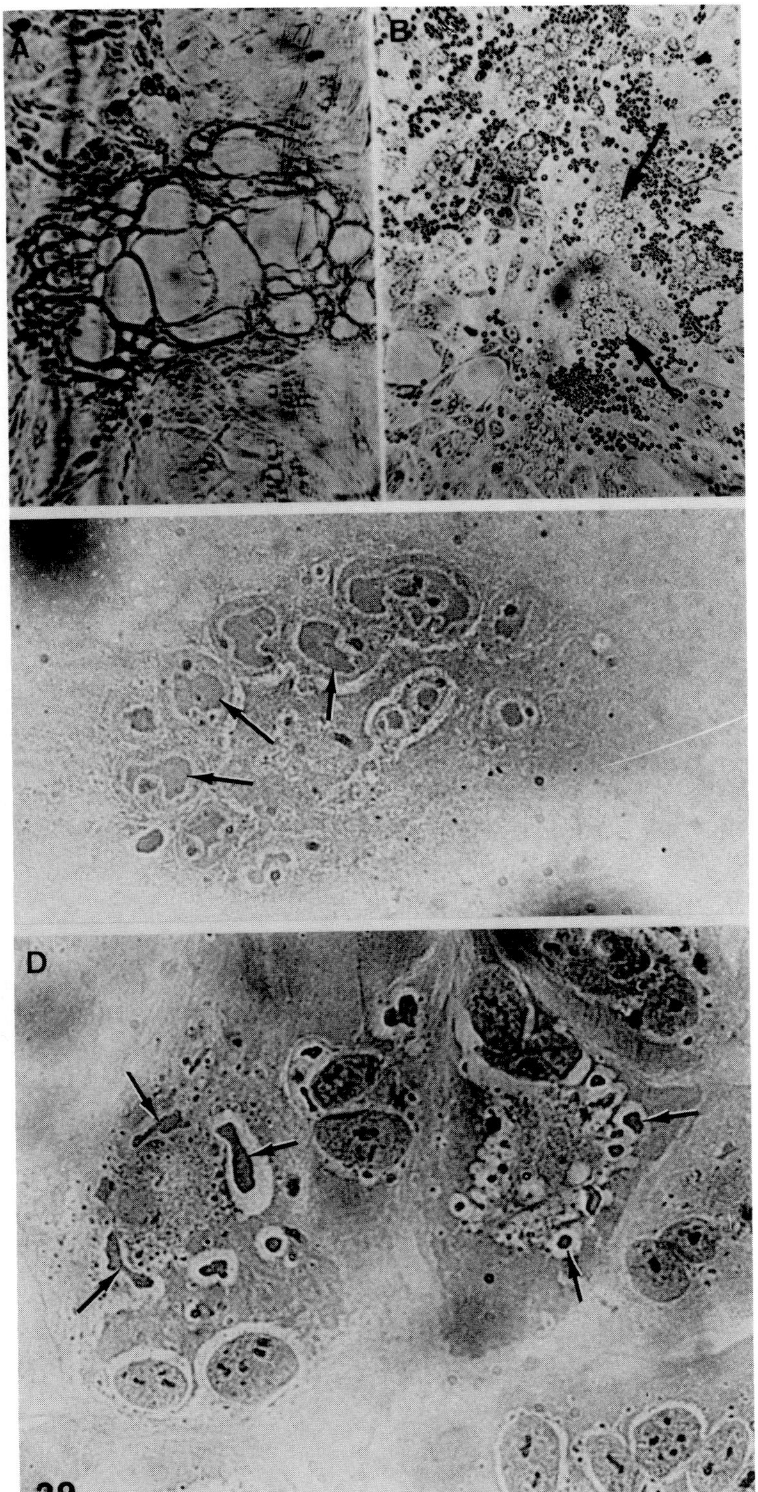

FIGURE 29. Rhesus monkey kidney cells and human kidney cells infected with measles virus. [(A and B) Unfixed, unstained, magnification × 100; (C and D) hematoxylin and eosin stained, magnification × 400.] (A) RhMK cells infected with measles virus showing cytopathic effect due to syncytia formation. (B) Measles virus infected RhMK cells showing positive hemadsorption with monkey erythrocytes. (Note: large syncytial cells, indicated by arrows.) (C) Measles virus infected human kidney cells. (Note: intranuclear eosinophilic inclusions indicated by arrows.) (D) Same preparation as C; illustrating intracytoplasmic inclusions (indicated by arrows).

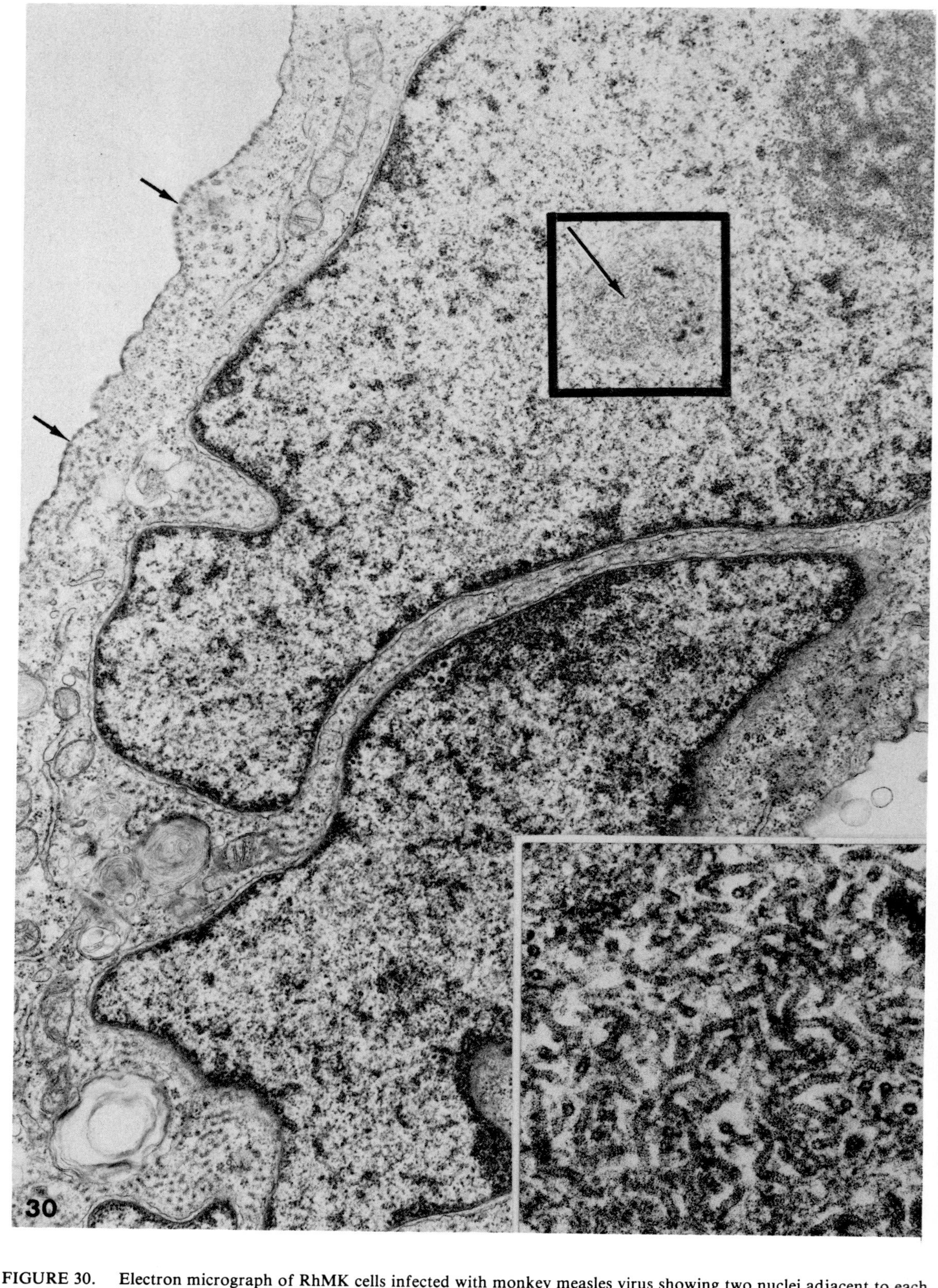

FIGURE 30. Electron micrograph of RhMK cells infected with monkey measles virus showing two nuclei adjacent to each other. A small intranuclear inclusion consisting of helical nucleocapsids is seen (arrow) in one of the nuclei. Note: plasma membrane at the upper left was altered due to virus infection (short arrows). (Magnification × 19,680.) Insert shows the helical nucleocapsid in the nucleus shown above at a higher magnification. (Magnification × 100,800.)

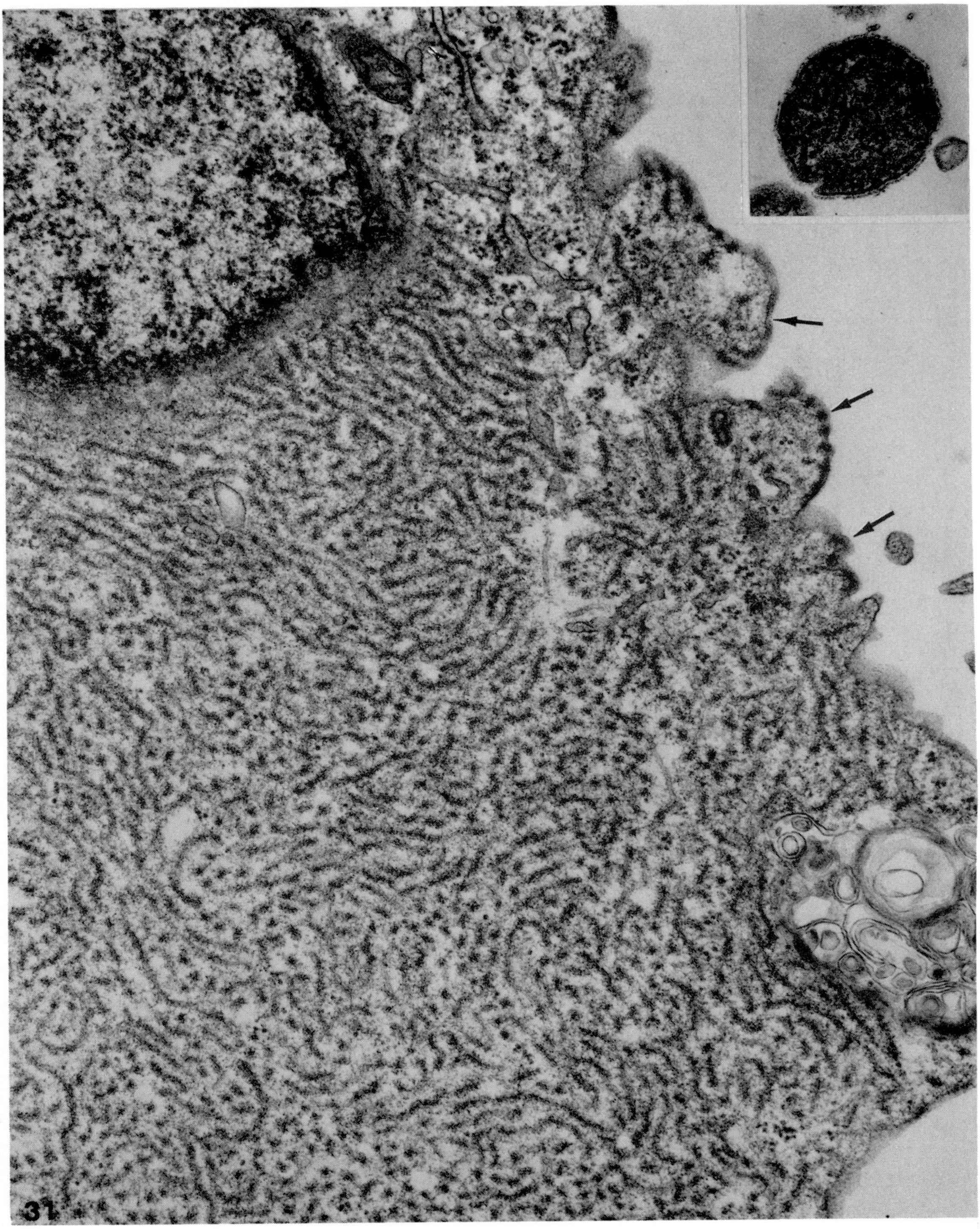

FIGURE 31. Same preparation as Figure 30. The cytoplasm of an infected cell was filled with measles virus nucleocapsids. Note: the plasma membrane was altered (arrows). (Magnification × 30,000.) Insert shows an extracellular measles virus particle. (Magnification × 48,000.)

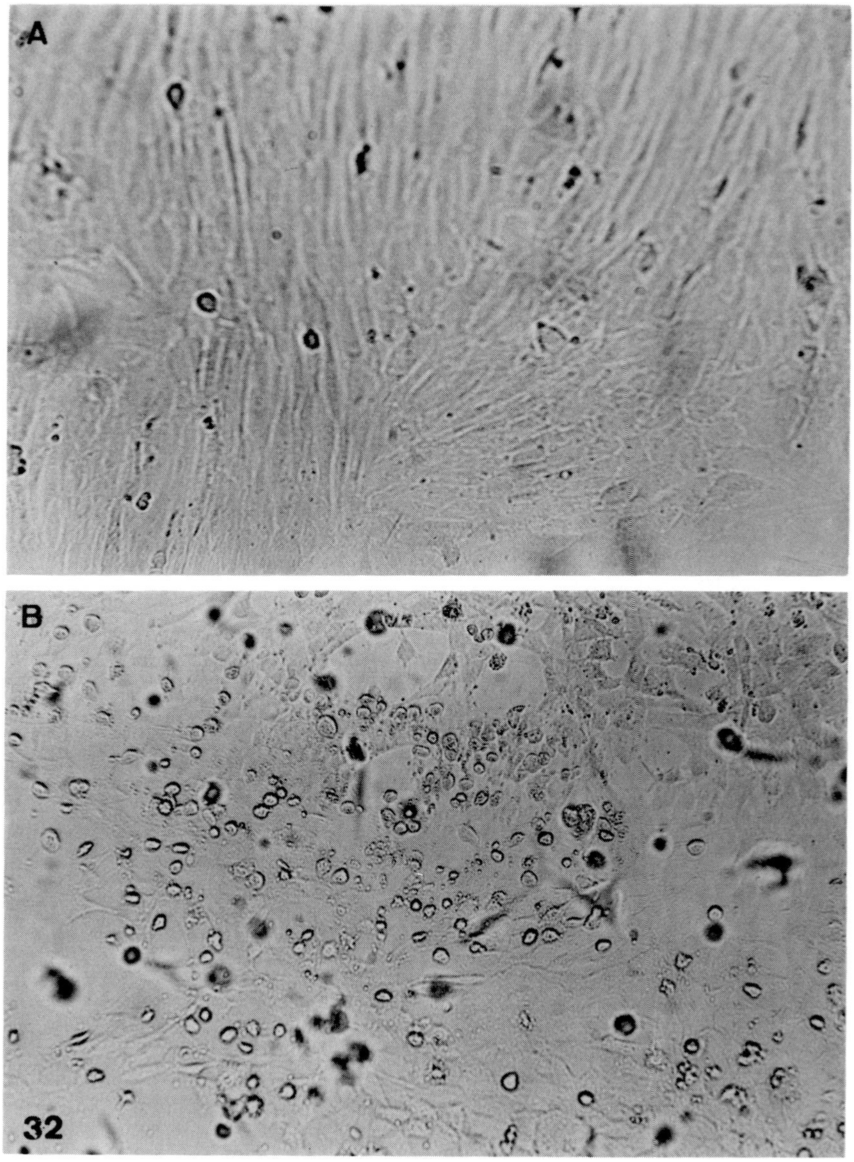

FIGURE 32. Chicken embryo (CE) cells infected with vesicular stomatitis virus (VSV). (Unfixed, unstained preparations; magnification × 100.) (A) Uninfected CE cells and (B) VSV infected CE cells, showing cytopathic effect of cell rounding.

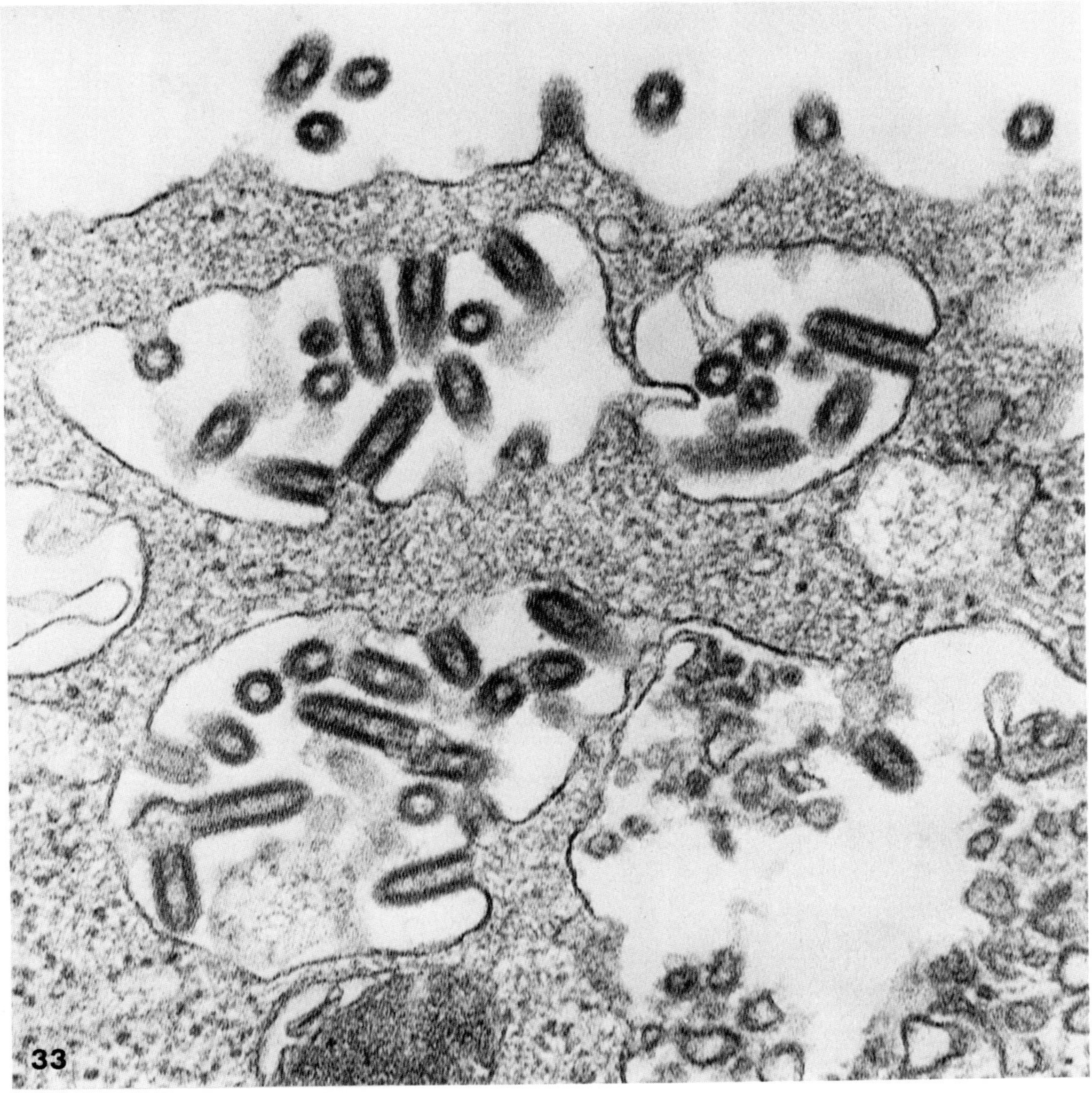

FIGURE 33. VSV-infected CE cell. Bullet-shaped virus particles (65 × 190 nm) are seen in the cytoplasmic vacuoles and extracellular space. Virus particles budding into cytoplasmic vacuoles and at plasma membrane are common. (Magnification × 100,800.)

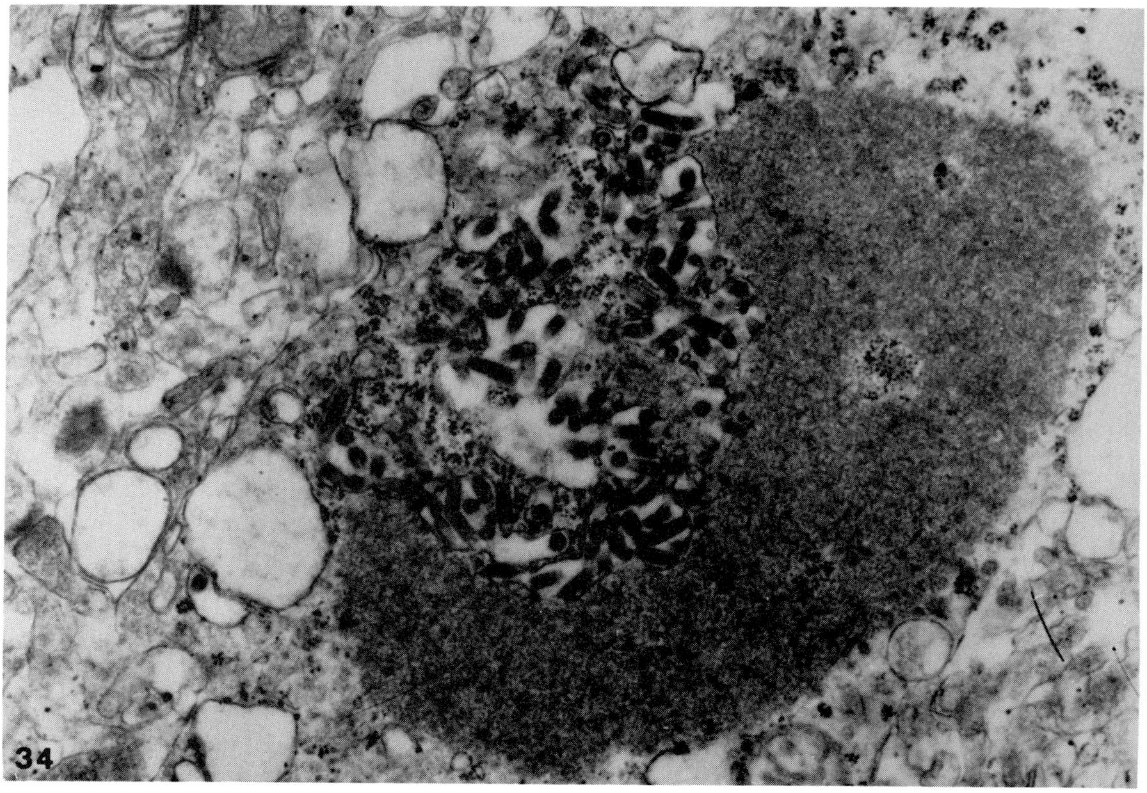

FIGURE 34. Electron micrograph of rabies virus-infected fox hippocampus. Bullet- and rod-shaped particles were present in an intracellular inclusion in the cytoplasm of a neuron. (Magnification × 30,000.) (From Dierks, R. E., Murphy, F. A., and Harrison, A. K., *Am. J. Pathol.*, 54, 251, 1969. With permission.)

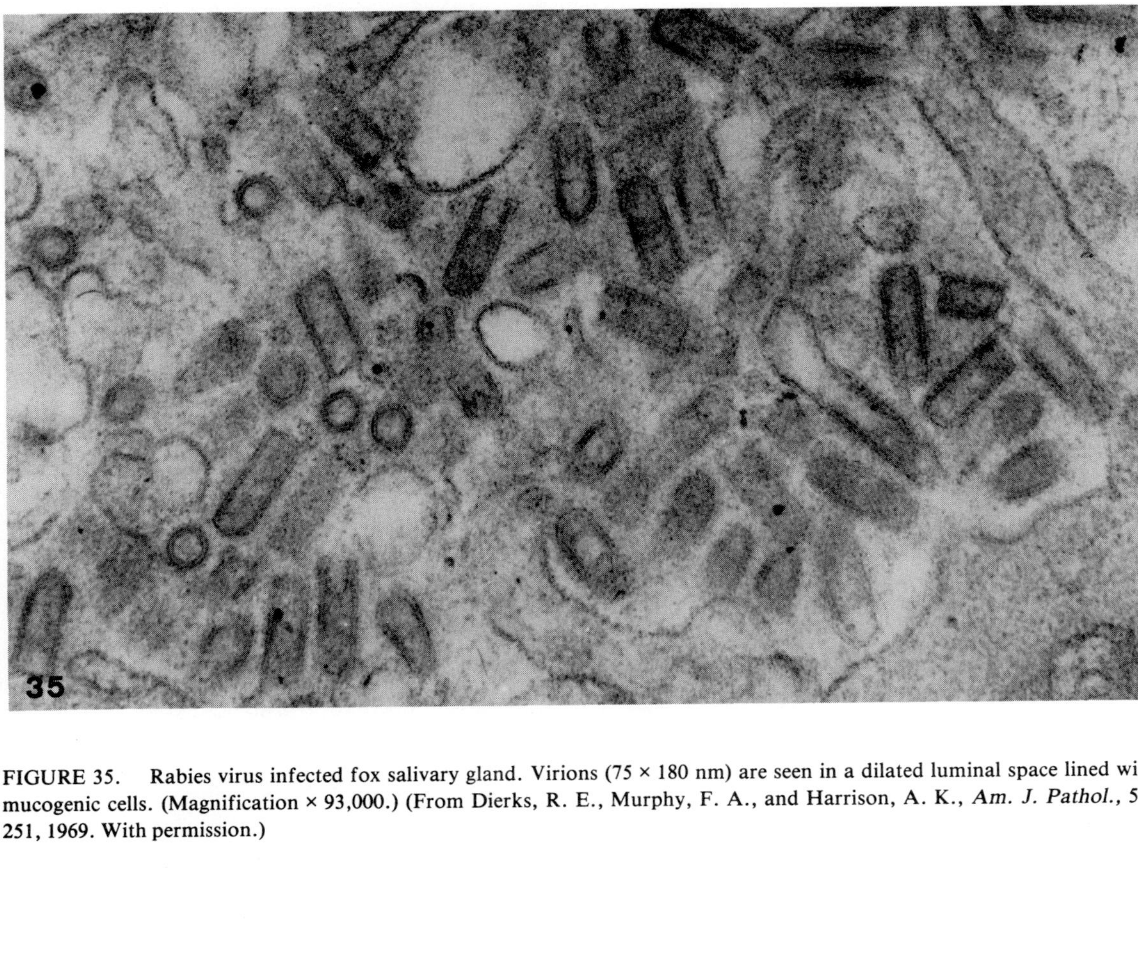

FIGURE 35. Rabies virus infected fox salivary gland. Virions (75 × 180 nm) are seen in a dilated luminal space lined with mucogenic cells. (Magnification × 93,000.) (From Dierks, R. E., Murphy, F. A., and Harrison, A. K., *Am. J. Pathol.*, 54, 251, 1969. With permission.)

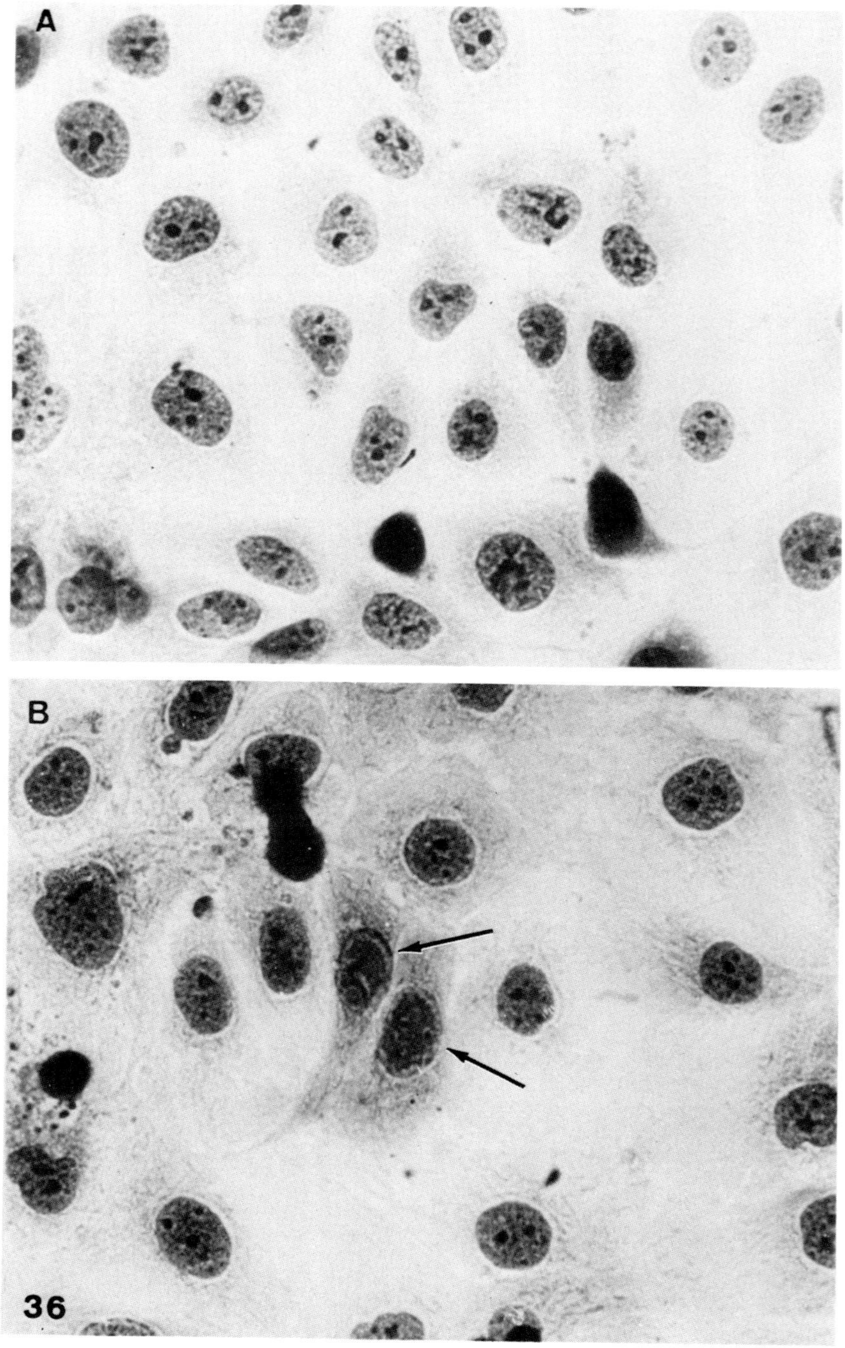

FIGURE 36. NB cells (cell line derived from SV40 transformed human embryonic kidney cells) infected with H-1 virus. (Hematoxylin and eosin stained; magnification × 400.) (A) Uninfected NB cells, and (B) H-1 virus-infected NB cells, 24 hr after infection, showing intranuclear basophilic inclusions (arrows).

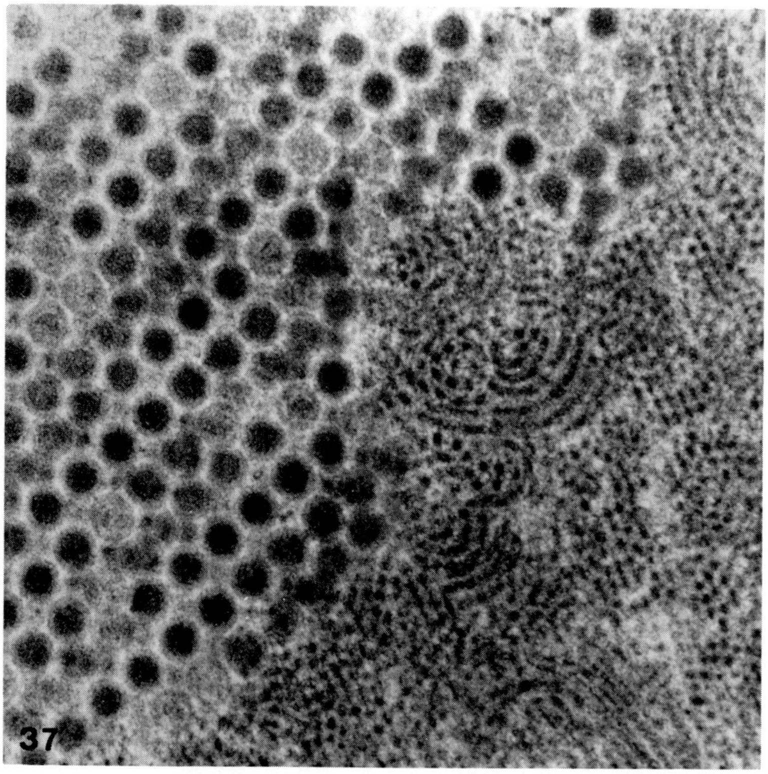

FIGURE 37. Electron micrograph of a cell infected with adeno-associated virus (AAV) in the presence of adenovirus replication. (Magnification × 57,000.) Upper left: adenovirus aggregate; lower right: AAV aggregate (Courtesy of C. J. Henry and M. Slifkin)

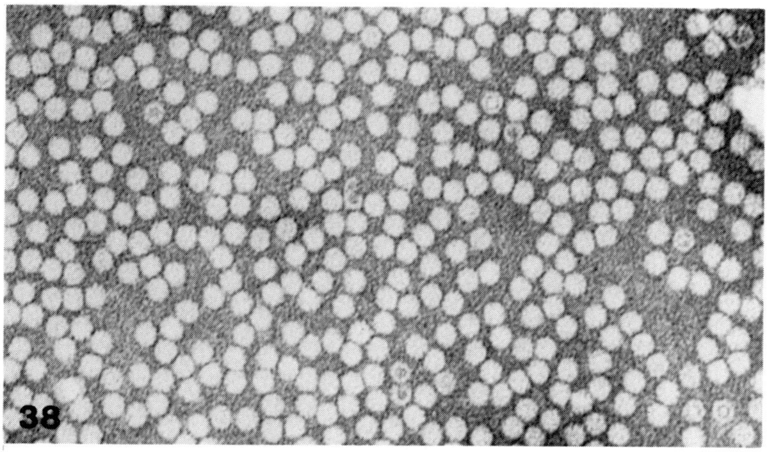

FIGURE 38. Purified H-1 virus stained with uranyl acetate. (Magnification × 125,000.) (Courtesy of F. Al-Lami and H. W. Toolan.)

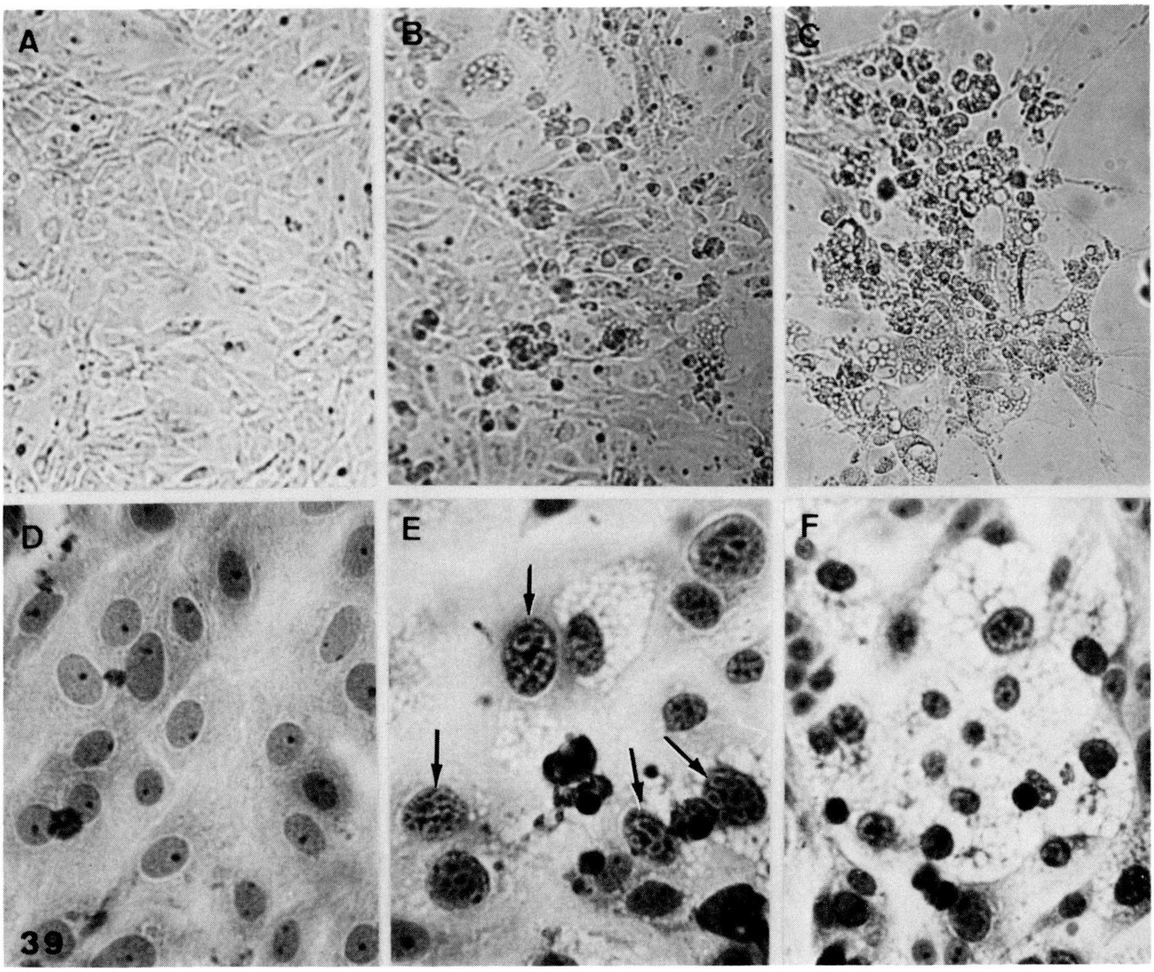

FIGURE 39. African green monkey kidney (AGMK) cells infected with SV40 [(A to C) Unfixed, unstained preparations; magnification × 100. (D to F) Hematoxylin and eosin stained; magnification × 400.] (A) Uninfected AGMK cells; (B) AGMK cells infected with SV40 showing early vacuolation; (C) AGMK cells infected with SV40 showing advanced CPE of vacuolation; (D) uninfected AGMK cell; (E) AGMK cells infected with SV40 showing intranuclear basophilic inclusions, indicated by arrows; and (F) SV40 infected AGMK cells showing intranuclear inclusions and extensive vacuolation in the cytoplasm.

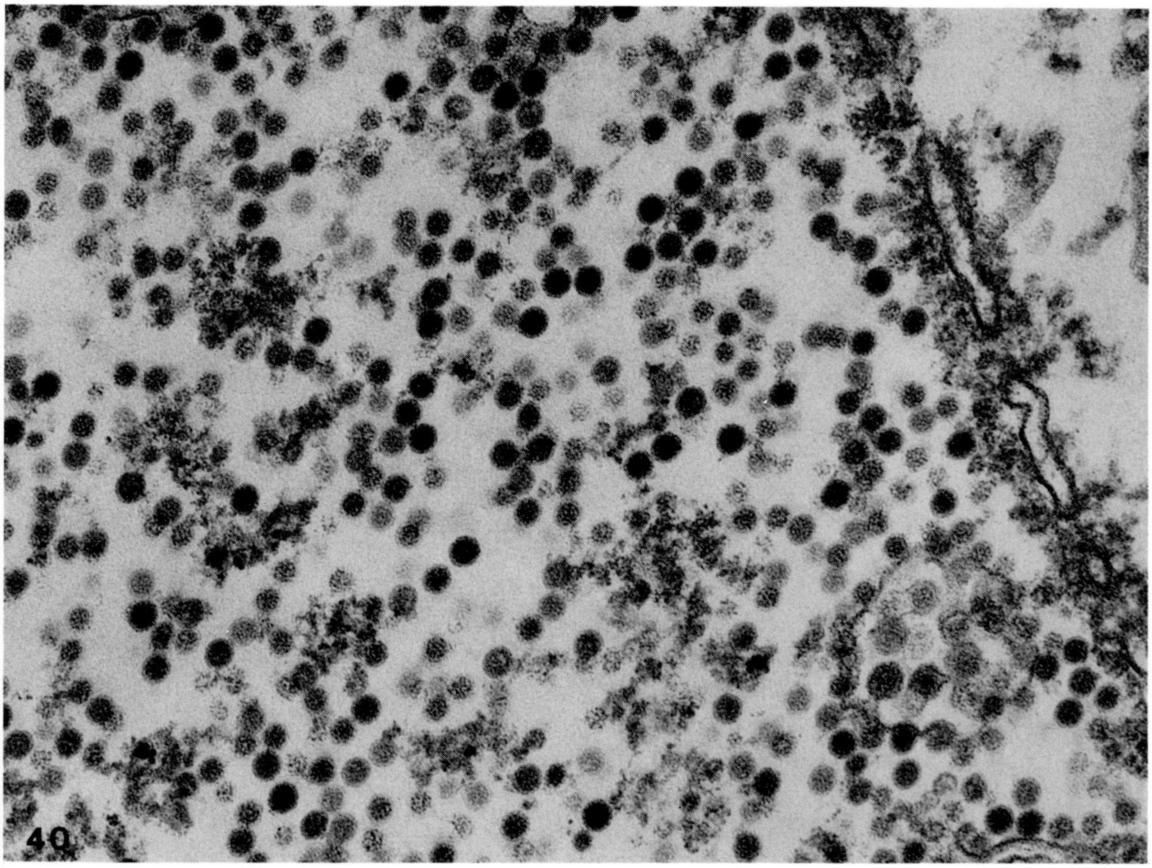

FIGURE 40. Electron micrograph of a SV40-infected AGMK cell. Many virus particles, 45 nm in diameter, are shown scattered throughout the nucleus. (Magnification × 88,000.)

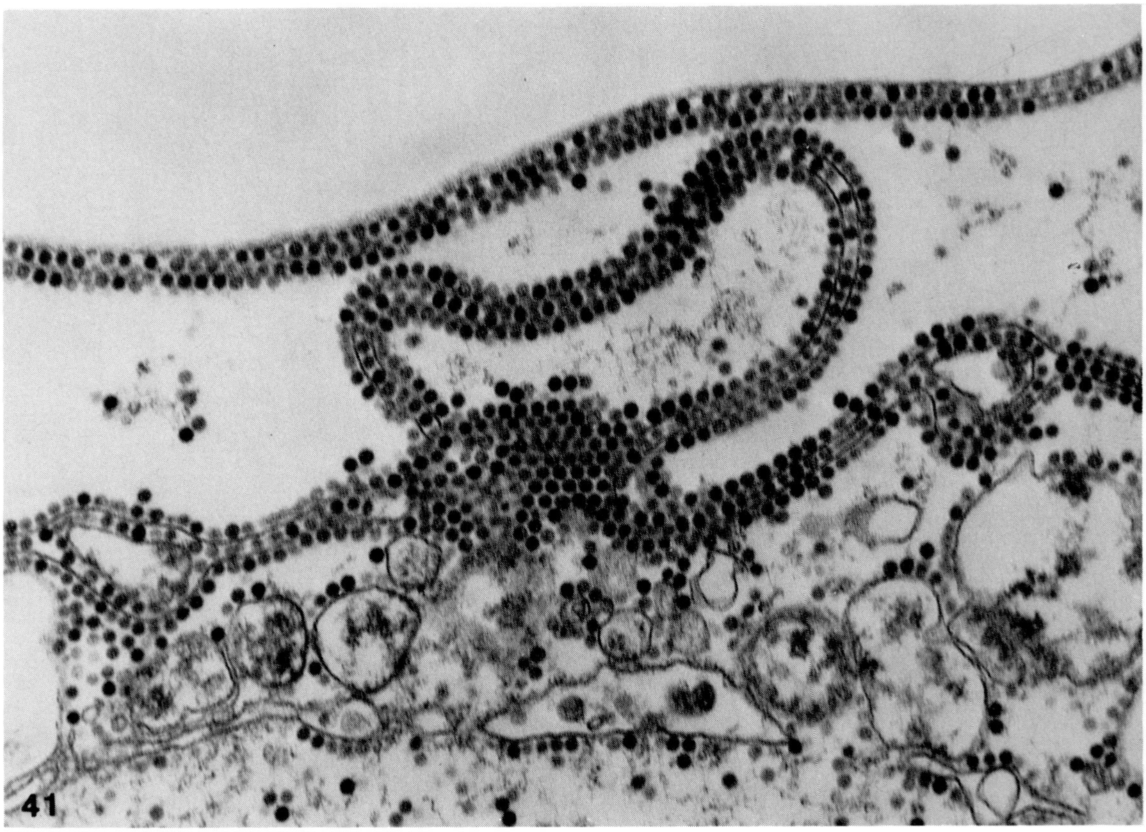

FIGURE 41. SV40 virions in association with membranes of a disrupted cell. (Magnification × 48,000.)

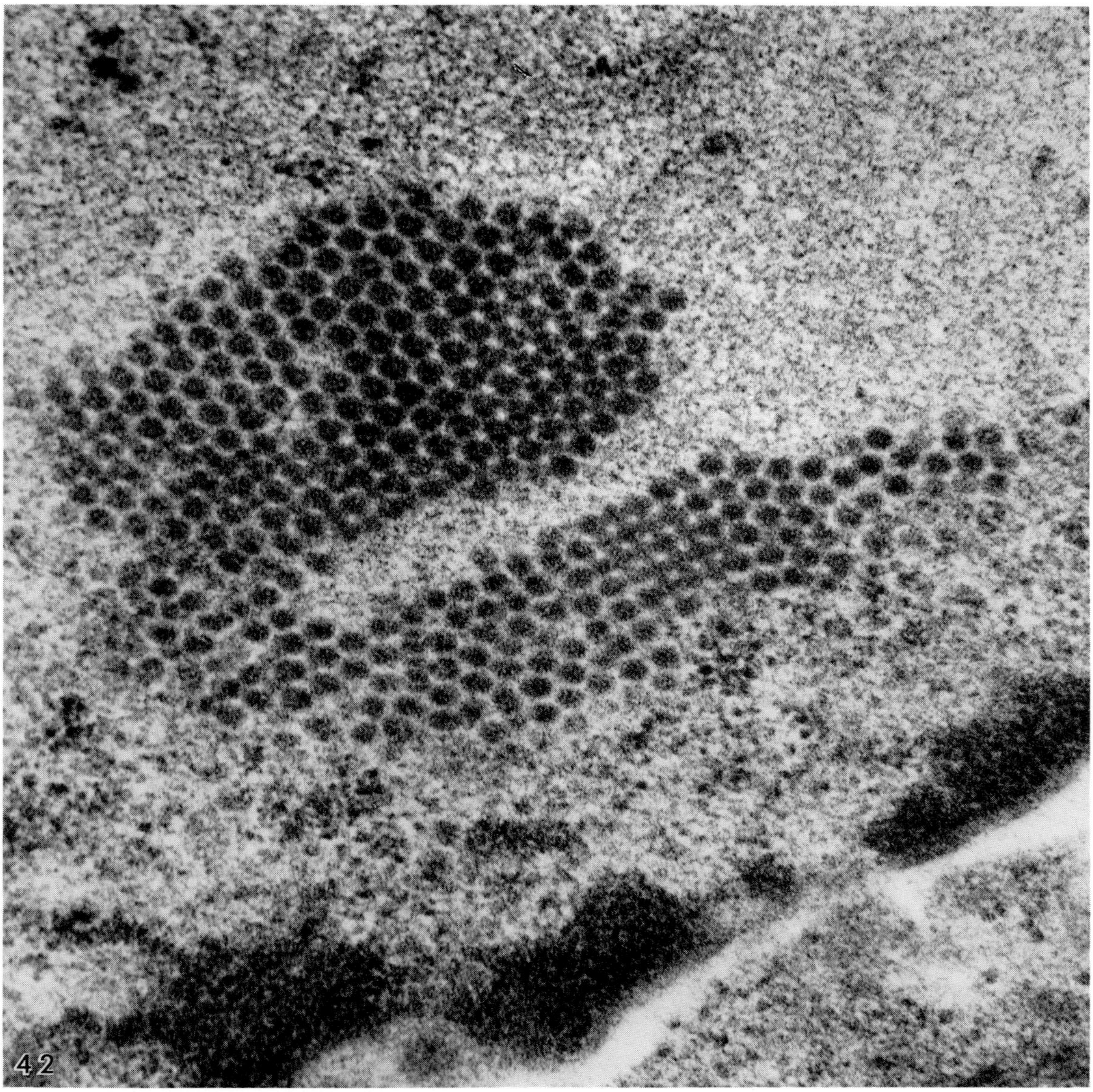

FIGURE 42. Electron micrograph of papilloma virus seen in human wart. Virus aggregate in crystalline arrays containing many virions, 50 nm diameter, in the nucleus. (Magnification × 79,000.) (Courtesy of S. Klaus.)

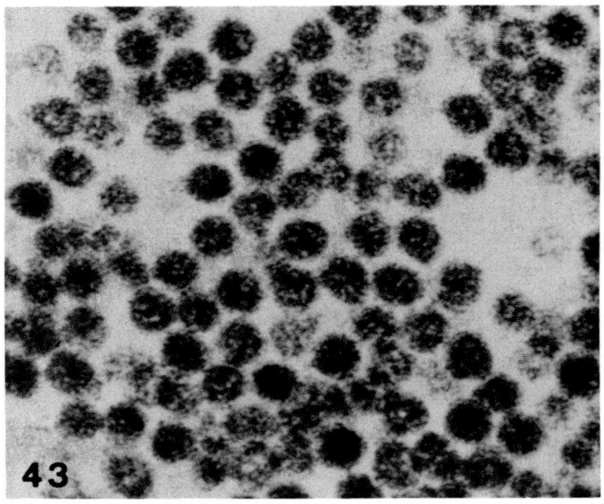

FIGURE 43. Human papovavirus seen in the nucleus of an enlarged epithelial cell from the urothelium. Virus particles, approximately 43 nm diameter. (Magnification × 100,000.) (From ZuRhein, G. M. and Varakis, J., *Lancet*, 2, 783, 1974. With permission.)

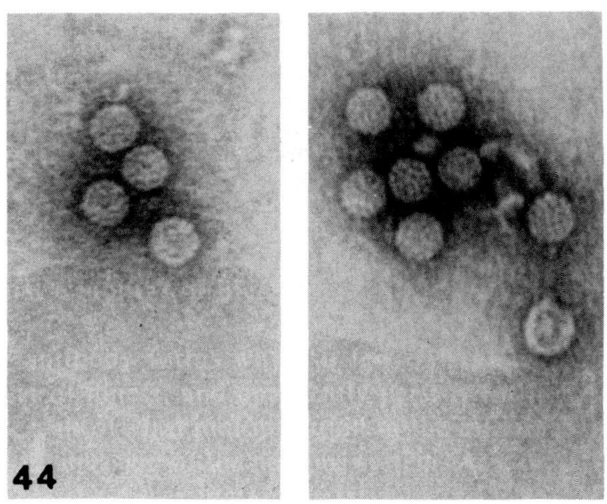

FIGURE 44. Papovavirus obtained from the urine of a renal-transplant patient. Stained with phosphotungstate. (Magnification × 110,000.) (Courtesy of Varakis, J. and ZuRhein, G. M.)

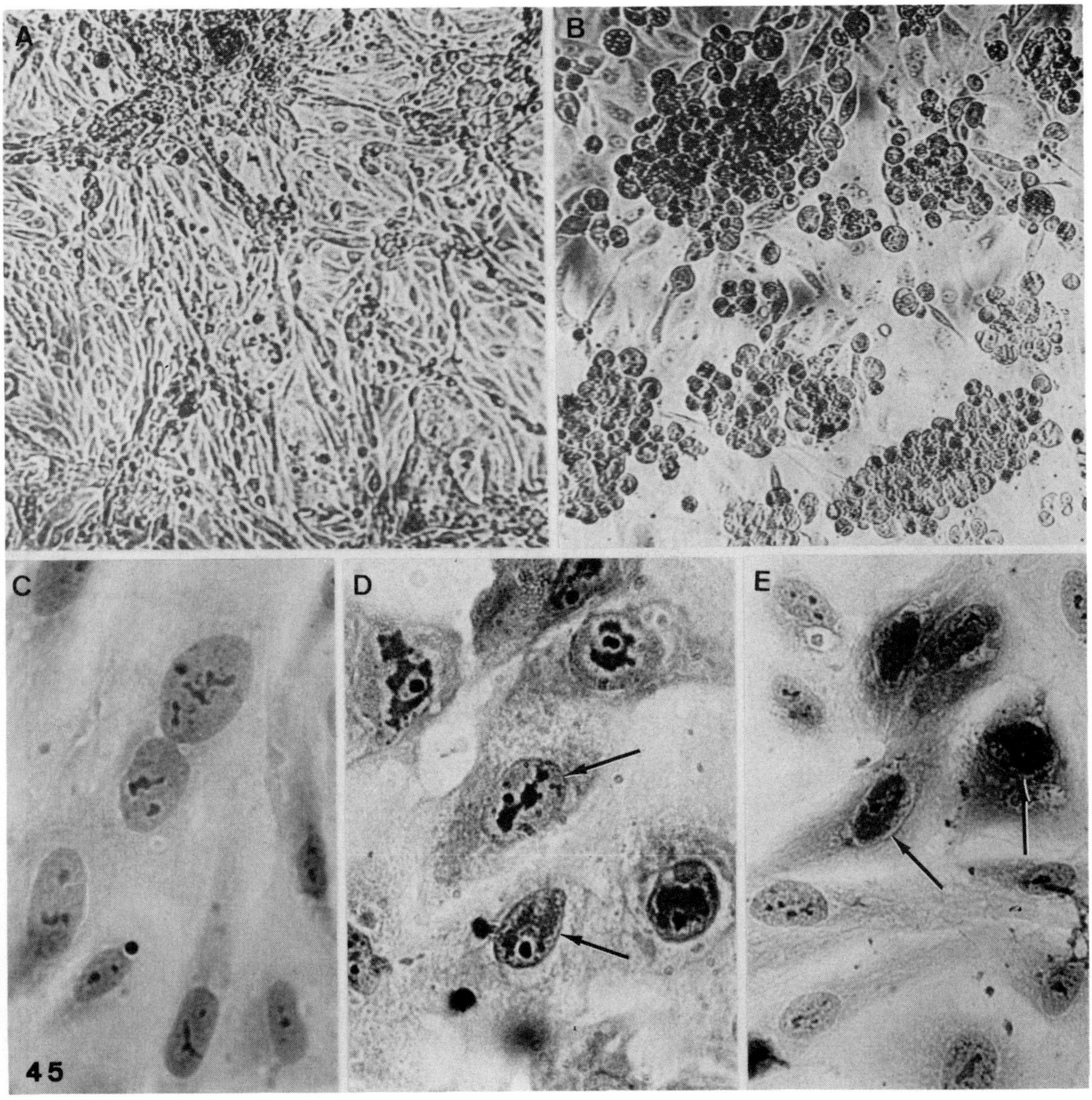

FIGURE 45. Human adenovirus infected human embryonic kidney (HEK) cell cultures. [(A and B) Unfixed, unstained; magnification × 100. (C to E) Hematoxylin and eosin stained; magnification × 400.] (A) Uninfected cells; (B) adenovirus type 3 infected HEK cells showing grape-like clusters of typical adenovirus induced CPE; (C) uninfected HEK cells; (D) adenovirus type 2 infected HEK cells; note: intranuclear basophilic inclusions (arrows); and (E) adenovirus type 7 infected HEK cells; Note: intranuclear basophilic inclusions (arrow). [(A) and (B) Modified from Hsiung, G. D., *Diagnostic Virology,* Yale University Press, 1973, 120. With permission.]

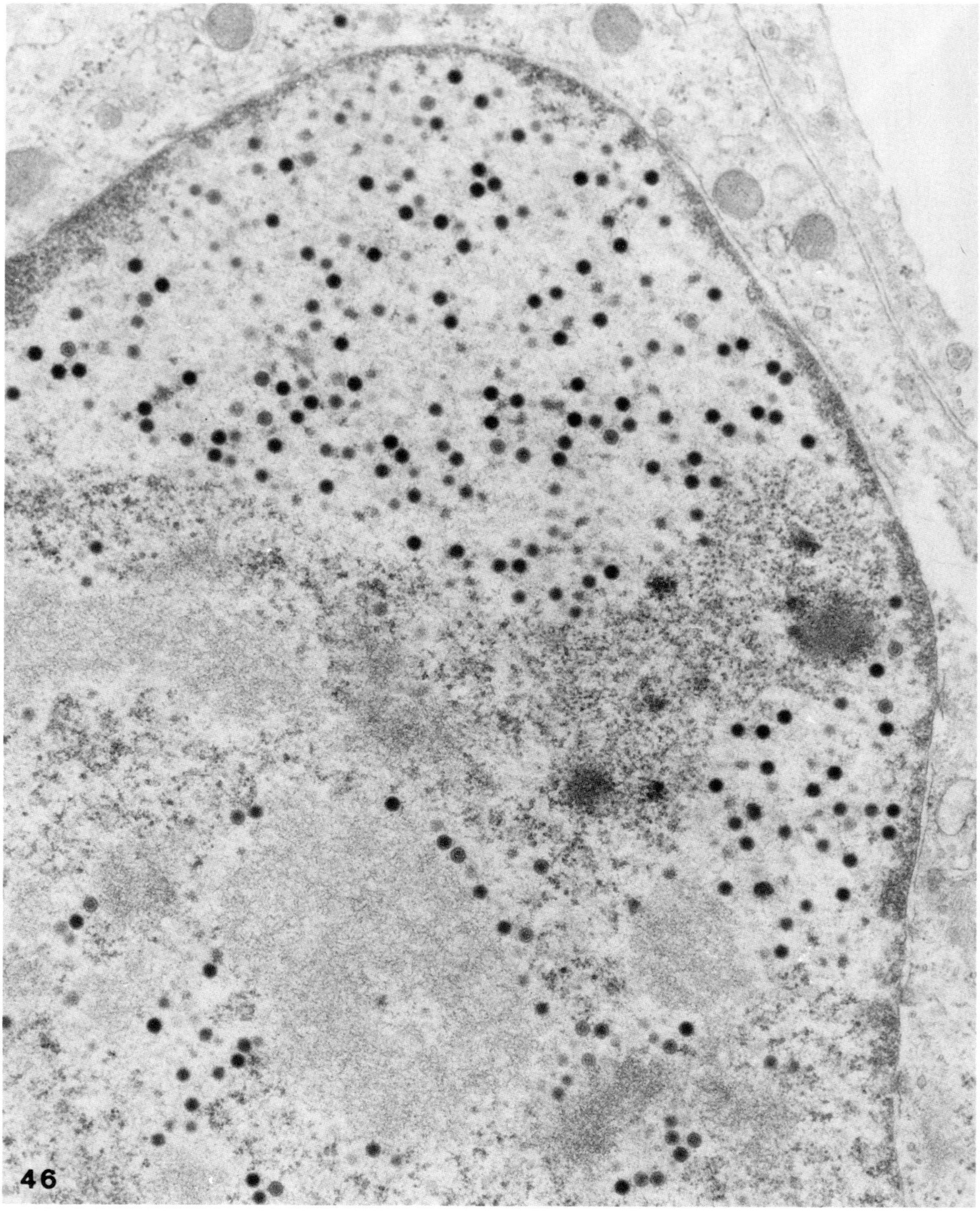

FIGURE 46. Electron micrograph of human adenovirus type 2 in HEK cells. Many spherical virus particles, 70 nm diameter, are scattered throughout the nucleus. (Magnification × 30,000.)

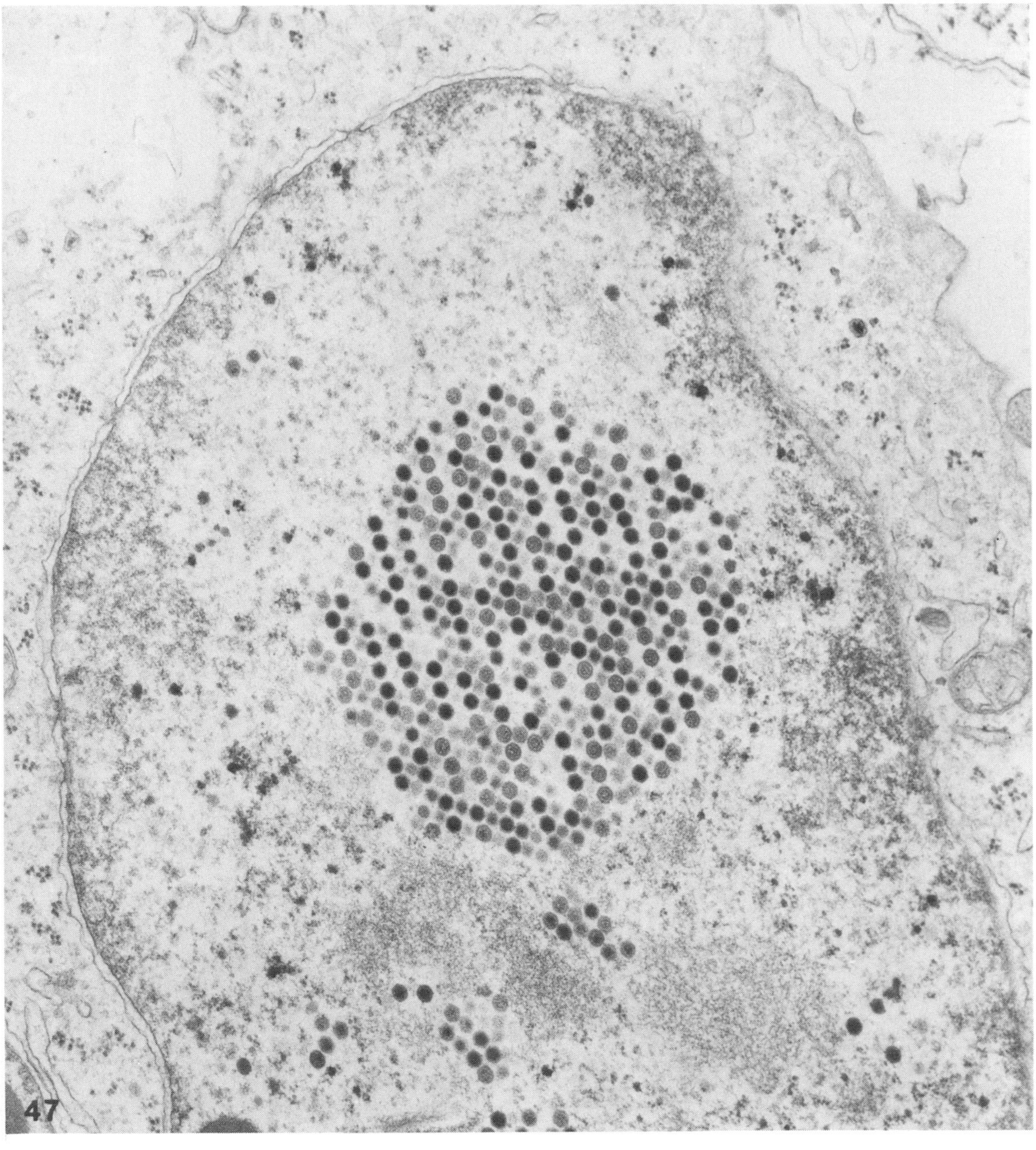

FIGURE 47. Human adenovirus type 7 in HEK cells. (Note: virus particles in crystalline arrays in the nucleus.) (Magnification × 30,000.)

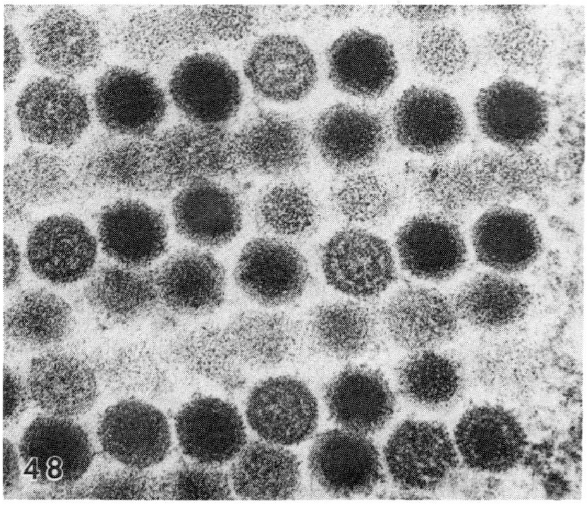

FIGURE 48. Higher magnification of a thin section of adenovirus type 7 crystal showing internal structure of viruses. (Magnification × 100,800.)

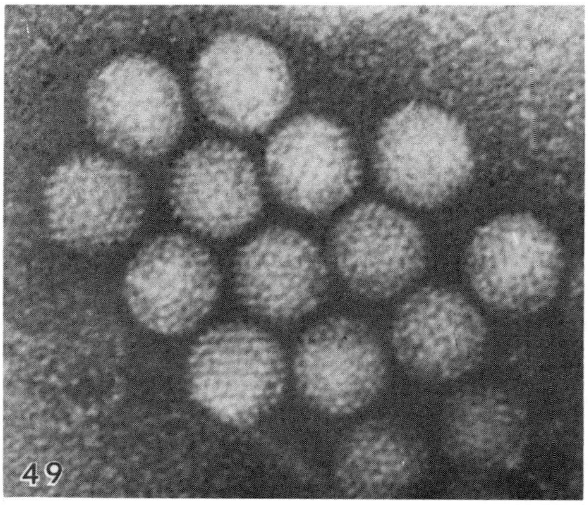

FIGURE 49. Adenovirus type 2 stained with uranyl acetate. (Magnification × 150,000.) (Courtesy of Al-Lami, F. and Ledinko, N.)

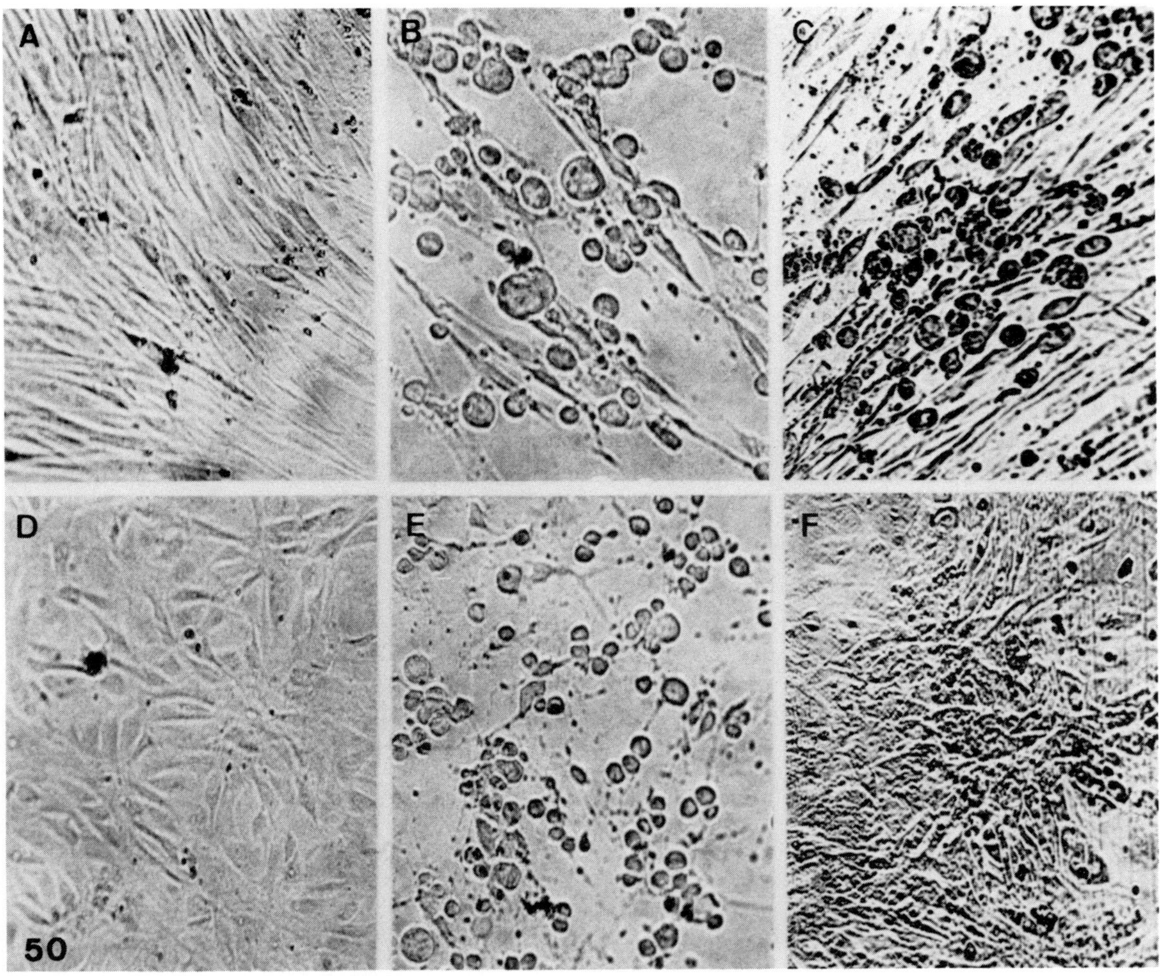

FIGURE 50. Human fibroblast, WI-38 and rabbit kidney (RK) cells infected with herpes simplex virus (HSV) or cytomegalovirus (CMV). (Unfixed, unstained; magnification × 100.) (A) and (D) uninfected WI-38 cells and RK cells, respectively; (B) and (E) HSV-infected WI-38 cells and RK cells showing cell rounding characteristic of HSV cytopathic effect; and (C) and (F) human CMV-infected WI-38 cells and RK cells, typical CMV cytopathic effect only appears in WI-38 cells, not in RK cells.

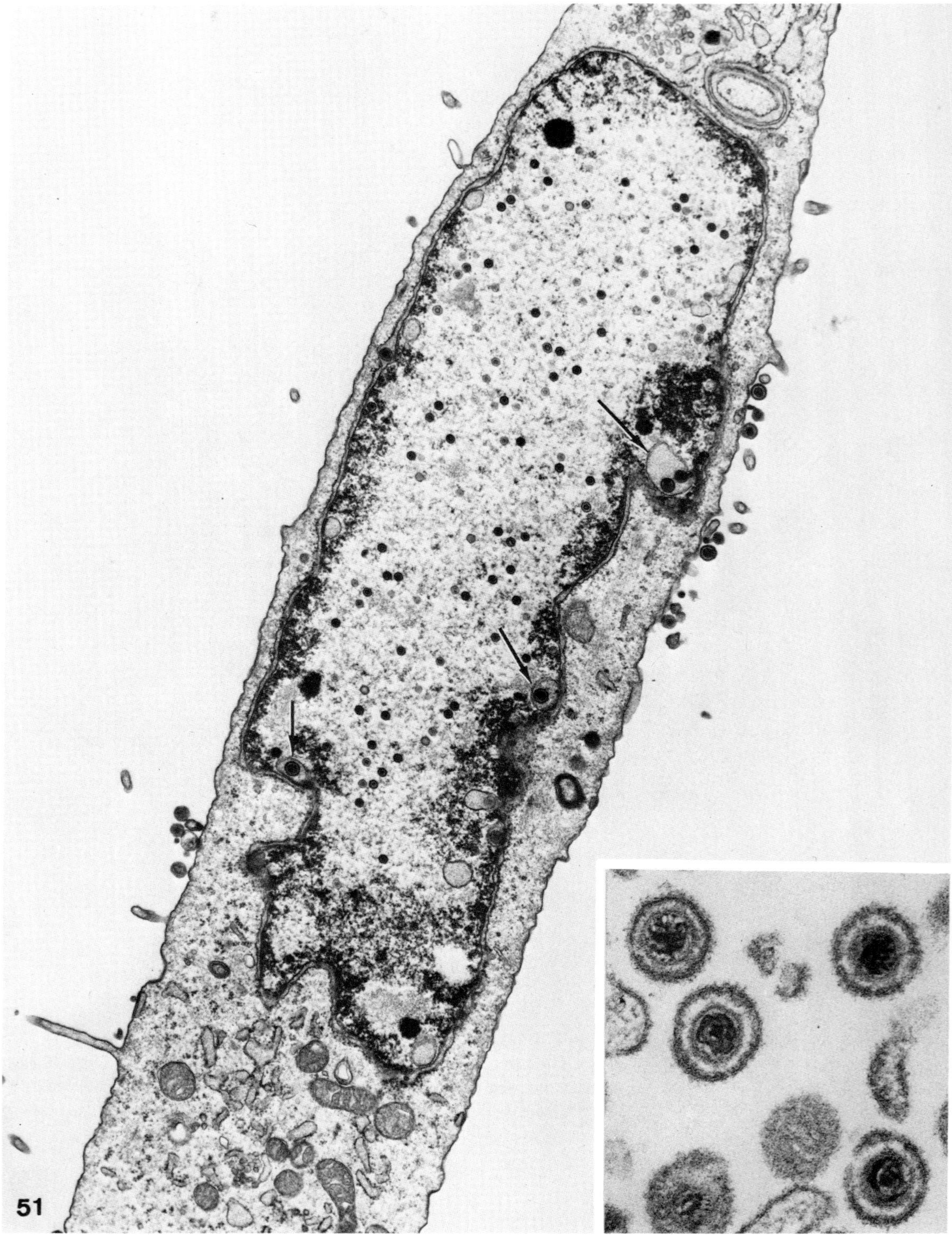

FIGURE 51. Electron micrograph of HSV type 1-infected RK cell. Nucleocapsids with or without dense cores are seen in the nucleus. Several enveloped mature virions are seen in membrane bound nuclear vacuoles (arrows) and at the extracellular space. (Magnification × 16,320, Insert: × 100,800.)

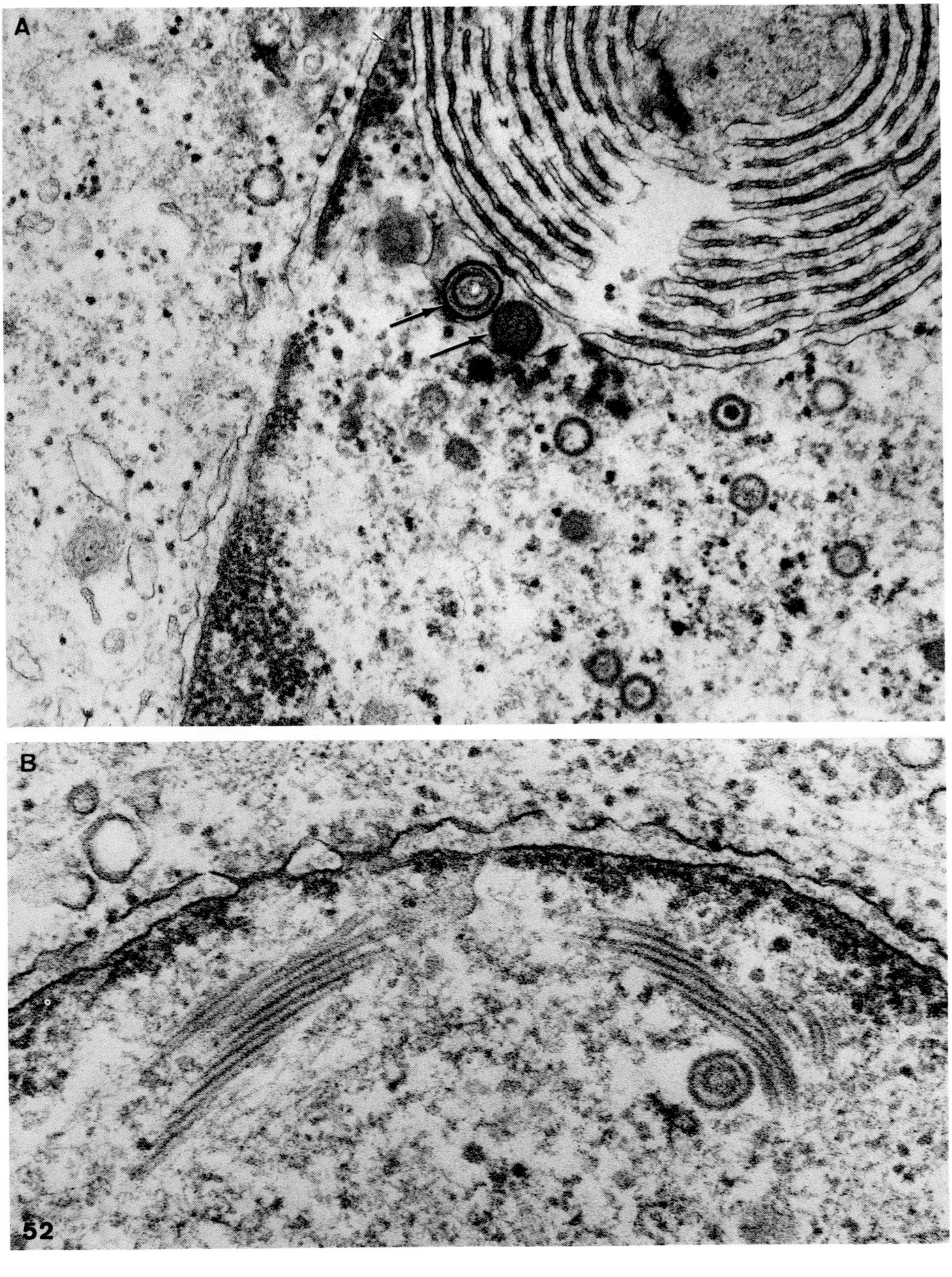

FIGURE 52. Electron micrographs of HSV type 2-infected RK cell. (Magnification × 100,800) (A) A portion of an infected cell showing nucleocapsids scattered throughout the nucleus, two enveloped mature virions (arrows) are seen in association with duplicated membrane induced by HSV infection. (B) A part of a HSV-infected nucleus showing filamentous structures specifically induced by HSV type 2.

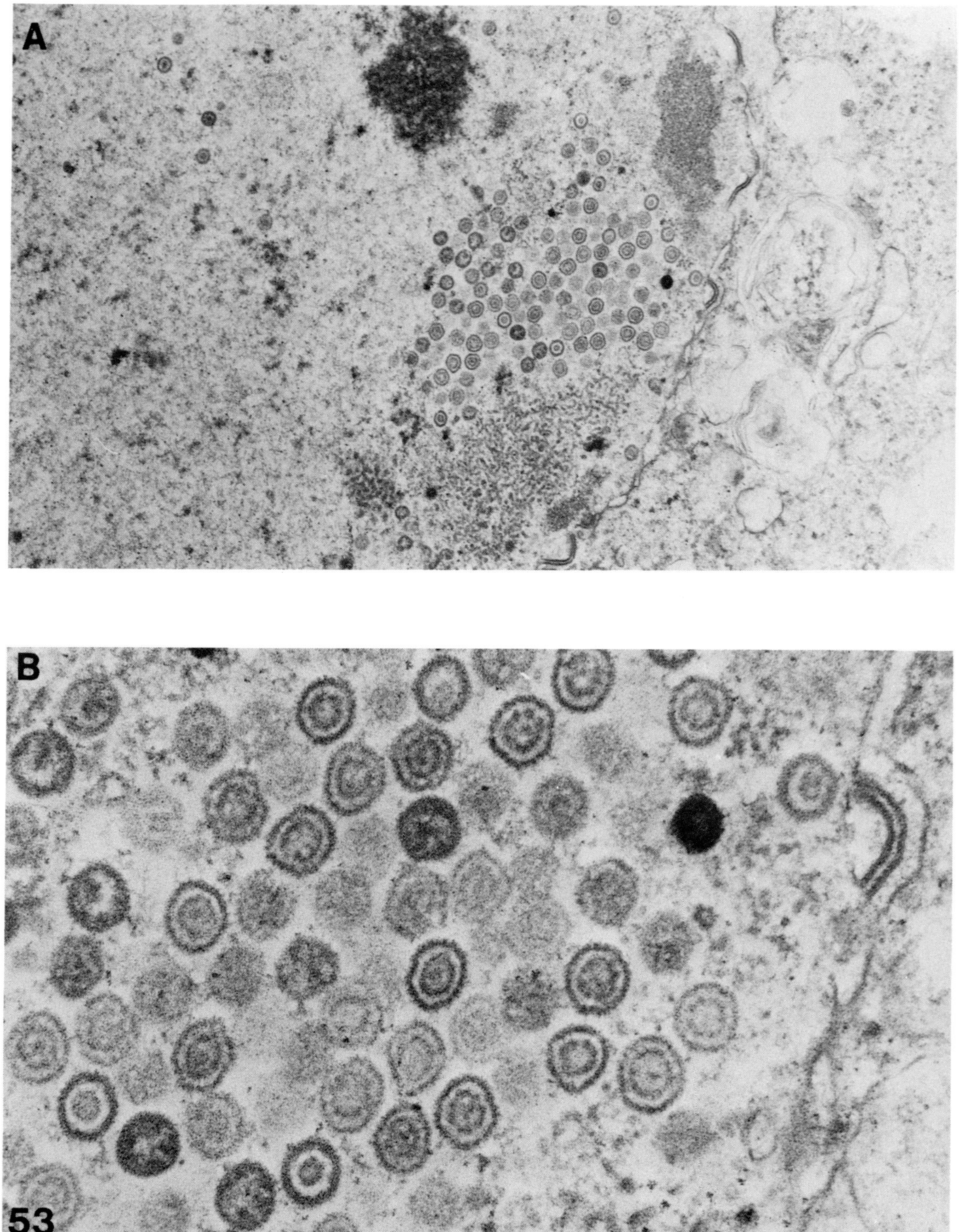

FIGURE 53. (A) HSV type 1 seen in a brain biopsy sample. Typical herpesvirus nucleocapsids were observed in the nucleus of an infected brain cell. (Magnification × 24,000.) (Courtesy of M. August.) (B) Higher magnification of nucleocapsids seen in (A). (Magnification × 100,800.) (Courtesy of M. August.)

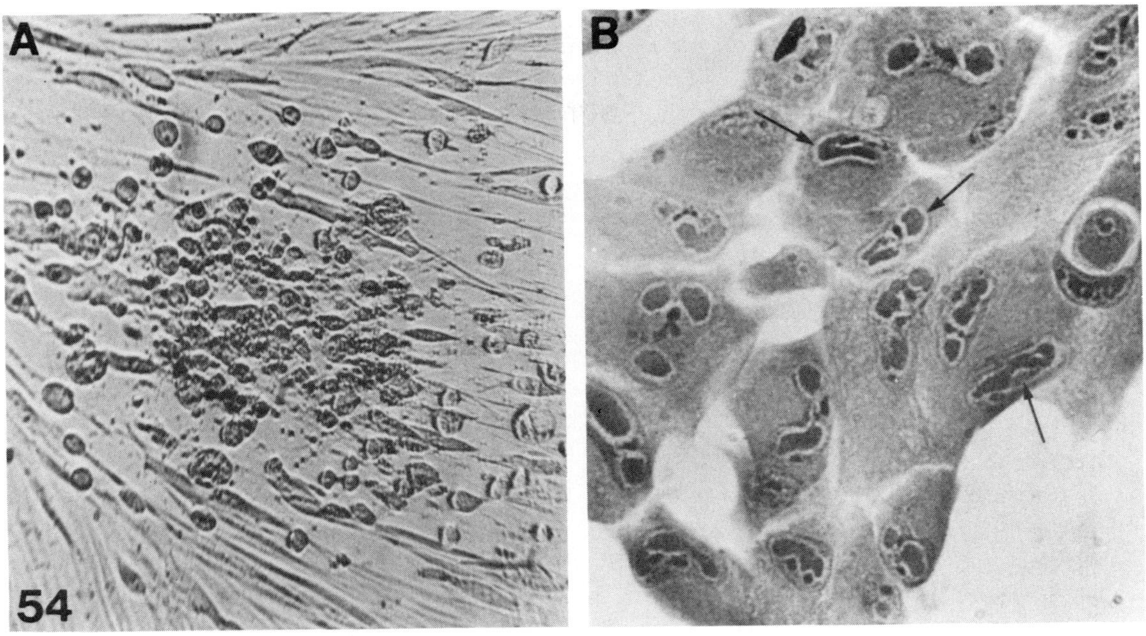

FIGURE 54. Human fibroblasts, WI-38 cells, infected with cytomegalovirus (AD169). (A) Typical cytopathic effect. (Unfixed, unstained; magnification × 100.) (B) Intranuclear inclusions (arrows). (Hematoxylin and eosin preparation; magnification × 400.)

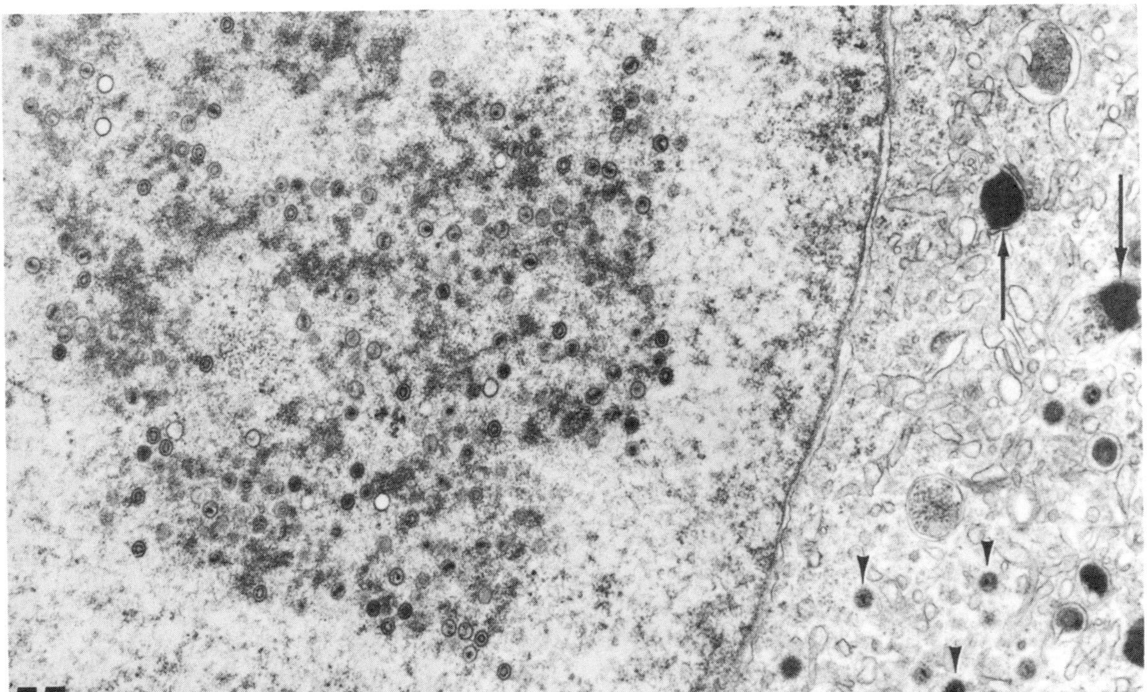

FIGURE 55. Electron micrograph of a human CMV, strain AD169, infected WI-38 cell. Intranuclear inclusion containing electron dense matrix and many nucleocapsids are seen in the nucleus. In the cytoplasm, several nucleocapsids coated with a layer of electron dense matrix (short arrows), and large dense bodies (long arrows) are seen scattered in the cytoplasm. (Magnification × 19,680.)

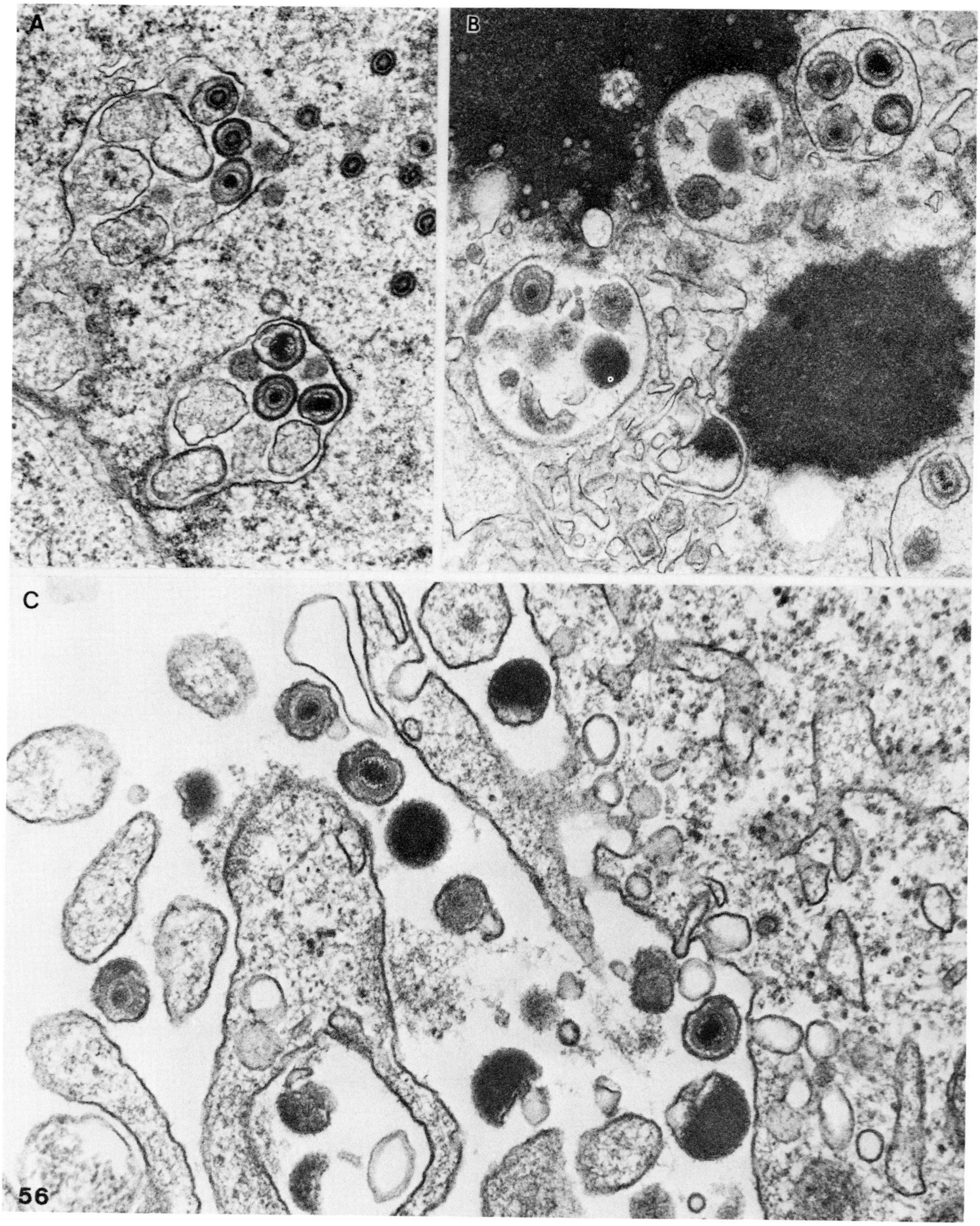

FIGURE 56. Electron micrographs of Human CMV infected WI-38 cells. [(A and B) Magnification × 48,000. (C) × 64,800.] (A) A portion of a nucleus showing viral nucleocapsids and enveloped virions; (B) a portion of cytoplasm containing enveloped virus particles in vacuoles near a large area of electron dense matrix; and (C) extracellular virus particles of CMV infected cell culture.

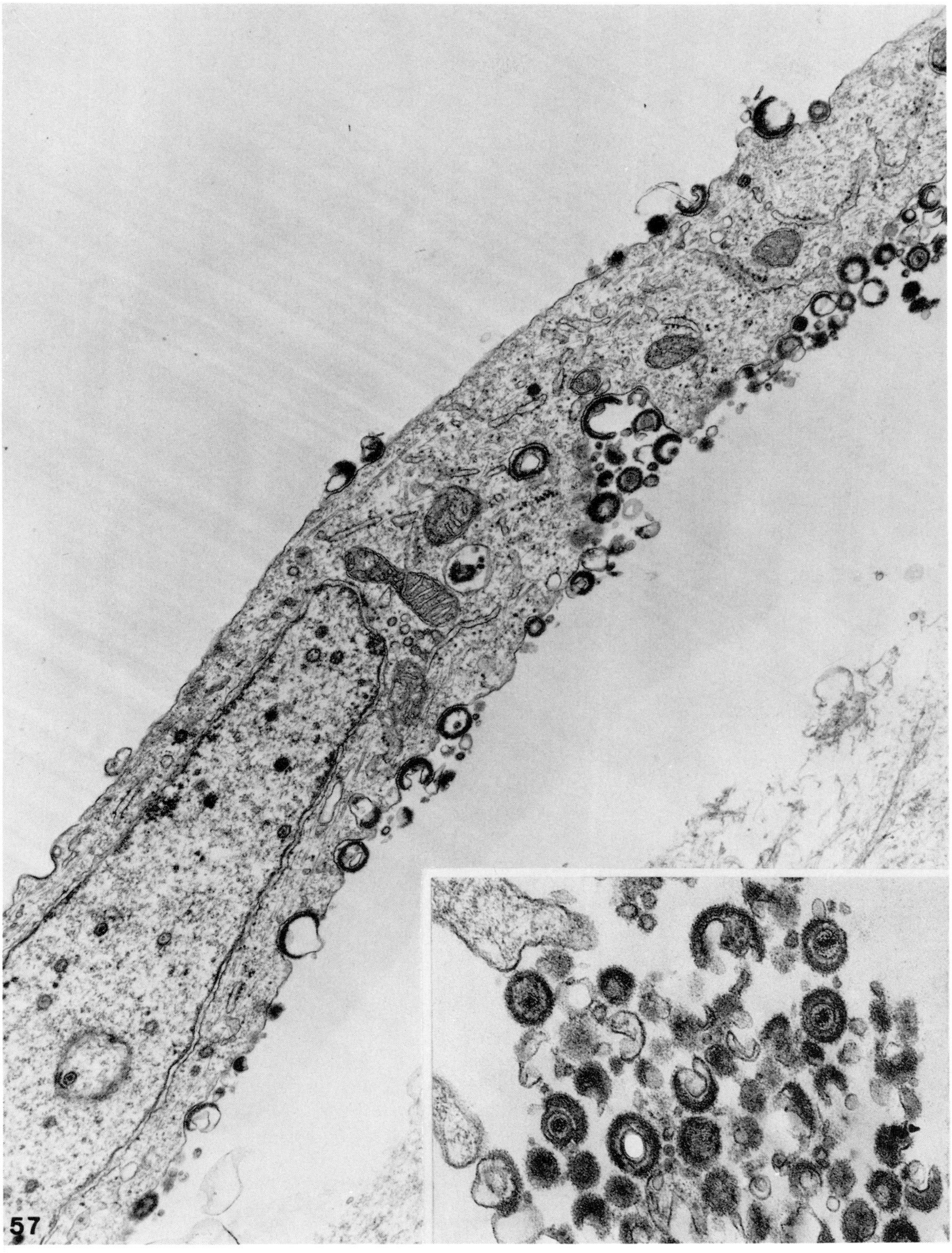

FIGURE 57. Electron micrograph of Varicella-Zoster (V-Z) virus-infected WI-38 cell showing nucleocapsids within the nucleus. Many extracellular virus particles are seen near the cell surface; note: many extracellular virus particles appeared to be distorted or partially distorted, a characteristic of V-Z virus in cell culture. (Magnification × 24,000, Insert: × 48,000.)

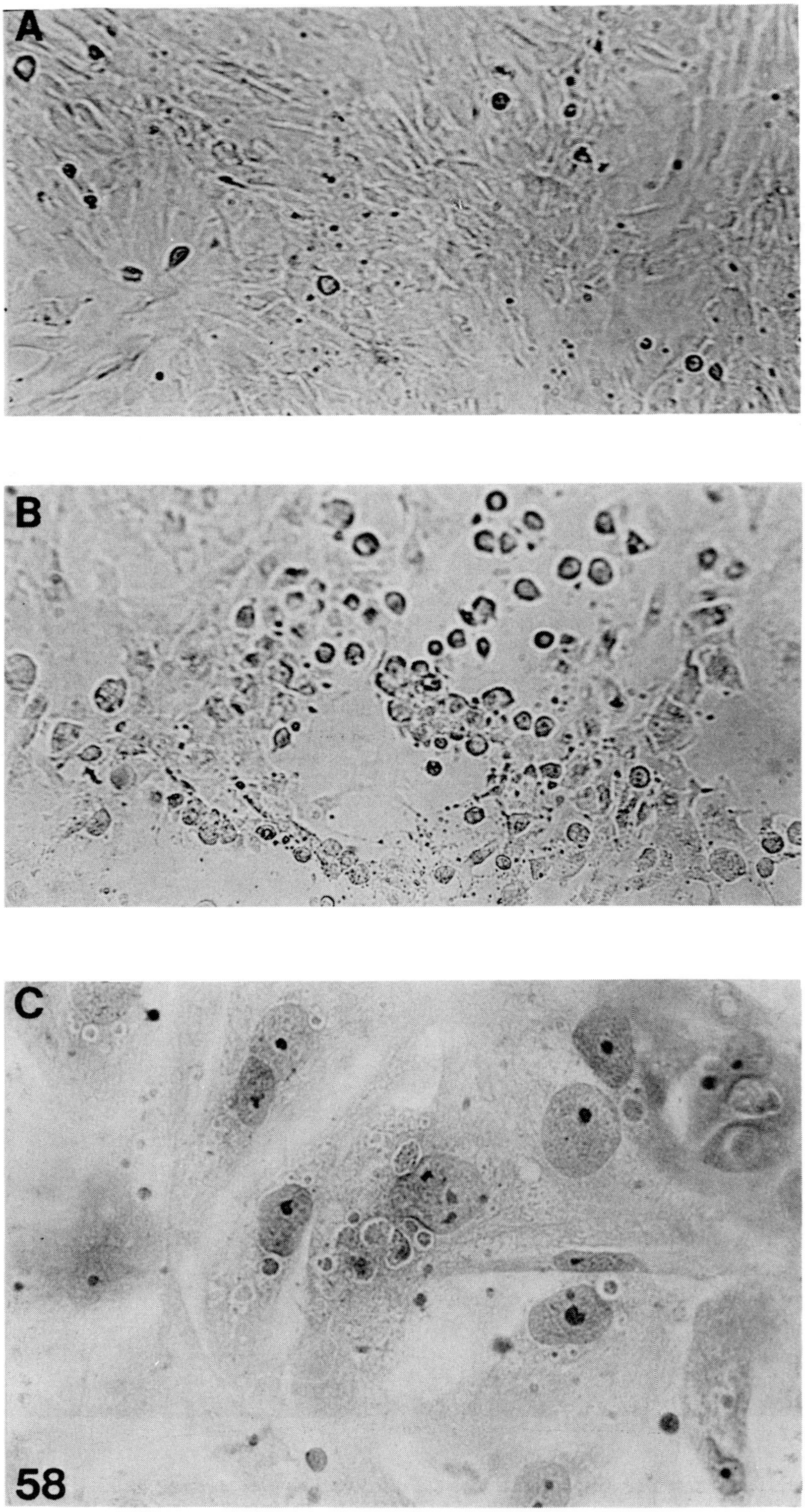

FIGURE 58. Nonprimate, guinea pig embryo fibroblast (GPEF) and primate rhesus monkey kidney (RhMK) cells infected with vaccinia virus. [(A and B) Unfixed, unstained; magnification × 100. (C) Hematoxylin and eosin; magnification × 100.] (A) Uninfected GPEF cells, (B) Vaccinia virus infected GPEF cells, showing typical cytopathic effect. Unfixed, unstained. (C) Vaccinia virus infected RhMK cells, showing intracytoplasmic eosinophilic inclusions (Guarnieri bodies).

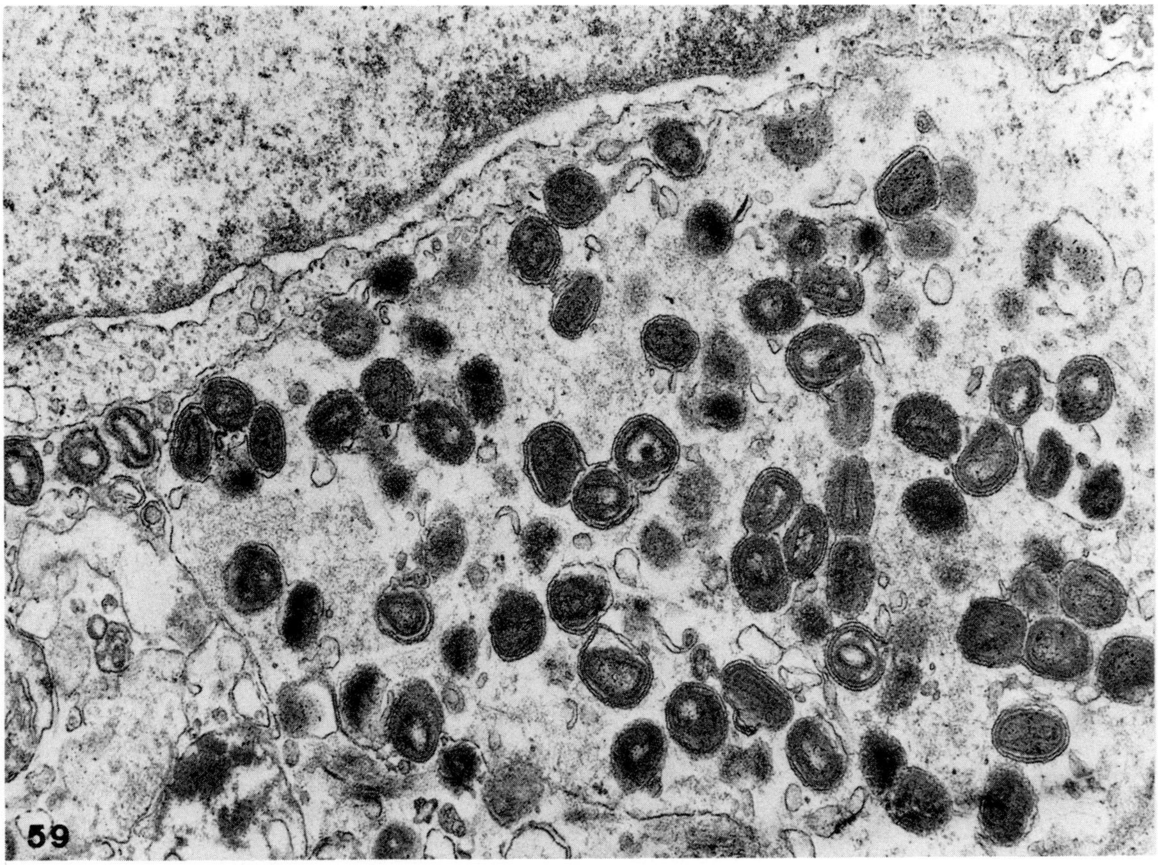

FIGURE 59. Electron micrograph of vaccinia virus infected RhMK cell showing intracytoplasmic inclusion containing many typical brick-shaped virions. (Magnification × 30,000.)

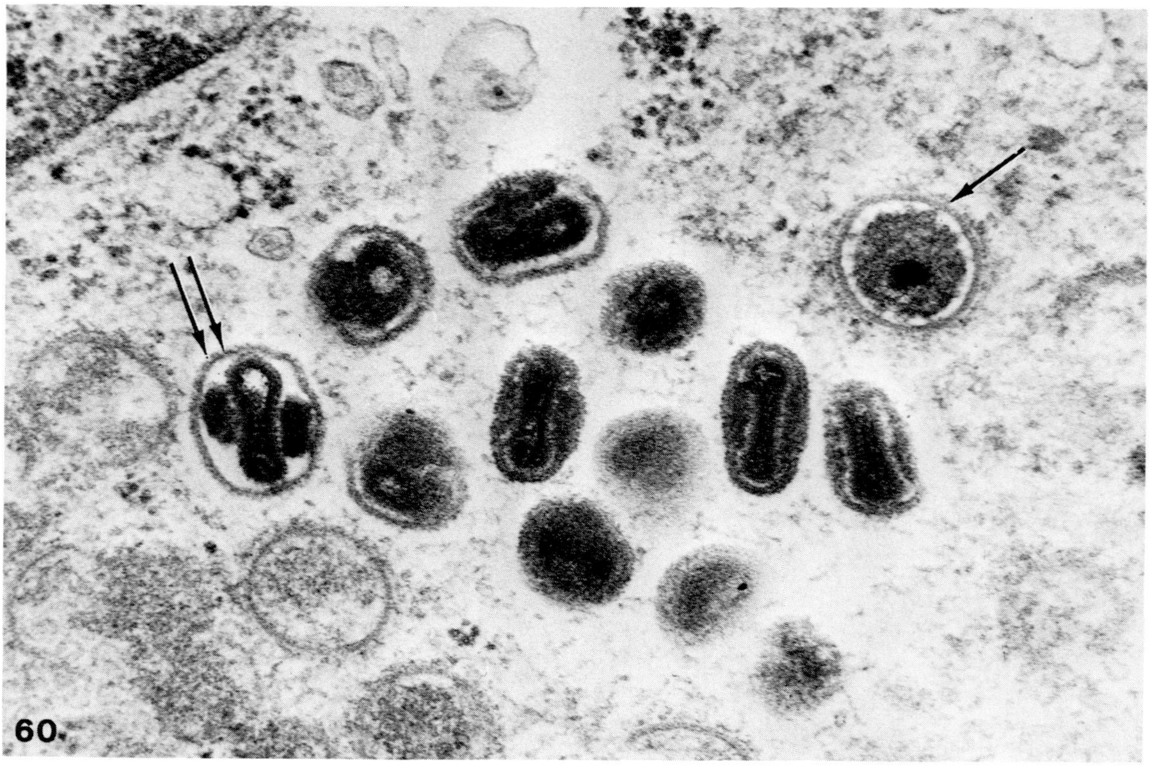

FIGURE 60. Molluscum contagiosum virus particles seen in tumor tissue taken directly from a patient with skin tumor. Spherical form containing electron dense material represents immature form of virus (arrow). A typical brick-shaped virus particle with dumbbell core and two lateral bodies is indicated by a double arrow. (Magnification × 64,800.) (Sample courtesy of D. Lowy.)

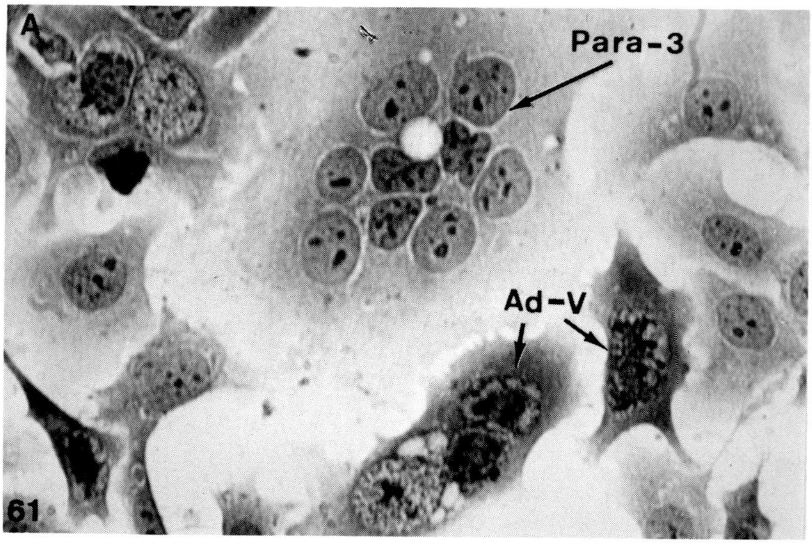

FIGURE 61. Mixed virus infection of Hep-2 cells with a clinical specimen containing both parainfluenza virus type 3 and adenovirus type 2. A multinucleated syncytial cell induced by parainfluenza virus, and intranuclear inclusions induced by adenovirus. (Hematoxylin and eosin stained; magnification × 400.) (From Hsiung, G. D., *Mt. Sinai J. Med.*, 44, 19, 1977. With permission.)

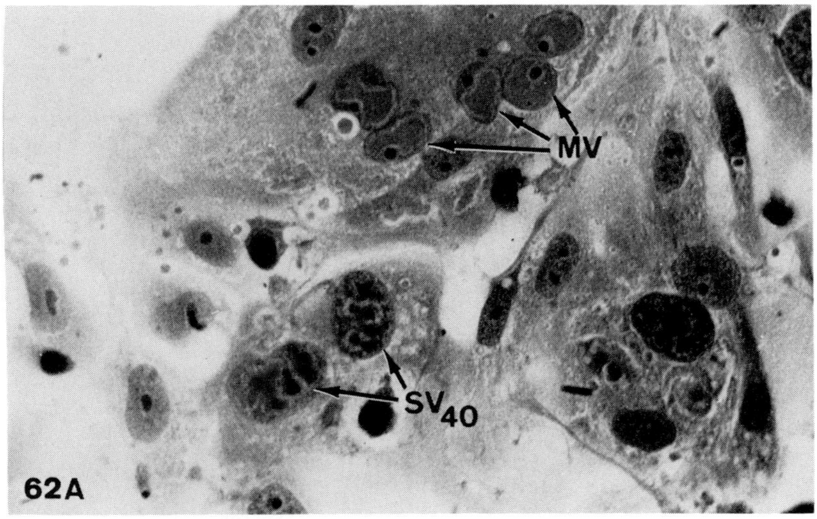

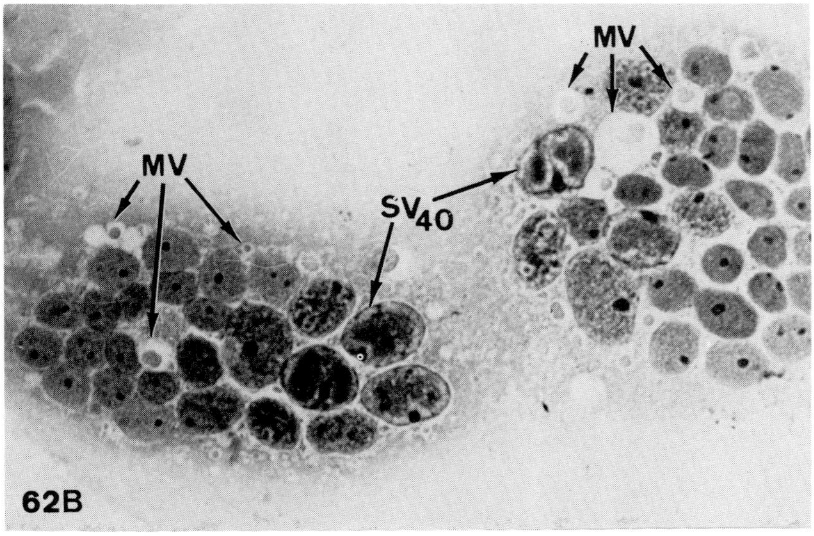

FIGURE 62. African green monkey kidney (AGMK) cells doubly infected with SV40 and measles virus. (Hematoxylin and eosin stained; magnification × 400.)
(A) A multinucleated syncytial cell containing many intranuclear eosinophilic inclusions due to measles virus infection. Several nuclei separate from the syncytium, contain intranuclear basophilic inclusions as a result of SV40 infection. (B) Two large syncytial cells; some nuclei contain intranuclear basophilic inclusions of SV40. Several intracytoplasmic inclusions due to measles virus infection can be seen.

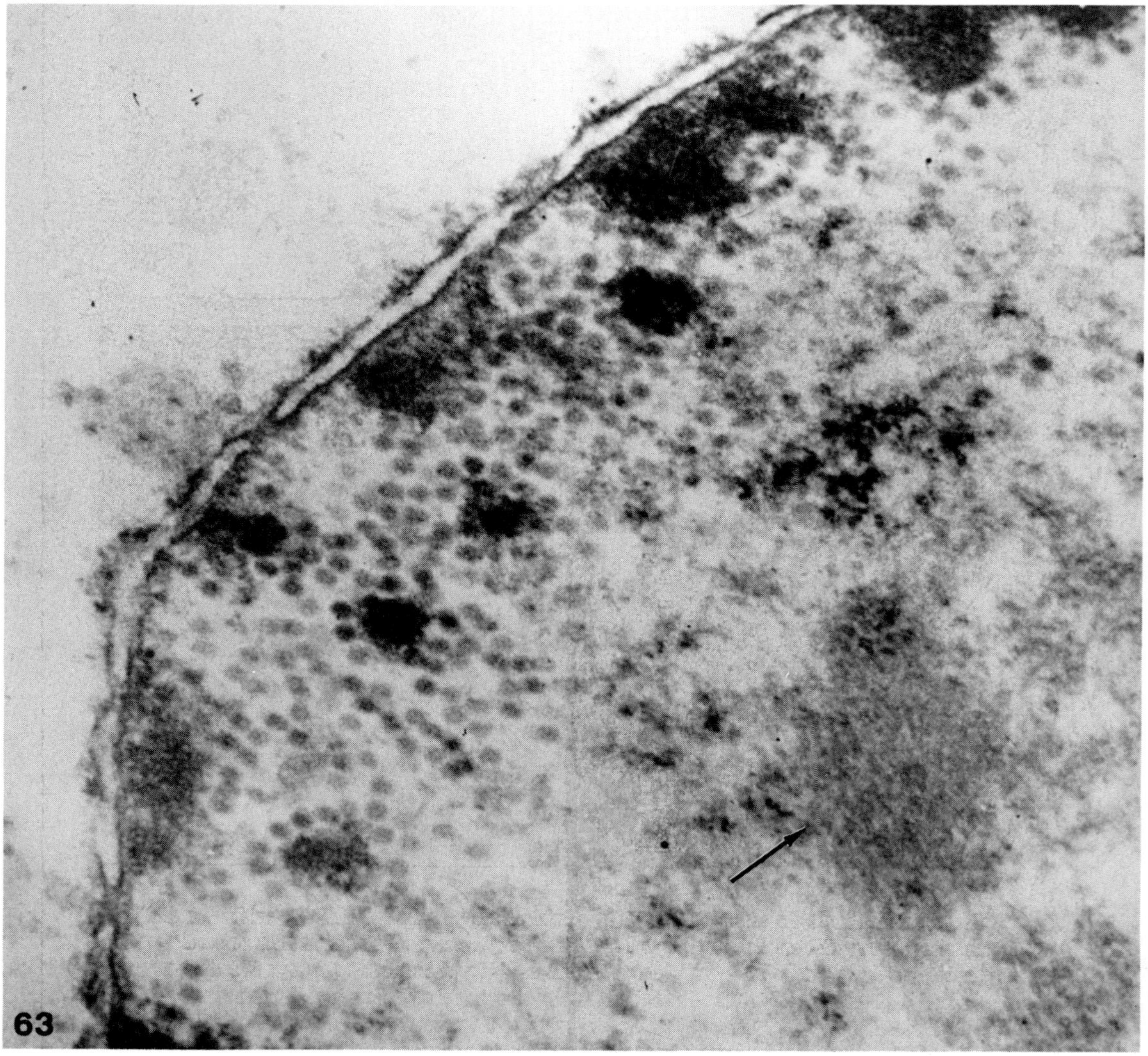

FIGURE 63. Electron micrograph of an AGMK cell doubly infected with SV40 and measles virus. Note: in the nucleus, both SV40 particles and helical nucleocapsids of measles virus (arrows) were present. (Magnification × 49,500.) (From Hsiung, G. D., Bensch, K. G., and Prose, P. H., *Nature (London)*, 215, 178, 1967. With permission.)

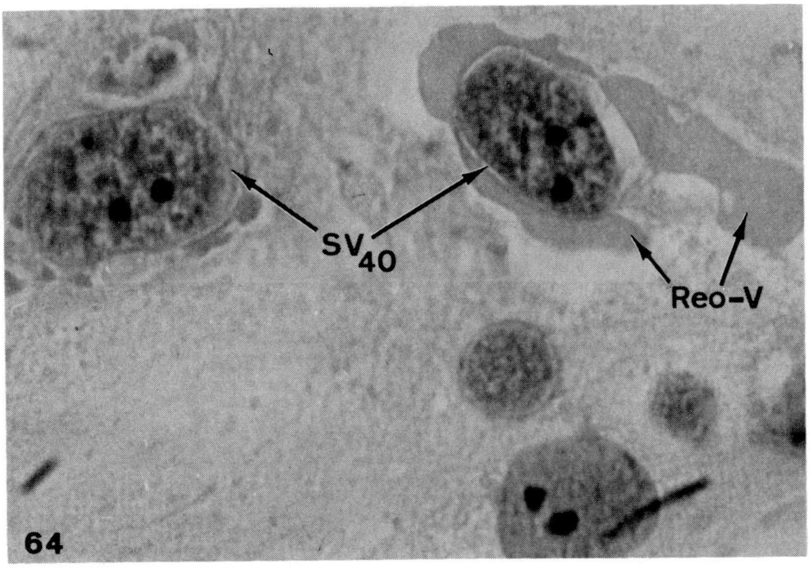

FIGURE 64. Mixed infection of AGMK cells with SV40 and reovirus. Typical intranuclear SV40 virus inclusions and intracytoplasmic inclusions of reovirus can be seen. (Hematoxylin and eosin stained; magnification × 100.)

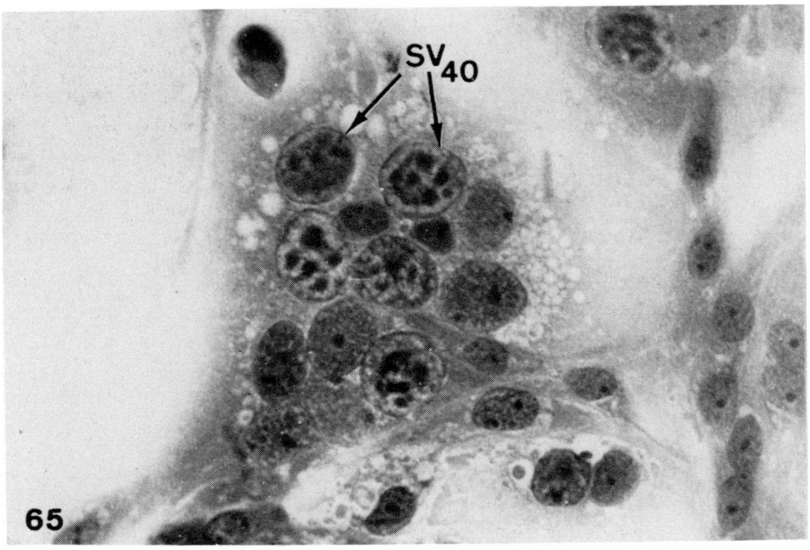

FIGURE 65. Mixed infection of rhesus monkey kidney cells (RhMK) with both SV40 and foamy virus. Note: intranuclear inclusions of SV40 virus in a multinucleated syncytial cell which resulted from the foamy virus infection. Also, large cytoplasmic vacuolation which is characteristic for foamy virus infection. (Magnification × 400.)

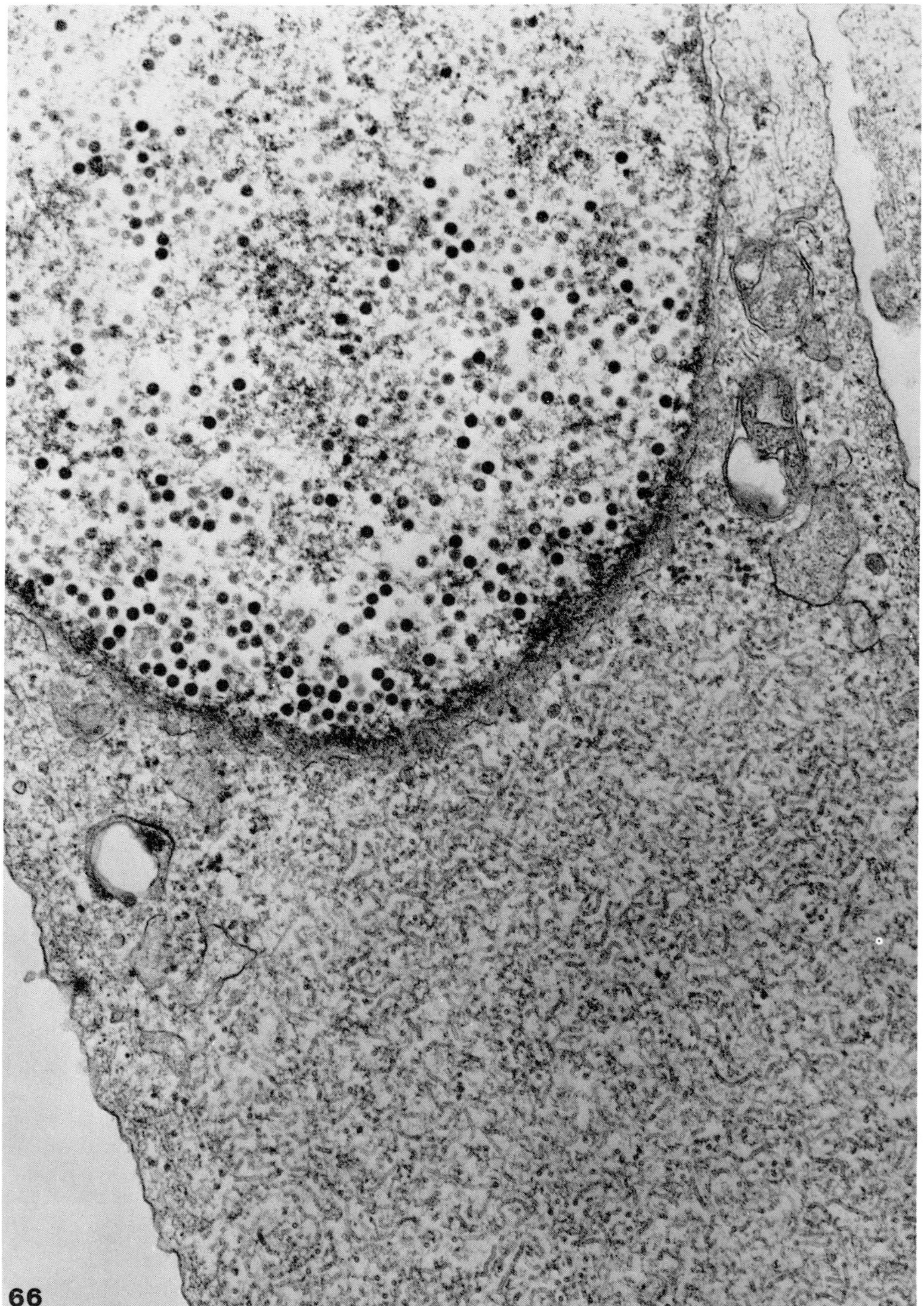

FIGURE 66. Electron micrograph of an AGMK cell doubly infected with SV40 and SV5 (parainfluenza virus type 5); SV40 virus is seen in the nucleus, and SV5 nucleocapsids are seen in the cytoplasm. (Magnification × 100,800.)

GENERAL REFERENCES

Maramorosch, K. and Kurstak, E., Eds., *Comparative Virology*, Academic Press, New York, 1971.

Dalton, A. J. and Haguenau, F., Eds. *Ultrastructure in Biological Systems vs. Ultrastructure of Animal Viruses and Bacteriophages: An Atlas*, Academic Press, New York, 1973.

Hsiung, G. D., *Diagnostic Virology. An Illustrated Handbook*, Yale University Press, New Haven, Conn., 1973.

Fenner, F. J., Classification and nomenclature of viruses, *Intervirology*, 7, 1, 1976.

Fenner, F. J., McAuslan, B. R., Mims, C. A., Sambrook, J. F., and White, D. O. *The Biology of Animal Viruses*, 2nd ed., Academic Press, New York, 1974.

Grist, N. R., Ross, C.A.C., and Bell, E. J. , *Diagnostic Methods in Clinical Virology*, 2nd ed., Lippencott, Philadelphia, 1974.

REFERENCES

Picornaviruses

1. **Godman, G. C.**, The cytopathology of enteroviral infection, *Int. Rev. Exp. Pathol.*, 5, 67, 1966.
2. **Dales, S., Eggers, H. J., Tamm, I., and Palade, G. E.**, Electron microscopic study of the formation of poliovirus, *Virology*, 26, 379, 1965.
3. **Jezéquel, A. M. and Steiner, J. W.**, Some ultrastructural and histochemical aspects of coxsackie virus-cell interactions, *Lab. Invest.*, 15, 1055, 1966.
4. **Hsiung, G. D.**, Applications of primary cell cultures in the study of animal viruses. III. Biological and genetic studies of enteric viruses of man (enteroviruses), *Yale J. Biol. Med.*, 33, 359, 1961.

Togaviruses

5. **Horzinek, M. C.**, The structure of togaviruses, *Progr. Med. Virol.*, 16, 109, 1973.
6. **Pfefferkorn, E. R. and Shapiro, D.**, Reproduction of togaviruses, in *Comprehensive Virology*, vol. 2. Fraenkel-Contrat, H. and Wagner, R. R., Eds., Plenum Press, New York, 1974, 171.
7. **Higashi, N.**, The togavirus family, in *Ultrastructure in Biological Systems vs. Ultrastructure of Animal Viruses and Bacteriophages: An Atlas*, Dalton, A. J. and Haguenau, F., Eds., Academic Press. New York, 1973, 173.

Rubella Virus

8. **von Bonsdorff, C. H. and Vaheri, A.**, Growth of rubella virus in BHK 21 cells: electron microscopy of morphogensis, *J. Gen. Virol.*, 5, 47, 1969.
9. **Edwards, M. R., Cohen, S. M., Bruno, M., and Deibel, R.**, Micromorphological aspects of the development of rubella virus in BHK-21 cells, *J. Virol.*, 3, 439, 1969.
10. **Holmes, I. H., Wark, M. C., and Warburton, M. F.**, Is rubella an arbovirus? II. Ultrastructural morphology and development, *Virology* 37,15, 1969.
11. **Oshiro, L. S., Schmidt, N. J., and Lennette, E. H.**, Electron microscopic studies of rubella virus, *J. Gen. Virol.*, 5, 205, 1969.

Reoviruses

12. **Joklik, W. K.**, Reproduction of reoviridae, in *Comprehensive Virology*, Vol. 2., Fraenkel-Conrat, H., and Wagner, R. R., Eds., Plenum Press, New York, 1974, 231
13. **Bishop, R. F., Davidson, G. P., Holmes, I. H., and Ruck, B. J.**, Virus particles in epithelial cells of duodenal mucosa from children with acute nonbacterial gastroenteritis, *Lancet*, 2, 1281, 1973.
14. **Flewett, T. H., Bryden, A. S., and Davies, H., Letter:** Virus particles in gastroenteritis, *Lancet*, 2, 1497, 1973.
15. **Kapikian, A. Z.**, Acute viral gastroenteritis, *Prevent. Med.*, 3, 535, 1974.
16. **Gomatos, P. J. and Tamm, I.**, Animal and plant virus with double-helical RNA, *Proc. Natl. Acad. Sci. U.S.A.*, 50, 878, 1963.
17. **Wood, H. A.**, Viruses with double-stranded RNA genomes, *J. Gen. Virol.*, 20 (Suppl.), 61, 1973.

Coronaviruses

18. **Almeida, J. D., and Tyrrell, D. A.,** The morphology of three previously uncharacterized human respiratory viruses that grow in organ culture, *J. Gen. Virol.*, 1, 175, 1967.
19. **Becker, W. B., McIntosh, K., Dees, J. H., and Chanock, R. M.,** Morphogenesis of avian infectious bronchitis virus and a related human virus (strain 229E), *J. Virol.*, 1, 1019, 1967.
20. **Hamre, D., Kindig, D. A., and Mann, J.,** Growth and intracellular development of a new respiratory virus, *J. Virol.*, 1, 810, 1967.
21. **McIntosh, K.,** Coronaviruses. A comparative review, *Curr. Top. Microbiol. Immunol.*, 63, 85, 1973.
22. **Oshiro, L. S., Schieble, J. H., and Lennette, E. H.,** Electron microscopic studies of coronavirus, *J. Gen. Virol.*, 12, 161, 1971.

Myxoviruses

23. **Waterson, A. P.,** Two kinds of myxovirus, *Nature* (London), 193, 1163, 1962.
24. **Compans, R. W. and Choppin, P. W.,** The structure and assembly of influenza and parainfluenza viruses, in *Comparative Virology*, Maramorosch, K. and Krustak, E., Eds., Academic Press. New York, 1971, 407.
25. **Howe, C. and Lee, L. T.,** Virus-erythrocyte interactions, *Adv. Virus Res.*, 17, 1, 1972.
26. **Choppin, P. W. and Compans, R. W.,** Reproduction of paramyxoviruses, in *Comprehensive Virology*, Vol. 4., Fraenkel-Conrat, H , and Wagner, R. R., Eds., Plenum Press, New York, 1975, 95.

Pseudomyoxviruses

27. **Armstrong, J. A., Pereira, H. G., and Valentine, R. C.,** Morphology and development of respiratory syncytial virus in cell cultures, *Nature* (London), 156, 1179, 1962.
28. **Jordan, W. S., Jr.,** Growth characteristics of respiratory syncytial virus, *J. Immunol.*, 88, 581, 1962.
29. **Mannweiler, K.,** Ultrastructural examinations of tissue cultures after infection with measles virus, *Arch. Gesamte Virusforsch*, 16, 89, 1965.
30. **Palacios, O.,** Cytochemical and fluorescent antibody studies on the growth of measles virus in tissue culture, *Arch. Gesamte Virusforsch.*, 16, 83, 1965.
31. **Waterson, A. P.,** Measles virus, *Arch. Gesamte Virusforsch.*, 16, 57, 1965.
32. **Richman, A. V., Pedreira, F. A., and Tauraso, N.M.,** Attempts to demonstrate hemagglutination and hemadsorption by respiratory syncytial virus, *Appl. Microbiol.*, 21, 1099, 1971.

Rhabdoviruses

33. **Howatson, A. F.,** Vesicular stomatitis and related viruses, *Adv. Virus Res.*, 16, 195, 1970
34. **Hummeler, K.,** Bullet-shaped viruses, in *Comparative Virology*, Maramorosch, K. and Kurstak, E., Eds., Academic Press, New York, 1971.
35. **Matsumoto, S.,** Rabies virus, *Adv. Virus Res.*, 16, 257, 1970.

Parvoviruses

36. **Rose, J. A.,** Parvovirus reproduction, in *Comprehensive Virology*, Vol. 3, Fraenkel-Conrat H., and Wagner, R. R., Eds.,Plenum Press, New York, 1974, 1.
37. **Toolan, H. W.,** The picodna virus, H. RV, and AAV, *Int. Rev. Exp. Pathol.*, 6, 135, 1968.
38. **Toolan, H. W.,** The parvoviruses, *Prog. Exp. Tumor Res.*, 16, 410, 1972.

Papovaviruses

39. **Melnick, J. L.,** Papova virus group, *Science* 135, 1128, 1962.
40. **Takemoto, K.K. and Mullarkey, M . F.,** Human papovavirus, BK strain: biological studies including antigenic relationship to simian virus 40, *J. Virol.*, 12, 625, 1973.
41. **Hsiung, C. D. and Gaylord, W. H.,** The vacuolating virus of monkeys. I. Isolation, growth characteristics, and inclusion body formation, *J. Exp. Med.*, 114, 975, 1961.
42. **Almeida, J. D., Howatson, A. F., and Williams, M. G.,** Electron microscope study of human warts: sites of virus producion and nature of the inclusion bodies, *J. Invest. Dermat.*, 38, 337, 1962.
43. **Howatson, A. F. and Almeida, J. D.,** an electron microscope study of polyoma virus in hamster kidney, *J. Biophys. Biochem. Cytol.*, 7, 753, 1960.
44. **Mattern, C.F.T., Takemoto, K.K., and Daniel, W.A.,** Replication of polyoma virus in mouse embryo cells: electron microscopic observations, *Virology*, 30, 242, 1966.
45. **Granboulan, N., Tournier, P., Wicker, R., and Bernard, W.,** An electron microscopic study of the development of SV40 virus, *J. Cell Biol.*, 17, 423, 1963.

Adenoviruses

46. **Ginsberg, H S. and Dingle, J. H.**, The adenovirus group, in *Viral and Rickettsial Infections of Man,* Horsfall, F. L., Tamm, I., Eds., Lippincott, Philadelphia, 1965, 860.
47. **Boyer, G. S., Denny, F. W. Jr., Miller, I., and Ginsberg, H.S.**, Correlation of production of infectious virus with sequential stages of cytological alteration in HeLa cells infected with adenoviruses types 5 and 7, *J. Exp. Med.,* 112, 865, 1960.
48. **Block, D. P., Morgan, C., Godman, G. C., Howe, C., and Rose, H. M.**, A correlated histochemical and electron microscopic study of the intranuclear crystalline aggregates of adenovirus (RI-APC virus) in HeLa cells, *J. Biophys. Biochem. Cytol.,* 3, 1, 1957.
49. **Morgan, C., Godman, G. C., Breitenfeld, P. M., and Rose, H. M.**, A correlative study by electron and light microscopy of the development of type 5 adenovirus. I. Electron microscopy, *J. Exp. Med.,* 112, 373, 1960.

Herpesviruses

50. **Roizman, B. and Furlong, D.**, The replication of herpesviruses, in *Comprehensive Virology,* Vol. 3, Fraenkel-Conrat, H. and Wagner, R. R., Eds. Plenum Press, New York, 1974, 229.
51. **Nii, S., Morgan, C., and Rose, H M.**, Electron microscopy of herpes simplex virus. II. Sequence of development, *J. Virol.,* 2, 517, 1968.
52. **Schwartz, J. and Roizman , B.**, Similarities and differences in the development of laboratory strains and freshly isolated strains of herpes simplex virus in Hep-2 cells: electron microscopy, *J. Virol.,* 4, 879, 1969.
53. **Siegert, R. S. and Falke, D.**, Electron microscopic investigation of the development of herpes virus hominis in culture cells, *Arch. Gesamte Virusforsch.,* 19, 230, 1966.
54. **Smith, J. D. and DeHarven, E.**, Herpes simplex virus and human cytomegalovirus replication in WI-38 cells. I. Sequence of viral replication, *J. Virol.,* 12, 919, 1973.
55. **Luse, S. A. and Smith, M. G.**, Electron microscopy of salivary gland viruses, *J. Exp. Med.,* 107, 623, 1958.
56. **Kanich, R. E. and Craighead, J. E.** Human cytomegalovirus infection of cultured fibroblasts. I. Cytopathologic effects induced by an adapted and a wild strain, *Lab. Invest.,* 27, 263, 1972.
57. **Kanich, R. E. and Craighead, J. E.**, Human cytomegalovirus infection of cultured fibroblasts. II. Viral replicative sequences of a "wild" and an "adapted" strain, *Lab. Invest.,* 27, 273, 1972.
58. **Iwasaki, Y., Furukawa, T , Plotkin , S., and Koprowski, N.**, Ultrastructural study on the sequence of human cytomegalovirus infection in human diploid cells, *Arch. Gesamte Virusforsch.,* 40, 311, 1973.
59. **Weller, T. H.**, Varicella-herpes-zoster virus, in *Viral and Rickettsial Infections of Man,* Horsfall, F. L. and Tamm, I., Eds., Lippincott, Philadelphia, 1965, 915.
60. **Cook, M. L. and Stevens , J. G.**, Replication of varicella-zoster virus in cell cultures: an ultrastructural study, *J. Ultrastruct. Res.,* 32, 334, 1970.
61. **Gershon, A., Cosio, L., and Brunell, P. A., Observations on the growth of varicella-zoster virus in human diploid cells,** J. Gen. Virol., *18, 21, 1973.*

Poxviruses

62. **Joklik, W. K.**, The poxviruses, *Ann. Rev. Microbiol.,* 22, 359, 1968,
63. **Dales, S. and Siminovitch, L.**, The development of vaccinia virus in Earle's L strain cells as examined by electron microscopy, *J. Biophys. Biochem. Cytol.,* 10, 475, 1961.
64. **Dales, S.**, The structure and replication of poxviruses as examplified by vaccinia, in *Ultrastructure in Biological Systems vs. Ultrastructure of Animal Viruses and Bacteriophages: An Atlas,* Dalton, A. J. and Haguenau, F., Eds., Academic Press, New York, 1974, 109.
65. **von Peters, D., Nielsen, G., Bayer, M. E.**, Reliability of the rapid electron microscope diagnosis: *variola,* Dtsch. Med. Wochenschr. 87, 2240, 1962.
66. **Cruickshank, J. G., Bedson, H. S., and Watson, D. H.**, Electron microscopy in the rapid diagnosis of smallpox, *Lancet,* 2, 527, 1966.

Mixed Virus Infections

67. **Ruckle, G.**, Studies with the monkey-intranuclear-inclusion-agent (MINIA) and foamy-agent derived from spontaneously degenerating monkey kidney cultures. I. Isolation and tissue culture behavior of the agent and identification of MINIA as closely related to measles virus, *Arch. Gesamte Virusforsch.,* 8, 139, 1958.
68. **Reissig, M.**, Double infection of cells in culture with measles and poliomyelitis viruses, *Ann. N.Y. Acad. Sci.,* 81, 17, 1959.
69. **O'Conor, G. T., Rabson, A. S., Berezesky, I. K., and Paul, F. J.**, Mixed infection with simian virus 40 and adenovirus 12, *J. Natl. Cancer Inst.,* 31, 903, 1963.

70. **Rabson , A. S., O'Conor, G. T., Paul, F. J., and Berezesky, I. K.,** Mixed infection with herpes simplex and simian virus 40, *Science,* 151, 1535, 1966.
71. **Hsiung, G. D., Bensch, K. G., and Prose, P. H.,** Electron microscopy of monkey kidney cells infected with both SV40 and measles virus, *Nature* (London), 215, 178, 1967.
72. **Hsiung, G. D.,** Latent virus infections in primate tissues with special reference to simian viruses, *Bacteriol. Rev.,* 32, 185, 1968.
73. **Smith, K. O.,** Adventitious viruses in cell cultures, *Progr. Med. Virol.,* 12, 302, 1970.
74. **Hsiung, G. D. and Fong, C.K.Y.,** Dual infection with endogenous herpes- and C-type viruses in cultured cells derived from normal and leukemic guinea pigs, *Proc. Soc. Exp. Biol. Med.,* 147, 635, 1974.

RNA Viruses

PICORNAVIRIDAE: ENTEROVIRUSES—POLIOVIRUSES

Joseph L. Melnick

Introduction

The enterovirus group, named in 1957,[1] brought together polioviruses, coxsackieviruses, and echoviruses, all of which inhabit the human alimentary tract. These viruses share a number of clinical, epidemiological, and ecological characteristics as well as physical and biochemical properties. Enteroviruses of human origin include the following:

- Polioviruses: types 1 to 3.
- Coxsackieviruses A: 23 types and several variants (coxsackieviruses A1 to A24; coxsackievirus type A23 is the same virus as echovirus 9).
- Coxsackieviruses B: types B1 to B6.
- Echoviruses: 31 types (types 1 to 33; echovirus 10 is now reclassified as reovirus type 1, and echovirus 28 as rhinovirus type 1A).
- Enterovirus types 68 to 71: these viruses would formerly have been classed as either coxsackievirus or echovirus types. As instances of overlapping properties between coxsackieviruses and echoviruses have accumulated, blurring the initial distinction made between these two groups, it was recommended that as new enterovirus serotypes were discovered they would simply be assigned sequential numbers. This has been the procedure for new enterovirus serotypes since 1969.

In 1963, the name picornavirus ("pico" = small; "rna" = ribonucleic acid genome) was introduced as a larger grouping which would include not only the enteroviruses, but also the rhinoviruses, because these viruses had fundamentally similar properties. As viral classification advanced on the basis of further knowledge of biophysical and biochemical characteristics, the International Committee on Nomenclature of Viruses (now the International Committee on Taxonomy of Viruses) officially assigned family status (Picornaviridae) to this larger taxon.[2]

Many enteroviruses cause diseases in man, ranging from severe and permanent paralysis to minor undifferentiated febrile illnesses. For all members of the group, however, subclinical infection is far more common than clinically manifest disease. The illnesses caused by coxsackieviruses, echoviruses, and the new numbered enterovirus types 68 to 71 range widely, from mild through very severe (discussed elsewhere). Polioviruses can cause the most severe disease of any for which enteroviruses are responsible, and these agents have received the most comprehensive study. In many instances, they have served as models in studies of other enteroviruses.

Many reviews and textbooks provide detailed information on the development of knowledge about the enteroviruses. Only a few of the historic advances in the understanding of poliomyelitis and the polioviruses will be mentioned here; for more information see Paul's history of poliomyelitis[3] and the author's recent chapter on enteroviruses.[4]

Crippling disease retrospectively recognizable as paralytic poliomyelitis was recorded in early antiquity, but the disease was described as a clinical entity only in the late 1700s and early 1800s. After increasingly severe epidemics began to appear in Europe and North America, studies were intensified. Experimental work became possible when poliomyelitis was successfully transmitted to monkeys. During the next 40 years, it was shown that the virus was regularly present in stools of patients; that subhuman primates could be infected not only by injection, but also by the alimentary route; and that strains could be adapted to growth in laboratory rodents, permitting expansion of laboratory studies. Significant antigenic differences among poliovirus strains were documented, resulting in their separation into three serotypes. A major landmark in modern virology was the discovery by Enders, Weller, and Robbins[5] in 1949 that poliovirus could be isolated and cultivated in cell cultures derived from tissues other than those of the nervous system. These investigators were awarded the Nobel Prize in 1954 for this development. The discovery of a way to grow poliovirus in cultures gave great impetus to the study of the disease and the virus, introducing a new era in polio research. Investigators could then develop relatively simple and rapid tests to detect and quantify the polioviruses and their antibodies; could obtain purer forms of the viruses; and could begin to analyze their biochemical constitution and physical structure as well as their biological behavior, including details of virus replication and the mechanisms by which they produce pathological effects. Furthermore, as cell cultures could be made from tissues of other than neural origin, virologists could also realistically begin to consider the possibility of developing polio vaccines. The availability of tissue cultures eliminated two important roadblocks hampering vaccine development that had existed as long as the monkey was the sole experimental host available for cultivating stocks of virus: nonneural tissue cultures would obviate the risk of allergic encephalitis, which was known to accompany injection of neural tissues, and the huge quantities of poliovirus stocks that would ultimately be required for preparation of vaccines could be cultivated economically and under controlled conditions in tissue cultures. This development also opened the possibility that poliovirus strains deliberately attenuated by propagation under specified conditions through many generations in cells widely different from human spinal cord cells (where virulent polioviruses do their damage) would serve as live virus vaccines.

Morphology

The polioviruses are typical of all the picornaviruses, especially the enteroviruses, in many fundamental morphological and chemical properties, in their reactions in culture, and in the methods used for isolating and identifying the virus. Type 1 poliovirus is the typespecies for the genus *Enterovirus.*[2]

The Picornaviridae[2] are small, ether-insensitive (not possessing a lipid-containing envelope) viruses with an RNA genome. Their nucleic acid is single stranded, has a molecular weight of 2.5×10^6, and constitutes approximately 30% of the particle mass. The nucleocapsid has cubic (icosahedral) symmetry; is naked, lacking essential lipid and carbohydrate; and is 20 to 30 nm in diameter (poliovirus particles are 28 nm in diameter). Virus maturation takes place in the cytoplasm. The family includes four genera, *Enterovirus, Rhinovirus, Cardiovirus,* and *Apthovirus.*

The enterovirus capsid is thought to be composed of 32 morphologic subunits, possibly in the form of a rhombic triacontahedron rather than a regular icosahedron. Rhinovirus capsid architecture appears to be similar. Infective nucleic acid has been extracted from several enteroviruses and rhinoviruses.

Crystalline aggregates of poliovirus particles may form in the cytoplasm of cells which support the growth of the virus (Figure 1). Figure 1 is an electron micrograph of part of the cytoplasm of a monkey kidney cell infected with attenuated type 1 poliovirus. Poliovirus particles may be observed directly and enumerated by electron microscopy (Figure 2). By counting representative electron microscope fields in which the particles have been sedimented, it is possible to compute the concentration of virus per milliliter of suspension. With appropriate staining and high-resolution electron microscopy, fine structure of the particle can be studied.

Physicochemical Properties

Like other enteroviruses, the polioviruses have a buoyant density in cesium chloride

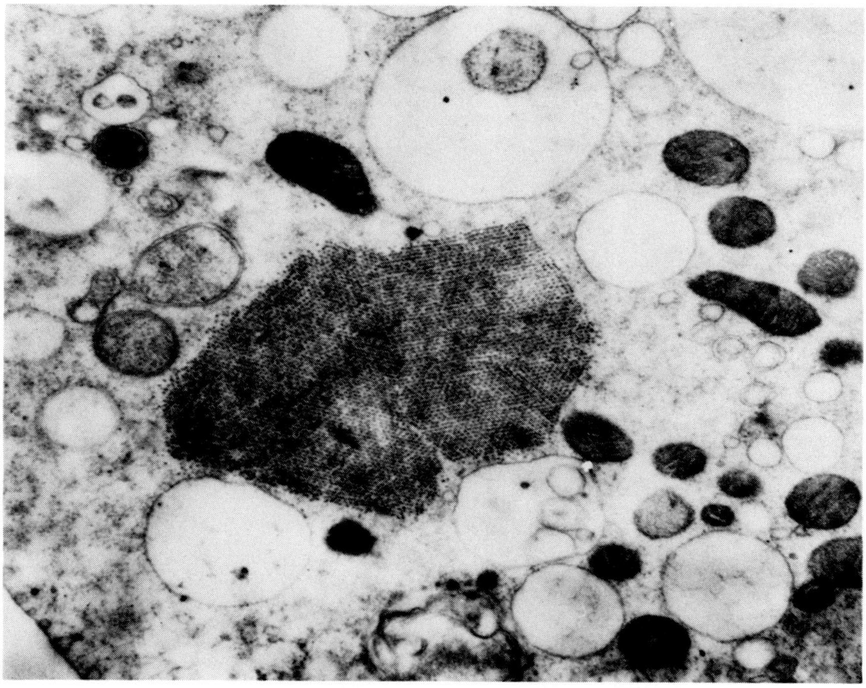

FIGURE 1. Ultrathin section of cytoplasm of a monkey kidney cell in culture, crystalline array. Two large virus crystals with a clear crystal interface are discernible. Numerous cellular vacuoles and well-preserved mitochondria are present. Magnification approximately × 37,000. (From Mayor, H. D. and Jordan, L., *Virology*, 16, 325, 1962. With permission.)

of 1.34 gm/ml and are stable at acid pH (3.0). They have a sedimentation coefficient of 158S and a particle weight of 8.4×10^6.

Virus infectivity is destroyed rapidly in aqueous suspension at temperatures of 45 to 55°C, but is stabilized against thermal inactivation by the presence of molar $MgCl_2$. The virus is able to withstand temperatures approximately 5°C higher when suspended in milk or milk products than when suspended in water; adequate pasteurization of milk destroys the virus. Virus infectivity is inactivated by ultraviolet (UV) light; ionizing radiation inactivities infectivity rapidly (target size equals the whole virion), but inactivates the complement-fixing antigen slowly (small target size). Visible light in the presence of acridine dyes (either as an admixture or incorporated into the virion during replication) will destroy infectivity.

Polioviruses also are sensitive to formaldehyde, hydroxylamine, and phenol (which denatures protein and releases infectious RNA). Urea inactivates poliovirus by degrading virions to procapsid substructures, while guanidine completely reduces capsid structures into soluble polypeptides. If no extraneous matter is present, polioviruses may be inactivated by low concentrations of chlorine (0.1 ppm); however, much higher chlorine concentrations are required if the virus is to be destroyed in the feces.[8]

Virus Replication

The replication of poliovirus provides useful insights into its survival and transmission, in vivo and in vitro host range, and relatedness to other enteroviruses. The essential theme of viral replication is that specific messenger RNA (mRNA) must be transcribed from the viral nucleic acid for successful expression and duplication of genetic information. Once this is accomplished, viruses use cell components to translate the

FIGURE 2. (A) Electron micrograph of field of purified poliovirus prepared for particle counting. Negatively charged virus particles are sedimented at high speed directly on positively charged grids which were prepared by coating collodion-covered grids with a thin layer of aluminum. Grids with poliovirus are shadowed with chromium and examined under the electron microscope. By counting the number of particles per field, the number of particles per milliliter of the original virus suspension can be calculated. Magnification × 90,000. (From Smith, K. O. and Melnick, J. L., *J. Immunol.*, 89, 279, 1962. With permission.)

mRNA. In the replication of poliovirus, all steps are independent of host DNA and occur in the cell cytoplasm.

Poliovirus attaches to specific cell receptors. The presence of these receptors controls the cell susceptibility to virus infection. It has recently been shown that human chromosome 19 contains the genetic information for this site. As diagramed in Figure 3, polioviruses adsorb to cells at specific cell receptor sites (Step 1), losing in the process one virus polypeptide (VP 4) which may, therefore, be important in adsorption. The sites are specific for interactions between the virus coat and the cell. Evidence of this specificity is the fact that intact poliovirus infects only primate cells in culture, whereas the isolated RNA will also infect nonprimate cells (rabbit, guinea pig, or chick) and complete one cycle of multiplication. Multiple cycles of infection are not observed in nonprimate cells because the resulting progeny possess protein coats and will again infect only primate cells. After attachment, the virus particles are taken into the cell by viropexis (similar to pinocytosis) (Step 2), and the viral RNA is uncoated (Step 3). The single stranded genomic RNA can then serve as its own messenger RNA. This messenger RNA is translated (Step 4), resulting in the formation of an RNA polymerase — a protein which has not yet been identified. The RNA polymerase catalyzes the production of a replicative intermediate (RI), a partially double-stranded molecule consisting of a complete RNA strand and numerous partially completed strands (Step 5). At the same time, inhibitors of cellular RNA and protein synthesis of unknown origin and nature are produced. It is not yet absolutely clear how synthesis of plus and minus strands of RNA comes about; the mechanism has been completely elucidated for only the plus strands but a similar mechanism for minus strands appears likely.

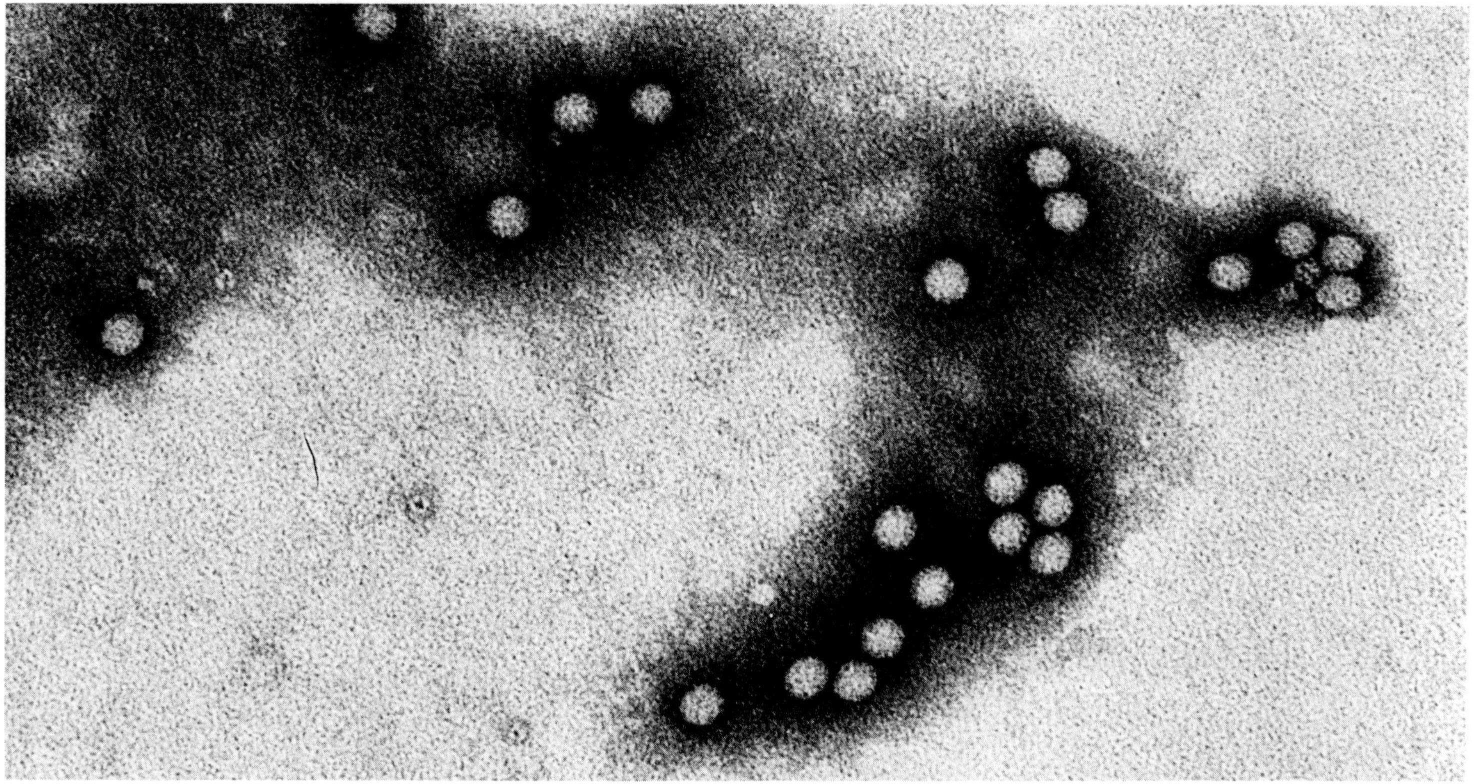

FIGURE 2. (B) Several fields of poliovirus prepared for high-resolution electron microscopy to show fine structure of the poliovirus. Negatively stained with phosphotungstic acid. Magnification X 237,000. (From McGregor, S. and Melnick, J.L. unpublished.)

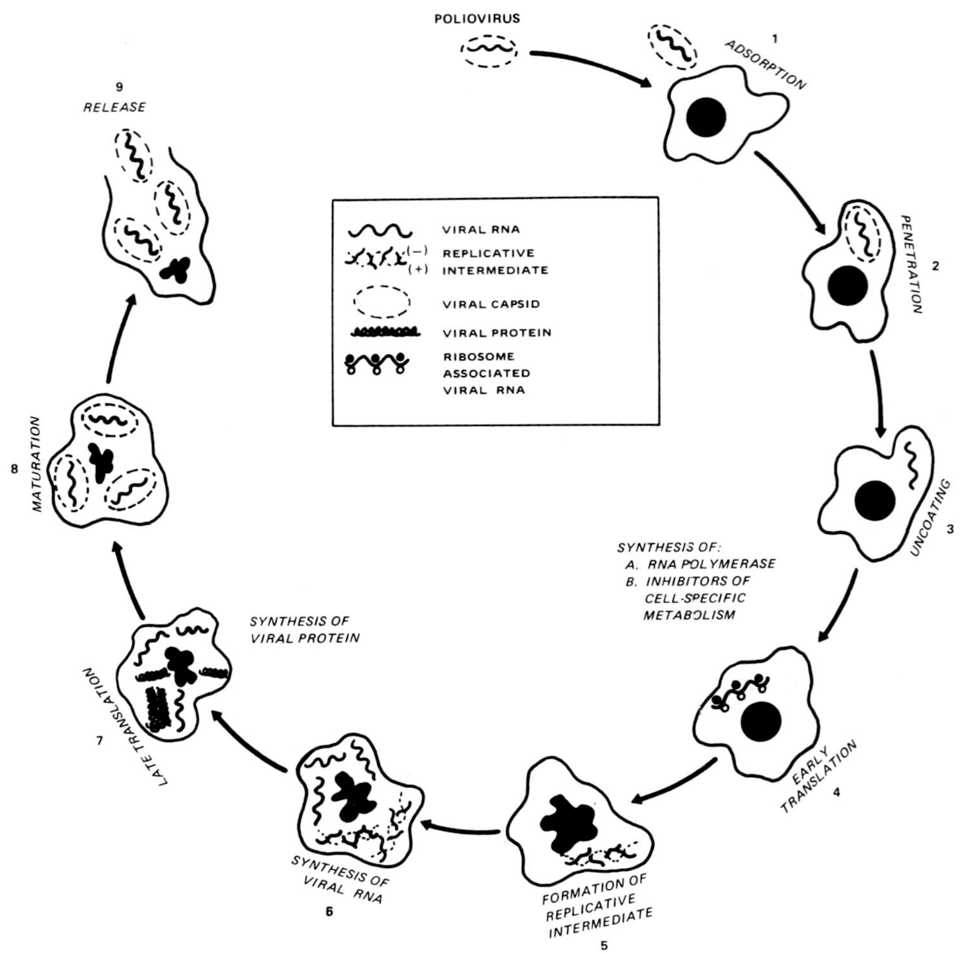

FIGURE 3. Replication of poliovirus.

Here the RI consists of one complete minus strand and many small pieces of newly synthesized plus-strand RNA (Step 6). Another form of RNA is produced in the infected cell. The replicative form (RF) consists of two complete RNA strands, one plus and one minus.

The single plus-strand RNA is made in large amounts and may perform any one of the following three functions: (1) It may serve as messenger RNA for synthesis of structural proteins; (2) it may serve as template for continued RNA replication; or (3) it may become encapsidated, resulting in mature progeny virions. The synthesis of viral capsid proteins (Step 7) is initiated at approximately the same time as RNA synthesis.

The entire poliovirus genome acts as its own mRNA, forming a polysome of approximately 350S, and is translated to form a single large polypeptide which is subsequently cleaved to produce the various viral capsid polypeptides. Thus, the poliovirus genome serves as a polycistronic messenger molecule. Poliovirus contains four polypeptides: VP 1 (3.5×10^4 mol wt), VP 2 (2.8×10^4 mol wt), VP 3 (2.4×10^4 mol wt) and VP 4 (6×10^3 mol wt). Polyacrylamide gel electrophoresis analysis has also been performed on virus particles devoid of RNA that have been separated from infectious virus by density gradient centrifugation. These particles were found to contain only VP 1 and VP 3, as well as a new polypeptide, VP 0. It is believed that this RNA-free particle (or procapsid) may be important in the regulation of virus functions and is a precursor of the

Table 1
PROTOTYPE STRAINS OF POLIOVIRUS[a]

Type	Prototype strain	Geographic origin	Illness in person yielding prototype virus	Investigator	Year
1	Brunhilde	Maryland	Paralytic polio	Howe and Bodian	1939
2	Lansing	Michigan	Fatal paralytic polio	Armstrong	1937
3	Leon	California	Fatal paralytic polio	Kessel	1937

[a] Established in 1951 through a collaborative study by the Committee on Typing, National Foundation for Infantile Paralysis.[9] For oral poliovirus vaccine, the following attenuated strains are used: type 1, LSc; type 2, P712; and type 3, Leon. The origin and history of these and many other strains of polioviruses are given in detail by Hahn.[10]

infectious virus and that, as the RNA becomes associated with the procapsid in some as yet unknown fashion (Step 7), VP 0 is cleaved to yield VP 2 and VP 4. The mechanism of cleavage is unknown, but the possibility exists that one of the three procapsid polypeptides (VP 0, VP 1, or VP 3) may possess an appropriate enzyme activity. Completion of encapsidation (Step 8) produces mature virus particles which are then released when the cell undergoes lysis (Step 9).

Guanidine in concentrations greater than 1 mM and 2-(α-hydroxybenzyl)-benzimidazole inhibit poliovirus multiplication in tissue culture. Guanidine acts by inhibiting the release of newly made viral RNA from the replicative complex.

Virus Strains and Antigenic Properties

The prototype strains of the three antigenic serotypes of poliovirus are listed in Table 1. Many of the classic strains of poliovirus that over the years have been isolated and passaged in the laboratory are listed elsewhere (see Reference 10).

Complement-fixing antigens are known for each serotype. They may be prepared from tissue culture or from infected central nervous system (CNS) tissue. Inactivation of the virus by formalin, heat, or UV light liberates a soluble complement-fixing antigen. This antigen is cross-reactive and fixes complement with heterotypic poliomyelitis antibodies. A type-specific precipitin reaction occurs when virus in sufficient concentration is used with immune animal or convalescent human sera. Two type-specific antigens are contained in poliovirus preparations and can be detected by precipitin and complement fixation (CF) tests. They are the D or N (native) and C or H (heated) antigens.[11,12] The D form can be converted to the C form by heating. The difference between the two is that the D form represents full particles containing RNA and the C form represents empty particles. Further, one polypeptide is lost during the conversion.

Clinical Features

Poliomyelitis is an acute infectious disease which in its serious form affects the CNS. The destruction of motor neurons in the spinal cord results in flaccid paralysis. However, only a small number of infections are clinically apparent.

*Pathogenesis and Pathology**

The mouth is the portal of entry of the virus, and primary multiplication takes place at the sites of viral implantation in the oropharynx or intestines. Two or more enteroviruses may propagate simultaneously in the alimentary tract, but under many circum-

* See References 4 and 13 for recent reviews and literature citations.

stances multiplication of one virus may interfere with growth of the heterologous type. Interference with the growth ("take") of live poliovaccine by other concurrent enterovirus infections is now well established.

Virus becomes widely disseminated via the bloodstream, but viremia is seldom detected in natural human infection because it occurs very early, before significant illness is observed. Antibodies to the virus appear early in the infection; they are usually present by the time paralysis appears. Viremia is also associated regularly with type 2 oral vaccination. Free virus is usually present between days 2 and 5 after vaccination, and virus is bound to antibody for an additional few days. Bound virus is detected by acid treatment, which inactivates the antibody and liberates active virus. These findings led to the view that the virus first multiplies in the tonsils, lymph nodes of the neck, Peyer's patches, and small intestine (Figure 4). The CNS may then be invaded by way of the circulating blood.

Poliovirus invades certain types of nerve cells and, in the process of its intracellular multiplication, may damage or completely destroy these cells. The anterior horn cells of the spinal cord are most prominently involved, but in severe cases the intermediate gray ganglia and even the posterior horn and dorsal root ganglia are often affected. Lesions are found as far forward as the hypothalamus and thalamus. In the brain, the reticular formation, vestibular nuclei, cerebellar vermis, and deep cerebellar nuclei are most often affected. The cortex is virtually spared, with the exception of the motor cortex along the precentral gyrus.

As the chief site of action of poliovirus is in the neuron, the changes which occur in peripheral nerves and voluntary muscles are secondary to destruction of the nerve cell. Changes occur rapidly in nerve cells, from mild chromatolysis to neuronophagia and complete destruction. Cells which are not killed but lose their function temporarily as a result of edema may recover completely. Inflammation occurs secondary to the attack on the nerve cells; the focal and perivascular infiltrations are chiefly lymphocytes, with some involvement of polymorphonuclear cells, plasma cells, and microglia. In addition to pathologic changes in the nervous system, hyperplasia and inflammatory lesions of lymph nodes and Peyer's patches and other lymph follicles in the intestinal tract are frequently observed.

Immunity

Immunity is permanent to the poliovirus type causing the infection. Passive immunity is transferred from mother to offspring. The maternal antibodies gradually disappear during the first 6 months of life. Passively administered antibody lasts only 3 to 5 weeks.

Virus-neutralizing antibody forms within a few days after exposure to the virus, often before the onset of illness, and apparently persists for life. Its formation early in the infection is a result of viral multiplication in the intestinal tract and deep lymphatic structures before invasion of the nervous system. As antibodies must be present in the blood to prevent the dissemination of virus to the CNS and are not effective after this has already occurred, immunization is of value only if it precedes the onset of symptoms referable to the nervous system.

A decrease in resistance to poliovirus accompanies removal of tonsils and adenoids. Preexisting secretory antibody levels in the nasopharynx decrease sharply following operation (especially in young male children) without any change in antibody levels in the serum. Local antibody levels remain low or absent for as long as 7 months. In seronegative children, nasopharyngeal antibody response to poliovaccine develops significantly later and to lower titers in children previously tonsillectomized than in those with intact tonsils. Thus, surgery of this type may eliminate a valuable source of immunocompetent tissue of importance in resistance to poliovirus.

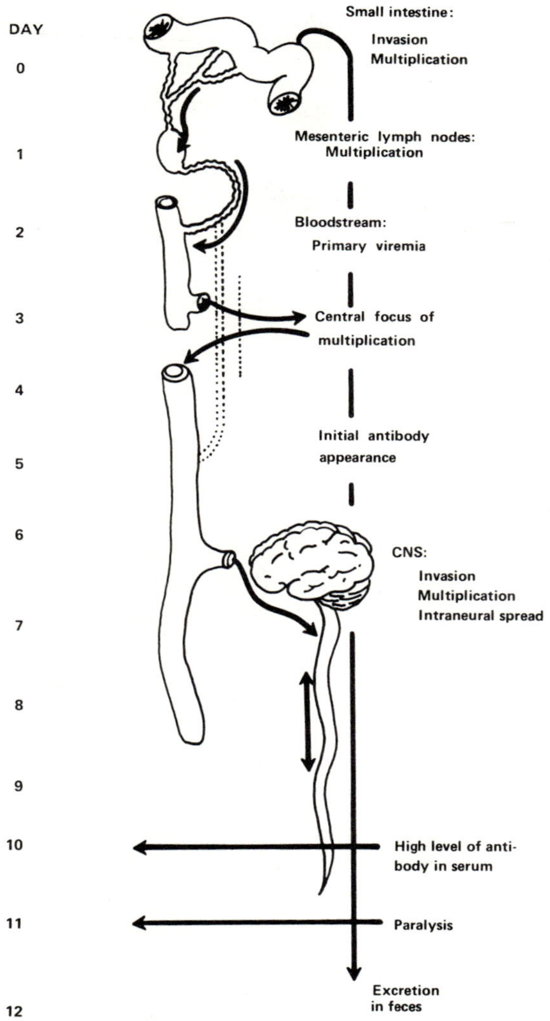

FIGURE 4. Schematic illustration of the pathogenesis of poliomyelitis. Virus enters by way of the alimentary tract and multiplies locally at the initial sites of viral implantation (tonsils, Peyer's patches) or the lymph nodes which drain these tissues and begins to appear in the throat and in the feces. Secondary virus spread occurs by way of the bloodstream to other susceptible tissues, namely, other lymph nodes, brown fat, and the central nervous system (CNS). Within the CNS the virus spreads along nerve fibers. If a high level of multiplication occurs as the virus spreads through the CNS, motor neurons are destroyed and paralysis occurs. The shedding of virus into the environment does not depend on secondary virus spread to the CNS. Secondary spread to the CNS is readily interrupted by the presence of antibodies induced by previous infection or vaccination.

Clinical Findings

When an individual susceptible to infection is exposed to poliovirus, one of the following responses may occur: (1) inapparent infection without symptoms, (2) mild (minor) illness, (3) aseptic meningitis, or (4) paralytic poliomyelitis. As the disease progresses one response may merge with a more severe form, sometimes resulting in a biphasic course: a minor illness followed first by a few days free of symptoms and then by the major, severe illness. Only approximately 1% of infections result in recognized clinical illness. The incubation period, defined as the time from exposure to onset of any symptoms, is usually between 7 and 14 days, but it may be short as 3 days or as long as 35 days.

Abortive poliomyelitis — This is the commonest form of the disease. The patient has only the minor illness characterized by fever, malaise, drowsiness, headache, nausea, vomiting, constipation, or sore throat in various combinations. The patient recovers in a few days. Even during an epidemic, the diagnosis of abortive poliomyelitis cannot be made with assurance except when the virus is isolated or antibody development is measured.

Nonparalytic poliomyelitis (aseptic meningitis) — In addition to the above symptoms and signs, the patient with the nonparalytic form presents stiffness and pain in the back and neck. The disease lasts 2 to 10 days, and recovery is rapid and complete. In a small percentage of cases the disease advances to paralysis. Poliovirus is only one of many viruses which produce aseptic meningitis.

Paralytic poliomyelitis — The predominating sign is flaccid paralysis resulting from lower motor neuron damage. However, incoordination secondary to brain stem invasion may occur, and there may also be painful spasms of nonparalyzed muscles. The amount of damage and destruction varies from case to case. Muscle involvement is usually maximal within a few days after the paralytic phase begins. The maximal recovery usually occurs within 6 months, but may take longer.

Laboratory Diagnosis

One cannot easily discuss diagnosis with respect to poliovirus alone in view of the common occurrence of other enteroviruses in stools of patients with symptoms resembling all but the most severe paralytic manifestations of poliovirus infection. Detailed descriptions of standard laboratory principles and procedures for investigation of enterovirus infections are available (see References 4 and 14 to 17). Background information and details on development of procedures utilizing monkeys as well as tissue cultures may be found in the second and third editions of *Diagnostic Procedures for Virus and Rickettsial Diseases*, American Public Health Association, 1956 and 1964.[16,17]

Choice of Specimen for Virus Isolation

Poliovirus is regularly present in the throat and stools before and after the onset of illness and can be isolated from rectal swabs or feces collected for several weeks. The likelihood of recovering the virus decreases as the disease runs its course. Virus has been found in approximately 80% of patients during the first 2 weeks of illness, but in only 25% during the third 2-week period. No permanent carriers are known. Although cerebrospinal fluid (CSF) frequently yields virus in cases of aseptic meningitis due to coxsackieviruses or echoviruses, detection of virus in CSF is rare when this syndrome is due to a poliovirus.

In fatal cases, poliovirus should be looked for in the cervical and lumbar enlargements of the spinal cord, in the medulla, and in the colon contents. If paralysis has lasted 4 to 5 days, it is difficult to recover the virus from the cord.

Not all agents recovered from feces and oropharynx are enteroviruses. Others which might be recovered are reoviruses, adenoviruses, and rhinoviruses as well as measles, mumps, rubella, and herpes simplex viruses. Many of these agents produce characteristic cytopathic effects (CPE) which at once differentiate them from enteroviruses. The ubiquity of attenuated vaccine strains of poliovirus in most developed nations presents problems for diagnostic laboratories, as the vaccine virus may be excreted by persons suffering from an unrelated disease; only in exceedingly rare instances has the vaccine itself been observed to be the cause of poliomyelitis.

Flow Chart for Isolation and Identification of Poliovirus

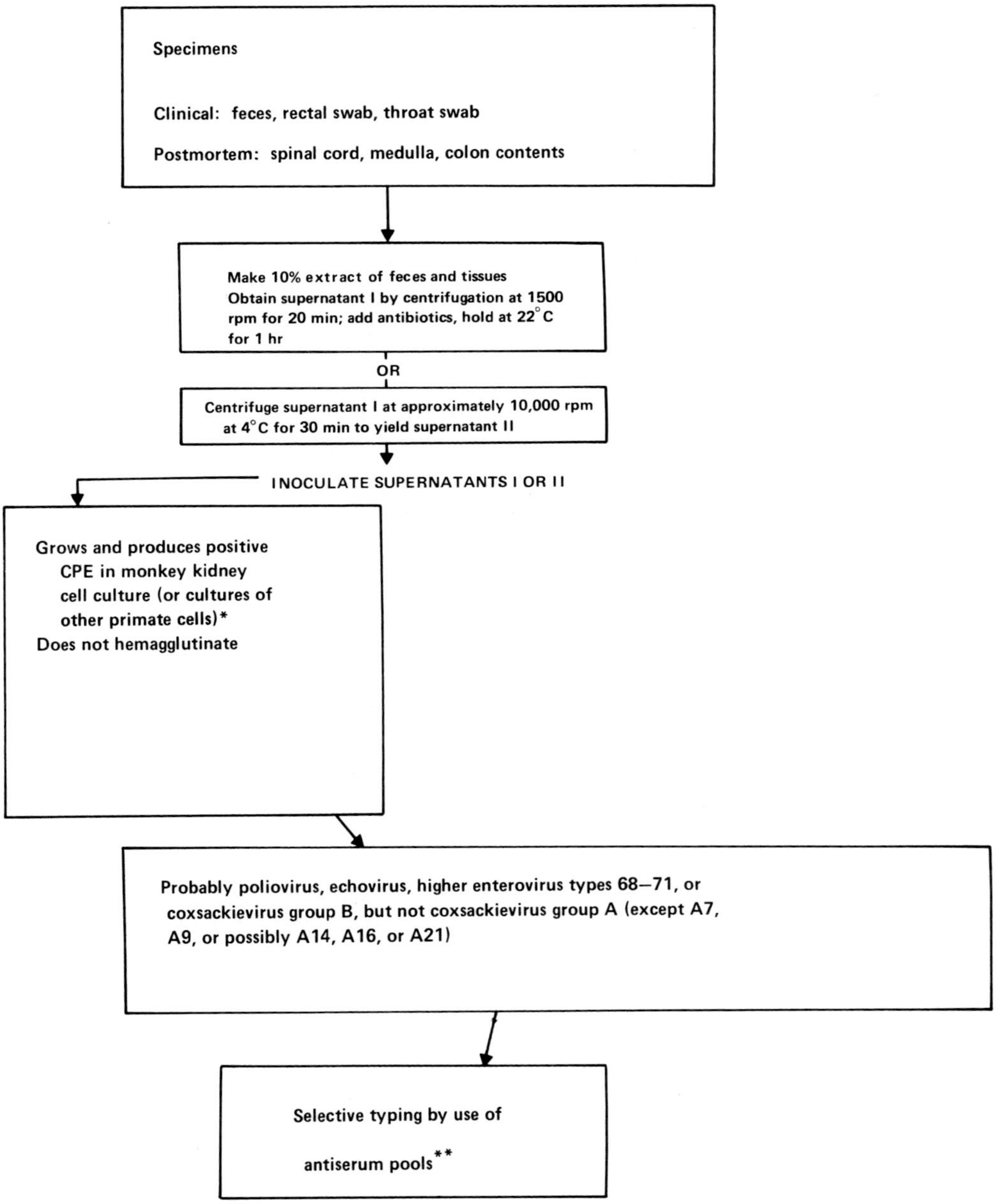

* Most of the common enterovirus serotypes which are cytopathogenic can be recovered by inoculation of specimens into primary cultures of rhesus monkey kidney cells, but at least one human tissue culture system (if available) should also be included.

** Identification of enterovirus isolates has been considerably simplified by the development of internationally standardized hyperimmune equine antisera [18,19] which are now incorporated into combination antiserum pools constituted in such a way that a given antiserum appears either in one, two, or three pools. An unknown enterovirus may be identified by its pattern of neutralization by the pool or pools containing its homotypic antiserum. Standardized antisera to the three poliovirus serotypes are included in the first set of such pools to be prepared and lyophilized (pools A to H). These eight pools contain antisera to 42 enteroviruses. A further set of seven enterovirus antiserum pools, J to P, has also been prepared, incorporating coxsackievirus A equine antisera developed in the program described above.[20] Sets of these pools are available from the Research Resources Branch, National Institute of Allergy and Infectious Diseases, National Institutes of Health, Bethesda, Maryland. However, it is prudent to conserve these reagent pools; therefore, if an epidemic due to a single serotype is in progress or if only poliovirus isolates must be identified, it may be sufficient to use a single monovalent hyperimmune serum or only the three polio antisera.

A number of laboratory methods have been developed in efforts to differentiate between wild poliovirus strains and strains derived from live polio vaccines. Efforts also have been made to develop in vitro tests that would distinguish virulent strains from attenuated strains. While several methods can provide clues, all must be interpreted with caution, for none of the in vitro tests can unfailingly make distinctions between virulence and attenuation. Thus far, the monkey neurovirulence test is the only one that regularly differentiates attenuated from virulent strains, but it does not indicate the relatedness of strains (e.g., it does not distinguish between virulent wild strains and vaccine-derived strains that have gained in virulence). Results of intratypic serodifferentiation (antigenic marker) tests may provide information to distinguish vaccine-derived strains from wild strains, but, again, interpretations must be approached with caution, for vaccine strains may show altered antigenicity on multiplication in the intestinal tract. The longer the period of multiplication, the more a change is likely to occur. Other marker tests that provide information on characteristics of a poliovirus strain include those based on growth in culture at elevated temperatures (approximately 40°C) or under high and low concentrations of bicarbonate. These tests are known as the rct test (for reproductive capacity at elevated temperatures) and the d test (for delayed growth under overlay medium containing low bicarbonate). Usually, but certainly not always, virulent strains are rct and d positive.

Photographs for Virus Recognition and Characterization in Cell Culture

Enteroviruses generally grow best in primate epithelial cells, and all three types of poliovirus grow well in cultures of monkey or human kidney or continuous cell lines such as HeLa. Growth of poliovirus in cultured cells is associated with a characteristic cytopathic effect.[14,17] Infected cells round up, show shrinkage and marked nuclear pyknosis, become refractile, and eventually degenerate and fall off the glass surface. Figure 5A shows normal monkey kidney cells in an uninfected culture which has been neither fixed nor stained; Figure 5B shows poliovirus CPE in approximately 25% of the cells (i.e., a CPE score of 1+). Figure 5C shows an advanced stage of CPE, 4+, in which almost 100% of the cells are affected.

Figure 6 illustrates the sequence of cytological changes induced in cultures of monkey kidney epithelial cells by poliovirus, followed through a single cycle of virus multiplication.[21] The rounded, pycnotic cell, customarily used as an index of the cytopathic response, was found only during the last stages of the infective process, i.e., approximately 8 hr after virus inoculation; however, by use of fixing and staining, alterations could be detected in cells as early as 4 hr after virus inoculation.

In order to evaluate the morphological data on a quantitative basis, the cells were classified into seven different types depending on the degree of cytopathic change, and differential cell counts were made on a number of areas in each culture. The morphological characteristics of the different cell types, as observed in preparations fixed in Zenker, Carnoy, or Bouin fluid and stained with hematoxylin and eosin, are given below.

- Type I. Normal cells with pale cytoplasm containing mitochondria and fat droplets. The nucleus exhibited a fine chromatin network and contained one or two nucleoli.
- Type II. Cells with typical early nuclear changes. There was loss of chromatin in the central zone of the nucleus and, in most cases, condensation of chromatin-like material about the nuclear membrane. One or more small acidophilic intranuclear inclusions were present. These inclusions differed from the early stages of herpes inclusions in that they were darker and their shape did not follow as closely the contours of the chromatin-free central areas, the poliomyelitis inclusions being more rounded and smaller. The cytoplasm appeared normal.

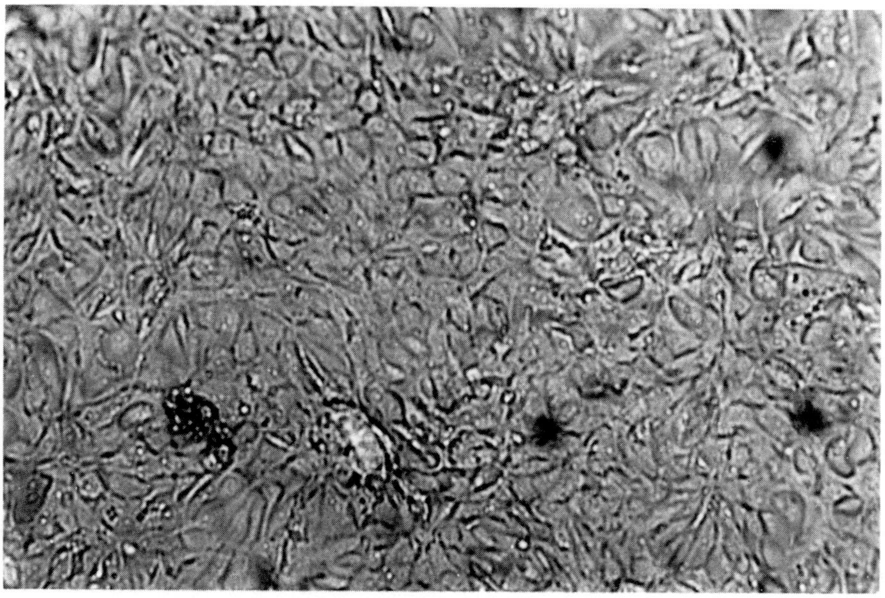

A

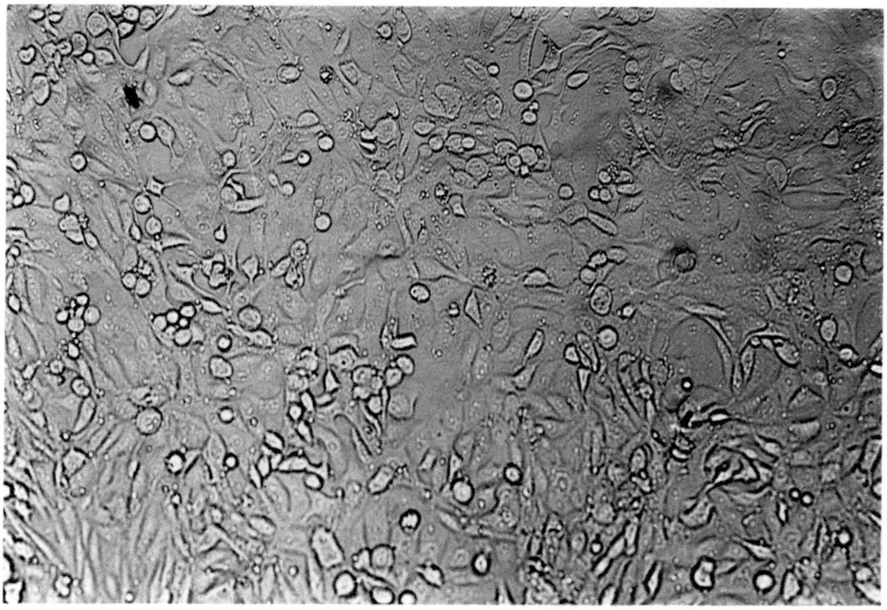

B

FIGURE 5. (A) Monolayer of normal unstained monkey kidney cells in culture. (B) Unstained monkey kidney cell culture showing early stage of cytopathic effects (CPE) typical of poliovirus infection. Approximately 25% of the cells in the culture show CPE indicative of virus multiplication (1+ CPE score). (C) Unstained monkey kidney cell culture illustrating more advanced cytopathic effect (3+ to 4+ CPE). Almost 100% of the cells are affected, and most of the cell sheet has come loose from the wall of the culture tube. Magnification × 200.

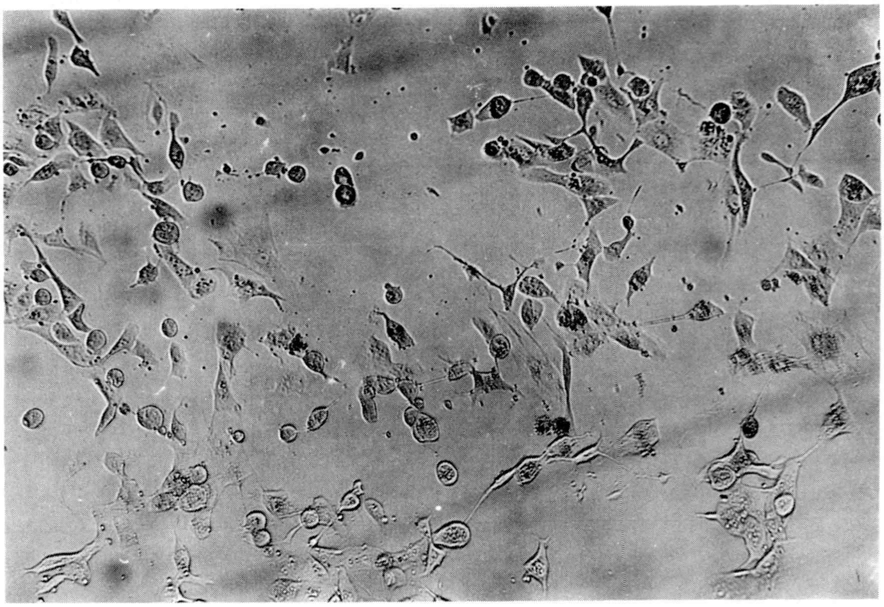

FIGURE 5(C).

- Type III. The nucleus contained one or more acidophilic inclusions and nucleoli were still present. The nucleus appeared wrinkled and somewhat distorted. No cytoplasmic changes were seen.
- Type IV. Cytoplasmic changes occurred consisting of the appearance of an eosinophilic paranuclear mass. The nuclear changes were essentially the same as those observed in type III cells, but the distortion of the nucleus was more marked.
- Type V. The cells began to pull away from the glass, and a great increase in the basophilia of the peripheral zone of the cytoplasm was observed. The cells did not appear fully rounded, but under the phase microscope their outlines appeared much sharper than those of uninfected cells.
- Type VI. The cells appeared fully rounded with an eccentric nucleus pushed aside by the eosinophilic cytoplasmic mass. In many cells of this type, the nucleolus and one or more acidophilic inclusions could be distinguished in the nucleus.
- Type VII. The cells were characterized by an acidophilic cytoplasm, nuclear pyknosis, and fragmentation of the nuclear chromatin (karyorrhexis).

Another cellular alteration observed was the presence of characteristic vacuoles in the periphery of the cytoplasm in some cells in the late stages of infection. These vacuoles appeared to arise from hyaline areas in the vicinity of the nucleus, moving later to the peripheral zone of the cytoplasm, and they were seen in some of the type V and VI cells. They could be followed under the phase microscope, but their short life and the difficulty of distinguishing them in some cases after staining prevented their inclusion in the differential cell counts. It has been suggested that these vacuoles play an important role in the mechanism of virus release.

The results of the differential cell counts in this experiment have been plotted in Figure 6. Appearance of infective virus within the cells occurred at approximately the same time that the first nuclear alterations were observed. In this experiment it was observed that virus release occurred just before the first type VI cells were seen. In other experiments the time of appearance of infective virus inside the cells usually

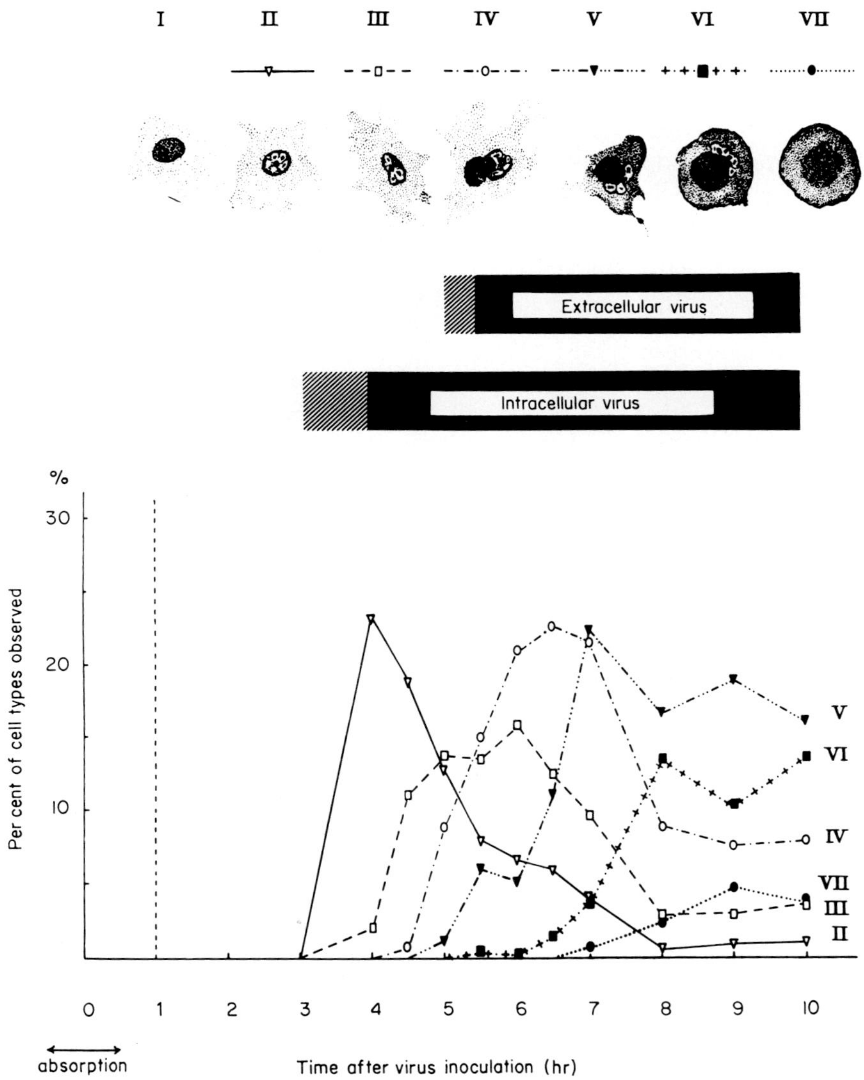

FIGURE 6. Schematic diagram illustrating successive morphological changes occurring in monkey kidney cells in culture after inoculation with poliovirus correlated with the appearance of new infective virus within the cells during a single cycle of virus multiplication. The percentage of each cell type was determined by differential cell counts conducted with fixed cells. (From Reissig, M., Howes, D. W., and Melnick, J. L., *J. Exp. Med.*, 104, 289, 1956. With permission.)

coincided with the time of appearance of type II cells (with intranuclear inclusions), but poliovirus release occurred at variable times, always before the rounded type VI cells appeared. It has been suggested that external factors which vary the permeability of the cell membrane may play a role in the release process.

Under the phase microscope, the first changes clearly observed were those corresponding to type V. When the same cell was followed throughout the cycle of virus growth, type IV cells could be distinguished in many cases. Intranuclear inclusions could not be seen with the phase microscope, and the distorted nucleus and the paranuclear mass were often masked by a cluster of fat droplets and mitochondria. However, the phase microscope was found to especially useful for detecting the characteristic vacuoles which appeared in the last stages. These vacuoles originated in hyaline perinuclear areas and underwent constant changes in shape and location until they finally developed into clearly defined vacuoles in the peripheral zone of the cytoplasm. The span of existence of the vacuoles was brief; they disappeared from the cell within approximately 1 hr of their formation so that the rounded cell of the last stage no longer contained them. The phase microscope also revealed that some of the type V and VI cells presented long cytoplasmic projections, frequently with dense thick nodules; they were never seen in the final stage.

Transition from one cell type to the next occurred rapidly and the whole sequence of changes ending with death of the cell was completed in approximately 7 hr after virus inoculation. Infective virus was formed in the cells some time before they rounded, and virus had already been released from cells before the typical pycnotic changes were seen.

Serological Tests

For any purpose except a serologic survey, paired serum specimens are required. The first sample must be taken as early as possible in the course of the illness or infection and the second 3 to 4 weeks later.

Neutralizing antibodies which can be measured quantitatively and rapidly by tissue culture methods also appear early and are usually already detectable at the time of hospitalization. However, if the first specimen is taken sufficiently early, a rise in titer can be demonstrated during the course of the disease. Acute and convalescent sera are usually tested simultaneously, using varying dilutions of serum against a constant amount of the specific virus (Table 2). Serum titers are calculated on the basis of the

Table 2
CELL CULTURE NEUTRALIZATION TEST WITH PAIRED SERA OF PATIENT INFECTED WITH TYPE 1 POLIOVIRUS

		Cellular degeneration (cytopathic effect),[b] final serum dilution					50% serum titer	
Virus[a]	Serum (day after onset)	1:2	1:10	1:50	1:250	1:1250	Logarithm	Antilog
Type 1	1	000	+ + +	+ + +	+ + +	+ + +	0.7	5
	20	000	000	000	00 +	+ + +	2.5	320
Type 2	1	000	+ + +	+ + +	+ + +	+ + +	0.7	5
	20	000	000	0 + +	+ + +	+ + +	1.5	32
Type 3	1	+ + +	+ + +	+ + +	+ + +	+ + +	0	0
	20	+ + +	+ + +	+ + +	+ + +	+ + +	0	0
None	1	000						
	20	000						

[a] 100 $TCID_{50}$ doses of each virus used in test. $TCID_{50}$ = 50% tissue culture infective dose, i.e., virus dose capable of infecting 50% of the inoculated cultures.

[b] Readings are shown for three culture tubes receiving each virus/serum mixture; +, = one tube showing cytopathic changes; 0, = one tube shwing no cytopathic changes.

dilution of serum which neutralizes a given amount of virus, and a fourfold or greater rise in antibody titer is considered significant, i.e., indicative of an infection during the period covered. However, it should be noted that neutralizing antibody titers may already be high at the time of onset of clinical symptoms, making interpretation difficult; if neutralizing antibody titers are found to be equally high in both acute and convalescent specimens, the infection might have taken place either recently or many years before.

Complement-fixing (CF) antibodies appear during the course of an infection, but may disappear or drop to a low level within a few months. N and H complement-fixing antigens of poliovirus have already been described. In the course of poliomyelitis infection, H antibodies decline first. Early acute stage sera thus contain H antibodies only; 1 to 2 weeks later both N and H antibodies are present; in late convalescent sera, only N antibodies are present. Only first infection with poliovirus produces strictly type-specific complement fixation responses. Subsequent infections with heterotypic polioviruses recall or induce antibodies, primarily against the heat-stable antigenic components shared by all three types of poliovirus, i.e., against the poliovirus group antigen. Very high CF antibody levels against a poliovirus are strongly suggestive of a recent infection.

Type-specific viral precipitating antibodies develop in convalescence. By the microprecipitation test only 50% of poliomyelitis patients show a fourfold or greater rise in antibody titer (due to the early appearance and early decline of the precipitating antibody). Therefore, the microprecipitation test has been less useful than the CF and neutralization tests.

Pathogenicity in Experimental Animals

Rhesus *(Macaca mulatta)* and cynomolgus (*Macaca irus*) monkeys as well as other species (e.g., *Cercopithecus aethiops*, the African green or grivet monkey) can readily be infected with polioviruses by the intraspinal (the most sensitive), intracerebral, and intranasal routes. Furthermore, certain poliovirus strains may infect monkeys on intradermal or intramuscular injection; cynomolgus monkeys have been infected with some strains by way of the gastrointestinal tract. Feeding poliovirus can cause infection of chimpanzees and cynomolgus monkeys, with prolonged excretion of virus in the stools associated with viremia and the development of serum antibodies. Naturally-acquired infection and even paralysis has occurred among captive primates, especially chimpanzees. The use of monkeys in the polio laboratory has been described in detail, and the reader is referred to the report of Paul and Melnick[16] for detailed information.

Laboratory rodents are usually resistant to freshly isolated polioviruses, although strains belonging to all three types have been adapted to cotton rats and suckling and adult mice. These animals usually develop flaccid paralysis when inoculated intracerebrally or intraspinally. For details on the use of mice in neutralization tests, the reader is again referred to the chapter mentioned above.[16]

Review

Table 3 is a summary table for poliovirus. Few viral diseases have been the object of so intense and comprehensive an attack, with massive concern and financial support from both the private and the public sectors in the United States and abroad. Because of the wide-ranging national and international programs concerned with poliomyelitis, the literature of the field is voluminous. The following reviews and symposium reports are only a few examples from this wealth of information: International Poliomyelitis Conferences I to V, held 1949 to 1961;[22] World Health Organization Expert Committee on Poliomyelitis, 1954 and 1958; [23,24] International Conferences on Live Poliovirus Vaccines, 1959 and 1960; [25,26] and Paul's history of poliomyelitis.[3]

Table 3
SUMMARY TABLE FOR POLIOVIRUS

		Laboratory diagnosis				
			Isolation technique			
Virus prototype strains[a]	Diseases associated	Source of specimen for virus isolation	Cell culture	Animal	Identification of virus isolate	Serologic diagnosis
Type 1: Brunhilde Type 2: Lansing Type 3: Leon	Only approximately 1% of poliovirus infections are recognized clinically; human response to infection ranges from symptomless or trivial through mild malaise to aseptic meningitis to paralytic poliomyelitis	Stool, rectal swab, throat swab; in fatal cases, spinal cord, medulla, colon contents	Monkey kidney or human cell cultures	None used at present; monkeys were chief tool before availability of tissue culture	Neutralization test vs. enterovirus equine antiserum pools or monovalent polio antisera	Neutralization test in cell culture; fourfold antibody titer rise; complement fixation test vs. polio antigens

[a] These prototype strains are described in References 9 and 10.

GENERAL REFERENCES

Fishbein, M., Ed., *International Poliomyelitis Conferences. Poliomyelitis: Papers and Discussions,* J. B. Lippincott, Philadelphia, 1949, 1952, 1955, 1958, 1961.

Expert Committee on Poliomyelitis, *WHO Tech. Rep. Ser.,* No. 81, 1954.

Paul, J. R. and Melnick, J. L., Poliomyelitis, in *Diagnostic Procedures for Virus and Rickettsial Diseases,* 2nd ed., Francis, T., Jr., Ed., American Public Health Association, New York, 1956, 53.

Expert Committee on Poliomyelitis, *WHO Tech. Rep. Ser.,* No. 145, 1958.

Live Poliovirus Vaccines, 1st Int. Conf., Scientific Publ. No. 44, Pan American Health Organization, 1959.

Live Poliovirus Vaccines, 2nd Int. Conf., Sci. Publ. No. 50, Pan American Health Organization, Washington, D.C., 1960.

Melnick, J. L. and Wenner, H. A., Enteroviruses, in *Diagnostic Procedures for Viral and Rickettsial Infections,* 4th ed., Lennette, E. H. and Schmidt, N.J., Eds., American Public Health Association, New York, 1969, 529.

Paul, J. R., *A History of Poliomyelitis,* Yale University Press, New Haven, 1971.

Melnick, J. L., Agol, V. I., Bachrach, H. L., Brown, F., Cooper, P. D., Fiers, W., Gard, S., Gear, J. H. S., Ghendon, Y., Kasza, L., LaPlaca, M., Mandel, B., McGregor, S., Mohanty, S. B., Plummer, G., Rueckert, R. R., Schaffer, F. L., Tagaya, I., Tyrrell, D. A. J., Voroshilova, M., and Wenner, H. A., Picornaviridae, *Intervirology,* 4, 303, 1974.

Phillips, C. A., Enteroviruses, in *Manual of Clinical Microbiology,* 2nd ed., Lennette, E. H., Spaulding, E. H., and Truant, J. P., Eds., American Society for Microbiology, Washington, D.C., 1974, 723.

Melnick, J. L., Enteroviruses, in *Viral Infections of Humans: Epidemiology, Pathogenesis, and Control,* Evans, A. S., Ed., Plenum Press, New York, 1976, 163.

Melnick, J. L., Picornavirus family (enterovirus and rhinovirus groups), in *Review of Medical Microbiology,* 12th ed., Jawetz, E., Melnick, J. L., and Adelberg, E. A., Eds., Lange, Los Altos, Cal., 1976, 368.

REFERENCES

1. Committee on the Enteroviruses, National Foundation for Infantile Paralysis, The enteroviruses, *Am. J. Public Health,* 47, 1556, 1957.
2. **Melnick, J. L., Agol, V. I., Bachrach, H. L., Brown, F., Cooper, P. D., Fiers, W., Gard, S., Gear, J. H. S., Ghendon, Y., Kasza, L., LaPlaca, M., Mandel, B., McGregor, S., Mohanty, S. B., Plummer, G., Rueckert, R. R., Schaffer, F. L., Tagaya, I., Tyrrell, D. A. J., Voroshilova, M., and Wenner, H. A.,** Picornaviridae, *Intervirology,* 4, 303, 1974.
3. **Paul, J. R.,** *A History of Poliomyelitis,* Yale University Press, New Haven, 1971.
4. **Melnick, J. L.,** Enteroviruses, in *Viral Infections of Humans: Epidemiology, Pathogenesis, and Control,* Evans, A. S., Ed., Plenum Press, New York, 1976, 163.
5. **Enders, J. F., Weller, T. H., and Robbins, F. C.,** Cultivation of the Lansing strain of poliomyelitis virus in cultures of various human embryonic tissues, *Science,* 109, 85, 1949.
6. **Mayor, H. D. and Jordan, L. E.,** Formation of poliovirus in monkey kidney tissue culture cells, *Virology,* 16, 325, 1962.
7. **Smith, K. O. and Melnick, J. L.,** Electron microscopic counting of virus particles by sedimentation on aluminized grids, *J. Immunol.,* 89, 279, 1962.
8. **Trask, J. D., Melnick, J. L., and Wenner, H. A.,** Chlorination of human, monkey-adapted and mouse strains of poliomyelitis virus, *Am. J. Hyg.,* 41, 30, 1945.
9. Committee on Typing of the National Foundation for Infantile Paralysis, Immunologic classification of poliomyelitis viruses. I. A cooperative program for the typing of one hundred strains, *Am. J. Hyg.,* 54, 191, 1951.
10. **Hahn, E. E. A.,** Polioviruses, in *Strains of Human Viruses,* Majer, M. and Plotkin, S. A., Eds., S. Karger, Basel, 1972, 155.
11. **LeBouvier, G. L.,** Poliovirus D and C antigens: their differentiation and measurement by precipitation in agar, *Br. J. Exp. Pathol.,* 40, 452, 1959.
12. **Hummeler, K. and Hamparian, V. V.,** Studies on the complement fixing antigens of poliomyelitis. I. Demonstration of type and group specific antigens in native and heated viral preparations, *J. Immunol.,* 81, 499, 1958.

13. **Melnick, J. L.**, Picornavirus family (Enterovirus and Rhinovirus groups), in *Review of Medical Microbiology,* 12th ed., Jawetz, E., Melnick, J. L., and Adelberg, E. A., Eds., Lange, Los Altos, Cal., 1976, 368.
14. **Melnick, J. L. and Wenner, H. A.**, Enteroviruses, in *Diagnostic Procedures for Viral and Rickettsial Infections,* 4th ed., Lennette, E. H. and Schmidt, N. J., Eds., American Public Health Association, New York, 1969, 529.
15. **Phillips, C. A.**, Enteroviruses, in *Manual of Clinical Microbiology,* 2nd ed., Lennette, E. H., Spaulding, E. H., and Truant, J. P., Eds., American Society for Microbiology, Washington, D.C., 1974, 723.
16. **Paul, J. R. and Melnick, J. L.**, Poliomyelitis, in *Diagnostic Procedures for Virus and Rickettsial Diseases,* 2nd ed., Francis, T., Jr., Ed., American Public Health Association, New York, 1956, 53.
17. **Melnick, J. L.**, Tissue culture methods for the cultivation of poliomyelitis and other viruses, in *Diagnostic Procedures for Virus and Rickettsial Diseases,* 2nd ed., Francis, T., Jr., Ed., American Public Health Association, New York, 1956, 97.
18. **Melnick, J. L. and Hampil, B.**, WHO collaborative studies on enterovirus reference antisera. Fourth report, *Bull. WHO,* 48, 381, 1973.
19. **Melnick, J. L., Rennick, V., Hampil, B., Schmidt, N. J., and Ho, H. H.**, Lyophilized combination pools of enterovirus equine antisera: preparation and test procedures for the identification of field strains of 42 enteroviruses, *Bull. WHO,* 48, 263, 1973.
20. **Melnick, J. L., Schmidt, N. J., Hampil, B., and Ho, H. H.**, Lyophilized combination pools of enterovirus equine antisera: preparation and test procedures for the identification of field strains of 19 group A coxsackievirus serotypes, *Intervirology,* 8, 172, 1977.
21. **Reissig, M., Howes, D. W., and Melnick, J. L.**, Sequence of morphological changes in epithelial cell cultures infected with poliovirus, *J. Exp. Med.,* 104, 289, 1956.
22. Fishbein, M., Ed., *International Poliomyelitis Conferences. Poliomyelitis: Papers and Discussions,* J. B. Lippincott, Philadelphia, 1949, 1952, 1955, 1958, 1961.
23. **Expert Committee on Poliomyelitis,** WHO Tech. Rep. Ser., *No. 81, 1954.*
24. **Expert Committee on Poliomyelitis,** WHO Tech. Rep. Ser., *No. 145, 1958.*
25. *Live Poliovirus Vaccines, 1st Int. Conf.,* Sci. Publ. No. 44, Pan American Health Organization, Washington, D.C., 1959.
26. *Live Poliovirus Vaccines, 2nd Int. Conf.,* Sci. Publ. No. 50, Pan American Health Organization, Washington, D.C., 1960.

PICORNAVIRIDAE: ENTEROVIRUS — COXSACKIEVIRUSES*

Richard L. Crowell and Burton J. Landau

Introduction

Three decades have elapsed since the discovery of the coxsackieviruses.[1] During this time, differences between these viruses and their more celebrated relatives, the polioviruses, have become more clearly defined. The burgeoning body of literature concerning the coxsackieviruses emphasizes the fact that gratuitous extrapolation to the coxsackieviruses of data derived from studies of poliovirus may be inappropriate. In fact, insight provided by studies of coxsackievirus morphogenesis, substructure analysis, antigenic properties, and disease potential justify an updated compilation of the properties of these viruses. Assorted facets of coxsackieviruses and their diseases have been reviewed. Those who are interested in a more detailed consideration are directed to the general references at the end of this chapter.

Since the pioneering efforts of Dalldorf[1] and Gifford and Dalldorf[2] and the work of Godman and his associates[3,4] and Melnick and Godman,[5] who demonstrated the unique coxsackievirus-induced histopathology in newborn mice, a host of human infections involving many organ systems have been shown to be caused directly by one or another of the coxsackieviruses. In several human disease states such as chronic heart disease, diabetes mellitus, muscular dystrophy, polymyositis, congenital anomalies, etc., the etiologic relationship of a coxsackievirus is suspected, but additional data must be accumulated before a final decision about their relationship can be established.[6-9]

Although these agents play a major role in severe, acute, human diseases, no vaccine or effective chemotherapy is available. The large number of immunotypes (23 in group A and 6 in group B) has undoubtedly contributed to the lack of progress in this area. However, if certain of the immunotypes can be shown to be the cause of some of the more serious infections, development of a coxsackievirus vaccine to selected viruses would be feasible.[10-12]

Morphology

Coxsackieviruses of both group A and group B are characterized by a nonenveloped capsid which exhibits icosahedral symmetry in a T = 1 or 3 structure[13-15] surrounding a single-stranded RNA genome of approximately 2.5×10^6 daltons mol wt.[15] The total protein mass of the capsid is approximately 5×10^6 daltons,[14,16] giving a virion mass of approximately 7.5×10^6 daltons.[17] When stained for electron microscopy with uranyl acetate-lead citrate or phosphotungstic acid (PTA), the virions in purified preparations have been shown to be 25 to 28 nm in diameter[10,17,18] (see Figure 1A).[19] Virions aggregated in a crystalline lattice structure measure approximately 20 nm in diameter (Figure 1B).[18] Thus, slight variations in reported measurements of coxsackievirus virions[20,21] may reflect not only the different techniques in fixation, but also the form of the virions measured.

The capsid of the virion is comprised of four nonidentical polypeptides designated VP-1, VP-2, VP-3, and VP-4 in order of decreasing molecular weights as determined by sodium dodeyl sulfate-polyacrylamide gel electrophoresis.[14-16, 22] The molecular weights of these polypeptides for B3 virus as determined in our laboratory are as follows: VP-1, 29,000; VP-2, 25,000; VP-3, 21,000; and VP-4, 5,500 daltons (see Figure 2A).[14,16] These four polypeptides occur in equimolar ratios with 60 monomers of each

* The studies cited from the authors' laboratory were supported by U.S. Public Health Service Research Grant AI-03771 from the National Institute of Allergy and Infectious Diseases.

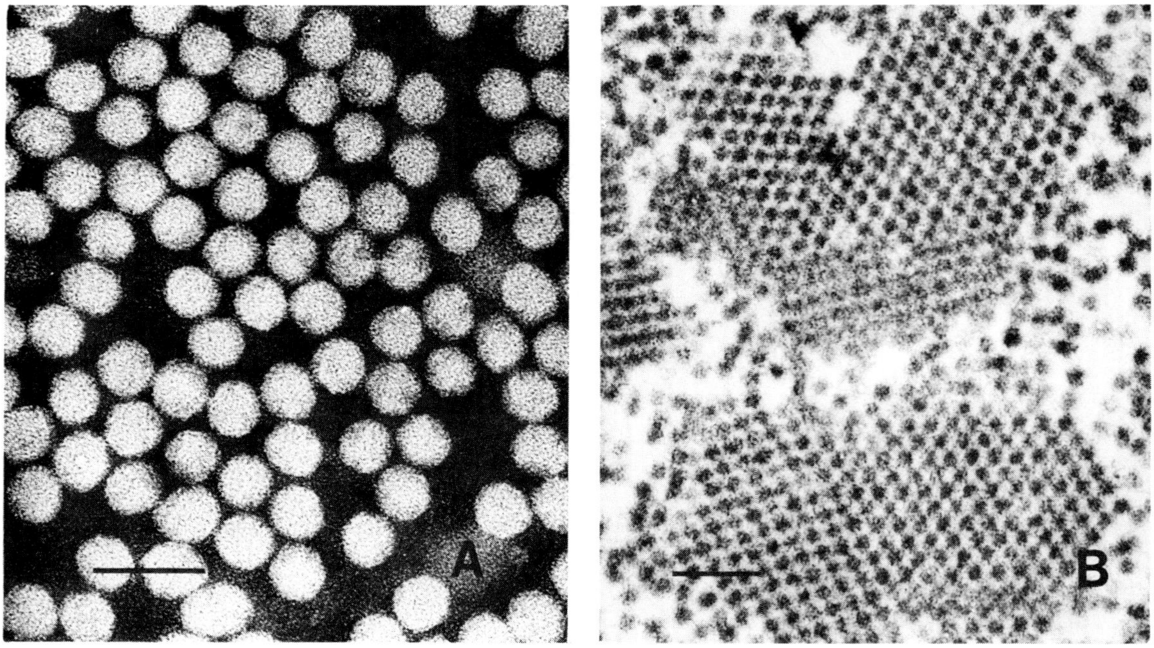

FIGURE 1. Electron micrographs of coxsackievirus B3 virions and procapsids. (A) A preparation of purified coxsackievirus B3 virions. The virions were diluted 1:2 with Tris buffer, pH 9.0, containing EDTA and DTT with no salt. Full capsids with an average diameter of 25 to 26 nm were seen. The bar represents 50 nm. (B) Crystalline arrays of B3 virions. The virions aggregated in the lattice measured 20 nm in diameter, with a center-to-center distance of 25 nm. Virions not aggregated in the lattice measured 25 nm in diameter. The bar represents 100nm. ([A] From Crowell, R. L., in *Cell Membrane Receptors for Viruses, Antigens and Antibodies, Polypeptide Hormones and Small Molecules,* Miles Int. Symp. Ser. No. 9, Beers, R. F., Jr. and Bassett, E. G., Eds., Raven Press, New York, 1976, 179. With permission. [B] From Roesing, T. G., Toselli, P. A., and Crowell, R. L., *J. Virol.,* 15, 654, 1975. With permission.)

per virion.[14,15] The provirions and procapsids (Figure 2B) have in common a precursor polypeptide designated VP-0, which has a molecular weight of 38,000 to 40,000 daltons. Recent studies have shown that coxsackievirus B3 virions can be dissociated by a procedure which permits recovery of substructures on sucrose gradients.[14] Electron micrographs of these substructures are illustrated in Figure 3, and some of the properties of these substructures are given in Table 1. The finding that the VP2-containing substructures, either alone or in combination with the other virion polypeptides, induce high titers of type-specific neutralizing antibodies to B3 virus makes feasible the prospect of a subunit vaccine.[10-12] The arrangement of these substructures at the vertices of the virion (Figure 4) has been proposed.[14]

Physicochemical Properties

Composition

As do the polioviruses, the coxsackieviruses lack lipid[23] and, despite an occasional report to the contrary,[26,27] probably lack carbohydrate moieties in the capsid.[15,24,25] No evidence exists for the presence of virion-containing, receptor-destroying enzyme (RDE).[28] The possibility that peptidase activity for cleaving the virion polypeptides during morphogenesis may be an intrinsic activity of the capsid polypeptides has not been excluded.[29,30]

The base compositions (Table 2) of the RNA of coxsackieviruses A9 and A10 are similar to each other, but differ from those found for polioviruses in that these coxsackieviruses contain more guanine.[17,31] On the other hand, the RNA base composition of coxsackievirus B1 resembles that of the polioviruses.[31,32] The presence of polyadenylic acid (poly A)[33] in the RNA of coxsackieviruses remains to be determined.

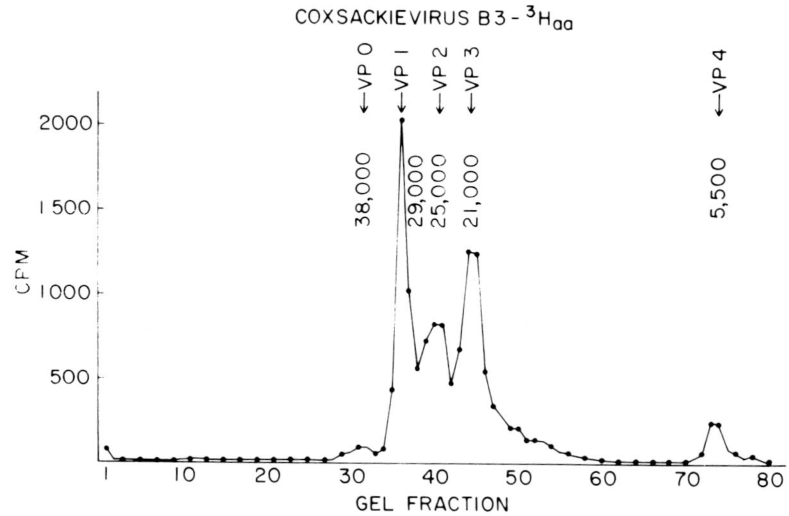

A

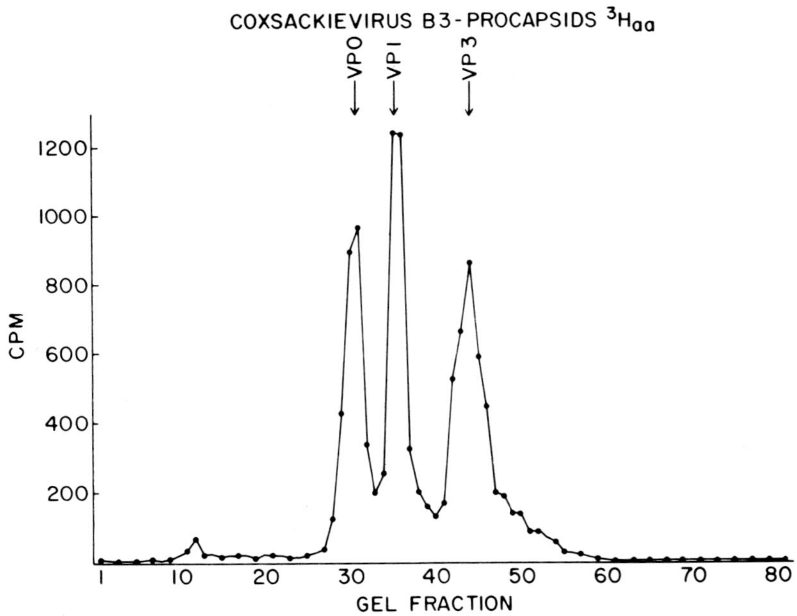

B

FIGURE 2. Sodium dodecyl sulfate-polyacrylamide gel electropherograms of the capsid polypeptides of coxsackievirus B33-Haa virions and procapsids. (A) Virions. The polypeptides with their respective molecular weights are indicated. The values were estimated by comparison with known poliovirus T1 polypeptides by co-electrophoresis.[16] (B) Procapsids. (From Philipson, L., Beatrice, S. T., and Crowell, R. L., *Virology*, 54, 69, 1973. With permission.)

Density

The buoyant density of coxsackievirus virions in cesium chloride gradients is 1.33 to 1.34 g/cc.[21,34] However, recent findings have revealed a small proportion of coxsackievirus B5 virions banded at a density of 1.44 g/cc.[34] The significance of the heavy particles is unknown, but may relate to the binding of cesium ions with virion proteins or RNA in this small fraction of virus.[15,34] In potassium tartrate gradients,

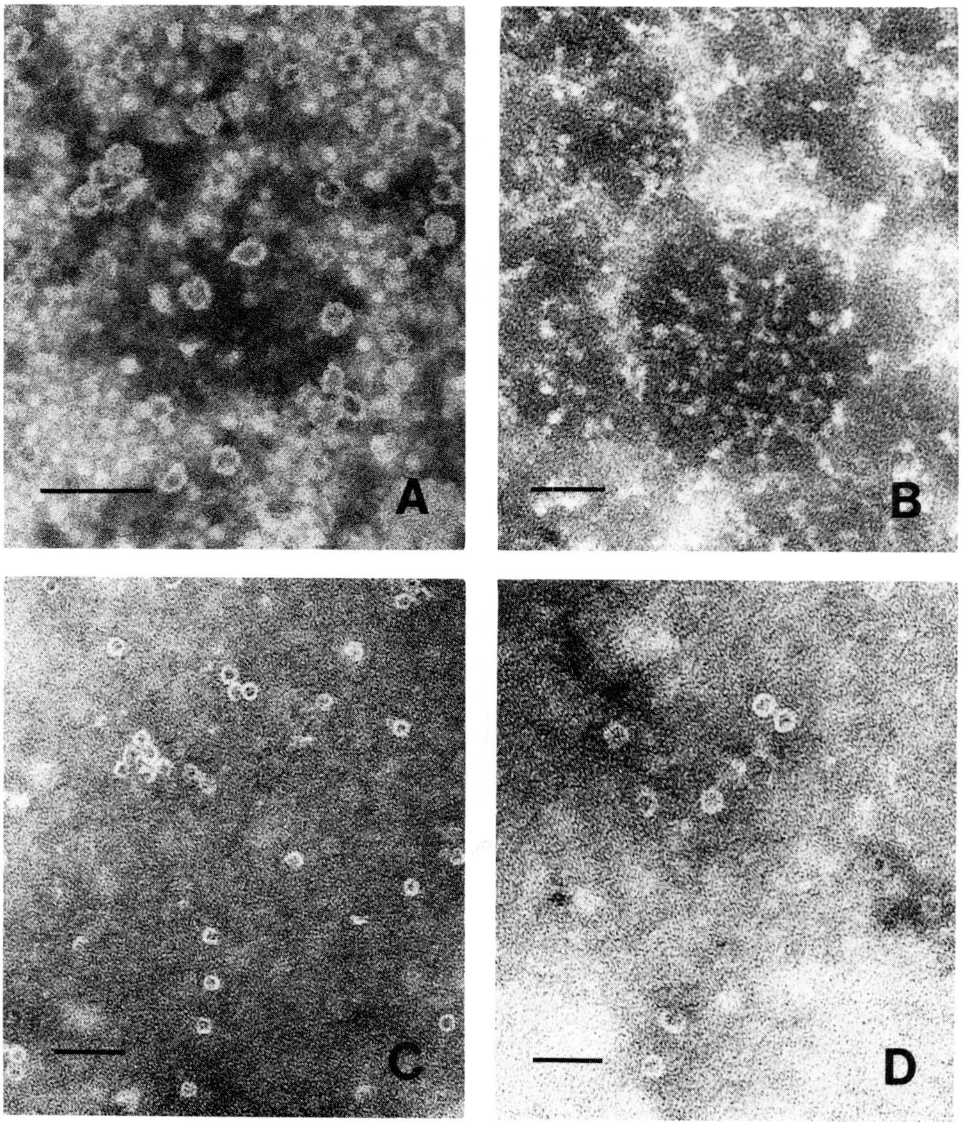

FIGURE 3. Electron micrographs of coxsackievirus B3 substructures obtained from urea-degraded virions following fractionation on sucrose gradients. (A) Preparation of urea-degraded B3 virions. The bar represents 100 nm. (B) Preparation of VP-2 substructures. These amorphous substructures have an average diameter of 5 to 6 nm. The bar represents 25 nm. (C) Preparation of VP-1, -3 substructures. These ring-like substructures have an average diameter of 9 to 10 nm. The bar represents 40 nm. (D) Higher magnification of the preparation shown in (C). The bar represents 25 nm. (From Beatrice.[10] Electron micrographs prepared by B. A. Zajac.)

the density of coxsackievirus B2 has been found to be 1.22,[21] which is in marked contrast to that of B3 virus.[16] In sucrose gradients, the sedimentation value has been found to be 150 to 160S.[14,17,34,35]

Virus Replication in Cell Cultures

The initial event in the interaction of coxsackieviruses with susceptible cells is attachment of virions to the receptors, which are located on the microvilli and adjacent areas of the plasma membrane, as can be seen in Figure 5. Studies using proteolytic enzyme

Table 1
PROPERTIES OF COXSACKIEVIRUS B3 VIRIONS AND SUBSTRUCTURES[10]

Virus structures[a]	Polypeptide content (VP)	S value	Capacity to induce neutralizing antibodies[b]
Virions	(1, 2, 3, 4)	150	+
Substructures	(1, 3), (2)	20, 5	+
	(2), (4)	5, <5	+
	(2)	5	+
	(1, 3)	20	−

[a] See Figures 1 and 3 for electron micrographs of the virus structures.
[b] +, High titers of neutralizing antibodies produced in rabbits immunized with the designated virus structure; −, no detectable neutralizing antibodies produced.

inactivation of these receptors followed by subcellular fractionation have shown that the receptors are limited to the external surface of cells.[36,37] The results of many studies[28,38-46] have revealed that there are specific receptors for the several genera of picornaviruses (Table 3).[19] It has been found that the three poliovirus immunotypes compete for a common receptor that is distinct from the receptor species which binds the six group B coxsackieviruses. It is also evident that members of the group A coxsackieviruses have another receptor(s). There also appear to be other receptors for echoviruses and human rhinoviruses.[36,39,43] Despite the generalization that receptors are virus specific, recent findings have shown that the adenovirus type 2 receptor is similar to that of coxsackievirus B3. Furthermore, the human rhinovirus type 14 (HRV-14) receptor is similar to that for coxsackievirus A21.[47] Significant differences in the early events of infection, which encompass virus attachment, penetration, and uncoating, may be a reflection of these diverse types of receptors for the different families of picornaviruses.[19,43,48]

Studies which focus on the early events of virus infection have revealed a complex sequence of events which is not fully understood. This sequence is depicted schematically in Figure 6. The events[49-51] in the replication of coxsackieviruses occurring in the cytoplasm of infected cells presumably mimic those identified for polioviruses and are outlined schematically in Figure 7. Figure 8 depicts a typical growth curve of coxsackievirus A5 in human rhabdomyosarcoma (RD) cells. Figure 9 shows the accumulation in human amnion cells of intracytoplasmic virions of coxsackievirus B5 in crystalline arrays, which are distinguishing features of coxsackieviruses and other enteroviruses in infected cells.[52-54] Tubular aggregates of B5 virus appearing in the cytoplasm have recently been described.[55] Their significance is unknown.

During infection, viral antigens can be detected in the cell cytoplasm by the use of immunofluorescent (Figure 9B) or immunoperoxidase tests. A hallmark of coxsackievirus infection of susceptible cells in culture is the characteristic cytopathic effect (CPE)[56] illustrated in Figure 10A. Formation of CPE has permitted an infectivity assay by the plaque method (Figure 11) for many of the coxsackieviruses, as will be discussed in the section on laboratory diagnosis.

Inhibition of Viral Replication

A number of factors have been shown to inhibit or reduce the replication of virus. These factors include such diverse elements as host cell physiology,[57-60] presence of antiviral agents,[61,62] the virus interference phenomenon,[63] and presence of antibodies to the virus[64] and to the receptors on cells.[39,63-67]

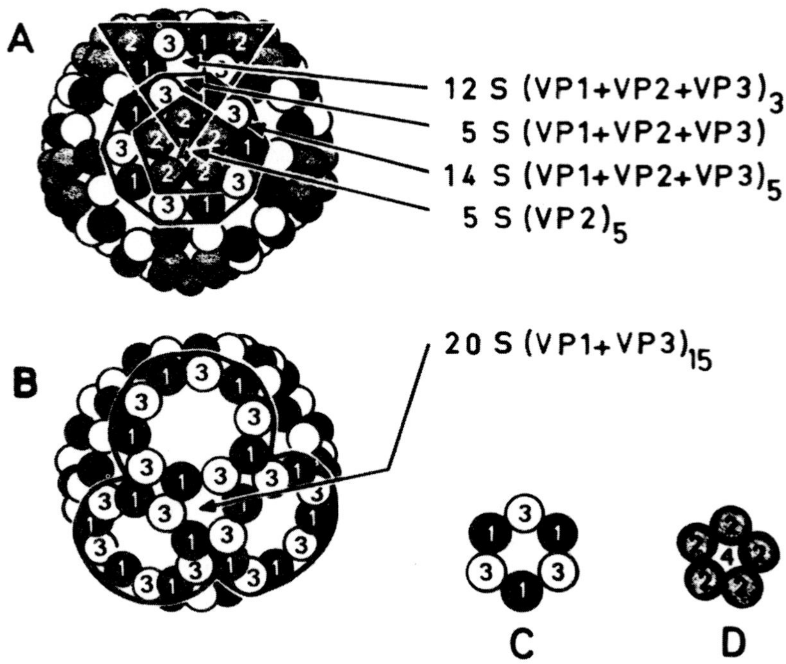

FIGURE 4. Hypothetical structure of picornaviruses as suggested from an analysis of urea-degraded virions and procapsids of coxsackievirus B3. (A) View of the virion along the fivefold axis, showing the individual polypeptides VP-1, VP-2, VP-3, and VP-4 marked by numbers. The cleavage planes for the 14S and 5S degradation structures observed by Dunker and Rueckert for mouse-Elberfeld virus[137] and the 12S structure observed by Talbot and Brown for foot-and-mouth disease virus,[138] as well as the 5S structure observed for coxsackievirus B3 comprising five VP-2 units, are indicated. The five polypeptides of VP-4 at the vertices are labeled with a single 4 in the figure. (B) The pentamers at the vertices have been removed, and the 20S structure containing pentadecameric VP-1 and VP-3 found for coxsackievirus B3 is outlined. (C) A schematic view of the hexamer of picornavirus containing VP-1 and VP-3. Note the close resemblance between the model and the ring-like structures observed in Figure 3C. (D) A schematic view of the pentamer of picornaviruses with the 5VP-2 unit surrounding five polypeptides of VP-4 (marked as a single 4). The latter are released from the pentamer as monomeric units in the initial event in the degradation of the virion. Note the relatively amorphous substructures of VP-2 in Figure 3B. (From Philipson, L., Beatrice, S. T., and Crowell, R. L., *Virology,* 54, 69, 1973. With permission.)

Although it is well recognized that the general state of health of cells in culture is important for optimal yields of virus, a number of more specific inhibitors of virus replication have been described.[61,68-72] The extensive studies of Tamm and Eggers[71,75] and Eggers and Tamm[73,74] concerning hydroxybenzimidazole (HBB) have indicated inhibition of coxsackieviruses at the level of RNA synthesis.

Virus Prototype Strains and Variants

Group A and B coxsackieviruses comprise 29 immunotypes. The prototype viruses for each immunotype are referred to in Tables 4 and 5. These viruses have served as reference strains for the typing of new isolates obtained from a variety of human diseases or in the course of screening specimens from normal subjects. However, experience has revealed that recent isolates[76-80] may not readily be categorized into the designated types because of antigenic variation. The significance of this observation

Table 2
COMPARISON OF BASE COMPOSITION OF RIBONUCLEIC ACIDS OF PURIFIED COXSACKIEVIRUSES OF GROUPS A AND B TO POLIOVIRUSES

	Moles per 100 mol nucleotides[a]				
Virus	A	U	G	C	Refs.
Coxsackie A9	27	25	28	20	17,31
Coxsackie A10	28	23	28	21	17,31
Coxsackie B1	29	24	24	23	31,32
Polio (3 types)	29	25	24	22	31
Polio 1	30	24	23	23	32

[a] A, adenine; U, uracil; G, guanine; C, cytosine.

remains to be determined. Perhaps the real issue is whether these new isolates represent "prime" strains which are not neutralized by antisera prepared against the prototype strains[77,81] or whether they are intratypic variants with a narrower range of antigens.[79] The selective pressures which may account for the origin of intratypic variants include those factors which interact with the surface of the virions. Numerous studies have shown that virus variants may be selected from parental viral populations, as evidenced by their resistance to antibodies[82] and naturally occurring inhibitors in serum.[83] The recent finding that the VP-2-containing substructure of coxsackievirus B3 induces neutralizing antibodies[10,11] suggests that virus variants may arise by the occurrence of faulty polypeptide cleavages at that site during virus maturation.

Shortly after the introduction of monolayer cell cultures for virus isolation and the development of techniques for the plaque assay, it became apparent that the coxsackieviruses produce a variety of plaque types under different environmental conditions and in different cell strains. In general, the range of plaque morphology produced by different virus strains includes plaques that are large or small in diameter, have sharp or diffuse borders, and appear clear or turbid.[62] Many studies have established that the type of plaque morphology produced by a given virus strain reflects the constituents in the agar overlay, e.g., primarily agar-sulfated polysaccharides;[83-85] inhibitors in the serum;[62] ion species, especially $MgCl_2$ and their concentrations[86] and strain or type of cultured cells employed in the plaque assay.

An example of the important role played by the type of cells used in culture for the selection of virus variants from clinical specimens has been documented experimentally for coxsackievirus B4 in the report of Choppin and Eggers.[83] It is noteworthy that the virus variant that was selected by the tissue cultures from a mixed population of antibody-sensitive and -resistant virions was antibody sensitive. Thus, attempts to identify a tissue culture isolate as the etiologic agent of a given disease syndrome in man may, at least on occasion, be a variant of the real pathogen, and conclusions drawn to correlate tissue culture isolates with the true virus variant which may be causing the human disease syndrome should be tempered with caution.

Although small-plaque mutants of many viruses have been found generally to be less virulent in animal hosts than are their large-plaque counterparts, this relationship does not always hold. The subject of plaque mutants was reviewed extensively in 1966.[62] The identification of in vitro markers of virus strains has great importance for the future selection of virus variants for the development of potentially useful live-virus vaccines.

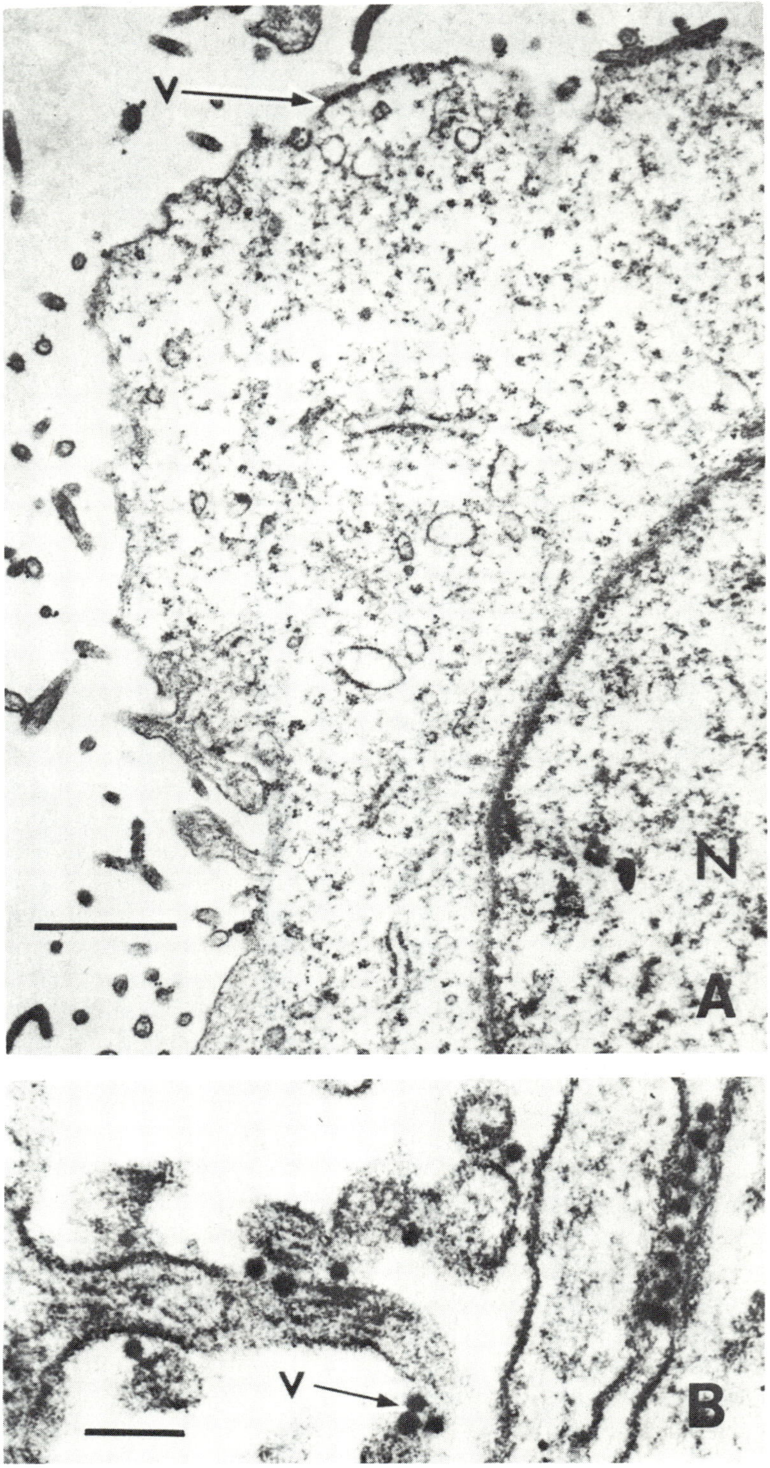

FIGURE 5. Electron micrographs of a HeLa cell preparation with receptor-saturating amounts of coxsackievirus B3 located at the cell surface. (A) The virions (V) are limited to discrete areas of the external surface of the plasma membrane and the microvilli. The cell nucleus is labeled (N). The bar represents 1.0 μm. (B) Higher magnification of a preparation similar to that shown in (A). The bar represents 100 nm. (From Roesing, T. G., Toselli, P. A., and Crowell, R. L., *J. Virol.*, 15, 654, 1975. With permission.)

Table 3
EVIDENCE FOR A CLOSE CORRELATION AMONG THE RECEPTORS FOR PICORNAVIRUS GENERA AND THEIR RESPECTIVE TAXONOMIC DESIGNATION AS BASED ON INTERFERENCE OF VIRUS ATTACHMENT[19,28,40,41,47]

Virus types used to saturate receptors	Interference of attachment of virus immunotypes					
	Polio 1—3	Cox B1—6	Cox A21	HRV2	HRV14	E6
Polio 1—3	+	−	−	−	−	−
Coxsackie B1—6	−	+	−[a]	−	−	−
Coxsackie A21	−	−	+	−	+	NT
HRV 2	−	−	−	+	−	NT
HRV 14	−	−	+	−	+	NT

Note: HRV, human rhinovirus; Cox, coxsackievirus; E6, echovirus type 6. In brief, HeLa cells were saturated with the interfering virus and the cells were washed and tested for capacity to attach a second type of virus (either homologous or heterologous). (+) Denotes interference of attachment of the challenge virus, whereas (−) denotes normal attachment of the second virus. NT, not tested.

[a] In addition, coxsackieviruses A13, A15, and A18 were not inhibited by coxsackievirus B5.

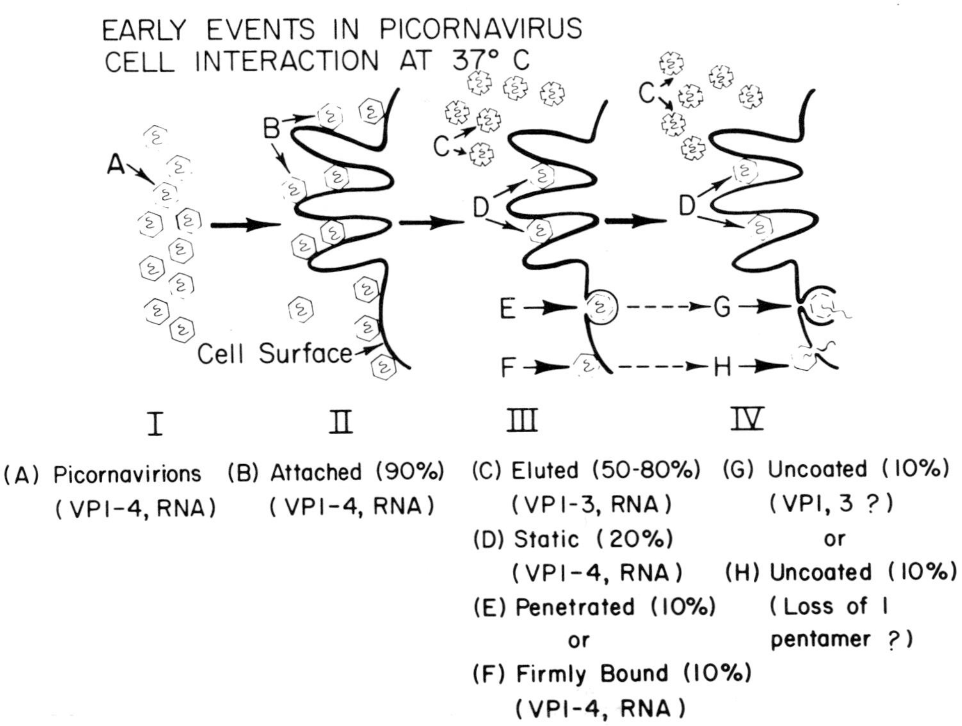

FIGURE 6. A schematic portrayal of the early events in picornavirus-cell interaction at 37°C. A population of virions containing a full complement of capsid polypeptides (VP-1 to VP-4) enclosing the RNA is shown in step I. Following attachment to the cell surface (step II), populations of the virions are known to be eluted, remain stationary at the cell surface, be engulfed, or be firmly bound (step III). The final sequence (step IV) suggests that virions may lose all or as few as one of their pentameric polypeptides from within a vacuole or at the cell surface to permit emergence of the virion RNA (uncoating) initiating productive infection. The percentages indicated in parentheses give the approximate distribution of the attached virions. (From Crowell, R. L., in *Cell Membrane Receptors for Viruses, Antigens and Antibodies, Polypeptide Hormones and Small Molecules*, Miles Int. Symp. Ser. No. 9, Beers, R. F., Jr. and Bassett, E. G., Eds., Raven Press, New York, 1976, 179. With permission.)

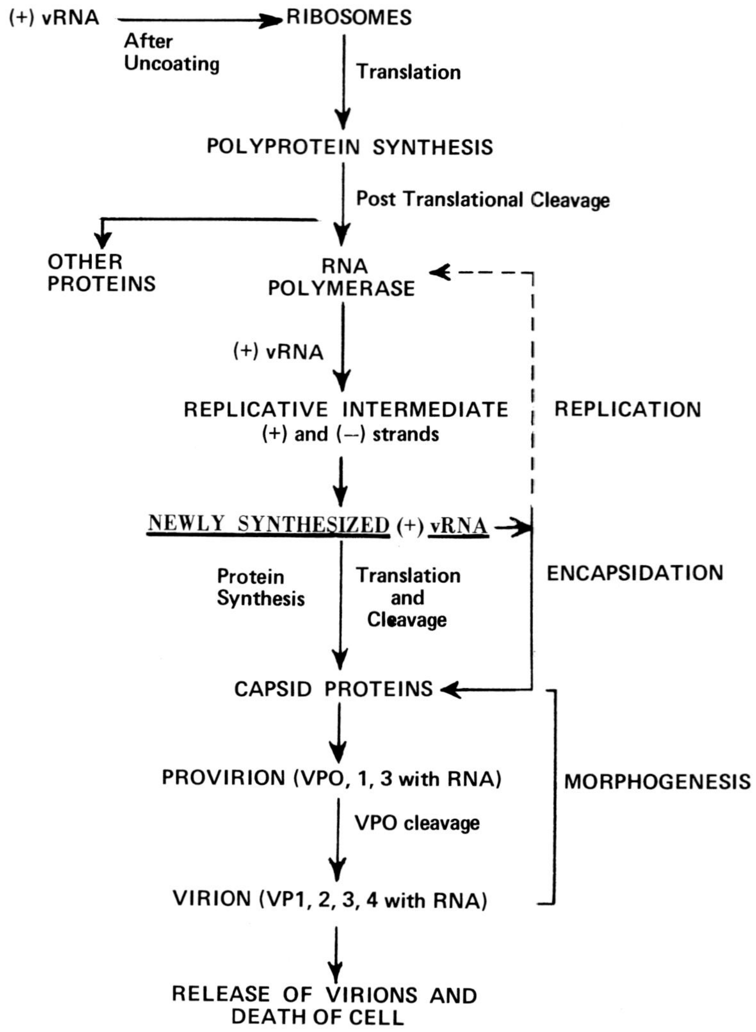

FIGURE 7. Events in the intracytoplasmic replication of coxsackieviruses. See text for details.

Examples of distinctive markers other than morphology and virulence used to identify coxsackievirus variants include temperature sensitivity,[87-90] resistance to heat,[91] drug dependence,[62,74] the ability to agglutinate erythrocytes,[92,93] and the capacity to attach to selected adsorbents.[94] Although it is unlikely to occur in nature, the potential of transcapsidation of coxsackievirus RNA by the protein coat of another virus in dual infections provides for the production of a different type of variant.[95] It is evident that coxsackieviruses are mutable and represent an array of variants, some of which may partially account for the many different types of human disease syndromes associated with these agents.

Clinical Features

Coxsackievirus infections in man have been documented and reviewed extensively.[7,96-103] Human infections with coxsackieviruses may result in a wide variety of syndromes including aseptic meningitis, myositis, pleurodynia, herpangina, myocarditis, pericarditis, pancreatitis, nephritis, orchitis, exanthems, gastroenteritis, acute respiratory disease, and congenitally acquired disease. Inapparent infections are common and may serve to maintain the virus in the population between epidemics. The

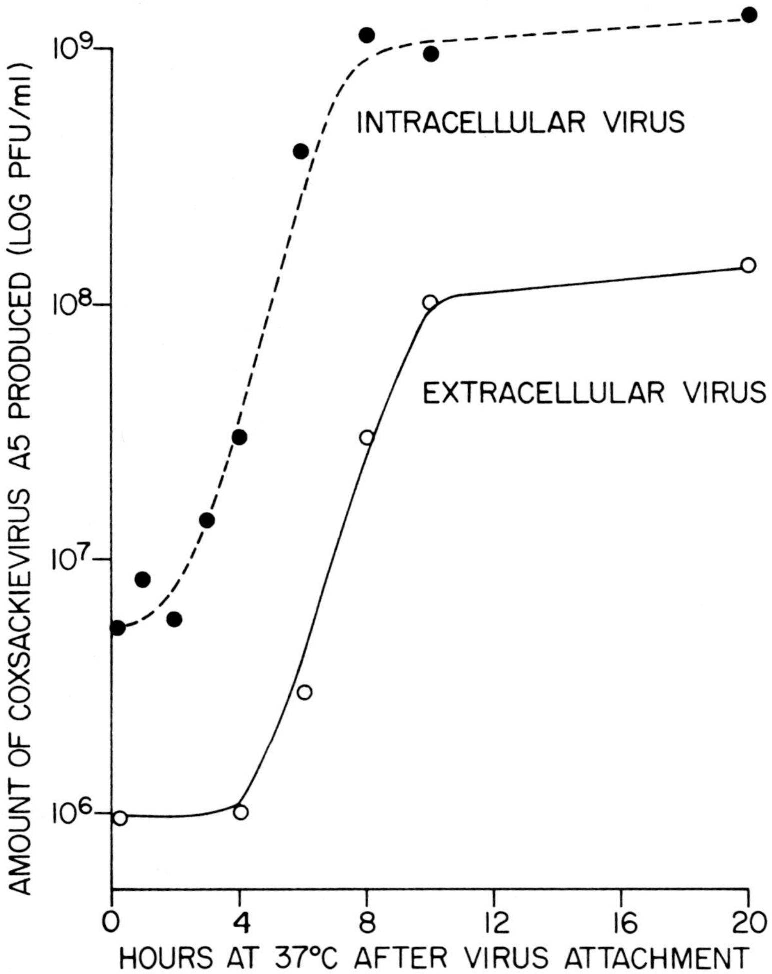

FIGURE 8. Production of coxsackievirus A5 by RD cells in replicate monolayer cultures. The cells received a virus multiplicity of 20 PFU/cell, and virus attachment was permitted at 37°C for 1 hr. The monolayers were washed, overlaid with medium, and incubated at 37°C for virus production. At intervals, replicate plates were sampled for determination of extracellular and intracellular amounts of virus by plaque assay. As is typical of picornavirus replication, virus accumulated intracellularly prior to release of virus into the fluid phase. (Courtesy of T. A. Stoneback.[136])

length of the incubation period has been shown to vary between 2 and 14 days, with an average of between 3 and 5 days.

Virus enters the body by way of the oral and respiratory routes (Figure 12). The presence of coxsackievirus in raw sewage, contaminated water, shellfish, and the feces of infected persons is well documented.[104,105] Infectious virus may persist on hard surfaces and enter the mouth via the fingers, etc. Spread of virus from person to person is also likely to occur. Following ingestion, coxsackieviruses replicate in the lymphoid tissue associated with the pharynx and/or gut, as do other enteroviruses. Virus spreads to the regional lymphatics, and a viremia and fever ensue. Presumably, the hematogenous spread of the virus distributes the virus to a variety of target organs which are affected by coxsackievirus infection. The pantropic effects of these viruses may be determined by the presence of specific viral receptor sites on cells of the central nervous system, heart, liver, pancreas, respiratory tract, and skin and mucous membranes.

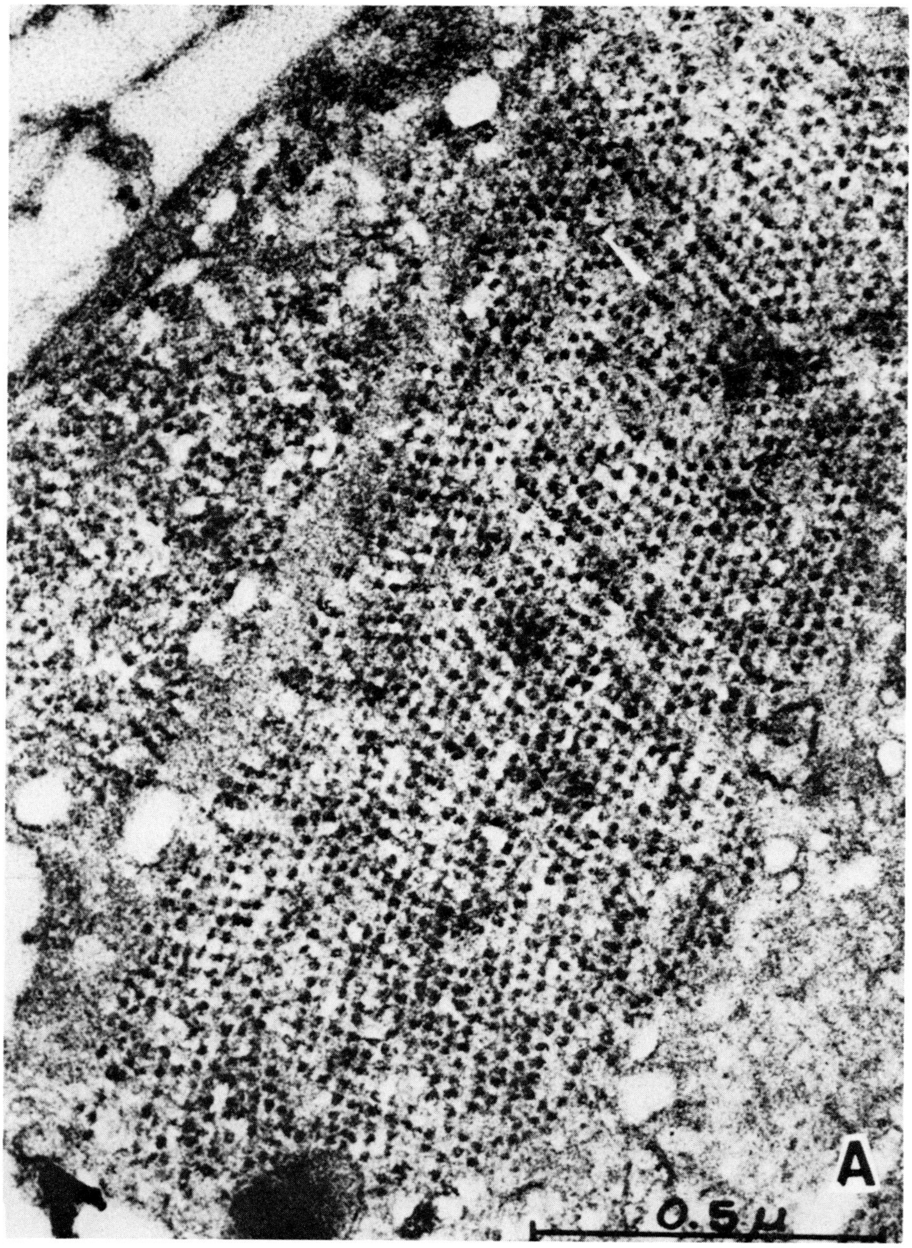

FIGURE 9. Evidence of coxsackievirus infection of cells is provided by the visualization of intracytoplasmic crystalline arrays of virions and by the detection of viral antigens by immunofluorescence. (A) Electron micrograph of coxsackievirus B5 in FL amnion cells. (B) Coxsackievirus A13 antigen in the cytoplasm of an infected myoblast and myotube as revealed by indirect immunofluorescence. Magnification × 400. ([A] From Stuart, D. C. and Fogh, J., *Virology,* 13, 177, 1961. With permission. [B] From Goldberg, R. J. and Crowell, R. L., *J. Virol.,* 7, 759, 1971. With permission.)

Why some individuals develop one or more syndromes or have inapparent infection is poorly understood.

The dramatic effects of acute coxsackievirus-induced disease may overshadow a potentially greater problem, i.e., the capacity of these viruses to cause chronic human

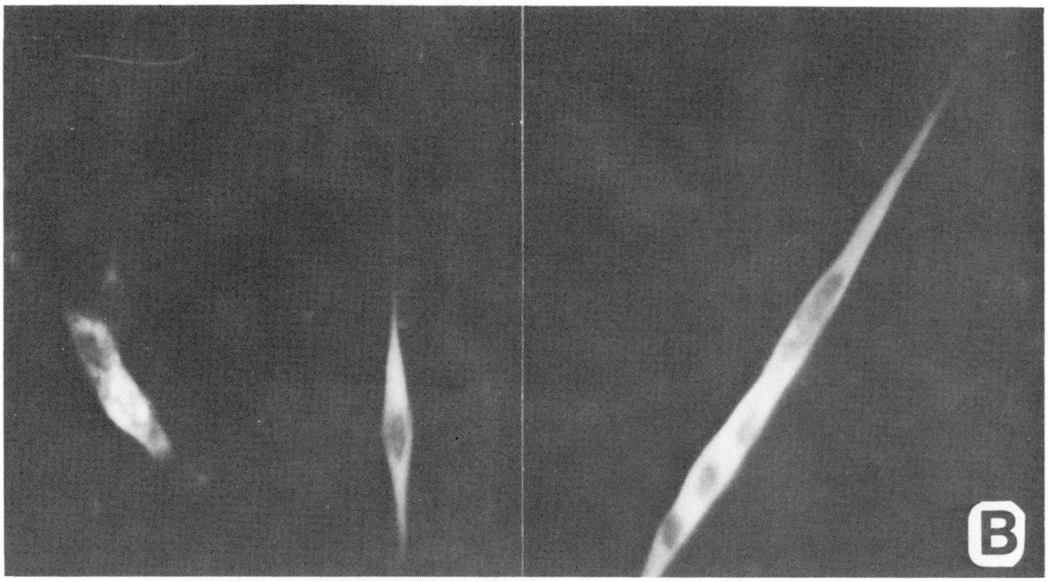

FIGURE 9B

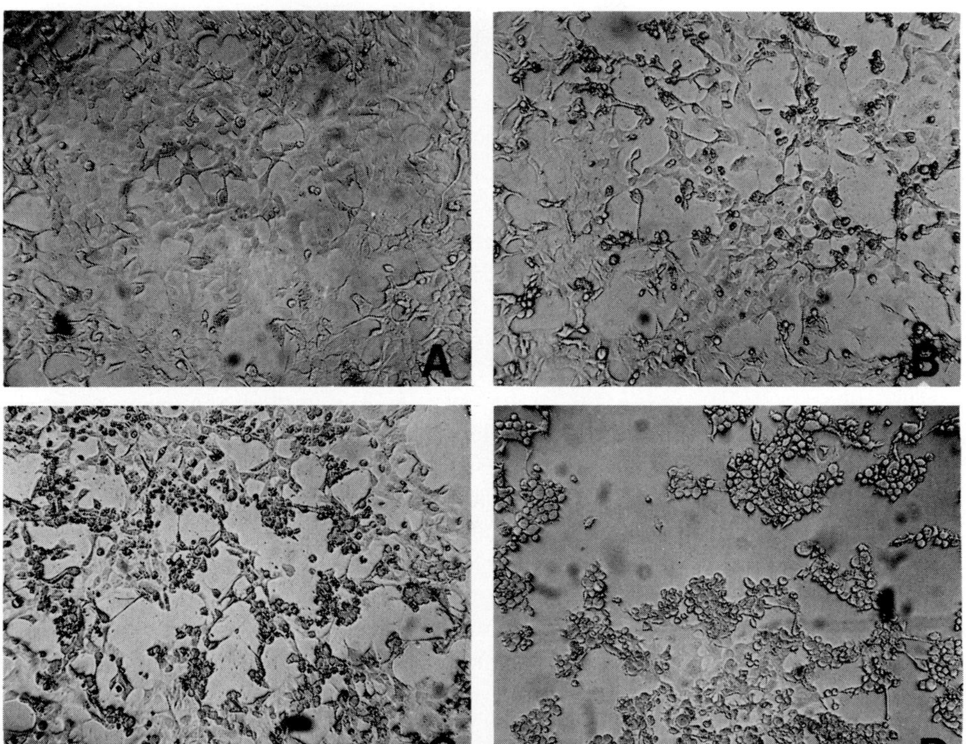

FIGURE 10. Cytopathic effect of coxsackievirus B3 in monolayer cultures of HeLa cells and the protective effect of homotypic antibody. (A) Normal-appearing HeLa cells protected from the effects of virus by homotypic antibody. The photograph was made 3 days after inoculation of the culture with virus and antibody. (B to D) Appearance of progressive degenerative effects of virus on HeLa cells photographed 48 to 72 hr after inoculation of B3 virus. Unstained preparations; magnification × 50. (From Crowell, R. L. and Syverton, J. T., *J. Immunol.*, 74, 169, 1955. With permission.)

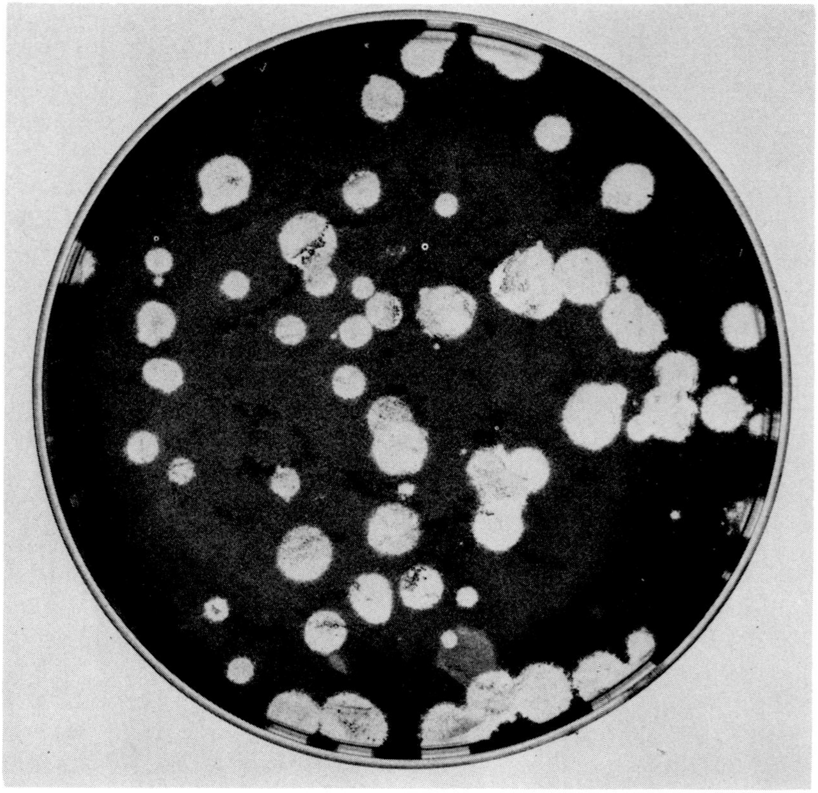

FIGURE 11. Appearance of plaques produced by coxsackievirus B3 on monolayer cultures of HeLa cells at 48 hr postinfection.

disease. These chronic diseases may be sequelae of prior acute disease or inapparent infection.[106] Two of the more prominent diseases currently under scrutiny are diabetes mellitus[101,107-112] and myocarditis.[99,113-117] The major problem remaining is definition of the mechanisms by which chronic disease occurs. Many attempts to recover infectious virus from humans and animals suffering from chronic disease have been unsuccessful, suggesting that additional factors such as immunologic injury may produce the disease. Consequently, indirect and less definitive evidence of a coxsackievirus etiology has been utilized to make these associations. Such evidence includes seroepidemiologic studies; viral isolation from random surveys of presumably normal individuals; screening of autopsy and biopsy material for virus-specific antigen by indirect immunologic techniques; and prior clinical history of coxsackievirus infection, even if it is often lacking in laboratory verification. Nevertheless, the volume of reports, together with results of experiments on animals, present a convincing basis for association of coxsackievirus with chronic disease and justify further study of this subject.

Laboratory Diagnosis

Virus Isolation and Immunotyping

The laboratory diagnosis of coxsackievirus infections is essential in order to establish these viruses as etiologic agents of the diverse syndromes referred to above. The laboratory diagnostic procedures have been reviewed extensively in recent years.[118-120] A laboratory diagnosis ideally consists of the recovery of a virus, its identification by serological techniques, and demonstration in the patient of a corresponding fourfold or greater rise in antibody titer over the course of the infection.[121] Occasionally, a new

Table 4
RESPONSE OF SELECTED CULTURE SYSTEMS TO PROTOTYPE STRAINS OF GROUP A COXSACKIEVIRUSES[a]

		Established cell lines			Primary or secondary cell cultures		
Virus type	Prototype strain	RD[124,126]	ML[124]	KB or HeLa[123]	Human amnion[123,126,130]	Rhesus monkey kidney[123,126]	Human embryonic kidney[123,126]
1	Tompkins	−	−	−	−	−	−
2	Fleetwood	+	−	−	+	−	−
3	Olson	+	−	−	+	−	−
4	Ala 4-CDC	+	−	−	−/+	−	−
5	Swartz	+	−	−	−	−	−
6	Gdula	+	−	−	−	−	−
7	AB IV	+	+	+	+	+	−/+
8	Donovan	+	−	−	+	−	−
9	Wiederhold	+	−	−	+	+	+
10	Kowalik	+	+	−	+	+	−
11	Belgium-1	+	−	+	+	−	+
12	Texas-12	+	−	−	+	−	+
13	Flores	−	+	+	+	−	+
14	G14	+	−	+	+	+	+
15	G9	+	+	+	+	−	+
16	G10	+	−	+	+	+	−/+
17	G12	+	−	−	+	−	−/+
18	G13	+	+	+	+	−	+
19	Dohi	−	−	−	−	−	−
20	IH-35	−	+	+	+	−	+
21	Kuykendall	+	−	+	+	−	+
22	Chulman	−	−	−	−	−	−
24	Joseph	−	+	+	+	−	−/+

[a] +, Growth of virus or cytopathic effect (CPE); −, absence of growth or CPE.

Table 5
RESPONSE OF SELECTED CULTURE SYSTEMS TO PROTOTYPE STRAINS OF GROUP B COXSACKIEVIRUSES[a]

		Established cell lines		Primary or secondary cell cultures		
Virus type	Prototype strain	RD[b]	HeLa KB, etc.[56,59]	Human amnion[125]	Rhesus monkey kidney[126]	Human embryonic kidney[127]
B1	Conn-5	+	+	+	+	+
B2	Redmond	−	+	+	+	+
B3	Nancy	+	+	+	+	+
B4	JVB	−	+	+	+	+
B5	Faulkner	+	+	+	+	+
B6	Schmitt	+	+		+	+

[a] +, Growth of virus or cytopathic effect (CPE); −, absence of growth or CPE.
[b] Only the hemagglutinating strains of virus grew.[135]

Table 6
COMPARATIVE EFFECTS OF POLYIONS AND $MgCl_2$ on PLAQUE PRODUCTION BY COXSACKIEVIRUSES[40,62,83,86,136]

		Agar additive[a]		
Virus type	Strain	Dextran sulfate	DEAE-D[b]	$MgCl_2$
A5	Swartz	−	−	+
A7	Habel	NC/−	+	NC/+
A9	Grigg	+/−	NC	+
B1	Conn-5	−	NC	NC/+
B2	Ohio-1	−	NC	+
B3	Nancy	NC	NC	+
B4	Powers	NC	+	+
B5	Faulkner	+	NC/+	+
B6	Schmitt	NC	NC	+

[a] +, Enhancement; −, inhibition; NC, no change in the efficiency of plating.
[b] DEAE-D, diethylaminoethyl dextran.

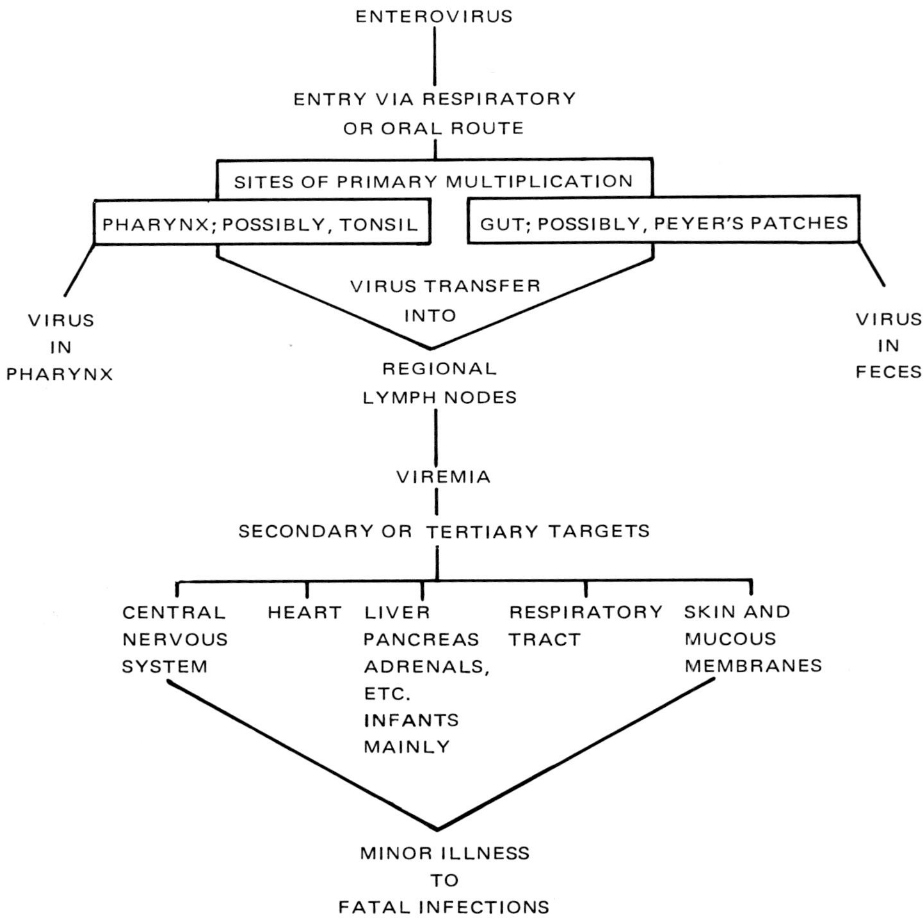

FIGURE 12. Schematic diagram depicting essential features of the replication and spread of enteroviruses in the body. (From Bodian, D., *Science*, 122, 105, 1955. With permission.)

virus immunotype is recovered, making a diagnosis more difficult.[122] Frequently, fulfillment of all criteria is not possible. An acute-phase serum sample may contain a high titer of antibody; consequently, an increase in antibody titer is not readily detected in the convalescent-phase serum sample. An equally difficult problem confronts the clinician in obtaining laboratory evidence to support a clinical impression of chronic disease, as virus is seldom recovered from specimens obtained late in the course of disease.

A protocol similar to that schematically presented in Figure 13 is usually followed for virus isolation. The clinical specimens from which coxsackieviruses are most likely to be recovered are listed in Table 7. The newborn mouse is the definitive animal host for the recovery of group A coxsackieviruses from clinical specimens. The use of these animals is often supplemented or replaced with susceptible tissue cultures. The tissue cultures used for isolation of some of the group A (Table 4) and group B coxsackviruses (Table 5) are of human or monkey origin.

As an extension of studies[52] which showed that differentiating cultures of primary mouse fetal muscle cells support the growth of previously uncultured viruses of group A, a human rhabdomyosarcoma cell line (RD cells) has been shown to be capable of supporting the growth of 16 of the 23 immunotypes of group A.[124,126] The RD cells, together with other cell cultures (mainly primary human amnion and primary monkey kidney cells), permit the recovery of 21 of 23 group A immunotypes. Although these findings make possible a screening technique for virus isolation more convenient than use of the newborn mouse, the mouse is usually more sensitive.[126] Further studies are needed to find a cell system which will support the growth of every member of this large group of viruses. The application of an "enhancer preparation"[128] may facilitate the detection in tissue culture of some of the less cytopathic viruses.

In marked contrast to the group A coxsackieviruses, the viruses of group B grow readily in many commonly available primary and established cell lines. Utilization of immunoelectron microscopy to detect and identify viruses in specimens which may not yield virus in cell cultures is receiving increased attention.[129] It is apparent that new methods and approaches are needed. Assignment of viral isolates to a specific immunotype is best accomplished by the neutralization test. This procedure has been facilitated by the availability of standardized pools of typing serums.[130,131] (For a detailed account of the use of these standardized antiserum pools for the typing of enteroviruses, see References 118, 119, and 132.) Identification of recent coxsackievirus isolates is complicated by the emergence of "prime" strains.[76,77,79,80] Thus, the standardized serum pools for typing may not contain antibodies to these antigenic variants.

Serodiagnosis

A number of methods have been employed to aid in the serologic diagnosis of coxsackievirus infections. A comprehensive review of these procedures has recently been compiled.[118] Despite the variety of immunologic techniques that have been adapted for this purpose, no rapid, simple, and sensitive screening test is currently available to efficiently catalogue the numerous coxsackieviruses that play major roles in human disease.

The choice of serodiagnostic test is influenced by the fact that, in contrast to other virus groups, these viruses have not yet been shown to possess a group antigen which can be used reliably for detecting antibody. Furthermore, a few members of the group A coxsackieviruses cannot be grown in tissue culture, thus reducing the universal applicability of the neutralization tests conducted in vitro. Lastly, the ubiquity of coxsackievirus infection in the general population suggests that most individuals have measurable antibody levels to some immunotypes in the absence of overt disease, and an immune response to heterologous immunotypes may be misleading. The laboratory

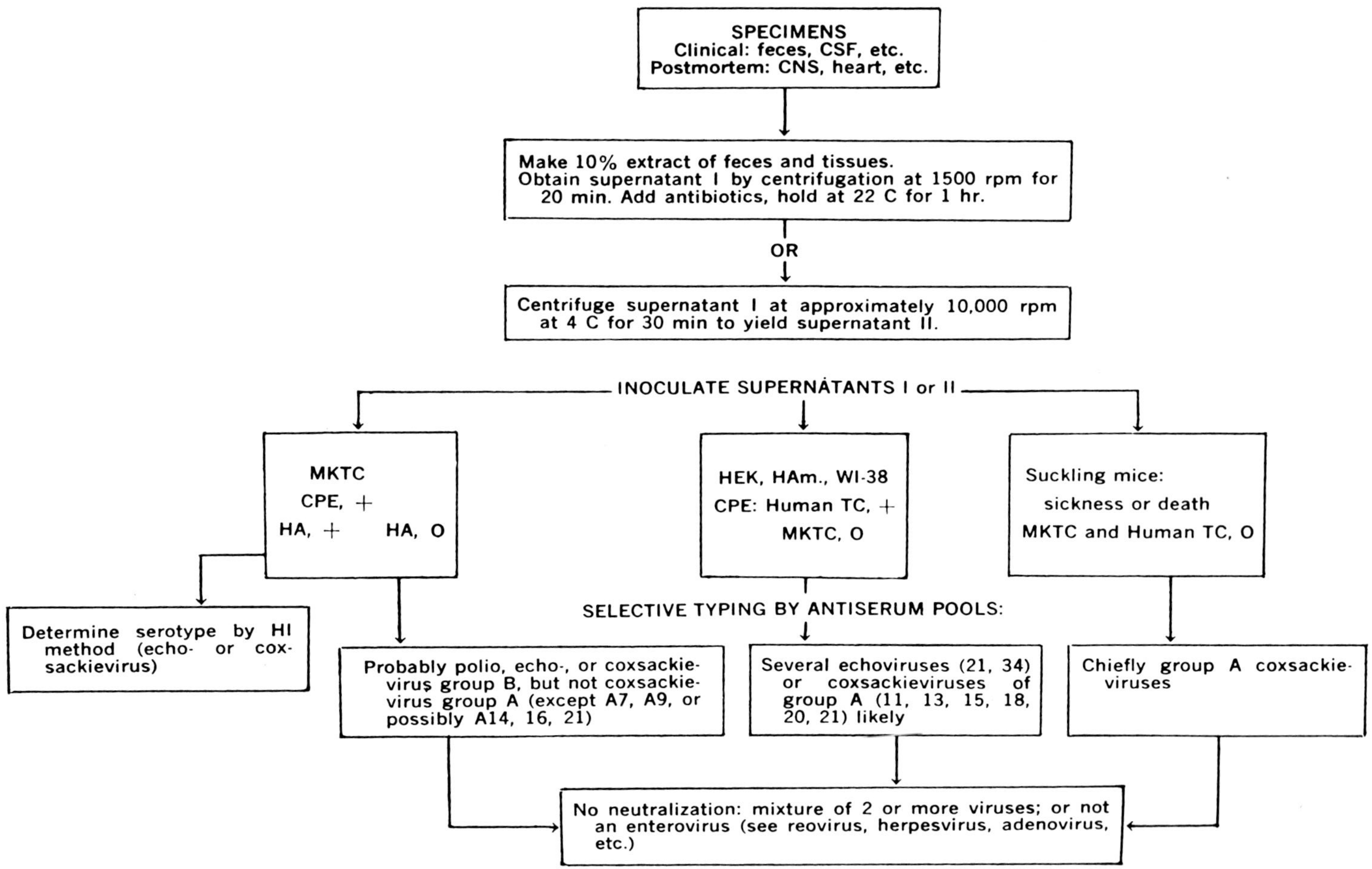

Code:
MKTC = Rhesus or African green monkey kidney tissue cultures
HEK = Human embryonic kidney tissue cultures.
HAm. = Human amnion tissue cultures.
WI-38 = Human embryonic lung tissue cultures.

FIGURE 13. Flowchart of procedures for isolation and identification of enteroviruses. (From Melnick, J. L. and Wenner, H. A., in *Diagnostic Procedures for Viral and Rickettsial Infections*, 4th ed., Lennette, E. H. and Schmidt, N. J., Eds., American Public Health Association, New York, 1969, 529. With permission.)

Table 7
COLLECTION OF CLINICAL SPECIMENS FOR VIRUS STUDIES

Specimen	Amount	Suspension agent[a]	Kind of container	Transport	Comments
Feces					
Stool	4—8 g	None	Ointment box	Freeze (≤0°C)	
Rectal swab	2 swabs[b]	BSS	Tube		Swabs should show visible fecal particles
Pharynx					
Gargle	10—15 ml	BSS	Tube	Freeze (≤0°C)	
Swab	2 swabs[b]	BSS	Tube		
Blood					
Whole	≥10 ml	None	Tube	Do not freeze	
Serum	5—10 ml	None	Tube	Freeze	
Cerebrospinal fluid	≥ 3 ml	None	Tube	Freeze	
Vesicle fluid	0.2 ml[c]	BSS	Tube	Freeze	
Pericardial fluid, ascitic fluid	≥ 1 ml	None	Tube	Freeze	
Urine	10 ml	None	Tube	Freeze	
Autopsy					
Blood	≥ 5 ml	None	Tubes and/or sterile containers	Freeze	Use sterile instruments for collection of body fluids and tissues
Cerebrospinal fluid	≥ 3 ml	None		Freeze	
Effusion fluids	≥ 3 ml or 3 g	None		Freeze	
Visceral tissues	≥ 3 ml or 3 g	None		Freeze	

Note: Place all samples in "sterile" containers. An exception may be made for feces, which may be collected in previously unused, clean containers.

[a] BSS, balanced salt solutions (e.g., Hanks' or Earle's BSS). When these are unavailable, any sterile nutrient (peptone) broth may be used.

[b] BSS is added in small amounts to prevent drying.

[c] Collect by aspiration with a tuberculin syringe. Cleanse the area with ether or acetone. Arrange for the fastest available transportation of specimens. If wet ice is used in transport, specimens must be in watertight containers.

From Wenner, H. A. and Ray, G. C., in *Brennemann's Practice of Pediatrics,* Vol. 2, Kelley, V. Ed., Harper & Row, New York, 1974, 1. With permission.

worker choosing among a variety of serodiagnostic procedures must initially determine which method(s) facilitates a presumptive diagnosis by ruling out other agents, e.g., complement fixation (CF) tests for group antigens for adenoviruses.

The serologic procedures used (other than the neutralization test, which is the standard reference test) usually reflect the investigator's interest and the availability of reagents, animals, or tissue cultures. The choice of test is influenced by whether the virus immunotype can cause agglutination of red blood cells; grow in tissue cultures; and produce high-titer antigens detectable by CF, immunodiffusion, immunoelectrophoresis, immunoelectron microscopy, or latex agglutination. Purified viral preparations, if available, reduce the number of nonspecific reactions which are inherent in all of these tests. Finally, the role of heat-labile and heat-stable inhibitors present in human sera must be considered in interpreting the results of any of these tests. The routine inclusion of known positive and negative control sera in each set of serologic tests is essential.

Pathogenicity in Experimental Animals

Early studies of the morbid anatomy of coxsackievirus infections of mice have been reviewed thoroughly by Dalldorf and Melnick.[96] Interest in the coxsackieviruses currently centers around the realization that these viruses may cause a variety of human diseases, including those of the cardiovascular system and the pancreas. These latter disease syndromes have been well established in experimental studies of virus-infected animals.

The coxsackieviruses are known to infect and cause disease in newborn and adult mice and a variety of other vertebrate species, including monkeys, chimpanzees, rats, squirrels, dogs, swine, and bats. Nevertheless, their natural host appears to be man. No known significant animal or arthropod reservoir exists in nature, although antibodies or inhibitors to some of the coxsackieviruses have been found in the serum of cows,[96] swine,[133] and rats.[96] Because studies of the pathogenicity of the coxsackievirus group in humans are almost always retrospective, experimental animal model systems have been used to define the histopathologic response following infection with these viruses. Animal responses to infection by these viruses include generalized myositis, paralysis, encephalitis, myocarditis, degeneration of fat tissue, pancreatitis, hepatitis, nephritis, enteritis, involvement of the reticuloendothelial system, and congenital anomalies.

GENERAL REFERENCES

Dalldorf, G. and Melnick, J. L., Coxsackie viruses, in *Viral and Rickettsial Infections of Man,* 4th ed., Horsfall, F. L. and Tamm, I., Eds., J.B. Lippincott, Philadelphia, 1965, 474.

Melnick, J. L. and Wenner, H. A., Enteroviruses, in *Diagnostic Procedures for Viral and Rickettsial Infections,* 4th ed., Lennette, E. H. and Schmidt, N. J., Eds., American Public Health Association, New York 1969, 529.

Sergiescu, D., Horodniceanu, F., and Aubert-Combiescu, A., The use of inhibitors in the study of picornavirus genetics, *Prog. Med. Virol.,* 14, 123, 1972.

Abelmann, W. H., Viral myocarditis and its sequalae, *Annu. Rev. Med.,* 24, 145, 1973.

Gear, J. H. S. and Measroch, V., Coxsackievirus infection of the newborn, *Prog. Med. Virol.,* 15, 42, 1973.

Lerner, A. M. and Wilson, F. M., Virus myocardiopathy, *Prog. Med. Virol.,* 15, 63, 1973.

Levintow, L., The reproduction of picornaviruses, in *Comprehensive Virology,* Vol. 2, Frankel-Conrat, H. and Wagner, R. R., Eds., Plenum Press, New York, 1974, 209.

Wenner, H. A. and Ray, C. G., Diseases associated with coxsackie and echoviruses, in *Brennemann's Practice of Pediatrics,* Vol. 2, Kelley, V., Ed., Harper & Row, New York, 1974, 1.

Craighead, J. E., The role of viruses in the pathogenesis of pancreatic disease and diabetes mellitus, *Prog. Med. Virol.,* 19, 161, 1975.

Rueckert, R. R., On the structure and morphogenesis of picornaviruses, in *Comprehensive Virology,* Vol. 6, Frankel-Conrat, H. and Wagner, R. R., Eds., Plenum Press, New York, 1976, 131.

REFERENCES

1. **Dalldorf, G.**, The coxsackie group of viruses, *Science,* 110, 594, 1949.
2. **Gifford, R. and Dalldorf, G.**, The morbid anatomy of experimental coxsackie virus infection, *Am. J. Pathol.,* 27, 1047, 1951.
3. **Godman, G. C., Bunting, H., and Melnick, J. L.**, The histopathology of coxsackievirus infection in mice. I. Morphologic observations with four different viral types, *Am. J. Pathol.,* 28, 223, 1952.

4. **Godman, G. C., Bunting, H., and Melnick, J. L.**, The histopathology of coxsackievirus infections in mice. II. Histochemical observations on the lesions in muscle and fat, *Am. J. Pathol.*, 28, 583, 1952.
5. **Melnick, J. L. and Godman, G. C.**, Pathogenesis of coxsackie virus infection. Multiplication of virus and evolution of the muscle lesion in mice, *J. Exp. Med.*, 93, 247, 1951.
6. **Drachman, D. B., Weiner, L. W., Price, D. L., and Chase, J.**, Experimental arthrogryposis caused by viral myopathy, *Arch. Neurol.* (Chicago), 33, 362, 1976.
7. **Brown, G. C. and Karunas, R. S.**, Relationship of congenital anomalies and maternal infection with selected enteroviruses, *Am. J. Epidemiol.*, 95, 207, 1972.
8. **Pearson, C. M.**, Editorial: Myopathy with viral-like structures, *N. Engl. J. Med.*, 292, 641, 1975.
9. **Tang, T. T., Sedmak, G. V., Siegesmund, K. A., and McCreadie, S. R.**, Chronic myopathy associated with coxsackievirus type A9. A combined electron microscopical and viral isolation study, *N. Engl. J. Med.*, 292, 608, 1975.
10. **Beatrice, S. T.**, Immunogenic and Antigenic Properties of Coxsackievirus B3 Virions and Substructures, *Ph.D. thesis*, Hahnemann Medical College, Philadelphia, 1976.
11. **Beatrice, S. T. and Crowell, R. L.**, Induction of type-specific neutralizing and precipitating antibodies to coxsackievirus B3 by the viral polypeptide, VP2, *Abstr. Annu. Meet. Am. Soc. Microbiol.*, p. 211, 1976.
12. **Ginsberg, H. S.**, Subunit viral vaccines, in *Viral Immunology and Immunopathology*, Notkins, A. L., Ed., Academic Press, New York, 1975, 317.
13. **Dunker, A. K.**, Structure of isometric viruses containing nonidentical polypeptide chains, *J. Virol.*, 14, 878, 1974.
14. **Philipson, L., Beatrice, S. T., and Crowell, R. L.**, A structural model for picornaviruses as suggested from an analysis of urea-degraded virions and procapsids of coxsackievirus B3, *Virology*, 54, 69, 1973.
15. **Rueckert, R. R.**, On the structure and morphogenesis of picornaviruses, in *Comprehensive Virology*, Vol. 6, Frankel-Conrat, H. and Wagner, R. R., Eds., Plenum Press, New York, 1976.
16. **Crowell, R. L. and Philipson, L.**, Specific alterations of coxsackievirus B3 eluted from HeLa cells, *J. Virol.*, 8, 509, 1971.
17. **Mattern, C. F. T.**, Some physical and chemical properties of coxsackieviruses A9 and A10, *Virology*, 17, 520, 1962.
18. **Roesing, T. G., Toselli, P. A., and Crowell, R. L.**, Elution and uncoating of coxsackievirus B3 by isolated HeLa cell plasma membranes, *J. Virol.*, 15, 654, 1975.
19. **Crowell, R. L.**, Comparative generic characteristics of picornavirus-receptor interactions, in *Cell Membrane Receptors for Viruses, Antigens and Antibodies, Polypeptide Hormones and Small Molecules*, Miles Int. Symp. Ser. No. 9, Beers, R. F., Jr., and Bassett, E. G., Eds., Raven Press, New York, 1976, 179.
20. **Frommhagen, L. H.**, The separation and physiocochemical properties of the C and D antigens of coxsackievirus, *J. Immunol.*, 95, 818, 1965.
21. **Jamison, R. M. and Mayor, H. D.**, Comparative study of seven picornaviruses of man, *J. Bacteriol.*, 91, 1971, 1966.
22. **Cords, C. E., James, C. G., and McLaren, L. C.**, Alteration of capsid proteins of coxsackievirus A 13 by low ionic concentrations, *J. Virol.*, 15, 244, 1975.
23. **Noll, H. and Younger, J. S.**, Virus-lipid interactions. II. The mechanism of adsorption of lipophilic viruses to water-insoluble polar lipids, *Virology*, 8, 319, 1959.
24. **Drzeniek, R. and Bilello, P.**, Absence of glycoproteins in poliovirus particles, *J. Gen. Virol.*, 25, 125, 1974.
25. **Mak, T. W. and Rueckert, R. R.**, Inactivation of mouse Elberfeld virus by periodate, *Intervirology*, 4, 129, 1974.
26. **Lerner, A. M., Bailey, E. J., and Tillotson, J. R.**, Enterovirus hemmagglutination: inhibition by several enzymes and sugars, *J. Immunol.*, 95, 1111, 1966.
27. **Tillotson, J. R. and Lerner, A. M.**, Effect of periodate oxidation on hemagglutinating and antibody-producing capacities of certain enteroviruses and reoviruses, *Proc. Natl. Acad. Sci. U.S.A.*, 56, 1143, 1966.
28. **Crowell, R. L.**, Specific cell-surface alteration by enteroviruses as reflected by viral-attachment interference, *J. Bacteriol.*, 91, 198, 1966.
29. **Butterworth, B. E. and Korant, B. D.**, Characterization of the large picornaviral polypeptides produced in the presence of zinc ion, *J. Virol.*, 14, 282, 1974.
30. **Kiehn, E. D. and Holland, J. J.**, Synthesis and cleavage of enterovirus polypeptides in mammalian cells, *J. Virol.*, 5, 358, 1970.
31. **Schaffer, F. L.**, Physical and chemical properties and infectivity of RNA from animal viruses, *Cold Spring Harbor Symp. Quant. Biol.*, 27, 89, 1962.

32. **Holland, J. J.**, Altered base ratios in RNA synthesized during enterovirus infection of human cells, *Proc. Natl. Acad. Sci. U.S.A.*, 48, 2044, 1962.
33. **Spector, D. H. and Baltimore, D.**, Requirement of 3′-terminal poly (adenylic acid) for the infectivity of poliovirus RNA, *Proc. Natl. Acad. Sci. U.S.A.*, 71, 2983, 1974.
34. **Rowlands, D. J., Shirley, M. W., Sanger, D. V., and Brown, F.**, A high density component in several vertebrate enteroviruses, *J. Gen. Virol.*, 29, 223, 1975.
35. **Schaffer, F. L. and Frommhagen, L. H.**, Similarities of biophysical properties of several human enteroviruses as shown by density gradient ultracentrifugation of mixtures of the viruses, *Virology*, 25, 662, 1965.
36. **McLaren, L. C., Scaletti, J. V., and James, C. G.**, Isolation and properties of enterovirus receptors, *Wistar Inst. Symp. Monogr.*, 8, 123, 1968.
37. **Zajac, I. and Crowell, R. L.**, Location and regeneration of enterovirus receptors of HeLa cells, *J. Bacteriol.*, 89, 1097, 1965.
38. **Zajac, I. and Crowell, R. L.**, Effect of enzymes on the interaction of enteroviruses with living HeLa cells, *J. Bacteriol.*, 89, 574, 1965.
39. **Axler, D. A. and Crowell, R. L.**, Effect of anticellular serum on the attachment of enteroviruses to HeLa cells, *J. Virol.*, 2, 813, 1968.
40. **Crowell, R. L.**, Specific viral interference in HeLa cell cultures chronically infected with Coxsackie B5 virus, *J. Bacteriol.*, 86, 517, 1963.
41. **Crowell, R. L. and Syverton, J. T.**, The mammalian cell-virus relationship. VI. Sustained infection of HeLa cells by coxsackie B3 virus and effect on superinfection, *J. Exp. Med.*, 113, 419, 1961.
42. **Levitt, N. H. and Crowell, R. L.**, Comparative studies of the regeneration of HeLa cell receptors for poliovirus T1 and coxsackievirus B3, *J. Virol.*, 1, 693, 1967.
43. **Lonberg-Holm, K. and Philipson, L.**, Early interaction between animal viruses and cells, *Monogr. Virol.*, 9, 1, 1974.
44. **Quersin-Thiry, L. and Nihoul, E.**, Interactions between cellular extracts and animal viruses. II. Evidence for the presence of different inactivators corresponding to different viruses, *Acta Virol.* (Engl. Ed.), 5, 283, 1961.
45. **Soloviev, V. D., Krispin, T. I., Zaslavsky, V. G., and Agol, V. I.**, Mechanism of resistance to enteroviruses of some primate cells in tissue culture, *J. Virol.*, 2, 553, 1968.
46. **Zajac, I. and Crowell, R. L.**, Differential inhibition of attachment and eclipse activities of HeLa cells for enteroviruses, *J. Virol.*, 3, 422, 1969.
47. **Lonberg-Holm, K., Crowell, R. L., and Philipson, L.**, Unrelated animal viruses share receptors, *Nature, London* 259, 679, 1976.
48. **Hall, L. and Rueckert, R. R.**, Infection of mouse fibroblasts by cardioviruses: premature uncoating and its prevention by elevated pH and magnesium chloride, *Virology*, 43, 152, 1971.
49. **Baltimore, D.**, Expression of animal virus genomes, *Bacteriol. Rev.*, 35, 235, 1971.
50. **Jezequel, A. M. and Steiner, J. W.**, Some ultrastructural and histochemical aspects of coxsackie virus-cell interactions, *Lab. Invest.*, 15, 1055, 1966.
51. **Levintow, L.**, The reproduction of picornaviruses, in *Comprehensive Virology*, Vol. 2, Fraenkel-Conrat, H. and Wagner, R. R., Eds., Plenum Press, New York, 1974, 109.
52. **Landau, B. J., Crowell, R. L., Boclair, C. W., and Zajac, B. A.**, The permissiveness of differentiating mouse muscle-cell cultures to infection by group A coxsackievirus types 1 and 5, *Proc. Soc. Exp. Biol. Med.*, 141, 753, 1972.
53. **Hecker, W., Meyer, J., Boeni, R., and Bienz, K.**, Pinocytotic uptake and intralysosomal crystal formation of coxsackievirus A9 in monkey kidney cells. An electron microscopic autoradiographic study, *Arch. Gesamte Virusforsch.*, 46, 167, 1974.
54. **Stuart, D. C. and Fogh, J.**, Micromorphology of FL cells infected with polio and coxsackie viruses, *Virology*, 13, 177, 1961.
55. **Shahrabadi, M. S. and Morgante, O.**, Tubular aggregation of coxsackievirus B5 in HeLa cells, *Virology*, 80, 434—440, 1977.
56. **Crowell, R. L. and Syverton, J. T.**, The viral range in vitro of a malignant, human epithelial cell (strain HeLa, Gey). IV. The cytopathogenicity of C viruses, *J. Immunol.*, 74, 169, 1955.
57. **Buthala, D. A.**, Effect of carbon dioxide concentration on ability of coxsackie A21 to form plaques, *J. Bacteriol.*, 86, 1356, 1963.
58. **Came, P. E. and Crowell, R. L.**, Studies of resistance of fetal mouse tissues in culture to coxsackie group A viruses, *Virology*, 23, 542, 1964.
59. **Khesin, Y. E. and Amchenkova, A. M.**, On the relationship between the enzymatic composition and the susceptibility of the cell to viruses, *J. Gen. Virol.*, 15, 105, 1972.
60. **Zuschek, F.**, Influence of potassium on coxsackie B2 virus growth in tissue cultures, *Proc. Soc. Exp. Biol. Med.*, 109, 540, 1962.
61. **Sergiescu, D., Horodniceanu, F., and Aubert-Combiescu, A.**, The use of inhibitors in the study of picornavirus genetics, *Prog. Med. Virol.*, 14, 123, 1972.

62. **Takemoto, K. K.**, Plaque mutants of animal viruses, *Prog. Med. Virol.*, 8, 314, 1966.
63. **Huang, A. S.**, Defective interfering viruses, *Annu. Rev. Microbiol.*, 27, 101, 1973.
64. **Rager-Zisman, B. and Allison, A. C.**, Effects of immunosuppression on coxsackie B3 virus infection in mice, and passive protection by circulating antibody, *J. Gen. Virol.*, 19, 339, 1973.
65. **Habel, K., Hornibrook, J. W., Gregg, N. C., Silverberg, R. J., and Takemoto, K. K.**, The effect of anticellular sera on virus multiplication in tissue culture, *Virology*, 5, 7, 1958.
66. **Much, D. H. and Zajac, I.**, The effect of an antireceptor serum on mammalian cell lines, *J. Gen. Virol.*, 23, 205, 1974.
67. **Quersin-Thiry, L.**, Action of anticellular sera on virus infections. I. Influence on homologous tissue cultures infected with various viruses, *J. Immunol.*, 81, 253, 1958.
68. **Galabov, A., Shindarov, L., Vassilev, G., and Vassileva, R.**, Inhibitor effect of *N*-phenyl-*N'*-aryl or alkythiourea derivatives on the multiplication of some picornaviruses (coxsackie B1, echo 19, and foot-and-mouth disease virus) in cell cultures, *Arch. Gesamte Virusforsch.*, 38, 159, 1972.
69. **Paget, C. J., Kisner, K., Stone, R. L., and Delong, D. C.**, Heterocyclic substituted ureas. I. Immunosuppression and virus inhibition by benzimidazolureas, *J. Med. Chem.*, 12, 1010, 1969.
70. **Shannon, W. M., Arnett, G., and Schabel, F. M., Jr.**, 3-Deazauridine: inhibition of ribonucleic acid virus-induced cytopathogenic effects in vitro, *Antimicrob. Agents Chemother.*, 2, 159, 1972.
71. **Tamm, I. and Eggers, H. J.**, Specific inhibition of replication of animal viruses. Chemical inhibitors are helping to elucidate virus-specific processes in virus reproduction, *Science*, 142, 24, 1963.
72. **Yamazi, Y., Takahashi, M., and Todome, Y.**, Inhibition of poliovirus by effect of a methylthiopyrimidine derivative, *Proc. Soc. Exp. Biol. Med.*, 133, 674, 1970.
73. **Eggers, H. J. and Tamm, I.**, On the mechanism of selective inhibition of enterovirus multiplication by 2-(α-hydroxybenzyl)-benzimidazole, *Virology*, 18, 426, 1962.
74. **Eggers, H. J. and Tamm, I.**, Coxsackie A9 virus: mutation from drug dependence to drug independence, *Science*, 148, 97, 1965.
75. **Tamm, I. and Eggers, H. J.**, Differences in the selective virus inhibitory action of 2-(α-hydroxybenzyl)-benzimidazole and guanidine. HCl, *Virology*, 18, 439, 1962.
76. **Abraham, A.**, Isolation of two enteroviruses, possible "prime" strains of coxsackie A type 17 virus, *Proc. Soc. Exp. Biol. Med.*, 125, 195, 1967.
77. **Fujimiya, Y., Suto, T., Morita, M., and Ishida, N.**, Antigenic analyses of sequential isolates of coxsackie A16 virus from 1964 to 1970 in Japan, *Jpn. J. Microbiol.*, 18, 463, 1974.
78. **Harris, T. J. R., Doel, T. R., and Brown, F.**, Molecular aspects of the antigenic variation of swine vesicular disease and coxsackie B5 viruses, *J. Gen. Virol.*, 35, 299, 1977.
79. **Richter, F. A., Macpherson, L. W., Campbell, J. B., and Labzoffsky, N. A.**, Studies on intratypic variants of coxsackie B1 virus, *Arch. Gesamte Virusforsch.*, 38, 77, 1972.
80. **Schmidt, N. J. and Lennette, E. H.**, Antigenic variants of coxsackievirus type A24, *Am. J. Epidemiol.*, 91, 99, 1970.
81. **Tagaya, I. and Tachibana, K.**, Epidemic of hand-foot- and mouth disease in Japan, 1972—1973: difference in epidemiologic and virologic features from the previous one, *Jpn. J. Med. Sci. Biol.*, 28, 231, 1975.
82. **Voiculescu, C., Dinculescu, M., Gaicu, N., Georgescu, I., Epuran, E., Bruker, J., and Radu, J.**, Use of type specific serum for differentiation of the rates of adsorption on cultured cell receptors, of several strains of enteroviruses (polio 3, echo 1 and 7, coxsackie A9, B1, and B5), *Stud. Cercet. Inframicrobiol.*, 19, 33, 1968.
83. **Choppin, P. W. and Eggers, H. J.**, Heterogeneity of coxsackie B4 virus: two kinds of particles which differ in antibody sensitivity, growth rate, and plaque size, *Virology*, 18, 470, 1962.
84. **Koroleva, G. A., Lashkevich, V. A., Voroshilova, M. K., and Frolova, M. P.**, The effect of dextran sulfate upon plaque formation and virulence of coxsackie A7, A14, and A16, viruses, *Arch. Gesamte Virusforsch.*, 23, 175, 1968.
85. **Wallis, C. and Melnick, J.**, Mechanism of enhancement of virus plaques by cationic polymers, *J. Virol.*, 2, 267, 1968.
86. **Wallis, C., Morales, F., Powell, J., and Melnick, J. L.**, Plaque enhancement of enteroviruses by magnesium chloride, cysteine and pancreatin, *J. Bacteriol.*, 91, 1932, 1966.
87. **Gevaudan, P., Charrel, J., Pieroni, G., and Gevaudan, M. J.**, Nouveau variant froid de coxsackievirus B5. III. Pouvoir protecteur et aptitude a induire de l'interferon, *C. R. Seances Soc. Biol. Paris*, 164, 1805, 1970.
88. **Gevaudan, P., Pieroni, G., Gevaudan, M. J., and Charrel, J.**, La maturation du coxsackievirus. II. Intervention d'une rupture nucleasique thermolimitative, *Ann. Inst. Pasteur Paris*, 122, 115, 1972.
89. **Voicoulesco, C., Georgesco, I., and Epuran, E.**, Variable paralytogenic ability of coxsackie virus strain A of various types and origins on 16-day-old mice. Correlation with certain temperature markers, *Arch. Roum. Pathol. Exp. Microbiol.*, 26, 599, 1967.
90. **Trousdale, M. D., Paque, R. E., and Gauntt, C. J.**,Isolation of coxsackievirus B3 temperature-sensitive mutants and their assignment to complementation groups, *Biochem. Biophys. Res. Commun.*, 76, 368, 1977.

91. **Yamamato, S. and Shingu, M.**, Some properties of echo and coxsackie viruses in tissue culture and variations by heat, *Kurume Med. J.*, 9, 70, 1962.
92. **Berquist, K. R. and Fritts, S. P.**, Hemagglutinating activity of enteroviruses recovered in two primary cell lines and a stable line, *Appl. Microbiol.*, 16, 703, 1968.
93. **Kern, J. and Rosen, L.**, Identification of enteroviruses by hemagglutination-inhibition, *J. Bacteriol.*, 91, 1936, 1966.
94. **Shirobokov, V. P.**, Differentiation of coxsackieviruses based on the character of adsorption onto bentonite, *Acta Virol.* (Engl. Ed.), 12, 185, 1968.
95. **Holland, J. J. and Cords, C. E.**, Maturation of poliovirus RNA with capsid protein coded by heterologous enteroviruses, *Proc. Natl. Acad. Sci. U.S.A.*, 51, 1082, 1964.
96. **Dalldorf, G. and Melnick, J. L.**, Coxsackie viruses, in *Viral and Rickettsial Infections of Man*, 4th ed., Horsfall, F. L. and Tamm, I., Eds., J. B. Lippincott, Philadelphia, 1965, 474.
97. **Abelmann, W. H.**, Viral myocarditis and its sequelae, *Annu. Rev. Med.*, 24, 145, 1973.
98. **Gear, J. H. S. and Measroch, V.**, Coxsackievirus infection of the newborn, *Prog. Med. Virol.*, 15, 42, 1973.
99. **Lerner, A. M. and Wilson, F. M.**, Virus myocardiopathy, *Prog. Med. Virol.*, 15, 63, 1973.
100. **Wenner, H. A. and Ray, C. G.**, Diseases associated with coxsackie and echo viruses, in *Brennemann's Practice of Pediatrics*, Vol. 2, Kelley, V., Ed., Harper & Row, New York, 1974, 1.
101. **Craighead, J. E.**, The role of viruses in the pathogenesis of pancreatic disease and diabetes mellitus, *Prog. Med. Virol.*, 19, 161, 1975.
102. **Kibrick, S.**, Current status of coxsackie and ECHO viruses in human disease, *Prog. Med. Virol.*, 6, 27, 1964.
103. **Sanford, J. P.**, Coxsackie virus and echo virus infections, in *Infectious Diseases*, Hoeprich, P. D., Ed., Harper & Row, New York, 1972, 133.
104. **Gilbert, R. G., Rice, R. C., Bouwer, H., Gerba, C. P., Wallis, C., and Melnick, J. L.**, Wastewater renovation and reuse: virus removal by soil filtration, *Science*, 192, 1004, 1976.
105. **Hill, W. F., Jr., Jakubowski, W., Akin, E. W., and Clarke, N. A.**, Detection of virus in water: sensitivity of the tentative standard method for drinking water, *Appl. Environ. Microbiol.*, 31, 254, 1976.
106. **Sells, C. J., Carpenter, R. L., and Ray, C. G.**, Sequelae of central-nervous-system enterovirus infections, *N. Engl. J. Med.*, 293, 1, 1975.
107. **Coleman, T. J., Taylor, K. W., and Gamble, D. R.**, The development of diabetes following coxsacki B virus infection in mice, *Diabetologia*, 10, 755, 1974.
108. **Huff, J. C., Hierholzer, J. C., and Farris, W. A.**, An "outbreak" of juvenile diabetes mellitus: consideration of a viral etiology, *Am. J. Epidemiol.*, 100, 277, 1974.
109. **Imrie, C. W., Ferguson, J. C., and Sommerville, R. G.**, Coxsackie and mumpsvirus infection in a prospective study of acute pancreatitis, *Gut*, 18, 53, 1977.
110. **Samantray, S. K., Christopher, S., Mukundan, P., and Johnson, S. C.**, Lack of relationship between viruses and human diabetes mellitus, *Aust. N.Z. J. Med.*, 7, 139, 1977.
111. **Nelson, P. G., Pyke, D. A., and Gamble, D. R.**, Viruses and the aetiology of diabetes: a study in identical twins, *Br. Med. J.*, 4, 249, 1975.
112. **Yoon, J. W. and Notkins, A. L.**, Virus-induced diabetes mellitus. VI. Genetically determined host differences in the replication of encephalomyocarditis virus in pancreatic beta cells, *J. Exp. Med.*, 143, 1170, 1976.
113. **Burch, G. E., Sun, S. C., Chu, K. C., Sohal, R. S., and Colcolough, H. L.**, Interstitial and coxsackievirus B myocarditis in infants and children. A comparative histologic and immunofluorescent study of 50 autopsied hearts, *JAMA*, 203, 1, 1968.
114. **Grist, N. R. and Bell, E. J.**, A six-year study of coxsackievirus B infections in heart disease, *J. Hyg.*, 73, 165, 1974.
115. **Hirschman, S. Z. and Hammer, G. S.**, Coxsackie virus myopericarditis. A microbiological and clinical review, *Am. J. Cardiol.*, 34, 224, 1974.
116. **Lerner, A. M., Wilson, F. M., and Reyes, M. P.**, Enteroviruses and the heart (with special emphasis on the probable role of coxsackievirus group B, types 1—5). II. Observations in humans, *Mod. Concepts Cardiovasc. Dis.*, 44, 11, 1975.
117. **Schmidt, N. J., Magoffin, R. L., and Lennette, E. H.**, Association of group B coxsackieviruses with cases of pericarditis, myocarditis or pleurodynia by demonstration of immunoglobulin M antibody, *Infect. Immun.*, 8, 341, 1973.
118. **Schmidt, N. J. and Lennette, E. H.**, Advances in the serodiagnosis of viral infections, *Prog. Med. Virol.*, 15, 244, 1973.
119. **Melnick, J. L. and Wenner, H. A.**, Enteroviruses, in *Diagnostic Procedures for Viral and Rickettsial Infections*, 4th ed., Lennette, E. H., and Schmidt, N. J., Eds., American Public Health Association, New York, 1969, 529.

120. **Hsiung, G. D.**, Laboratory diagnosis of viral infections: general principles and recent developments, *Mt. Sinai J. Med.*, 44, 1, 1977.
121. **MacWilliam, K. M. and Cooper, M. A.**, Antibody levels in human sera measured by the fluorescent-antibody technique against the coxsackie B virus types 1—5 grown in HEp-2 cells compared with results obtained by neutralization, *J. Clin. Pathol.*, 27, 825, 1974.
122. **Melnick, J. L., Tagaya, I., and von Magnus, H.**, Enteroviruses, 69, 70, and 71, *Intervirology*, 4, 369, 1974.
123. **Wenner, H. and Lenahan, M. F.**, Propagation of group A coxsackieviruses in tissue culture. II. Some interactions between virus and mammalian cells, *Yale J. Biol. Med.*, 34, 421, 1961.
124. **Crowell, R. L. and Goldberg, B.**, Propagation and Assay of Group A Coxsackieviruses in RD cells, *Abstr. Annu. Meet. Am. Soc. Microbiol.*, p. 208, 1974.
125. **Lehmann-Grube, F. and Syverton, J. T.**, Pathogenicity for suckling mice of coxsackie viruses adapted to human amnion cells, *J. Exp. Med.*, 113, 811, 1961.
126. **Schmidt, N. J., Ho, H. H., and Lennette, E. H.**, Propagation and isolation of group A coxsackieviruses in RD cells, *J. Clin. Microbiol.*, 2, 183, 1975.
127. **Cooney, M. K.**, Relative efficiency of cell cultures for detection of viruses, *Health Lab. Sci.*, 10, 294, 1973.
128. **Righthand, V. F. and Hughes, P. F.**, Isolation of a specific enhancer for echovirus 6 from uninfected permissive host cells, *Infect. Immun.*, 9, 134, 1974.
129. **Hughes, J. H., Gnau, J. M., Hilty, M. D., Chema, S., Ottolenghi, A. C., and Hamparian, V. V.**, Picornaviruses: rapid differentiation and identification by immune electromicroscopy and immunodiffusion, *J. Med. Microbiol.*, 10, 203, 1977.
130. **Melnick, J. L. and Hampil, B.**, WHO collaborative studies on enterovirus reference antisera (fourth report), *Bull. WHO*, 48, 381, 1973.
131. **Melnick, J. L., Schmidt, N. J., Hampil, B., and Ho, H. H.**, Lyophilized combination pools of enterovirus equine antisera: preparation and test procedures for the idenfication of field strains of 19 Group A coxsackievirus serotypes, *Interviology*, 8, 172, 1977.
132. **Wenner, H. A.**, The enteroviruses, *Am. J. Clin. Pathol.*, 57, 751, 1972.
133. **Brown, F., Wild, T. F., Rowe, L. W., Underwood, B. O., and Harris, T. J. R.**, Comparison of swine vesicular disease virus and coxsackie B5 virus by serological and RNA hybridization methods, *J. Gen. Virol.*, 31, 231, 1976.
134. **Goldberg, R. J. and Crowell, R. L.**, Susceptibility of differentiating muscle cells of the fetal mouse in culture to coxsackievirus A13, *J. Virol.*, 7, 759, 1971.
135. **Crowell, R. L. and Goldberg, R. B.**, unpublished data.
136. **Stoneback, T. A.**, Development of a Plaque Assay for Coxsackievirus A5 by Use of a Human Rhabdomyosarcoma Cell Line and Studies of Virus Attachment, M.S. thesis, Hahnemann Medical College, Philadelphia, 1975.
137. **Dunker, A. K. and Rueckert, R. R.**, Fragments generated by pH dissociation of ME-virus and their relation to the structure of the virion, *J. Mol. Biol.*, 58, 217, 1971.
138. **Talbot, P. and Brown, F.**, A model for foot-and-mouth disease virus, *J. Gen. Virol.*, 15, 163, 1972.

PICORNAVIRIDAE: ENTEROVIRUSES — ECHOVIRUSES

C.A. Phillips

Introduction

The echoviruses were first isolated in tissue culture during the search for poliovirus in stool specimens of patients.[1,2] This isolation followed the discovery by Enders et al.[3] that cultures of human and monkey cells could be used for the growth of polioviruses. The echoviruses were differentiated from the previously described polioviruses and coxsackieviruses because they were isolated from patients with no clinical disease and were nonpathogenic for newborn mice. Being without clinical manifestation of disease, they were originally referred to as "orphan" viruses, but the name evolved to echoviruses (or ECHOviruses) because they were, in fact, enteric cytopathogenic human orphan viruses.[4] In recent years, strains of echoviruses have been shown to be pathogenic for newborn mice and capable of producing a variety of clinical syndromes in man, ranging from fatal meningoencephalitis to the common cold. On the other hand, some strains of coxsackieviruses have been shown to be nonpathogenic for mice and have been isolated from clinically asymptomatic patients. It has recently been suggested that all new candidate echoviruses, as well as new coxsackieviruses, be classified as enteroviruses.[5] Although the name echovirus is now an anachronism, it is retained for its historical importance; echoviruses 1 to 34 will retain the designation, but there will be no new echoviruses.

Morphology

The capsid is composed of 32 morphologic subunits in the form of a rhombic triacontrahedron (Figure 1).

Physicochemical Properties

The echoviruses are small particles 20 to 28 nm in diameter, have an RNA core, are ether and acid stable, and are stabilized to 1 hr of heat at 50°C by molar $MgCl_2$ and other salts of divalent cations. They have a density of approximately 1.34 g/ml in cesium chloride.

Virus Replication

Replication of echoviruses takes place in the cell cytoplasm and is independent of host DNA. Virus is adsorbed to cells at specific receptor sites, taken into the cell, and the viral RNA is uncoated. Specific messenger RNA (mRNA) is transcripted from viral nucleic acid, and the single-stranded genomic RNA serves as its own mRNA. This mRNA is then translated, leading to the formation of RNA polymerase, an enzyme necessary for formation of the replicative mRNA. This is double-stranded consisting of a parental strand and a complementary strand. At the same time, RNA polymerase stimulates the formation of inhibitors of both cellular RNA synthesis and cellular protein synthesis. From the double-stranded RNA, single-stranded viral RNA molecules are formed, which may serve as mRNA for the synthesis of structual proteins or the template for continued RNA replication or may become encapsulated and form mature progeny virions. Mature virus particles are finally released when cell lysis occurs.

Pathogenesis

Pathogenesis for echoviruses is similar to that for other enteroviruses. Virus enters the mouth and primary replication takes place in the tonsils, neck lymph nodes, Peyer's patches, and small intestine. Spread of virus may occur either via the blood stream or along peripheral nerve axons to the central nervous system (CNS).

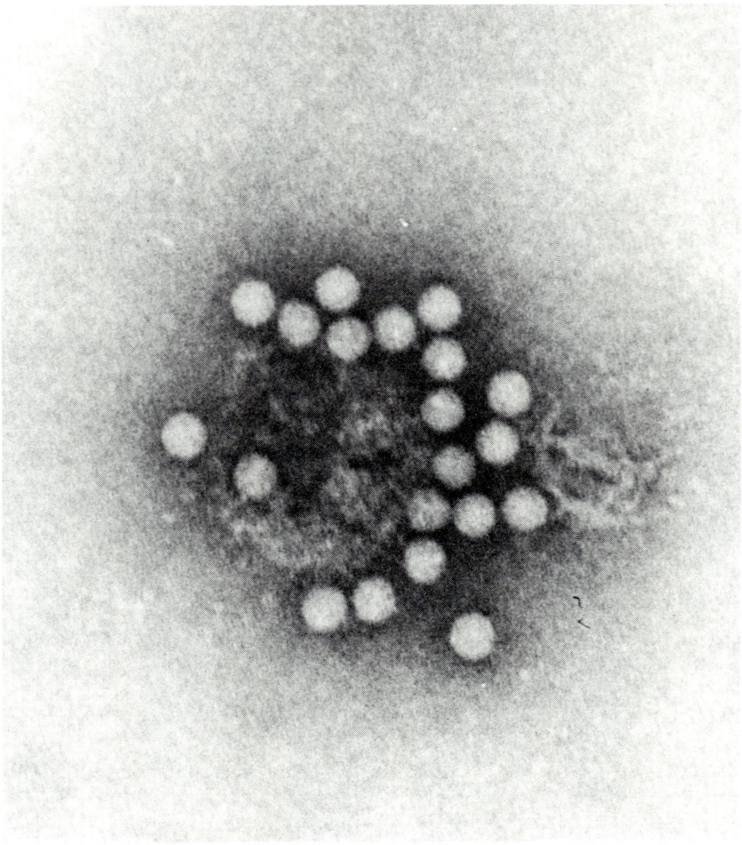

FIGURE 1. Phosphotungstic acid-stained preparation of echovirus type 34. Scale: 1 mm = 4 nm (particles, 24 nm). Magnification × 250,000. (Courtesy of Dr. J. P. Brunschwig, Baylor College of Medicine.)

Virus Strains and Antigenic Properties

The echoviruses are a subgroup of the enteroviruses, which also include the polioviruses and coxsackieviruses. The major difference separating the echoviruses from the other enteroviruses is their lack of pathogenicity for newborn mice, rabbits, and monkeys. To date, 34 echoviruses have been described (Table 1), but only 30 distinct serotypes exist (echovirus 1 is closely related to echovirus 8, echovirus 10 has been reclassified as reovirus 1, echovirus 28 has been reclassified as rhinovirus type 1A, and echovirus 34 is a prime strain of coxsackievirus A24). As has been noted above, no further serotypes will be added to the echovirus group, as inconsistent mouse pathogenicity led to the recommendation in 1969 that new enterovirus candidates be designated as enteroviruses and numbered as such (e.g., enteroviruses 68, 69, 70, and 71 rather than echoviruses 35, 36, 37, and 38).

Clinical Features

Persons infected with echoviruses may manifest no disease or may have a variety of clinical syndromes (Table 1) including paralysis; aseptic meningitis; encephalitis; acute respiratory tract infection; pandemic conjunctivitis; and rashes of all descriptions, including hemorrhagic rash resembling meningococcemia, morbilliform rash resembling

Table 1
ECHOVIRUS ANTIGENIC TYPES AND ASSOCIATED CLINICAL DISEASES

Echovirus type	Virus strain	Diseases associated (for details see virus infection section)	Original investigator	Refs.
1	Farouk	None	Melnick	1, 4, 6
2	Cornelis	Aseptic meningitis	Melnick	1, 4, 6
3	Morrisey	Aseptic meningitis	Melnick	1, 4, 6
4	Pesascek	Aseptic meningitis	Melnick	1, 4, 6
5	Noyce	Aseptic meningtitis	Melnick	1, 4, 6
6	D'Amori	Aseptic meningitis	Melnick	1, 4, 6
6′	Cox	None	Ramos-Alvarez, Sabin	7, 4
6″	Burgess	Aseptic meningitis	Melnick	6, 4
7	Wallace	None	Ramos-Alvarez, Sabin	7, 4
9	Hill	None	Ramos-Alvarez, Sabin	7, 4
11	Gregory	None	Ramos-Alvarez, Sabin	7, 4
12	Travis	None	Hammon, Ludwig	8, 9, 4
13	Del Carmen	None	Hammon, Ludwig	8, 9, 4
14	Tow	Aseptic meningitis	Melnick	10
15	CH 96-51	None	Ormsbee, Melnick	11
16	Harrington	Aseptic meningitis	Kibrick, Enders	10
17	CHHE-29	None	Ramos-Alvarez, Sabin	7, 10
18	Metcalf	Diarrhea	Ramos-Alvarez, Sabin	7, 10
19	Burke	Diarrhea	Ramos-Alvarez, Sabin	7, 10
20	JV-1	Fever	Rosen	12
21	Farina	Aseptic meningitis	Enders, Kibrick	13
22	Harris	Diarrhea	Sabin	13
23	Williamson	Diarrhea	Sabin	13
24	De Camp	Diarrhea	Sabin	13
25	JV-4	Diarrhea	Rosen	13
26	Coronel	None	Hammon	14
27	Bacon	None	Hammon	14
29	JV-10	None	Rosen	13
30	Bastianni	Aseptic meningitis	Duncan, Plager, Lennette	15
31	Caldwell	Aseptic meningitis	Wenner, Lennette, von Magnus	13
32	PR-10	Aseptic meningitis	Branche	13
33	Toluca-3	None	Rosen, Kern	16
34	DN-19	Infectious diarrhea	Melnick	17

rubeola, and scarlatiniform rash resembling scarlet fever. In addition, echoviruses may play a role in the etiology of hemolytic uremic syndrome, pancreatitis, and diarrhea.

Collection and Storage of Specimens

Specimens for echovirus study should be collected as soon after the onset of clinical symptoms as possible. Viruses are most readily isolated from feces or rectal swabs and may be recovered from this source for up to 28 days following the onset of clinical symptoms. They may also be isolated from throat swabs or washings for 7 to 10 days following the onset of clinical symptoms.

Fecal specimens are the most useful specimens for study and should be collected in 4- to 8-g amounts in leakproof containers (ice cream cartons or screw-cap jars may be used). They may be held for several hours at 4°C or they may be frozen (−20°C or colder) for longer periods of time until tested. If stool is unavailable, swabs may be obtained. Rectal swabs should be obtained on moist, sterile, cotton-tipped applicators, placed in an appropriate carrier medium such as veal infusion broth with 0.5% crystallized bovine plasma albumin or 1% fetal calf serum, and transported to the laboratory as soon as possible. Ideally, they should be kept at 4°C during transportation. If tissue culture inoculation is to take place within 4 hr of collection, swabs should be held at 4°C. Otherwise, they should be frozen at −20°C or colder until inoculated. Any balanced salt solution (BSS) such as Melnick's medium A or B or Hanks' or Earle's media may be used in place of veal infusion broth.

Throat swabs are handled in a similar manner. A charcoal medium has been developed for use in transporting specimens from the field to the laboratory.[18] Swabs placed in the medium remain stable at ambient or refrigerator temperatures (4°C) for up to 72 hr. This medium may be very helpful in field studies when specimens are inoculated within 72 hr of collection.

If throat washings are to be tested, they should be collected by having the patient gargle with an isotonic solution such as phosphate-buffered saline. Patients should then expectorate this material into a sterile vessel for subsequent inoculation.

In cases of CNS disease, cerebrospinal fluid (CSF) specimens are most desirable and should be collected in a sterile tube within 5 days of onset of clinical symptoms. They should be kept at 4°C if they are to be inoculated within 4 hr or frozen at −20°C or colder if they are to be kept for a longer period of time. Vesicular fluid, when available, may be collected in a sterile 1-ml syringe with a 26-gauge needle and eluted into a 2-ml aliquot of BSS, or it may be obtained on a cotton-tipped applicator and handled as noted above for other swabs. Post-mortem tissue specimens may occasionally be sent for examination; these should be placed in BSS and frozen until inoculated.

Serum specimens should be collected as soon after onset of clinical symptoms as possible and again at approximately 14 to 28 days after onset of symptoms. Blood should be placed in sterile centrifuge tubes, rimmed, and held at 4°C until clot retraction is adequate (2 to 24 hr). After centrifugation, serum is used for antibody determinations. If virus isolation attempts are to be made from blood, heparinized blood should be used.

Laboratory Diagnosis (See Figure 2)

Isolation in Tissue Culture

Isolation of echoviruses in tissue culture is the method used for diagnosing infection. Most laboratories use a combination of two tube culture systems including primary monkey kidney plus one human tissue culture system from among WI-38, HEK, HeLa, and human amnion cells. This author prefers to use rhesus monkey kidney and WI-38 cells for routine echovirus isolation studies. Tube cultures of monkey kidney, WI-38, HEK, and other cell lines are commercially available from several sources (Flow Laboratories, Rockville, Maryland; Mogul Diagnostics, Grand Island, New York; and Microbiological Associates, Bethesda, Maryland). For most laboratories it is probably most convenient to obtain cultures from one or more of these sources.

As feces is the specimen most likely to yield echoviruses, the following is offered as an example of how a specimen should be handled in the laboratory:

1. A 4- to 8-g specimen is made into a 10% suspension in cold, sterile Hanks' BSS. The suspension is agitated vigorously and allowed to settle at 4°C. The supernatant is then clarified by centrifugation at 2500 rpm for 10 min, separated, and treated with 1000 units of potassium penicillin G and 1000 μg of streptomycin sulfate per milliliter or gentamicin in a final concentration of 200 μg/ml. After incubation at 25°C for 1 hr, the specimen is again clarified by centrifugation (2500 rpm for 1 hr at 4°C) and is ready for inoculation.
2. Tissue culture tubes of primary rhesus monkey kidney and WI-38 cells containing

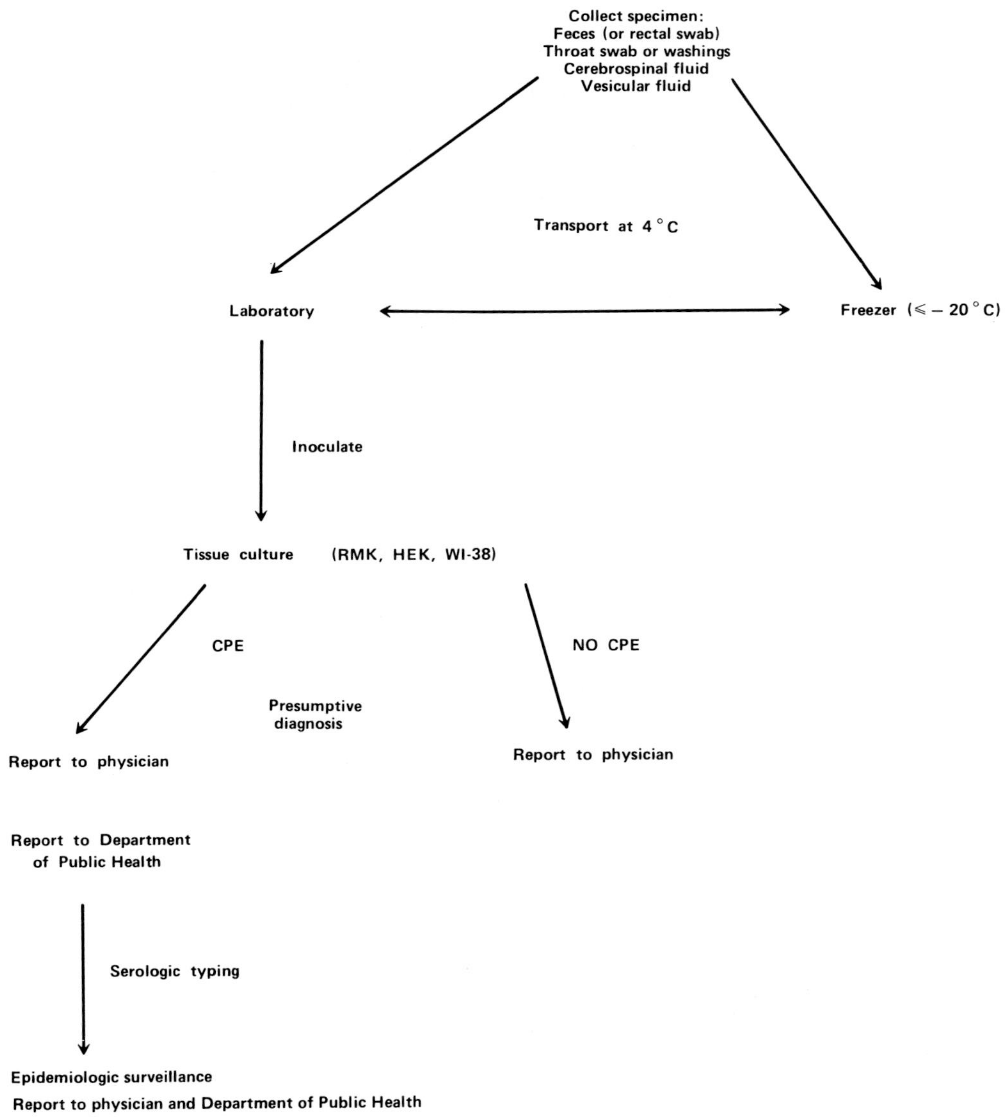

FIGURE 2. Flow chart for virus isolation and identification. CPE, cytopathic effect.

0.9 ml of fresh maintenance medium (Melnick's medium B for monkey and Eagle's for WI-38) with 2% fetal bovine serum (inactivated at 56°C for 30 min), $NaHCO_3$, supplemental L-glutamine for WI-38 cells, and either 100 units of potassium penicillin G and 100 μg of streptomycin sulfate per milliliter or gentamicin in a final concentration of 200 μg per ml are inoculated aseptically with 0.1 ml of the specimen.

3. Tubes are placed in an incubator and held stationary at 37°C.
4. Tubes are observed on day 1 for evidence of early cytopathic effect (CPE) or toxicity and thereafter at least every other day for the appearance of typical CPE. This CPE is evidenced by rounding up and shrinkage of cells and nuclear pyknosis. The cells may become refractile. This CPE usually begins at the periphery of the cell sheet and eventually involves the entire culture (Figures 3 to 8). (Note that if cell toxicity is observed [usually evident within 24 hr], a subpassage should be made as soon as possible.)

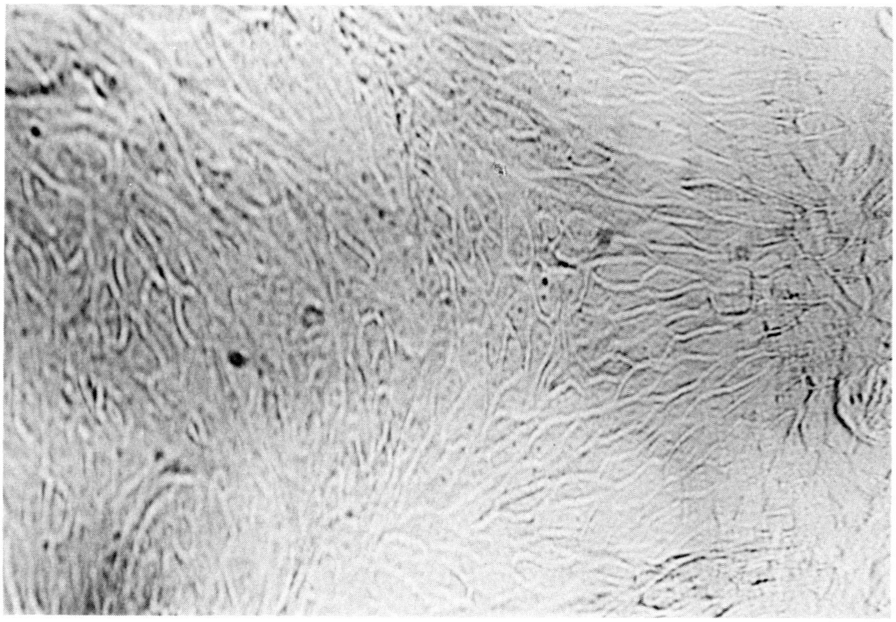

FIGURE 3. Monolayer of uninoculated rhesus monkey kidney cells.

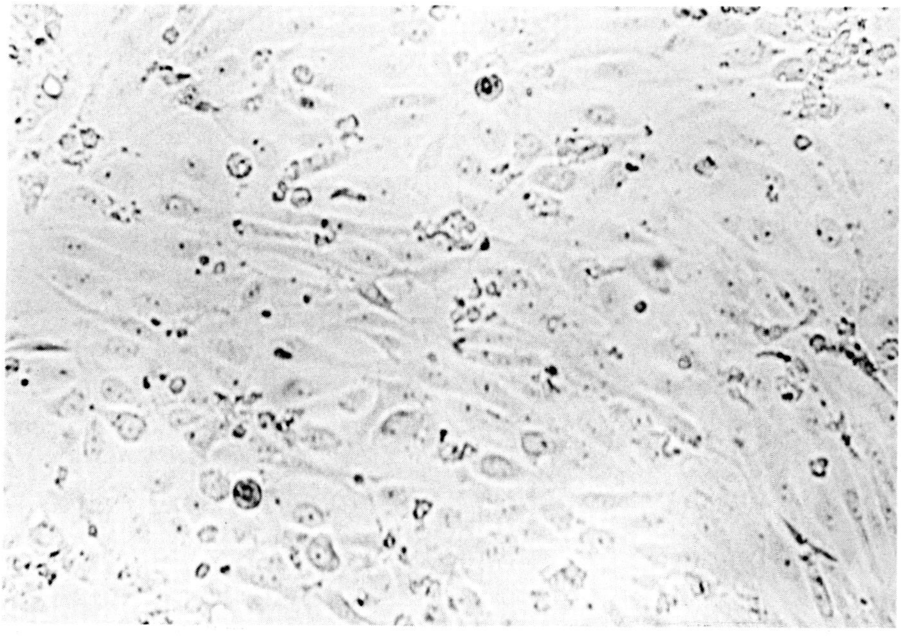

FIGURE 4. Rhesus monkey kidney cells infected with echovirus type 11. Note that infected cells are rounded. Magnification × 50.

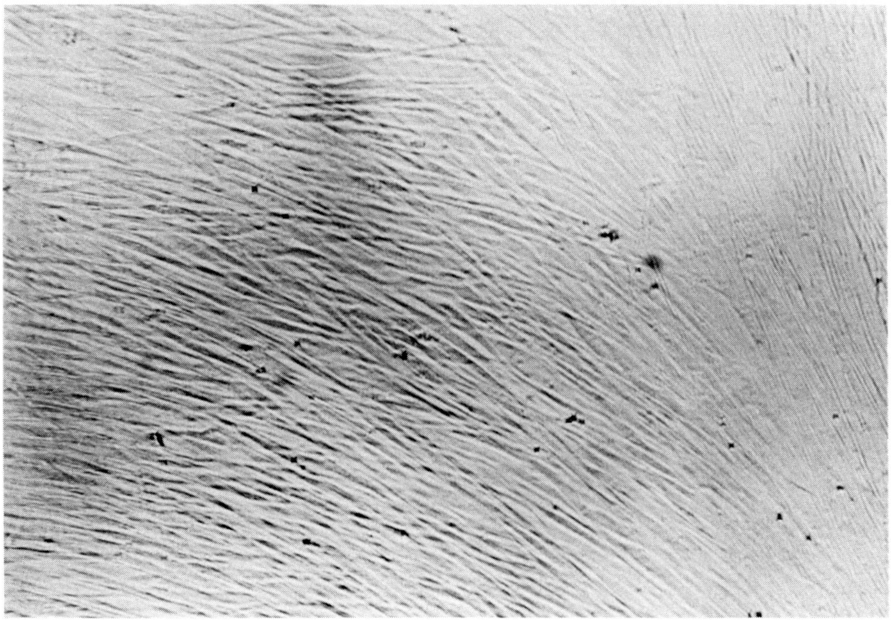

FIGURE 5. Monolayer of uninoculated human diploid cells (WI-38).

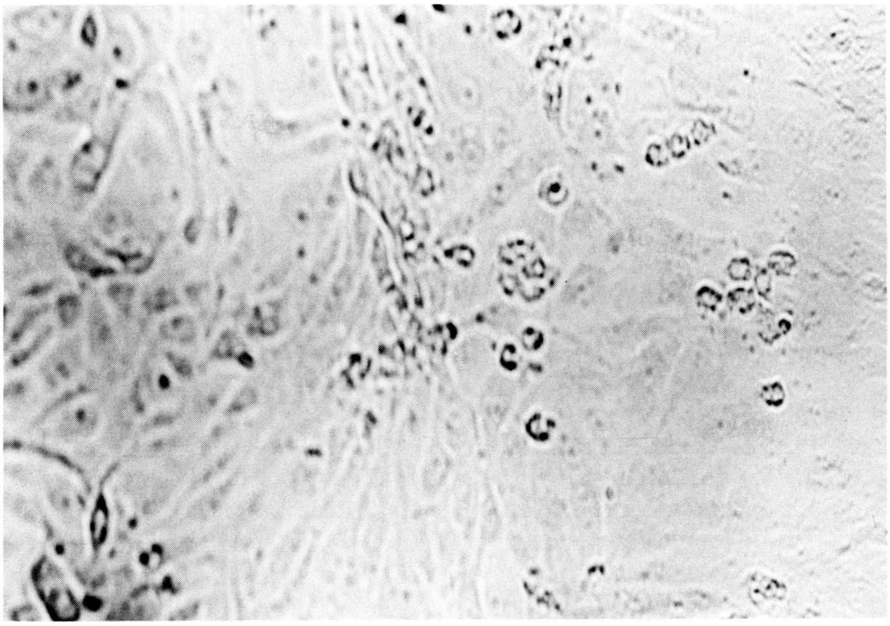

FIGURE 6. Rhesus monkey kidney cells infected with echovirus type 11. Rounded, infected cells can be seen. Magnification × 500.

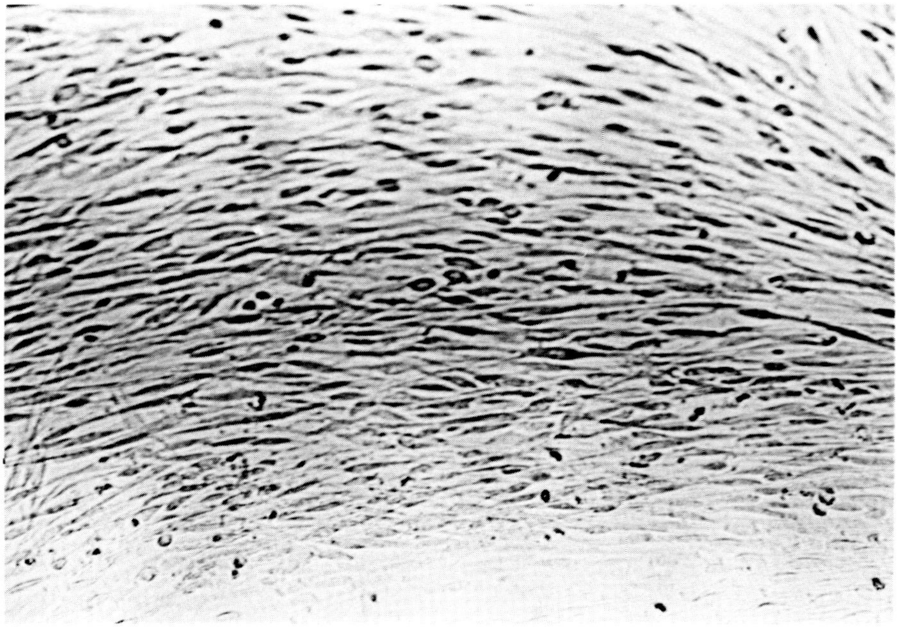

FIGURE 7. WI-38 cell culture infected with echovirus type 11. Cells have become rounded and necrotic and are beginning to fall off the tube wall. Magnification × 50.

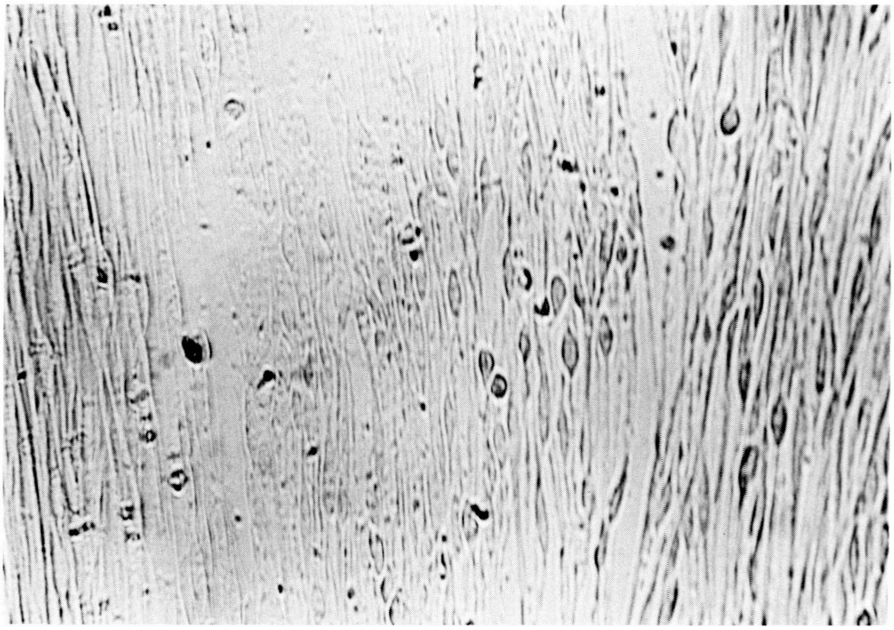

FIGURE 8. WI-38 cells infected with echovirus type 11, showing dendritic shapes as well as rounded infected cells. These are characteristic of the cytopathic effect of picornaviruses in WI-38 cells. Magnification × 500.

5. Cell cultures are observed at least 14 days before discarding. Cell sheets which remain in good condition may be kept for longer periods of observation or passed into new cultures for observation for an additional time period. If cultures are observed for an extended length of time, media should be changed approximately every 4 days.
6. Isolates may be passed to increase volume and infectivity titer before identification is attempted, or the original harvest may be of sufficient quantity to type the isolate directly.

Identification of Isolates

It is usually possible to make a presumptive diagnosis of echoviral disease based on the clinical history of the patient, the time of year at which the specimen was obtained, the tissue culture system in which the isolate grew, and the characteristic CPE. Reporting a presumptive diagnosis of viral infection without waiting for the specific typing of the isolate is of real value to the physician in the clinical management of the patient.

Specific identification of the serotype rests with serum neutralization, and this is most readily accomplished by use of intersecting pools of hyperimmune sera. These may be constructed in many ways, but the most economical one is the Lim Benyesh-Melnick scheme.[19] The Research Resources Branch (RRB), National Institute of Allergy and Infectious Diseases, National Institues of Health, now has available for all virus diagnostic laboratories dried equine antisera pools consisting of 42 equine antisera combined into eight pools which will correctly identify 37 enteroviruses. Detailed directions for use of the pools are supplied by the RRB when pools are shipped. Combination pools are very convenient and, as noted above, are available through the RRB. However, it is prudent to conserve these reagents. Therefore, if an epidemic due to a single serotype is in progress, use of the single hyperimmune serum to identify most isolates is clearly indicated. It should also be noted that it may not be in the best interest of the patient, the community, or the laboratory to attempt to identify the specific serotype of all isolates. At the present time, as no specific echovirus therapy is available, it is probably sufficient for the physician to know simply that an echovirus has been isolated. If a virus isolate cannot be identified, it may represent a new enterovirus, a mixture of viruses, or a virus other than an enterovirus. If there is uncertainty about the group to which an isolate belongs, determination of size, nucleic acid, ether stability, and acid lability should determine whether or not the isolate is an enterovirus.[20]

Serologic Diagnosis

No rapid or inexpensive serologic screening procedure is presently available to detect antibody titer rises to the echoviruses. The neutralization test is accurate and type specific, but requires the use of many antigens in tissue culture systems to screen completely for rises in antibody titer. Serologic diagnosis is feasible only when a virus isolate from the patient is available, an epidemic due to a single serotype is present, or a clinical picture such as pleurodynia clearly implicates a small number of echoviruses such as echovirus type 8 or the group B coxsackieviruses.

Paired sera are collected from the patient as soon after the onset of symptoms as possible (acute serum) and 4 to 8 weeks later (convalescent serum) and are diluted at 1:2, 1:8, 1:32, 1:128, and 1:512 with BSS or normal saline. Twofold dilutions (1:2, 1:4, 1:8, etc.) may be preferred by some laboratories. A 0.5-ml aliquot of each serum dilution is then mixed with an equal amount of virus diluted to contain approximately 100 $TCID_{50}$ per 0.1 ml, and the mixture is allowed to incubate for 2 hr at 37°C. Virus-serum mixtures in 0.2-ml amounts are then inoculated into the appropriate tissue culture system, usually monkey kidney tube cultures containing 0.9 ml of Melnick's B medium containing $NaHCO_3$ and antibiotics as noted above. Two tissue culture tubes are usually used for each dilution. A control virus titration to determine $TCID_{50}$ is always included. Virus is diluted in tenfold amounts and 0.1 ml of each dilution is inoculated into monkey kidney tissue culture tubes containing 0.9 ml of Melnick's B medium as noted above. Tubes are observed every 2 days for 8 to 10 days. Serum titers are calculated for the 50% endpoint per 0.1 ml of virus. A fourfold or greater rise in antibody titer is considered significant in making the diagnosis. It should be noted that neutralizing antibody may already be high at the time of onset of clinical symptoms, making interpretation of the test more difficult.

Other Serological Tests

Other serological tests which may be used include complement fixation (CF), hemagglutination inhibition (HI), and passive hemagglutination inhibition tests. The CF test has limited value in diagnosing echoviral disease because of major heterotypic cross-reactions among the antigens. Also, many individuals fail to develop homotypic CF antibody. The HI test is relatively easy to perform, and patients who become infected with an echovirus which hemagglutinates red blood cells (RBC) develop homotypic antibody. They may also develop heterotypic antibody, making the test somewhat nonspecific. However, the major drawback to the HI test is that only approximately one third of the known echoviruses agglutinate RBCs.

A new test originally developed to screen for rises in antibody to rhinoviruses has been found to be effective for detecting rises in antibody titer to the picornaviruses.[21] This test, a passive hemagglutination test, relies on the use of a coupling reagent, chromic chloride, to attach proteins to indicator erythrocytes. This has made possible the hemagglutination of RBCs by antigens which otherwise do not demonstrate this property. We have utilized this technique as a screening test, using paired sera collected during an epidemic of coxsackievirus B5, and have found it to be very useful. More work on this test is needed, especially with echoviruses, before it can be recommended for use in diagnostic laboratories.

Newer Methods for More Rapid Diagnosis

A new hope for rapid diagnosis rests with immunofluorescent staining of viral antigens.[22] A new technique which enables the rapid detection of viral antigens in cerebrospinal fluid (CSF) leukocytes is perhaps the most exciting recent development in this area.[23]

GENERAL REFERENCES

International Poliomyelitis Congress, *Poliomyelitis: Papers and Discussions Presented at the 1st Through 5th Int. Poliomyelitis Conf.*, Fishbein, M., Ed., J.B. Lippincott, Philadelphia, 1949, 1952, 1955, 1958, 1961.

Melnick, J. L., Advances in the study of the enteroviruses, *Prog. Med. Virol.*, 1, 59, 1958.

Paul, J. R., *A History of Poliomyelitis*, Yale University Press, New Haven, 1971.

Melnick, J. L., Enteroviruses, in *Viral Infections of Humans: Epidemiology, Pathogenesis, and Control*, Evans, A. S., Ed., Plenum Press, New York, 1976.

REFERENCES

1. **Melnick, J. L. and Aagren, K.,** Poliomyelitis and coxsackie viruses isolated from normal infants in Egypt, *Proc. Soc. Exp. Biol. Med.,* 81, 621, 1952.
2. **Robbins, F. C., Enders, J. F., Weller, T. H., and Florentino, G. L.,** Studies on the cultivation of poliomyelitis viruses in tissue culture. V. The direct isolation and serologic identification of virus strains in tissue culture from patients with nonparalytic and paralytic poliomyelitis, *Am. J. Hyg.,* 54, 286, 1951.
3. **Enders, J. F., Weller, T. H., and Robbins, F. C.,** Cultivation of the Lansing strain of poliomyelitis virus in cultures of various human embryonic tissues, *Science,* 109, 85, 1949.
4. Committee on the ECHO viruses, Enteric cytopathogenic human orphan (ECHO) viruses, *Science,* 122, 1187, 1955.
5. **Rosen, L., Melnick, J. L., Schmidt, N. J., and Wenner, H. A.,** Subclassification of enteroviruses and ECHOviruses type 34, brief report *Arch. Gesamte Virusforsch.,* 30, 89, 1970.
6. **Melnick, J. L., Shaw, E. W., and Curnen, E. C.,** A virus isolated from patients diagnosed as non-paralytic poliomyelitis or aseptic meningitis, *Proc. Soc. Exp. Biol. Med.,* 71, 344, 1949.
7. **Ramos-Alvarez, M. and Sabin, A. B.,** Characteristics of poliomyelitis and other enteric viruses recovered in tissue culture from healthy American children, *Proc. Soc. Exp. Biol. Med.,* 87, 655, 1954.
8. **Hammon, W. McD., Ludwig, E. H., Sather, G., and Yohn, D. S.,** Comparative studies on patterns of family infections with polioviruses and ECHO virus type 1 on an American military base in the Philippines, *Am. J. Public Health,* 47, 802, 1957.
9. **Hammon, W. McD., Yohn, D. S., Ludwig, E. G., Pavia, R. A., Sather, G. E., and McCloskey, L. W.,** A study of certain nonpoliomyelitis and poliomyelitis enterovirus infections. Clinical serologic associations, *JAMA,* 167, 727, 1958.
10. Committee on the Enteroviruses, National Foundation for Infantile Paralysis, The enteroviruses, *Am. J. Public Health,* 47, 1556, 1957.
11. **Ormsbee, R. A. and Melnick, J. L.,** Biologic and serologic characteristics of ECHO viruses from West Virginia, *J. Immunol.,* 79, 384, 1957.
12. **Rosen, L., Johnson, J. H., Huebner, R. J., and Bell, J. A.,** Observations on a newly recognized ECHO virus and description of an outbreak in a nursery, *Am. J. Hyg.,* 67, 300, 1958.
13. Panel for Picornaviruses, Picornaviruses: classification of nine new types, *Science,* 141, 153, 1963.
14. **Hammon, W. McD., Yohn, D. S., and Pavia, R. A.,** Isolation and characterization of prototype viruses ECHO-26, ECHO-27, and coxsackie B-6, *Proc. Soc. Exp. Biol. Med.,* 103, 164, 1960.
15. **Duncan, I. B. R.,** An outbreak of aseptic meningitis associated with previously unrecognized virus, *J. Hyg.,* 59, 181, 1961.
16. **Rosen, L. and Kern, J.,** Toluca-3, a newly recognized enterovirus, *Proc. Soc. Exp. Biol. Med.,* 118, 389, 1965.
17. **Melnick, J. L., Burkhardt, M., and Phillips, C. A.,** Isolation of a New Enterovirus from a Patient with Infantile Diarrhea, report to World Health Organization Virus Reference Laboratories, Moscow, 1966.
18. **Leibovitz, A.,** A transport medium for diagnostic virology, *Proc. Soc. Exp. Biol. Med.,* 131, 127, 1969.
19. **Lim, K. A. and Benyesh-Melnick, M.,** Typing of viruses by combinations of antiserum pools. Application to typing of enteroviruses (Coxsackie and ECHO), *J. Immunol.,* 84, 309, 1960.
20. **Hsiung, G. D.,** Use of ultrafiltration for animal virus grouping, *Bacteriol. Rev.,* 29, 477, 1965.
21. **Faulk, W. P., Vyas, G. N., Phillips, C. A., Fudenberg, H. H., and Chism, K.,** Passive hemagglutination test for anti-rhinovirus antibodies, *Nature New Biol.,* 231, 101, 1971.
22. **Sommerville, R. G.,** Rapid diagnosis of viral infections by immunofluorescent staining of viral antigens in leukocytes and macrophages, *Prog. Med. Virol.,* 10, 398, 1968.
23. **Taber, L. H., Mirkovic, R. R., Adam, V., Ellis, S. S., Yow, M. D., and Melnick, J. L.,** Rapid diagnosis of enterovirus meningitis by immunofluorescent staining of CSF leukocytes, *Intervirology,* 1, 128, 1973.

PICORNAVIRIDAE: RHINOVIRUSES

Charles J. Gauntt

Introduction

Rhinoviruses comprise the largest genus within the picornaviridae family. These viruses can most easily be differentiated from other viruses within this family on the basis of lability under acidic conditions. Rhinoviruses are the single most important genus of viruses which cause mild upper respiratory tract illness, particularly the common cold syndrome, in human beings. Additionally, the rhinoviruses are occasionally associated with more serious lower respiratory tract diseases in human beings, including asthma in children. Immunity against the more than 100 rhinovirus serotypes is type specific. Epidemiologic evidence suggests that new rhinovirus serotypes appear annually, thus presently assuring most human beings of repeated annual episodes of the common cold syndrome.

Morphology and Chemical Composition

Human rhinoviruses are naked icosahedral particles of 28 to 34 nm in diameter.[1-8] Crystalline arrays of virus particles have been observed in the cytoplasm of infected cells.[9-12] The protein capsid surrounding the single-stranded ribonucleic acid (RNA) genome is constructed of four polypeptides with estimated molecular weights of 45,000 to 25,000 and 8,000 daltons.[8,13] Virus particles may be constructed to contain 60 of each of the 4 polypeptides[13,14] or 60 of each of the 3 largest and 30 of the smallest polypeptides.[15] The two largest polypeptides are located at the capsid surface and the smallest is internal to the surface.[16]

RNA genomes of rhinoviruses have an approximate molecular weight of 2.2 to 2.8 $\times 10^6$ daltons[15,17-21] and account for 30% of the total particle mass.[7] Polyadenylic acid sequences are located at the 3′ terminus of virion RNA,[22,23] which has an unusually high adenine content.[21,24] Unlike enterovirus RNA, rhinovirus RNA appears to be highly fragile for unknown reasons.[25,26] The amino acid composition of rhinovirion polypeptides is not significantly different from polypeptides of other picornaviruses.[27] Rhinoviruses are distinguishable from enteroviruses on the basis of lability at pH 3[28,29] and a higher buoyant density of 1.38 to 1.41 g/ml at equilibrium in CsCl gradients.[30-33]

Virus Replication

Replication of rhinoviruses occurs in the cytoplasm of infected cells in tissue culture,[11] and single-step growth cycles require 10 to 15 hr, depending on the cell type.[31] Maximum yields of virus are produced at 33 to 35°C,[32,34] and yields are reduced by actinomycin D.[18,35] Attachment of virions to cells[36,37] and several definable steps in the process of penetration and uncoating of rhinovirus genomes have been identified.[31,38-41] Mechanisms of synthesis of rhinovirus RNA[7,18,19] and proteins[31,42-45] are similar to those described for other picornaviruses (see preceeding chapters). A schematic diagram of the replication of a rhinovirus is given in Figure 1.

Virus Strains

Rhinovirus strains are classified on the basis of neutralization tests in cell cultures using hyperimmune homotypic serum.[4] Reciprocal neutralization tests have identified strains which can be classified into 89 distinct serotypes and 1 subtype strain (see Table 1 and Reference 46). Information on requirements for designation of prototype strains,[4,47] methodology for identification of candidate strains,[47-50] and the reference

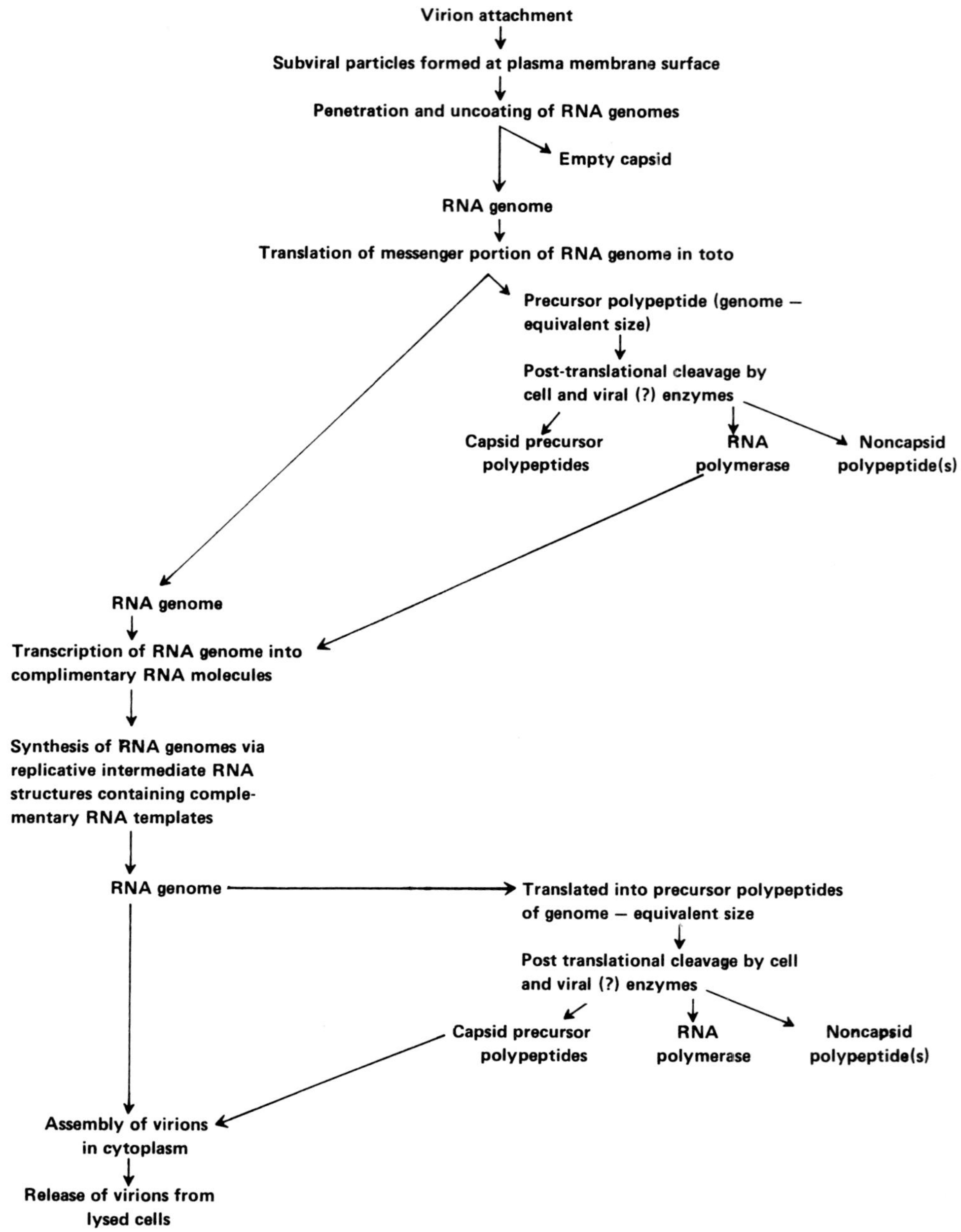

FIGURE 1. Replication of rhinovirus.

laboratories to which investigators may submit new candidate strains for inclusion into the rhinovirus numbering system has been published.[46] Reports of untyped strains are numerous,[46,51-55] and from these strains approximately 20 prototype strains will be selected for addition to the current taxonomic list.[56] Long-term epidemiological studies of the distribution of serotypes in a given geographic location suggest a shift in antigenic types; higher numbered, more recently isolated serotypes and untypeable strains replace previously discovered serotypes with lower numbers.[53-55,57] In agreement with epidemiological studies, extensive antigenic variations have been observed among serotype 51[58] and serotype 22[59] strains isolated in different years. It is also possible that a very large number of stable rhinovirus serotypes exist and have not circulated

Table 1
RHINOVIRUS SEROTYPES

Serotype	Diseases associated[a]	Laboratory diagnosis: Source of specimen	Laboratory diagnosis: Cell culture[b]	Ref. for prototype strain(s)
1A	Mild respiratory illness (Echo 28)	Nasal washings	PRMK	60, 61
1B	Common cold (B632)	Nasal washings	PHEK	62
2	Common cold (HGP)	Nasal washings	PHEK	63
3	Common cold (FEB)	Nasal washings	PHEK	63
4	Common cold (16/60)	Nasal washings	PHEK	64
5	Common cold (Norman)	Nasal washings	PHEK	65
6	Common cold (Thompson)	Nasal washings	PHEK	65
7	Acute URI (68-CV11)	Throat swab	HDC WI-10 and/or WI-26	1
8	Acute URI (MRH-CV12)	Throat swab	HDC WI-10 and/or WI-26	1
9	Acute URI (211-CV13)	Throat swab	HDC WI-10 and/or WI-26	1
10	Acute URI (204-CV14)	Throat swab	HDC WI-10 and/or WI-26	1
11	Acute URI (1-CV15)	Throat swab	HDC WI-10 and/or WI-26	1
12	Acute URI (181-CV16)	Throat swab	HDC WI-10 and/or WI-26	1
13	Mild acute afebrile URI (353)	Throat swab	PHEK	67
14	Mild acute afebrile URI (1,059)	Throat swab	PHEK	67
15	Mild acute afebrile URI (1,734)	Throat swab	PHEK	67
16	No illness in 2-year-old female (11,757)	Throat swab	PHEK	67
17	Mild acute afebrile URI (33,342)	Throat swab	PHEK	67
18	Bronchitis (5,986-CV17)	Throat swab	HDC WI-26	28
19	Bronchitis (6,072-CV18)	Throat swab	HDC WI-26	28
20	Acute URI (15-CV19)	Throat swab	HDC WI-26	28
21	Acute URI (47-CV21)	Throat swab	HDC WI-26	28
22	Acute URI (127-CV22)	Throat swab	HDC WI-26	28
23	Acute URI (5,124-CV24)	Throat swab	HDC WI-26	28
24	Acute URI (5,146-CV25)	Throat swab	HDC WI-26	28
25	Acute URI (5,426-CV26)	Throat swab	HDC WI-26	28
26	Acute URI (5,660-CV27)	Throat swab	HDC WI-26	28
27	Acute URI (5,870-CV28)	Throat swab	HDC WI-26	28
28	Acute URI (6,106-CV29)	Throat swab	HDC WI-26	28
29	No illness in 2-year-old child (5,582-CV30)	Throat swab	HDC WI-26	28
30	Acute respiratory illness (106F)	Nasal and throat swabs combined	SMK and SHK	68
31	Acute respiratory illness (140F)	Nasal and throat swabs combined	SMK and SHK	68
32	Afebrile URI (363)	Throat swab	PHEK	69
33	Afebrile URI (1,200)	Throat swab	PHEK	69
34	Acute respiratory illness (137-3)	Nasal and throat specimens combined	SHK	70
35	Acute respiratory illness (164A)	Nasal and throat swabs combined	SHK	70

Table 1 (continued)
RHINOVIRUS SEROTYPES

Serotype	Diseases associated[a]	Laboratory diagnosis: Source of specimen	Laboratory diagnosis: Cell culture[b]	Ref. for prototype strain(s)
36	Acute respiratory illness (342H)	Nasal and throat swabs combined	SHK	70
37	Acute respiratory illness (151-1)	Nasal and throat swabs combined	SHK	70
38	Acute respiratory illness (CH 79)	Throat swab	HEK	71
39	URI (209)	Nasal washing	HDC WI-26 or WI-38	72
40	URI (1,794)	Throat swab	HDC WI-26 or WI-38	72
41	URI (56,110)	Throat swab	HDC WI-26	72
42	URI (56,822)	Throat swab	HDC WI-26	72
43	URI (58,750)	Throat swab	HDC WI-26	72
44	No illness in 18-year-old male (71,560)	Throat swab	HDC WI-26	72
45	Common cold (Baylor 1)	Nasal and pharyngeal swabs	HDC WI-38	73
46	Common cold (Baylor 2)	Nasal and pharyngeal swabs combined	Human aorta (A-39) and HDC WI-38	73
47	Common cold (Baylor 3)	Nasal and pharyngeal swabs combined	Human aorta (A-39)	73
48	No illness in 5-year-old male (1,505)	Throat swab	HDC WI-26	74
49	Bronchiolitis (8,213)	Throat swab	HEK	74
50	Respiratory illness (A2 No. 58)	Gargling fluids	HDC WI-38	75
51	URI (F01-4,081)	Throat washings	HDC (LS)	4
52	URI (F01-3,772)	Throat washings	Human fetal kidney	4
53	URI (F01-3,928)	Throat washings	HDC (LS)	4
54	URI (F01-3,774)	Throat washings	HDC (LS)	4
55	Mild cold (WIS 315E)	Throat swab and gargling combined	HDC WI-38	76
56	URI (CH82)	Throat swab	PHEK	71
57	URI (CH47)	Throat swab	HDC WI-26	71
58	URI (21-CV20)	Throat swab	HDC WI-26	28
59	URI (611-CV35)	Throat swab	HDC WI-26	77
60	URI (2,268-CV37)	Throat swab	HDC WI-26	77
61	Bronchiolitis (6,669-CV39)	Throat swab	HDC WI-26	77
62	URI (1963M-CV40)	Throat swab	HDC WI-26	77
63	Bronchiolitis (6,360-CV41)	Throat swab	HDC WI-26	77
64	URI (6,258-CV44)	Throat swab	HDC WI-26	77
65	URI (425-CV47)	Throat swab	HDC WI-26	77
66	URI (1,983-CV48)	Throat swab	HDC WI-26	77
67	URI (1,857-CV51)	Throat swab	HDC WI-26	77
68	URI (F02-2317-Wood)	Nasopharyngeal swab	HDC (SLF)	78
69	No illness in 23-year-old male (F02-2513-Mitchinson)	Nasopharyngeal swab	HDC (SLF)	78
70	URI (F02-2547-Treganza)	Nasopharyngeal swab	HDC (SLF)	78
71	Respiratory illness (SF365)	Nasal and throat swabs combined	HDC WI-26	57
72	URI (K2207)	Gargling fluids	HDC WI-38	46

Table 1 (continued)
RHINOVIRUS SEROTYPES

Serotype	Diseases associated[a]	Laboratory diagnosis: Source of specimen	Laboratory diagnosis: Cell culture[b]	Ref. for prototype strain(s)
73	Acute respiratory illness (107E)	Nasal and throat swabs combined	SHK	70
74	Acute respiratory illness (328A)	Nasal and throat swabs combined	SHK	46
75	Acute respiratory illness (328F)	Nasal and throat swabs combined	SHK	79
76	Acute respiratory illness (H 00062)	Nasal and throat swabs combined	HDC WI-38	79
77	Cold in chronic bronchitic patient with exacerbation of bronchitic condition (130-63)	Nasal swab and/or throat wash	HDC WI-38 and human embryonic fibroblastic cells of lung and kidney origin	80
78	Bronchitis (2030-65)	Nasal and throat swabs combined	Human embryo kidney fibroblastic cells	81
79	Acute respiratory illness (101-1)	Nasal and throat swabs combined	SHK and human diploid cells of embryonic lung	70
80	No illness in 21-year-old male (277G)	Nasal and throat swabs combined	SHK	46
81	Acute respiratory illness (483F2)	Nasal and throat swabs combined	SHK	46
82	Acute respiratory illness (03647)	Nasal and throat swabs combined	HDC WI-38	46
83	Common cold (Baylor 7)	Nasal and throat swabs combined	PHEK	82
84	Acute respiratory illness (432D)	Nasal and throat swabs combined	SHK and HDC WI-38	79
85	URI (50-525-CV54)	Throat swab	HDC WI-26 or WI-38	83
86	URI (121564-Johnson)	Throat washing	HDC (SL)	46
87	URI (F02-3607-Corn)	Throat washing	PRMK	46
88	Rhinitis, pharyngitis and otitis media (CVD 01-0165-Dambrauskas)	Nasopharyngeal swab	HDC (SL)	46
89	URI (41467-Gallo)	Nasopharyngeal swab	HDC (SL)	46

[a] Original (prototype) strain designation in parentheses.
[b] PRMK, primary rhesus monkey kidney; PHEK, primary human embryonic kidney; HDC, human diploid cells: WI-10 of fetal kidney origin, WI-26 of fetal lung origin, WI-38 of embryonic lung origin, (LS) of fetal kidney origin, (SLF) of fetal lung origin, (SL) of fetal lung origin; SMK, secondary monkey kidney; SHK, secondary human kidney, several different sources; HEK, human embryonic kidney.

through areas under rhinovirus surveillance[51] or cannot be isolated by currently employed cell or organ culture methods.

Clinical Features

Rhinoviruses as a group account for the largest proportion (up to 30%) of all etiologic agents associated with acute upper respiratory tract illnesses such as the "common cold" in human beings.[53,71,84-89] Symptoms commonly ascribed to rhinovirus illnesses include sneezing, nasal discharge, nasal obstruction, coughing, hoarseness, pharyngeal discomfort, watery or burning eyes, headache, malaise, and chilliness.

Signs of a rhinovirus illness generally include the presence of rhinitis, pharygnitis, otitis, and conjunctivitis with no fever. The incubation period is usually 2 to 4 days, and the major symptoms and signs of a typical "cold" generally persist for a median of 7 days in young adults. Symptoms last for up to 2 weeks in 25% of cases.[51] The onset of illness closely correlates with the time of maximum virus shedding,[90,91] suggesting that direct injury to both ciliated and nonciliated nasal epithelial cells as a result of virus replication may be a mechanism by which rhinoviruses produce illnesses.[51,91] Virus shedding may continue for 2 weeks after onset of illness[90] and for up to 55 days in some patients.[53] Reports of rhinovirus infections of the lower respiratory tract and associated illnesses include tracheobronchitis,[92] acute exacerbations of chronic bronchitis,[80,93,94] bronchopneumonia,[59,95] pneumonia,[96] and exacerbations of chronic lung disease.[97] There is increasing evidence that rhinovirus infections may be associated with exacerbations of intermittent reversible obstructive airway disease (asthma) in children.[98-100] The possibility cannot be excluded that rhinoviruses are merely opportunistic pathogens in hosts infected by other viral or bacterial pathogens which cause the lower respiratory tract illnesses.[51] Rhinoviruses have been reported as opportunistic pathogens in a fatal case of pneumonia in an adult[101] and in two cases of "sudden death" in infants showing rhinovirus viremia.[102]

Rhinoviruses are spread by direct person-to-person contact by an unknown mechanism(s),[103] possibly via inoculation of the nose or conjunctiva by virus-contaminated fingers.[104] Transmission of rhinoviruses by aerosol droplets has been suggested, but this mode is not accepted by all workers.[91]

Laboratory Diagnosis

Laboratory diagnosis of rhinovirus infections is made by isolation of the virus from respiratory tract specimens in cell cultures including human fetal diploid cells,[99,105,106] HEp 2 cells,[98,99,106] and primary rhesus monkey kidney cells.[98,99,106] Organ cultures of human embryonic trachea or nasal epithelium have also been used. Nasal and pharyngeal swabs in combination,[70,73,79] pharyngeal swabs,[1,28,67,69] nasopharyngeal washings,[63-65,107] or throat gargling fluids[4,74,75] should be collected within 2 to 5 days after onset of symptoms.[51] These specimens should be inoculated into tissue or organ cultures within 2 to 3 hr of collection. Addition of proteins such as gelatin, heat-inactivated animal sera, or bovine serum albumin to the clinical specimen and holding the specimen at 2 to 4°C may be of value in preventing loss in infectivity during the interval between collection and inoculation.[108] If storage of specimens is necessary, they can be frozen at −70°C, but only under nonacidic conditions.[108] Tissue cultures inoculated with presumptive rhinovirus-containing specimens are maintained at pH 7.0 to 7.2 and incubated at 33°C in a roller drum.[108] Cytopathic effects (CPE) produced by rhinoviruses in tissue cultures are similar to CPE produced by other picornaviruses, i.e., discrete foci of rounded or oval cells which may become refractile and enlarged; many cells detach from the surface, and all cells in the foci eventually lyse (Figure 2). CPE may develop slowly if the inoculum contains little virus, but it generally appears within 2 weeks and almost all cells in the tissue culture eventually exhibit CPE. Some strains of rhinoviruses do not produce CPE in tissue culture cells, and three strains were detected only by an interference test in human embryonic kidney cells.[52] At least one untyped strain (HS) has not been propagated in tissue culture, and it has been isolated only in organ cultures.[2,112] CPE in organ cultures is indicated by decreasing ciliary activity and degeneration of epithelial cells (Figure 3). A flow diagram showing all successfully applied methods for the isolation and subsequent identification of an unknown virus isolate as a rhinovirus is shown in Figure 4. Most rhinoviruses in clinical specimens replicate well in tissue culture and can easily be identified as rhinoviruses on the basis of stability in ether and lability under acid conditions.

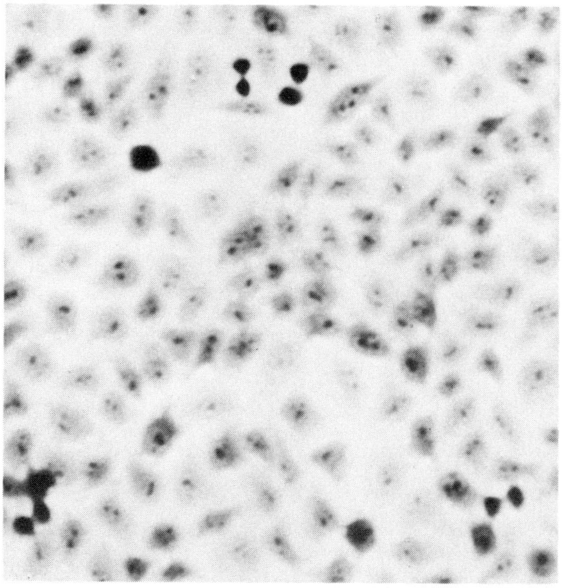

A

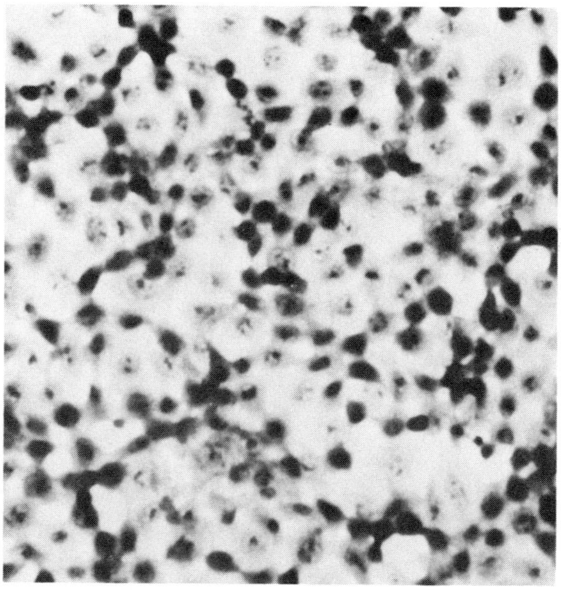

B

FIGURE 2. HeLa cell control (A) and HeLa cells inoculated with RV14 (B) and incubated for 15 hr at 34.5°C. Hematoxylin-eosin; total magnification × 315.

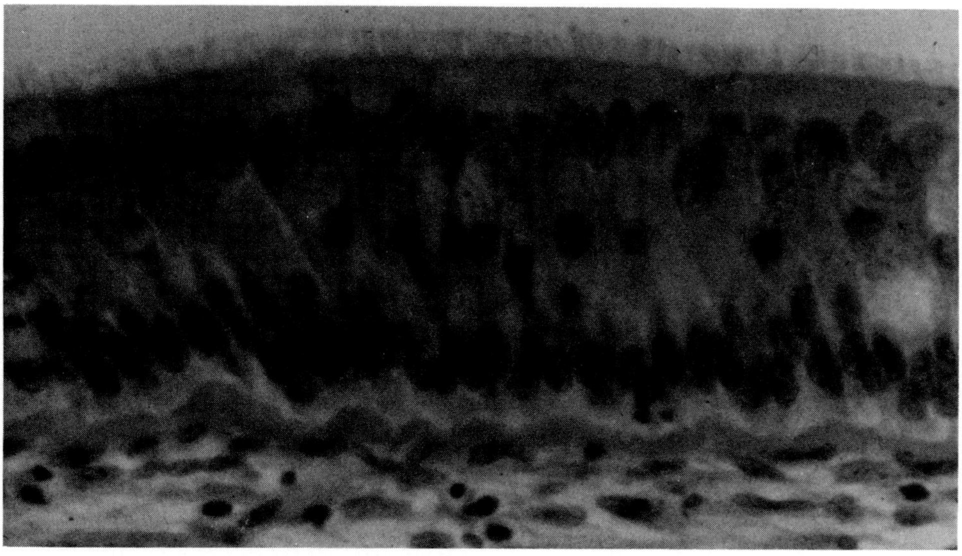

FIGURE 3. Organ culture of human respiratory tract epithelium (A) and similar organ culture previously infected with a rhinovirus (B). Note the shedding of degenerate surface epithelial cells. (From original Kodachrome by B. Hoorn and with the kind permission of B. Hoorn and D.A.J. Tyrrell.)

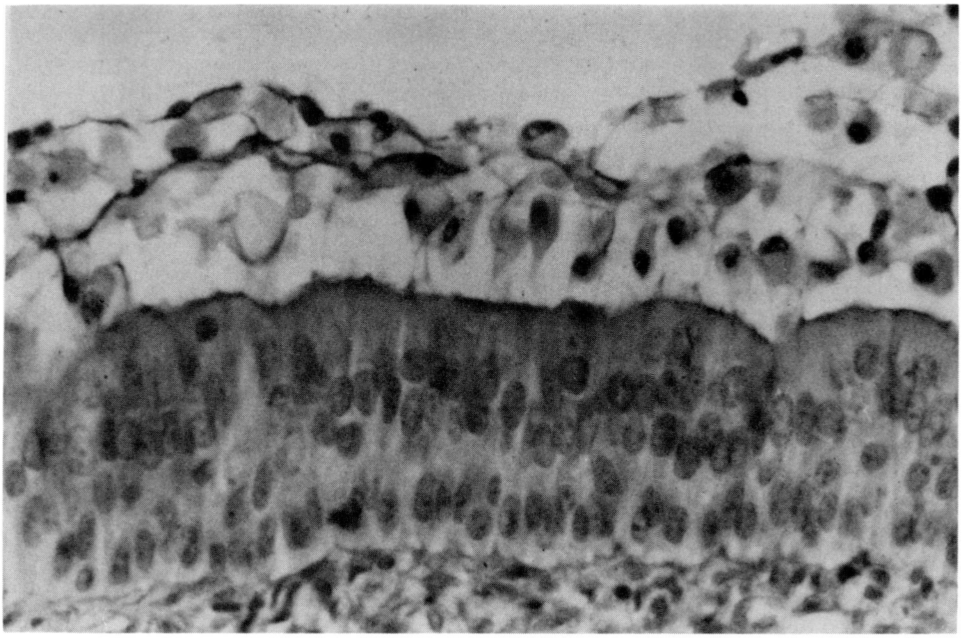

FIGURE 3(B)

Serologic diagnosis of rhinovirus infections by testing for rhinovirus antibody is possible only when the infecting strain of virus has been identified or propagated in cell or organ cultures. Specific antibody in tears, parotid saliva, nasal secretions, nasal washings, and serum has been measured by neutralization of CPE in cell cultures.[110] More sensitive plaque reduction methods for measuring antibody to rhinoviruses have been developed,[111,112] but they have not been used in epidemiologic studies. Specificity of the humoral immune response in rhinovirus infections is not resolved, as both homotypic[113] and heterotypic[114-116] antibody responses have been reported in experimental and natural infections in human beings. Heterotypic antibody responses have also been reported in animals immunized sequentially with one or two serotypes, suggesting relatedness in antigenic determinants among several groups of rhinovirus serotypes.[116-118] Complement fixation tests with rhinoviruses have been performed,[119,120] but are not practical for diagnostic tests for rhinovirus antibody due to heterotypic reactions with human sera. Assay for rhinovirus antibody by a hemagglutination inhibition test has been reported;[121] however, this technique has limitations in that only a few serotypes exhibit hemagglutination.[32] A passive hemagglutination test employing erythrocytes with chemically coupled rhinovirus proteins has also been used to assess rhinovirus antibody levels in human sera.[122]

Pathogenicity in Experimental Animals

Human rhinoviruses have thus far been shown to infect only chimpanzees[122-126] and gibbons.[127,128] Infection of human beings by rhinovirus strains of bovine,[129,130] equine,[131,132] or feline[133] origins has not been reported.

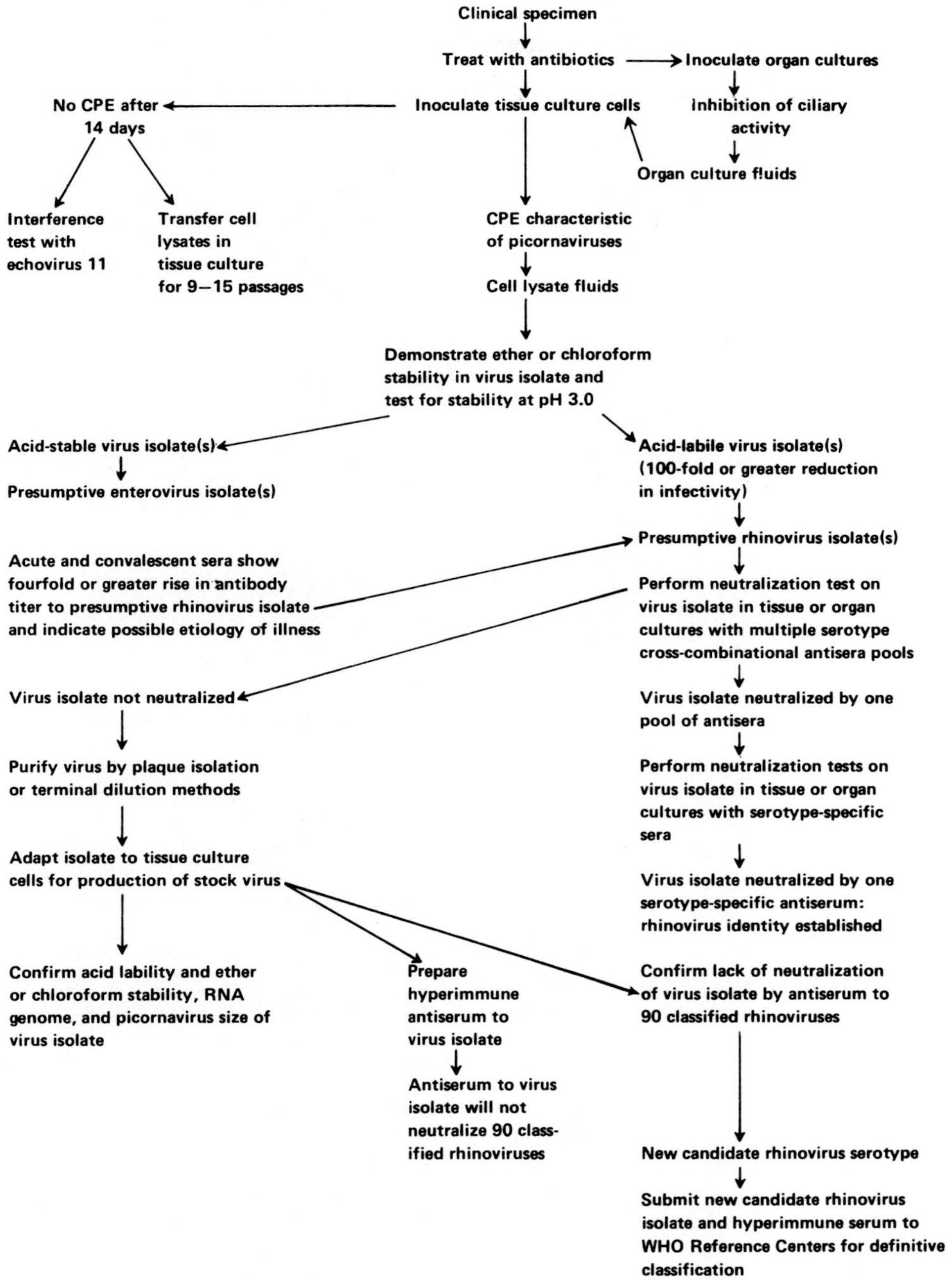

FIGURE 4. Flow chart for isolation and identification of rhinoviruses.

GENERAL REFERENCES

Gwaltney, J. M., Jr. and Jordan, W. S., Jr., Rhinoviruses and respiratory disease, *Bacteriol. Rev.*, 28, 409, 1964.

Andrewes, C. H., Rhinoviruses and common colds, *Annu. Rev. Med.*, 17, 361, 1966.

Hamre, D., Rhinoviruses, *Monogr. Virol.*, 1, 13, 1968.

Tyrrell, D. A. J., Rhinoviruses, *Virol. Monogr.*, 2, 69, 1968.

Stott, E. J. and Killington, R. A., Rhinoviruses, *Annu. Rev. Microbiol.*, 26, 503, 1972.

Kapikian, A. Z., Rhinoviruses, in *Strains of Human Viruses,* Majer, M. and Plotkin, S. A., Eds., S. Karger, New York, 193, 1972.

Hsiung, G. D., Rhinoviruses, in *Diagnostic Virology: An Illustrated Handbook,* Hsiung, G. D., Ed., Yale University Press, New Haven, 1973, 68.

Schieble, J. H., Rhinoviruses, in *Manual of Clinical Microbiology,* 2nd ed., Lennette, E. H., Spaulding, E. H., and Truant, J. P., Eds., American Society for Microbiology, Washington, D.C., 1974, 728.

Gwaltney, J. M., Jr., Rhinoviruses, *Yale J. Biol. Med.*, 48, 17, 1975.

Butterworth, B. E., Grunert, R. R., Korant, B. D., Lonberg-Holm, K., and Yin, F. H., Replication of rhinoviruses, *Arch. Gesamte Virusforsch.*, 51, 169, 1976.

REFERENCES

1. **Hamparian, V. V., Ketler, A., and Hilleman, M. R.,** Recovery of new viruses (coryzavirus) from cases of common cold in human adults, *Proc. Soc. Exp. Biol. Med.*, 108, 444, 1961.
2. **McGregor, S., Phillips, C. A., and Mayor, H. D.,** Purification and biophysical properties of rhinoviruses, *Proc. Soc. Exp. Biol. Med.*, 122, 118, 1966.
3. **Tyrrell, D. A. J. and Chanock, R. M.,** Rhinoviruses: a description, *Science,* 141, 152, 1963.
4. **Kapikian, A. Z. et al.,** Rhinoviruses: a numbering system, *Nature,* 213, 761, 1967.
5. **Phillips, C. A., Melnick, J. L., and Sullivan, L.,** Characterization of four new rhinovirus serotypes, *Proc. Soc. Exp. Biol. Med.*, 134, 933, 1970.
6. **Thomas, D. C., Conant, R. M., and Hamparian, V. V.,** Rhinovirus replication in suspension cultures of HeLa cells, *Proc. Soc. Exp. Biol. Med.*, 133, 62, 1970.
7. **Brown, F. and Hull, R.,** Comparative virology of the small RNA viruses, *J. Gen. Virol.*, (Suppl. 20), 43, 1973.
8. **Korant, B. D., Lonberg-Holm, K., Noble, J., and Stasny, J. T.,** Naturally occurring and artificially produced components of three rhinoviruses, *Virology,* 48, 71, 1972.
9. **Fiala, M.,** Plaque formation by 55 rhinovirus serotypes, *Appl. Microbiol.*, 16, 1445, 1968.
10. **Kawana, R. and Matsumoto, I.,** Electron microscopic study of rhinovirus replication in human fetal lung cells, *Jpn. J. Microbiol.*, 15, 207, 1971.
11. **Gauntt, C., Sauck, J., and Carlson, E.,** Growth of rhinovirus type 14; brief report *Arch. Gesamte Virusforsch.*, 41, 382, 1973.
12. **Blough, H. A., Tiffany, J. M., Gordon, G., and Fiala, M.,** The effect of magnesium on intracellular crystallization of rhinovirus, *Virology,* 38, 694, 1969.
13. **Medappa, K. C., McLean, C., and Rueckert, R. R.,** On the structure of rhinovirus 1A, *Virology,* 44, 259, 1971.
14. **Rueckert, R. R.,** Picornaviral architecture, in *Comparative Virology,* Maramorosch, K. and Kurstak, E., Eds., Academic Press, New York, 1971, 255.
15. **Stott, E. J. and Killington, R. A.,** Rhinoviruses, *Annu. Rev. Microbiol.*, 26, 503, 1972.
16. **Lonberg-Holm, K. and Butterworth, B. E.,** Investigation of the structure of polio- and human-rhinovirions through the use of selective chemical reactivity, *Virology,* 71, 207, 1976.
17. **Kolias, S. I. and Dimmock, N. J.,** Replication of rhinovirus RNA, *J. Gen. Virol.*, 20, 1, 1973.
18. **Gauntt, C. J.,** Synthesis of ribonucleic acids in KB cells infected with rhinovirus type 14, *J. Gen. Virol.*, 21, 253, 1973.
19. **Sethi, S. K.and Schwerdt, C. E.,** Studies on the biosynthesis and characterization of rhinovirus ribonucleic acid, *Virology,* 48, 221, 1972.
20. **Nair, C. N. and Lonberg-Holm, K. K.,** Infectivity and sedimentation of rhinovirus ribonucleic acid, *J. Virol.*, 7, 278, 1971.
21. **Brown, F., Newman, J. F. E., and Stott, E. J.,** Molecular weight of rhinovirus ribonucleic acid, *J. Gen. Virol.*, 8, 145, 1970.
22. **Nair, C. N. and Owens, M. J.,** Preliminary observations pertaining to polyadenylation of rhinovirus RNA, *J. Virol.*, 13, 535, 1974.
23. **MacNaughton, M. R. and Dimmock, N. J.,** Polyadenylic acid sequences in rhinovirus RNA species from infected human diploid cells, *J. Virol.*, 16, 745, 1975.
24. **Newman, J. F. E., Rowlands, D. J., and Brown, F.,** A physicochemical subgrouping of the mammalian picornaviruses, *J. Gen. Virol.*, 18, 171, 1973.
25. **Gauntt, C. J. and Griffith, M. M.,** Fragmentation of RNA in virus particles of rhinovirus type 14, *J. Virol.*, 13, 762, 1974.

26. **Dimmock, N. J.**, Differences between the thermal inactivation of picornaviruses at "high" and "low" temperatures, *Virology*, 31, 338, 1967.
27. **McGregor, S. and Mayor, H. D.**, Biophysical and biochemical studies on rhinovirus and poliovirus. II. Chemical and hydrodynamic analysis of the rhinovirion, *J. Virol.*, 7, 41, 1971.
28. **Ketler, A., Hamparian, V. V., and Hilleman, M. R.**, Characterization and classification of ECHO 28-rhinovirus-coryazvirus-agents, *Proc. Soc. Exp. Biol. Med.*, 110, 821, 1962.
29. **Hughes, J. H., Thomas, D. C., and Hamparian, V. V.**, Acid lability of rhinovirus type 14: effect of pH, time, and temperature, *Proc. Soc. Exp. Biol. Med.*, 144, 555, 1973.
30. **Rowlands, D. J., Sangar, D. V., and Brown, F.**, Buoyant density of picornaviruses in caesium salts, *J. Gen. Virol.*, 13, 141, 1971.
31. **Butterworth, B. E., Grunert, R. R., Korant, B. D., Lonberg-Holm, K., and Yin, F. H.**, Replication of rhinoviruses, *Arch. Gesamte Virusforsch.*, 51, 169, 1976.
32. **Stott, E. J. and Killington, R. A.**, Haemagglutination by rhinoviruses, *Lancet*, 1, 1369, 1972.
33. **Medappa, K. C. and Rueckert, R. R.**, Binding of cesium ions to human rhinovirus 14, *Abstr. Annu. Meet. Am. Soc. Microbiol.*, p. 207, 1974.
34. **Stott, E. J. and Heath, G. F.**, Factors affecting the growth of rhinovirus 2 in suspension cultures of L 132 cells, *J. Gen. Virol.*, 6, 15, 1970.
35. **Sethi, S. K.**, Kinetics of Rhinovirus Replication and Virus Specific Nucleic Acid Synthesis in HeLa Cells, Ph.D. thesis, Stanford University, 1971.
36. **Lonberg-Holm., K. and Korant, B. D.**, Early interaction of rhinoviruses with host cells, *J. Virol.*, 9, 29, 1972.
37. **Lonberg-Holm, K. and Philipson, L.**, Early interaction between animal viruses and cells, *Monogr. Virol.*, 9, 1, 1974.
38. **Noble-Harvey, J. and Lonberg-Holm, K.**, Sequential steps in attachment of human rhinovirus type 2 to HeLa cells, *J. Gen. Virol.*, 25, 83, 1974.
39. **Noble, J. and Lonberg-Holm, K.**, Interactions of components of human rhinovirus type 2 with HeLa cells, *Virology*, 51, 270, 1973.
40. **Korant, B. D., Lonberg-Holm, K., Yin, F. H., and Noble-Harvey, J.**, Fractionation of biologically active and inactive populations of human rhinovirus type 2, *Virology*, 63, 384, 1975.
41. **Lonberg-Holm, K. and Noble-Harvey, J.**, Comparison of *in vitro* and cell-mediated alteration of a human rhinovirus and its inhibition by sodium dodecyl sulfate, *J. Virol.*, 12, 819, 1973.
42. **McLean, C. and Rueckert, R. R.**, Picornaviral gene order: comparison of a rhinovirus with a cardiovirus, *J. Virol.*, 11, 341, 1973.
43. **Butterworth, B. E.**, A comparison of virus-specific polypeptides of encephalomyocarditis virus, human rhinovirus-1A and poliovirus, *Virology*, 56, 439, 1973.
44. **Butterworth, B. E. and Korant, B. D.**, Characterization of the large picornaviral polypeptides produced in the presence of zinc ion, *J. Virol.*, 14, 282, 1974.
45. **Korant, B. D., Kauer, J. C., and Butterworth, B. E.**, Zinc ions inhibit replication of rhinoviruses, *Nature*, 248, 588, 1974.
46. **Kapikian, A. Z.**, A collaborative report: rhinoviruses — extension of the numbering system, *Virology*, 43, 524, 1971.
47. **Conant, R. M. and Hamparian, V. V.**, Rhinoviruses: basis for a numbering system. I. HeLa cells for propagation and serologic procedures, *J. Immunol.*, 100, 107, 1968.
48. **Kenny, G. E., Cooney, M. K., and Thompson, D. J.**, Analysis of serum pooling schemes for identification of large numbers of viruses, *Am. J. Epidemiol.*, 91, 439, 1970.
49. **Conant, R. M. and Hamparian, V. V.**, Rhinoviruses: basis for a numbering system. II. Serologic characterization of prototype strains, *J. Immunol.*, 100, 114, 1968.
50. **Gwaltney, J. M., Jr.**, Micro-neutralization test for identification of rhinovirus serotypes, *Proc. Soc. Exp. Biol. Med.*, 122, 1137, 1966.
51. **Gwaltney, J. M., Jr.**, Rhinoviruses, *Yale J. Biol. Med.*, 48, 17, 1975.
52. **Olson, L. C., Willhight, M., and Buescher, E. L.**, Recovery and characterization of non-cytopathogenic rhinoviruses, *J. Gen. Virol.*, 17, 237, 1972.
53. **Fox, J. P., Cooney, M. K., and Hall, C. E.**, The Seattle virus watch. V. Epidemiologic observations of rhinovirus infections, 1965—1969, in families with young children, *Am. J. Epidemiol.*, 101, 122, 1975.
54. **Calhoun, A. M., Jordan, W. S., Jr., and Gwaltney, J. M., Jr.**, Rhinovirus infections in an industrial population. V. Change in distribution of serotypes, *Am. J. Epidemiol.*, 99, 58, 1974.
55. **Monto, A. S. and Cavallaro, J. J.**, The Tecumseh study of respiratory illness. IV. Prevalence of rhinovirus serotypes, 1966—1969, *Am. J. Epidemiol.*, 94, 280, 1972.
56. **Hamparian, V. V.**, personal communication.
57. **Gwaltney, J. M., Jr., Hendley, J. O., Simon, G., and Jordan, W. S., Jr.**, Rhinovirus infections in an industrial population. III. Number and prevalence of serotypes, *Am. J. Epidemiol.*, 87, 158, 1968.

58. **Stott, E. J. and Walker, M.**, Antigenic variation among strains of rhinovirus type 51, *Nature*, 224, 1311, 1969.
59. **Schieble, J. H., Lennette, E. H., and Fox, V. L.**, Antigenic variation of rhinovirus type 22, *Proc. Soc. Exp. Biol. Med.*, 133, 329, 1970.
60. **Pelon, W., Mogabgab, W. J., Phillips, I. A., and Pierce, W. E.**, A cytopathogenic agent isolated from naval recruits with mild respiratory illnesses, *Proc. Soc. Exp. Biol. Med.*, 94, 262, 1957.
61. **Pelon, W.**, Classification of the '2060' virus as ECHO 28 and further study of its properties, *Am. J. Hyg.*, 73, 36, 1961.
62. **Kendall, E. J. C., Bynoe, M. L., and Tyrrell, D. A. J.**, Virus isolations from common colds occurring in a residential school, *Br. Med. J.*, 2, 82, 1962.
63. **Tyrrell, D. A. J., Bynoe, M. L., Hitchcock, G., Pereira, H. G., and Andrewes, C. H.**, Some virus isolations from common colds. I. Experiments employing human volunteers, *Lancet*, 1, 235, 1960.
64. **Hobson, D. and Schild, G. C.**, Virological studies in matural common colds in Sheffield in 1960, *Br. Med. J.*, 2, 1414, 1960.
65. **Tyrrell, D. A. J. and Bynoe, M. L.**, Some further virus isolations from common colds, *Br. Med. J.*, 1, 393, 1961.
66. **Andrewes, C. H., Chapronieri, D. M., Gompels, A. E. H., Pereira, H. G., and Roden, A. T.**, Propagation of common-cold virus in tissue cultures, *Lancet*, 2, 546, 1953.
67. **Johnson, K. M., Bloom, H. H., Chanock, R. M., Mufson, M. A., and Knight, V.**, Acute respiratory diseases of viral etiology. VI. The newer enteroviruses, *Am. J. Public Health*, 52, 933, 1962.
68. **Connelly, A. P., Jr. and Hamre, D.**, Virologic studies on acute respiratory disease in young adults. II. Characteristics and serologic studies of three new rhinoviruses, *J. Lab. Clin. Med.*, 63, 30, 1964.
69. **Webb, P. A., Johnson, K. M., and Mufson, M. A.**, A description of two newly-recognized rhinoviruses of human origin, *Proc. Soc. Exp. Biol. Med.*, 116, 845, 1964.
70. **Hamre, D., Connelly, A. P., Jr., and Procknow, J. J.**, Virologic studies of acute respiratory disease in young adults. III. Some biologic and serologic characteristics of seventeen rhinovirus serotypes isolated, October 1960 to June 1961, *J. Lab. Clin. Med.*, 64, 450, 1964.
71. **Gwaltney, J. M., Jr., and Jordan, W. S., Jr.**, Rhinoviruses and respiratory disease, *Bacteriol. Rev.*, 28, 409, 1964.
72. **Mufson, M. A., Kawana, R., James, H. D., Jr., Gauld, L. W., Bloom, H. H., and Chanock, R. M.**, A description of six new rhinoviruses of human origin, *Am. J. Epidemiol.*, 81, 32, 1965.
73. **Phillips, C. A., Melnick, J. L., and Grim, C. A.**, Characterization of three new rhinovirus serotypes, *Proc. Soc. Exp. Biol. Med.*, 119, 798, 1965.
74. **Kapikian, A. Z., Mufson, M. A., James, H. D., Jr., Kalica, A. R., Bloom, H. H., and Chanock, R. M.**, Characterization of two newly recognized rhinovirus serotypes of human origin, *Proc. Soc. Exp. Biol. Med.*, 122, 1155, 1966.
75. **Mogabgab, W. J.**, Acute respiratory illnesses in university (1962—1966), military and industrial (1962—1963) populations, *Am. Rev. Respir. Dis.*, 98, 359, 1968.
76. **Dick, E. C., Blumer, C. R., and Evans, A. S.**, Epidemiology of infections with rhinovirus types 43 and 55 in a group of University of Wisconsin student families, *Am. J. Epidemiol.*, 86, 386, 1967.
77. **Hamparian, V. V., Leagus, M. B., and Hilleman, M. R.**, Additional rhinovirus serotypes, *Proc. Soc. Exp. Biol. Med.*, 116, 976, 1964.
78. **Schieble, J. H., Lennette, E. H., and Fox, V. L.**, Rhinoviruses: the isolation and characterization of three new serologic types, *Proc. Soc. Exp. Biol. Med.*, 127, 324, 1968.
79. **Hamre, D., Connelly, A. P., Jr., and Procknow, J. J.**, Virologic studies of actue respiratory disease in young adults. IV. Virus isolations during four years of surveillance, *Am. J. Epidemiol.*, 83, 238, 1966.
80. **Stott, E. J., Grist, N. R., and Eadie, M. B.**, Rhinovirus infections in chronic bronchitis: isolation of eight possibly new rhinovirus serotypes, *J. Med. Microbiol.*, 1, 109, 1968.
81. **Stott, E. J., Eadie, M. B., and Grist, N. R.**, Rhinovirus infections of children in hospital: isolation of three possible new rhinovirus serotypes, *Am. J. Epidemiol.*, 90, 45, 1969.
82. **Phillips, C. A., Melnick, J. L., and Grim, C. A.**, Rhinovirus infections in a student population: isolation of five new serotypes, *Am. J. Epidemiol.*, 87, 447, 1968.
83. **Mascoli, C. C., Leagus, M. B., Hilleman, M. R., Weibel, R. E., and Stokes, J., Jr.**, Rhinovirus infection in nursery and kindergarten children. New rhinovirus serotype 54, *Proc. Soc. Exp. Biol. Med.*, 124, 845, 1967.
84. **Jackson, G. G., Dowling, H. F., and Mogabgab, W. J.**, Infectivity and interrelationships of 2060 and JH viruses in volunteers, *J. Lab. Clin. Med.*, 55, 331, 1960.
85. **Tyrrell, D. A. J. and Bynoe, M. L.**, Cultivation of viruses from a high proportion of patients with colds, *Lancet*, 1, 76, 1966.
86. **Mufson, M. A., Webb, P. A., Kennedy, H., Gill, V., and Chanock, R. M.**, Etiology of upper-respiratory-tract illnesses among civilian adults, *JAMA*, 195, 1, 1966.

87. **Hilleman, M. R.**, Present knowledge of the rhinovirus group of viruses, *Curr. Top. Microbiol. Immuno.*, 41, 1, 1967.
88. **Gwaltney, J. M., Jr., Hendley, J. O., Simon, G., and Jordan, W. S., Jr.**, Rhinovirus infections in an industrial population. I. The occurrence of illness, *N. Engl. J. Med.*, 275, 1261, 1966.
89. **Cooney, M. K., Hall, C. E., and Fox, J. P.**, The Seattle virus watch. III. Evaluation of isolation methods and summary of infections detected by virus isolations, *Am. J. Epidemiol.*, 96, 286, 1972.
90. **Gwaltney, J. M., Jr., Hendley, J. O., Simon, G., and Jordan, W. S., Jr.**, Rhinovirus infections in an industrial population. II. Characteristics of illness and antibody response, *JAMA*, 202, 494, 1967.
91. **Douglas, R. G., Jr.**, Pathogenesis of rhinovirus common colds in human volunteers, *Ann. Otol. Rhinol. Laryngol.*, 79, 563, 1970.
92. **Glezen, W. P., Loda, F. A., Clyde, W. A., Jr., Senior, R. J., Sheaffer, C. I., Conley, W. G., and Denny, F. W.**, Epidemiologic patterns of acute lower respiratory disease of children in a pediatric group practice, *J. Pediatr.*, 78, 397, 1971.
93. **Eadie, M. B., Stott, E. J., and Grist, N. R.**, Virological studies in chronic bronchitis, *Br. Med. J.*, 2, 671, 1966.
94. **Stenhouse, A. C.**, Rhinovirus infection in acute exacerbations of chronic bronchitis: a controlled prospective study, *Br. Med. J.*, 3, 461, 1967.
95. **Cherry, J. D., Diddams, J. A., and Dick, E. C.**, Rhinovirus infections in hospitalized children. Provocative bacterial interrelationships, *Arch. Environ. Health* (Chicago), 14, 390, 1967.
96. **George, R. B. and Mogabgab, W. J.**, Atypical pneumonia in young men with rhinovirus infections, *Ann. Intern. Med.*, 71, 1073, 1969.
97. **McNamara, M. J., Phillips, I. A., and Williams, O. B.**, Viral and *Mycoplasma pneumoniae* infections in exacerbations of chronic lung disease, *Am. Rev. Respir. Dis.*, 100, 19, 1969.
98. **Minor, T. E., Dick, E. C., Baker, J. W., Ovellette, J. J., Cohen, M., and Reed, C. E.**, Rhinovirus and influenza type A infections as precipitants of asthma, *Am. Rev. Respir. Dis.*, 113, 149, 1976.
99. **Minor, T. E., Dick, E. C., DeMeo, A. N., Ovellette, J. J., Cohen, M., and Reed, C. E.**, Viruses as precipitants of asthmatic attacks in children, *JAMA*, 227, 292, 1974.
100. **Minor, T. E., Baker, J. W., Dick, E. C., DeMeo, A. N., Ovellette, J. J., Cohen, M., and Reed, C. E.**, Greater frequency of viral respiratory infections in asthmatic children as compared with their nonasthmatic siblings, *J. Pediatr.*, 85, 472, 1974.
101. **Craighead, J. E., Meier, M., and Cooley, M. H.**, Pulmonary infection due to rhinovirus type 13, *N. Engl. J. Med.*, 281, 1403, 1969.
102. **Urquhart, G. E. D. and Stott, E. J.**, Rhinoviraemia, *Br. Med. J.*, 4, 28, 1970.
103. **D'Alessio, D. J., Peterson, J. A., Dick, C. R., and Dick, E. C.**, Transmission of experimental rhinovirus colds in volunteer married couples, *J. Infect. Dis.*, 133, 28, 1976.
104. **Hendley, J. O., Wenzel, R. P., and Gwaltney, J. M., Jr.**, Transmission of rhinovirus colds by self-inoculation, *N. Engl. J. Med.*, 288, 1361, 1973.
105. **Rosenbaum, M. J., DeBerry, P., Sullivan, E. J., Pierce, W. E., Mueller, R. E., and Peckinpaugh, R. O.**, Epidemiology of the common cold in military recruits with emphasis on infections by rhinovirus types 1A, 2 and two unclassified rhinoviruses, *Am. J. Epidemiol.*, 93, 183, 1971.
106. **Wulff, H., Noble, G. R., Maynard, J. E., Feltz, E. T., Poland, J. D., and Chin, T. D. Y.**, An outbreak of respiratory infection in children associated with rhinovirus types 16 and 29, *Am. J. Epidemiol.*, 90, 304, 1969.
107. **Cate, T. R., Couch, R. B., and Johnson, K. M.**, Studies with rhinoviruses in volunteers: production of illness, effect of naturally acquired antibody, and demonstration of a protective effect not associated with serum antibody, *J. Clin. Invest.*, 43, 56, 1964.
108. **Schieble, J. H.**, Rhinoviruses, in *Manual of Clinical Microbiology*, 2nd ed., Lennette, E. H., Spaulding, E. H., and Truant, J. P., Eds., American Society for Microbiology, Washington, D.C., 1974, 728.
109. **Hoorn, B. and Tyrrell, D. A. J.**, A new virus cultivated only in organ cultures of human ciliated epithelium, *Arch. Gesamte Virusforsch.*, 18, 210, 1966.
110. **Douglas, R. G., Jr., Fleet, W. F., Cate, T. R., and Couch, R. B.**, Antibody to rhinovirus in human sera. I. Standardization of a neutralization test, *Proc. Soc. Exp. Biol. Med.*, 127, 497, 1968.
111. **Cate, T. R., Douglas, R. G., Jr., and Couch, R. B.**, Interferon and resistance to upper respiratory virus illness, *Proc. Soc. Exp. Biol. Med.*, 131, 631, 1969.
112. **Fiala, M.**, A study of the neutralization of rhinoviruses by means of a plaque-reduction test, *J. Immunol.*, 103, 107, 1969.
113. **Hamparian, V. V., Leagus, M. B., Hilleman, M. R., and Stokes, J., Jr.**, Epidemiologic investigations of rhinovirus infections, *Proc. Soc. Exp. Biol. Med.*, 117, 469, 1964.
114. **Fleet, W. F., Douglas, R. G., Jr., Cate, T. R., and Couch, R. B.**, Antibody to rhinovirus in human sera. II. Heterotypic responses, *Proc. Soc. Exp. Biol. Med.*, 127, 503, 1968.

115. **Fawzy, K. Y., Fox, J. P., Ketler, A., Brandt, C. D., Hall, C. E., and Haraway, A. W.,** The virus watch programs: a continuing surveillance of virus infection in metropolitan New York families. V. Observations in employed adults on etiology of acute upper respiratory disease and heterologous antibody response to rhinoviruses, *Am. J. Epidemiol.,* 86, 653, 1967.
116. **Monto, A. S. and Johnson, K. M.,** Serologic relationships of the B632 and ECHO-28 rhinovirus strains, *Proc. Soc. Exp. Biol. Med.,* 121, 615, 1966.
117. **Fenters, J. D., Gillum, S. S., Holper, J. C., and Marquis, G. S.,** Serotypic relationships among rhinoviruses, *Am. J. Epidemiol.,* 84, 10, 1966.
118. **Cooney, M. K., Wise, J. A., Kenny, G. E., and Fox, J. P.,** Broad antigenic relationships among rhinovirus serotypes revealed by cross-immunization of rabbits with different serotypes, *J. Immunol.,* 114, 635, 1975.
119. **Chapple, P. J., Head, B., and Tyrrell, D. A. J.,** A complement fixing antigen from an M rhinovirus, (brief report) *Arch. Gesamte Virusforsch.,* 21, 123, 1967.
120. **Dans, P. E., Forsyth, B. R., and Chanock, R. M.,** Density of infectious virus and complement-fixing antigens of two rhinovirus strains, *J. Bacteriol.,* 91, 1605, 1966.
121. **Reed, S. E. and Hall, T. S.,** Hemagglutination-inhibition test in rhinovirus infections of volunteers, *Infect. Immun.,* 8, 1, 1973.
122. **Faulk, W. P., Vyas, G. N., Phillips, C. A., Fudenberg, H. H., and Chism, K.,** Passive haemagglutination test for anti-rhinovirus antibodies, *Nature* [New Biology,] 231, 101, 1971.
123. **Shipkowitz, N. L., Bower, R. R., Schleicher, J. B., Aquino, F., Appell, R. N., and Roderick, W. R.,** Antiviral activity of a *bis*-benzimidazole against experimental rhinovirus infections in chimpanzees, *Appl. Microbiol.,* 23, 117, 1972.
124. **Dick, E. C.,** Experimental infection of chimpanzees with human rhinovirus types 14 and 43, *Proc. Soc. Exp. Biol. Med.,* 127, 1079, 1968.
125. **Dick, E. C. and Dick, C. R.,** A subclinical outbreak of human rhinovirus 31 infection in chimpanzees, *Am. J. Epidemiol.,* 88, 267, 1968.
126. **Boyle, J. J., Raupp, W. G., Stanfield, F. J., Haff, R. F., Dick, E. C., D'Alessio, D., and Dick, C. R.,** Progress in rhinovirus chemotherapy, *Ann. N.Y. Acad. Sci.,* 173, 477, 1970.
127. **Pinto, C. A. and Haff, R. F.,** Experimental infection of gibbons with rhinoviruses, *Nature* (London), 224, 1310, 1969.
128. **Dick, E. C. and Dick, C. R.,** Natural and experimental infections of nonhuman primates with respiratory viruses, *Lab. Anim. Sci.,* 24, 177, 1974.
129. **Mohanty, S. B., Lillie, M. G., and Albert, T. F.,** Experimental exposure of calves to a bovine rhinovirus, *Am. J. Vet. Res.,* 30, 1105, 1969.
130. **Ide, P. R. and Darbyshire, J. H.,** Studies with a rhinovirus of bovine origin. I. Growth *in vitro* of strain RS 3X, *Arch. Gesamte Virusforsch.,* 36, 166, 1972.
131. **Ditchfield, W. J. B.,** Rhinoviruses and parainfluenza viruses of horses, *J. Am. Vet. Med. Assoc.,* 115, 384, 1969.
132. **Plummer, G.,** An equine respiratory enterovirus. Some biological and physical properties, *Arch. Gesamte Virusforsch.,* 12, 694, 1963.
133. **Crandell, R. A.,** A description of eight feline picornaviruses and an attempt to classify them, *Proc. Soc. Exp. Biol. Med.,* 126, 240, 1967.

TOGAVIRIDAE: ALPHAVIRUS

Jordi Casals

Introduction

Alphavirus is a genus in the family Togaviridae.[1] The description of the taxon is based primarily on studies with two viruses, Semliki Forest virus and Sindbis virus, which have little or no clinical significance. To a lesser extent, the taxon description depends on studies of a few other agents, especially chikungunya, western equine encephalitis, and Venezuelan equine encephalitis viruses. Most of the remaining viruses of the set, which includes at least 20 distinct agents, have not been characterized biochemically or physically. Inclusion of the latter viruses in the genus is based on serological relationships: alphavirus is a synonym for arbovirus antigenic group A.[2] Whether there exist alphaviruses that are not antigenically part of group A is a moot question.

Morphology

The alphaviruses contain a single linear molecule of single-stranded RNA of 4×10^6 daltons mol wt and have a capsid with cubic symmetry 20 to 40 nm in diameter enclosed within a lipoprotein envelope with an overall diameter of 50 to 70 nm. The viruses in the genus are all serologically related antigenic group A and are unrelated to any other viruses. [3] The overall shape is spherical, with surface projections or spikes. Virus particle diameter is 50nm in ultrathin sections and between 60 and 80 nm is negatively stained preparations. The surface projections are between 6 and 12 nm long. The nucleocapsid has cubic symmetry with 32 capsomers (Figure 1). Figure 2 shows the location of viral nucleoids in the cytoplasm of a thin section of mouse brain infected with eastern equine encephalitis virus.[3a]

Physicochemical Properties

The single RNA molecule has a sedimentation coefficient of 42 to 49S, represents 4% of the weight of the virion, is infectious, and serves as messenger RNA. Virions contain three or four structural polypeptides and no enzymes; their density is 1.18 to 1.24, sedimentation coefficient 250-275S, and molecular weight 1×10^8 daltons. Virions are stable at pH between 6.5 and 9, are rapidly inactivated by heat at 56°C and by ultra-violet light, are sensitive to lipid solvents and detergents, and relatively resistant to trypsin.[4]

Virus Replication

Synthesis takes place exclusively in the cytoplasm in close connection with cytoplasmic reticulum membranes and plasma membrane. The viral RNA acts as messenger RNA; its information is translated directly into protein. The resulting large polypepide is subsequently cleaved into smaller fragments — two or three glycoproteins of 52,000, 49,000, and 10,000 mol wt, respectively, which are incorporated in the envelope and a nucleoprotein of 30,000 mol wt which is incorporated in the core. Nonstructural proteins include a number of large polypeptides, primarily precursors to viral proteins, and a virus-specified RNA polymerase.

The single-stranded viral genome replicates by a semiconservative process in which the 42S RNA viral or positive strand is transcribed by polymerase into the complementary or minus strand 42S RNA, which then serves as a template for the smaller-than-genome RNA species 26S 33S, and 38S and the progeny 42S RNA. Alphavirus replication is not affected by the presence of actinomycin D.[4]

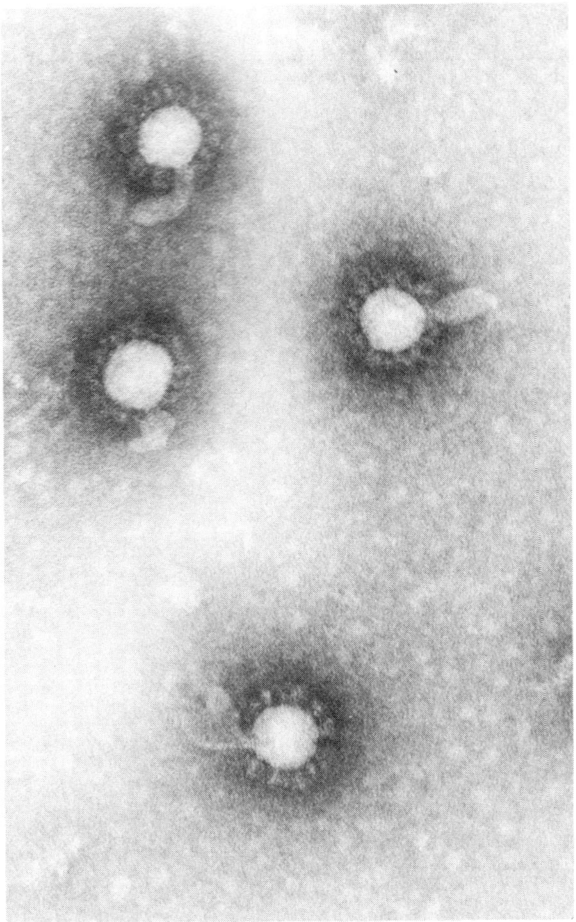

FIGURE 1. Venezuelan equine encephalitis virus in BHK-21 cells. Virions between 60 and 70 nm. Negative contrast strain; magnification × 191,000 (Courtesy of Dr. Frederick A. Murphy, Center for Disease Control, Atlanta, Ga.)

Maturation

Nucleocapsids assembled in the cytoplasm move toward the plasma membrane, causing the cellular glycoproteins in the cell membrane to be totally displaced and substituted by virus-specified glycoproteins. The virion is released by budding through the host cell plasma membrane. By this process the virion acquires its envelope. The lipids in the envelope are those of the host cell membrane.

In general, alphaviruses propagate with cytopathic effect (CPE) in a variety of cell cultures both under fluid medium and under an overlay: in the latter case, the viruses induce the formation of plaques. Among the most commonly used stable cells are African green monkey kidney (especially VERO line) and baby hamster kidney cells (BHK-21); primary chick embryo cell cultures are also susceptible to these viruses. *Aedes albopictus* and *A. aegypti* cells in culture[9] maintain replication of alphaviruses, but with no visible CPE; the virus can be detected in high titer in these cultures either by in vitro tests or by subpassage to another host.

Virus Types and Antigenic Properties

Viruses which may have morphological, biochemical, and physicochemical proper-

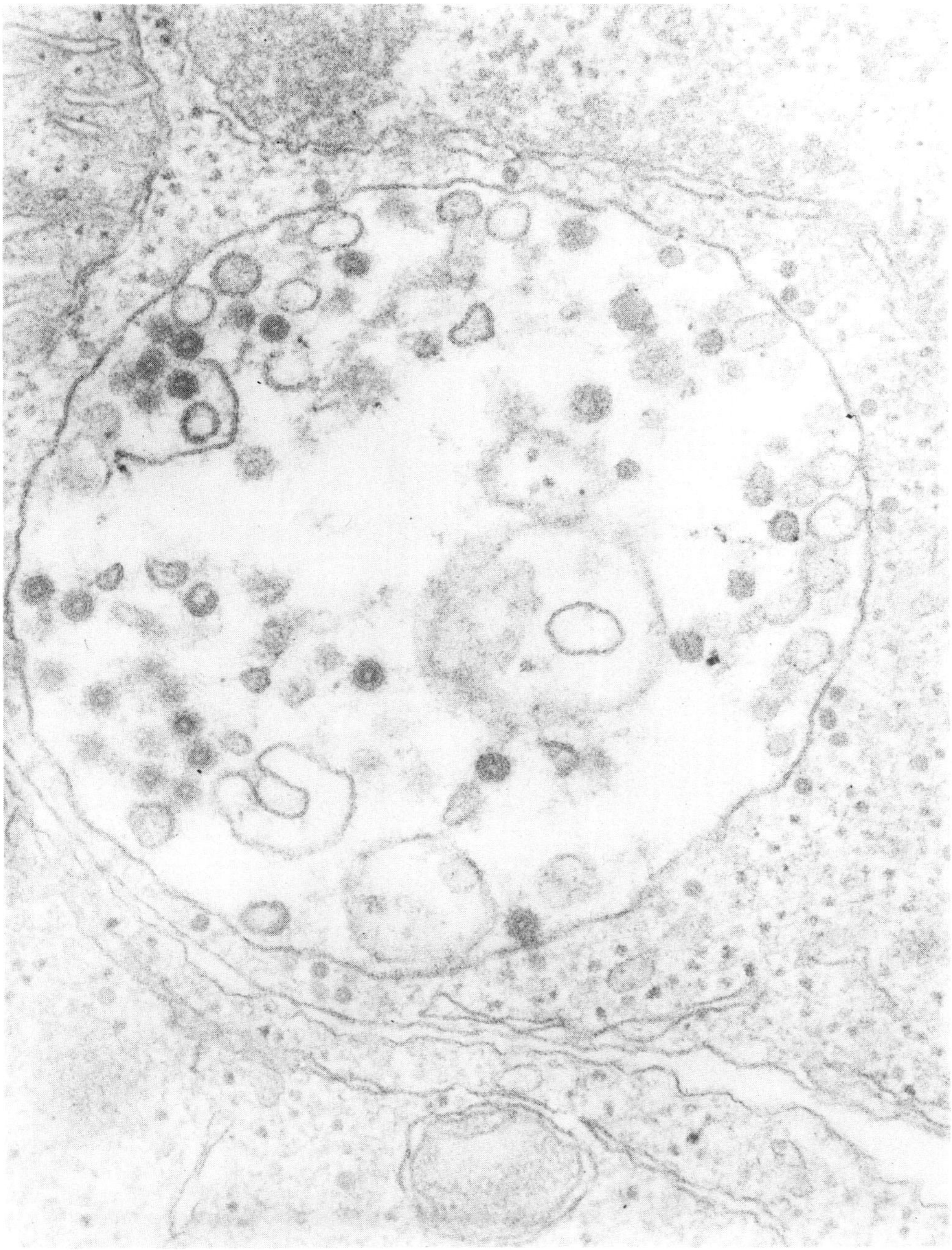

FIGURE 2. Thin section of mouse brain tissue infected with eastern equine encephalitis virus. Viral capsids are visible in the cytoplasm at the surface of vacuoles. Magnification × 91,000. (From Murphy, F. A. and Whitfield, S. G., *Exp. Mol. Pathol.*, 13, 131, 1970. With permission.)

ties of alphaviruses but are not serologically part of the arbovirus antigenic group A[2] are not alphaviruses by definition[1] and are excluded from further consideration.

The number of distinct antigenic specificities in the alphavirus virion is not presently known, nor is the number of nonstructural antigens. It appears that an envelope antigen, a large glycoprotein, is involved in virus neutralization. Core proteins are broadly cross-reactive and are responsible for the group demarcation. On the other hand, en-

Table 1
TOGAVIRUS: ALPHAVIRUS

		Laboratory diagnosis, isolation technique					
Virus type	Clinical manifestations	Source of specimen	Cell culture	Animal	Fluorescent antibody and/or electron microscopy	Serologic diagnosis[a]	Ref.
Chikungunya	Dengue-like, febrile with rash, no mortality	Blood	Possible; no reports	Newborn mouse	No reports	N, CF, HI; FAT less used	10
Eastern equine encephalitis	Encephalitis, high mortality	Brain, blood	HeLa; lower vertebrates, chick embryo	Mouse: newborn, weanling; wet chicks; chick embryos	No reports	N, CF, HI; FAT less used	11
Mayaro	Systemic, febrile, no mortality	Blood	Possible; no reports	Newborn mouse	No reports	N, CF, HI; FAT less used	12
O'nyong-nyong	Dengue-like, febrile with rash, no mortality	Blood	Possible; no reports	Newborn mouse	No reports	N, CF, HI	13
Ross River	Polyarthritis, seldom febrile, rash	Blood	Possible; no reports	Newborn mouse	No reports	N, CF, HI	14
Sindbis	Febrile with rash; no mortality	Blood, skin	Possible; no reports	Newborn mouse	No reports	N, CF, HI; FAT less used	15
Venezuelan equine encephalitis	Febrile, meningitis, encephalitis; low to moderate mortality	Blood, brain, throat washings	Duck and chick embryo; VERO, BHK-21, others	Mouse: newborn, weanling; chick embryos, guinea pig, rabbit	No reports	N, CF, HI; FAT increasingly used	16
Western equine encephalitis	Encephalitis, moderate mortality	Blood, brain, cerebro spinal fluid	Duck and chick embryo; hamster kidney; others	Mouse: newborn, weanling; chicks; chick embryos	No reports	N, CF, HI, FAT	17

Note: Human cases, natural and/or laboratory acquired, have been reported in small numbers with Everglades and Mucambo viruses.

[a] N, neutralization; CF, complement fixation; HI, hemagglutination inhibition; FAT, fluorescent antibody test.

velope proteins are subgroup or complex and, especially type specific; antisera against core and envelope do not cross-react.[5,6]

From a practical, diagnostic point of view it is useful to know that the degree of serological cross-activity among the alphaviruses varies among members of the group as well as with the test employed. Viruses are more closely related to their subgroup members than to the remaining members of the group. Based on results of the more universally used diagnostic serological tests — hemagglutination inhibition (HI), complement fixation (CF), and neutralization — a stepwise relationship has been established. The steps define relationships in order of increasing relatedness from the alphavirus genus at one extreme to individual viruses and subtypes at the other extreme.[7] In order of increasing restriction in definition these steps include group, subgroup or complex, type or virus, subtype, and variety. For the alphaviruses (group A), the group includes the following 20 agents: Aura, Bebaru, chikungunya, eastern equine encephalitis (EEE), Everglades, Getah, Mayaro, Middleburg, Mucambo, Ndumu, o'nyong-nyong, Pixuna, Ross River, Sagiyama, Semliki Forest, Sindbis, Una, Venezuelan equine encephiaitis (VEE), Western equine encephalitis (WEE), and Whataroa.

The subgroup or complex viruses are EEE (EEE); Semliki (Bebaru, chikungunya, Getah, Mayaro, o'nyong-nyong, Ross River, Sagiyama, Semliki Forest, and Una); VEE (Everglades, Mucambo, Pixuna, and VEE); and WEE (Aura, Sindbis, WEE, and Whataroa). Middleburg and Ndumu could presently be considered to be complexes that include one virus only, as is the case with EEE, but as little information is available concerning these two viruses, it is best to leave them as unattached viruses.

Clinical Features

Approximately half of the listed alphaviruses are known to infect man (Table 1); however, as far as is known, some cause disease only rarely and others do so only as a result of laboratory exposure. Viruses of medical importance (as judged either by the severity of the disease that they cause, the numbers of persons affected, or both) are chikungunya, EEE, o'nyong-nyong, Ross River, VEE (including other viruses in the complex), and WEE. Of lesser importance but still to be considered is Mayaro.

As is the case with arboviruses from other taxons, especially flaviviruses, the fully developed disease caused by alphaviruses is generally biphasic. Effective exposure is followed by local multiplication of the virus and is manifested clinically by ill-defined, generalized or systemic signs and symptoms including fever, malaise, headache, and body aches. After a short remission, the disease progresses to a second phase with viremia followed by localization of the virus in specific organs or tissues, resulting in characteristic syndromes depending on the tropism of the virus. EEE, VEE, and WEE cause encephalitis; chikungunya and o'nyong-nyong lead to generalized lymphadenopathy, rash, and muscular and joint pains; chikungunya can cause hemorrhagic manifestations of considerable intensity; Ross River virus appears to be the cause of a well-defined type of epidemic polyarthritis with skin rash.

Often, the infection does not progress beyond the first phase. With some viruses such as Mayaro, general systemic involvement with no particular localization appears to be the rule. Fully developed EEE virus encephalitis as observed in the United States is an extremely severe illness, with mortality between 20 and 50%. Survivors are likely to retain serious, incapacitating sequelae. WEE encephalitis is less severe, with mortality ranging from occasional deaths to up to 10%. Death can follow infection with VEE virus, but this does not often occur. Fatal outcome directly attributable to infection with o'nyong-nyong or chikungunya virus in Africa appears to be rare or unreported; the latter virus seems to have been associated with deaths following severe hemorrhagic manifestations in India.

Laboratory Diagnosis

Specific clinical diagnosis is not possible even in fully developed cases with organ localization. Laboratory diagnosis, which is the only specific procedure, is accomplished by isolation of the virus from the patient and identification and by demonstration of antibody development between early and late serum specimens. Ideally, both virus isolation and detection of antibodies should be accomplished; however, it is often impossible to isolate virus, and antibody determination is the only proof of specificity.

Specimen Collection

The clinical specimen of choice in a suspected alphavirus infection is blood serum. A sample of serum should be taken as soon as possible after onset (the acute or early sample). If the patient survives the disease, a second sample is taken between 10 to 12 and 20 to 22 days after onset (the convalescent or late sample). In fatal cases, usually as a result of encephalitis, and if an autopsy is allowed, specimens of brain tissue from several areas are collected. Virus isolation is attempted from the early serum and from brain tissue. Isolation of virus from the cerebrospinal fluid (CSF) has been successful, but only rarely. VEE has been isolated from throat washings during the early phase of the disease.

Virus Isolation and Identification

Virus isolation and identification is accomplished as follows:

1. Early serum, undiluted and in dilutions of 1:10 and 1:100, is inoculated intracerebrally (i.c.) into newborn mice or to cell cultures, among which VERO, BHK-21, and chick embryo fibroblasts are recommended. There has recently been interest in inoculating cell cultures derived from *Aedes* mosquitoes. Although the cells show no CPE, they support virus growth which can be demonstrated by serological tests or passage to other susceptible serum.
2. If viral isolation is successful, i.e., mice sicken or die or cells in culture show CPE, a stock suspension is prepared by collecting brain tissue from the mice or the fluid and cells from the tissue cultures. Usually, the mice are sick or the cells in cultures show CPE from 24 to 72 hr after inoculation.
3. Identification of the virus in the stock suspensions is carried out. Certain preliminary steps are necessary in order to rule out bacterial contaminants and toxic effects on the cells in culture due to the serum. Furthermore, it is necessary to ascertain that no indigenous viruses from the mice or the cell cultures are responsible for the observed effects. If the titer of the virus isolated is sufficiently high, the stocks prepared with the first mouse or tissue culture passages may be used directly for identification of the agent by using the stocks as antigens and testing them against a collection of alphavirus immune sera by the CF and flourescent antibody (FAT) tests and later by the neutralization test. It is most important to realize that a positive serological reaction (especially CF, FAT, and, when used, HI) against a serum is not necessarily diagnostic alone. The known cross-reactivity among these viruses necessitates that a positive result be supported by negative results or proportionally lower levels of positive results against other sera.

Electron Microscopy

The morphology and morphogenesis of a virus in cell cultures or mouse organs cannot be used as a specific diagnostic criterion. At best, the picture may give an indication that the virus involved is an alphavirus.

Serological Tests

The most generally used serological tests are neutralization , CF, HI, and, increasingly, FAT. Other tests such as indirect hemagglutination, ferritin precipitation, agar gel diffusion and precipitation, and radiommunoassay have also been applied, but less generally. The tests are employed for identification of virus isolates and diagnosis of virus infections through the detection of antibodies in sera.

A complete collection of immune reagents is required for identification of an isolate. For technical reasons, these reagents are usually prepared in mice and consist of either immune serum or immune ascitic fluid, the latter induced by inoculation of immunized mice with ascites-producing agents such as Freund's complete adjuvant or certain lines of mouse sarcoma. The homologous titers of the immune reagents in each of the various tests is determined. Of course, cross-reactions are anticipated among viruses of the group, especially among viruses of a subgroup or complex. Cross-reactions can be minimized by using sera from animals inoculated only once or twice with the virus; if the resulting antibody titer is too low, more immunizing injections are required, but they should be held to a minimum.

Serological tests are also used for presumptive specific diagnosis of infection in patients from whom virus is not isolated during the acute phase or in individuals tested in the course of seroepidemiological surveys. A collection of antigens of known potency is required for CF, HI, or neutralization tests to be performed in mice or cells in culture. For the diagnosis of a current infection, acute and convalescent samples of serum are tested simultaneously against several antigens. In situations in which only a given virus of the taxon is known to be active in a geographical area, this antigen alone is used. Conversion from negative in the acute sample to positive in the second sample or from a low titer to a titer at least four or eight times greater is considered diagnostic. The CF test is a simple and reliable method for the diagnosis of a current infection; it is more type specific than the HI test, but less type specific than the neutralization test. Furthermore, the development of CF antibodies appears to lag somewhat as compared with that of HI and neutralizing antibodies; therefore, conversions in the former test are often observed when the latter already show high titers in the early serum. FAT has been effectively used for identification of virus isolates in certain ecological situations in which only one group A virus, WEE, is enzootic or endemic. [8]

Pathogenicity in Experimental Animals

All currently known alphaviruses are pathogenic for newborn (from 0- to 4- or 5-days-old) Swiss mice on i.c. inoculation; disease and death follows from 2 to 4 or 5 days later, depending on the dose of virus and degree of adaptation to the host. Weanling and adult mice are lethally infected by i.c. inoculation of EEE, WEE, and VEE epidemic types and certain strains of Sindbis, Semliki Forest, and chikungunya viruses, but rarely by the other agents of the group. Guinea pigs and hamsters are lethally infected by i.c. inoculation of strains of EEE and WEE and epidemic strains of VEE; the response of these animals to inoculation of other viruses is less well established, although high titers of antibodies develop with some viruses, presumably indicating inapparent infection. Range of and response to inoculation of experimental animals as a final means of specific diagnosis for alphaviruses is of limited value.

Alphaviruses in general propagate with CPE in a variety of cell cultures both under fluid medium and under an overlay in the latter case with plaque formation. Among the most commonly used stable cells are African green monkey kidney (especially VERO line) and baby hamster kidney cells (BHK-21); primary chick embryo cell cultures are also susceptible to these viruses. *Aedes albopictus* and *A. aegypti* cells in culture[9] support replication of alphaviruses, but with no visible CPE; the virus can be detected in high titer in these cultures either by in vitro tests or by subpassage to another host.

General References

Casals, J., The arthropod-borne group of animal viruses, *Trans. N. Y. Acad. Sci.* (Ser. 2), 19, 219, 1957.

Wildy, P., Classification and nomenclature of viruses, *Monogr. Virol.*, No. 5, 1971.

Berge, T. O., Ed.,*International Catalogue of Arboviruses,* 2nd ed., Publ. No. (CDC) 75-8301, Department of Health, Education, and Welfare, Washington, D. C., 1975.

Fenner, F., The classification and nomenclature of viruses, *J. Gen. Virol.*, 31, 463, 1976.

Porterfield, J. S., Casals, J., Chumakov, M. P., Gaidamovich, S. Ya., Hannoun, C., Holmes, I. H., Horzinek, M. C., Mussgay, M., Oker-Blom, N., and Russell, P. K., Togaviridae , *Intervirology,* 9, 129, 1978.

REFERENCES

1. **Fenner, F.**, The classification and nomenclature of viruses, *J. Gen. Virol.*, 31, 463, 1976.
2. **Casals, J.**, Viruses: the versatile parasite: the arthropod-borne group of animal viruses, *Trans. N. Y. Acad. Sci.*, 19, 219, 1957.
3. **Wildy, P.**, Classification and nomenclature of viruses, *Monogr. Virol.*, No. 5, 1971.

3a. **Murphy, F. A. and Whitfield, S. G.**, Eastern equine encephalitis virus infection: electron microscopic studies of mouse central nervous system, *Exp. Mol. Pathol.*, 13, 131, 1970.

4. **Porterfield, J. S., Casals, J., Chumakov, M. P , Gaidamovich, S. Ya., Hannoun, C., Holmes, I. H., Horzinek, M. C., Mussgay, M., Oker-Blom, N., and Russell, P. K.**, Togaviridae, *Intervirology,* 9, 129, 1978.
5. **Dalrymple, J. M., Vogel, S. N., Teramoto, A. Y., and Russell, P. K.**, Antigenic components of group A arbovirus virions, *J. Virol.*, 12, 1034, 1973.
6. **Pedersen, C. E. and Eddy, G. A.**, Separation, isolation and immunological studies of the structural proteins of Venezuelan equine encephalomyelitis virus, *J. Virol.*, 14, 740, 1974.
7. **Berge, T. O., Ed.**, International Catalogue of Arboviruses, 2nd ed., Publ. No. (CDC) 75-8301, Department of Health, Education, and Welfare, Atlanta, 1975.
8. **Emmons, R. W., Dodero, D. V., Devlin, V., and Lennette, E. H** , Serologic diagnosis of Colorado tick fever. A comparison of complement-fixation, immunofluorescence, and plaque reduction methods, *Am. J. Trop. Med. Hyg.*, 18, 796, 1969.
9. **Singh, K. R. P.**, Cell cultures derived from larvae of *Aedes albopictus* (Skuse) and *Aedes aegypti* (L),*Curr. Sci.*, 36, 506, 1967.
10. **Ross, R. W.**, The Newala epidemic. III. The virus: isolation, pathogenic properties and relationship to the epidemic, *J. Hyg.*, 54, 177, 1956.
11. **Ten Broeck, C. and Merrill, M. H.**, A Serological difference between eastern and western equine encephalomyelitis virus, *Proc. Soc. Exp. Biol. Med.*, 31, 217, 1933.
12. **Anderson, C. R., Downs, W. G., Wattley, G. H., Ahin, N. W., and Reese, A. A.**, Mayaro virus: a new human disease agent. II. Isolation from blood of patients in Trinidad, B.W.I., *Am. J. Trop. Med. Hyg.*, 6, 1012, 1957.
13. **Williams, M. C and Woodall, J. P.**, O'nyong-nyong fever: an epidemic virus disease in East Africa. II. Isolation and some properties of the virus, *Trans. R. Soc. Trop. Med. Hyg.*, 55, 135, 1961.
14. **Doherty, R. L., Whitehead, R. H., and Gorman, B. M.**, The isolation of a third group A arbovirus in Australia, with preliminary observations on its relationship to epidemic polyarthritis, *Aust. J. Sci.*, 26, 183, 1963.
15. **Taylor, R. M., Hurlbut, H. S., Work, T. H , Kingston, J. R., and Frothingham, T. E.**, Sindbis virus: a newly-recognized arthropod-transmitted virus, *Am. J. Trop. Med. Hyg.*, 4, 844, 1955.
16. **Beck, C. E. and Wyckoff, R. W. G.**, Venezuelan equine encephalomyelitis, *Science,* 88, 530, 1938.
17. **Meyer, K. F., Haring, C. M., and Howitt, B.** , The etiology of epizootic encephalomyelitis in horses in the San Joaquin valley, *Science,* 74, 227, 1931.

TOGAVIRIDAE: FLAVIVIRUS

Jordi Casals

Introduction

Flavivirus is a genus in the family Togaviridae.[1] The description of this genus and the properties of the agents included have much in common with those of the alphavirus genus.

In order to characterize a virus, its genome and polypeptides must be investigated; however, with the exception of morphological studies, the description of this taxon is based on more or less detailed examination of only 10 to 12 of the member viruses: dengue 1; dengue 2; Japanese encephalitis (JE); St. Louis encephalitis (SLE); yellow fever (YF); Murray Valley encephalitis (MVE); Russian spring-summer encephalitis(RSSE); Banzi (BAN); Kunjin (KUN); and central European tick-borne encephalitis (CETBE) strains Kumlinge, Hypr, and Hanzalova. With the exception of some electron microscopy studies, little has been reported on the remaining viruses included in the taxon, which now numbers 55 distinct serotypes. Viruses are assembled in the taxon primarily on the basis of documented serological cross-reactivity.[1] Therefore, flavivirus is synonymous with antigenic group B,[2] but the correspondence between these two designations may not be complete. There are at least two new viruses which are considered to be flaviviruses but are antigenically unrelated to any of the group B arboviruses.[3,4] Because the current definition of flavivirus includes serological relationship, the taxonomic position of these two viruses is not clear.

The flaviviruses contain a single piece of linear, single-stranded RNA of 4×10^6 daltons mol wt. They have a capsid, probably of cubic symmetry, enclosed within a lipoprotein envelope with an overall external diameter of approximately 45 nm. Some flaviviruses have mosquitoes as invertebrate hosts, others have ticks, and some have no known arthropod hosts.

Morphology

The overall shape of the viruses is spherical; they have knob-like surface projections. Virions have a diameter smaller than that of the alphaviruses (between 35 and 45 nm). The terminal knobs of the projections are 2 nm in diameter. A bilayer membrane separated from the core by an electronlucent zone is perceived in thin sections. In other instances, virions appear to have a dense core surrounded by a translucent halo. The structural symmetry of the flaviviruses remains unresolved, although certain observations suggest that they are cubic[5] (Figure 1). Figure 2, a thin section of mouse brain tissue infected with St. Louis encephalitis virus, shows the accumulation of mature virions within cytoplasmic cisternae.[4a]

Physicochemical Properties

The single RNA molecule has a sedimentation coefficient of 42S, represents 8% of the weight of the virion, is infectious, and is presumed to serve as messenger RNA. Virions contain three polypeptides and no enzymes. Their density is 1.19 to 1.24 ; their sedimentation coefficient is 175 to 200S (dengue 2), 218S (CETBE), or 205S (Kunjin); and their molecular weight is 4.6×10^6 daltons (dengue 2). Virions are stable between pH 3 and 9; They are rapidly inactivated at 56°C; and they are sensitive to lipid solvents, detergents, and trypsin.[6]

Virus Replication and Maturation

Assembly takes place exclusively in the cytoplasm in close connection with cyto-

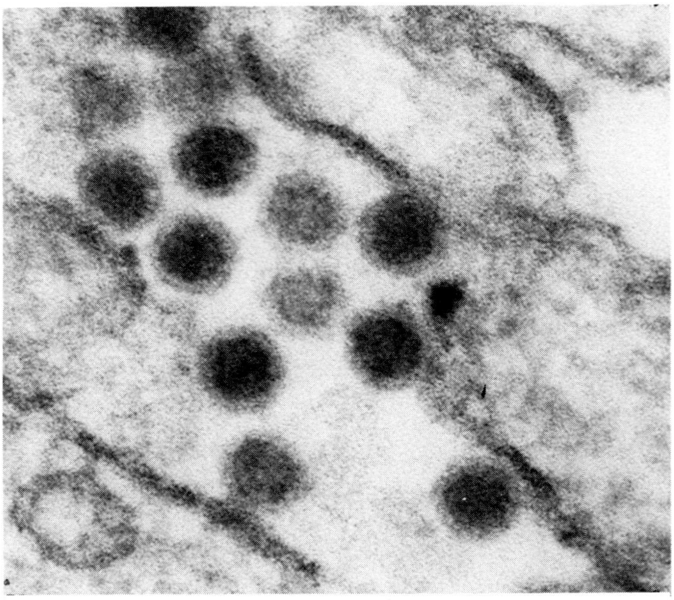

FIGURE 1. St. Louis encephalitis virus in the brain of infected newborn mice. Virions 38 nm in extracellular space. Lead citrate; magnification × 290,000 (Courtesy of Dr. Frederick A. Murphy, Center for Disease Control, Atlanta, Ga.)

plasmic reticulum membranes. The viral RNA codes for three structural polypeptides, two of which are in the envelope: a glycoprotein between 51,000 and 58,000 daltons mol wt and a nonglycoprotein of 9,000 daltons mol wt. The third polypeptide, of between 13,000 and 18,000 daltons mol wt, is a nucleoprotein in the core. Five nonstructural polypeptides have been reported in JE virus, with molecular weights of 10,500, 19,000, 45,000, 71,000, and 93,000 daltons.

The single-stranded viral RNA is transcribed by a virion-induced polymerase into complementary of minus-strand RNA, which serves as template for production of progeny RNA. Virus replication is not affected by actinomycin D.[6]

Viral synthesis takes place in the cytoplasm, and maturation occurs by budding on endoplasmic reticulum membranes without previous formation of distinct core particles. The mature virions accumulate in cytoplasmic cisternae.

Virus Types and Antigenic Properties

All antigenic group B viruses are serologically related. Whether viruses similar to established flaviviruses in all respects but lacking antigenic relationship should be included in the taxon has not been resolved.

There are at least two distinct antigenic moieties in the virion, a core protein and an envelope glycoprotein, and four or five distinct nonstructural antigens. The envelope antigen is involved in virus neutralization. Antisera prepared against core and envelope do not cross-react.

As is true with the alphaviruses, it is essential from a practical, diagnostic standpoint to know that the viruses in this taxon show varying degrees of cross-reactivity depending on the viruses and the test employed. Of the more generally used tests-hemagglutination inhibition (HI), complement fixation (CF), and neutralization — the HI test gives the most pronounced cross-reactions, neutralization (either intracerebral (i.c.) inoculation in mice or plaque reduction in cells in culture) gives the least cross-reactivity,

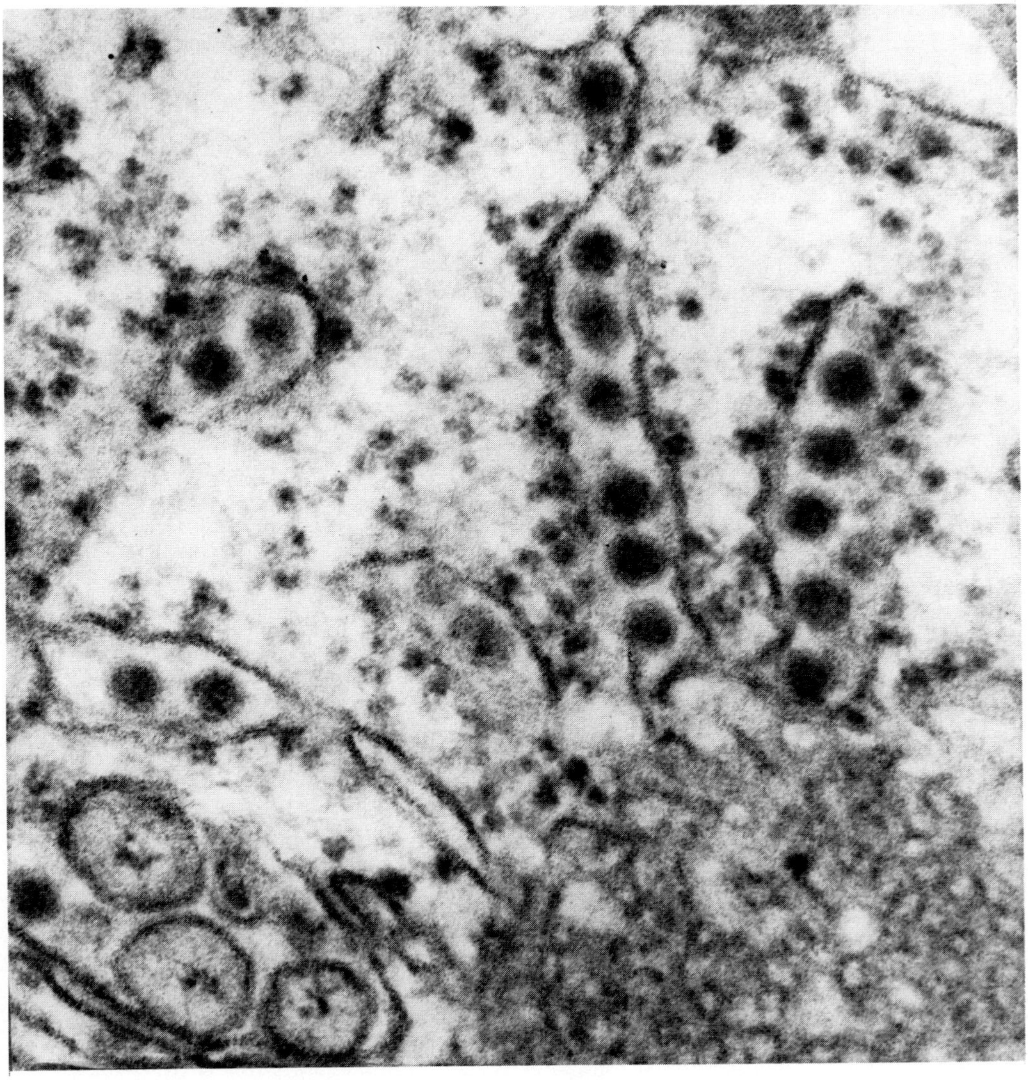

FIGURE 2. Thin section of mouse brain tissue infected with St. Louis encephalitis virus. Mature virions can be seen accumulating within the cisternae. Magnification × 146,000. (From Murphy, F. A., Harrison, A. K., Gary, G. W., Jr., Whitfield, S. G., and Forrester, F. T., *Lab. Invest.*, 19, 652, 1968. © U.S.-Canadian Division of the International Academy of Pathology. With permission.)

and the CF test is intermediate. According to the degree of cross-reactivity among the flaviviruses and their antisera, subgroups or complexes can be noted, and the viruses in a complex cross-react among themselves to a higher degree than with the other agents in the taxon.

The flavivirus taxon, or antigenic group B, is presently comprised of the following 55 agents:[7]

Alfuy, Apoi, Banzi, Batu Cave, Bouboui, Bukalasa bat, Bussuquara , Carey Island, Central European tick-borne encephalitis (CETBE) , Cowbone Ridge, Dakar bat, dengue 1, dengue 2, dengue 3, dengue 4, Edgehill, Entebbe bat, Ilheus, Israel turkey meningoencephalitis, Japanese encephalitis (JE), Jugra, Jutiapa, Kadam, Karshi, Kokobera, Koutango, Kunjin, Kyasanur Forest disease (KFD), Langat, Louping ill, Modoc, Montana myotis leukoencephalitis, Murray Valley encephalitis (MVE), Negishi,

Ntaya, Omsk hemorrhagic fever (OHF), Phnom-Penh bat, Powassan, Rio Bravo, Royal Farm, Russian spring-summer encephalitis (RSSE), Saboya, Sepik, Sokoluk, Spondweni, St. Louis encephalitis (SLE), Stratford, Tembusu, Tyuleniy, Uganda S, Usutu, Wesselsbron, West Nile (WN), yellow fever (YF), and Zika

The following subgroups or complexes have been identified:

Dengue: dengue types 1, 2, 3, and 4
Tick-borne complex: CETB, Kadam, KFD, Langat, Louping ill, Negishi, OHF, Powassan, Royal Farm, RSSE, and Tyuleniy
SLE-JE-WN complex: Edgehill, Ilheus, Israel turkey meningoencephalitis, JE, Kokobera, Kunjin, MVE, SLE, Stratford, Tembusu, Usutu, Wesselsbron, and WN

The remaining 27 viruses in the group cannot definitely be included in the above complexes or in others as yet undescribed; this may be attributable to insufficient information. Enumeration of subtypes and varieties will not be considered here.

Clinical Features

Infection producing clinical manifestation in man has been reported in association with approximately half the flaviviruses ; however, some cause disease in nature rarely or only as a result of laboratory infection. Importance of a virus may be measured by the severity of the disease produced, its prevalence, or both; The following viruses are important as causes of human illness: CETBE; dengues 1, 2, 3, and 4; JF; KFD; MVE; Negishi; OHF; Powassan; RSSE; SLE; WN; and YF (Table 1). A few others have been associated with disease less frequently; these viruses are Banzi, Bussuquara, Ilheus, Louping ill and Wesselsbron. Although diseases of domestic animals are outside the scope of this handbook, it may be noted that Wesselsbron virus has been associated with a disease of sheep in southern Africa, and the possibility exists that man may acquire infection from that source.

Diseases caused by flaviviruses have a biphasic course. After an incubation period which generally lasts between 5 and 10 days, there is a systemic phase with fever, malaise, and generalized vague signs and symptoms. Some infections do not progress beyond this point. In the fully developed disease, a short remission is followed by a second phase, characterized by manifestations due to localization of the virus in specific organs or tissues in addition to general signs and symptoms. Depending on the organ, the following syndromes are observed: meningoencephalitis, associated with CETBE virus; encephalitis, caused by JE, MVE, Negishi, Powassan, RSSE, SLE, and, occasionally, WN viruses; hemorrhagic fever with or without shock, caused by superinfection with a dengue serotype of a person previously infected by a different dengue virus; hemorrhagic fever with severe hepatitis, due to yellow fever virus; hemorrhagic fever, caused by KFD and OHF viruses; dengue syndrome, uncomplicated, caused by dengue types, 1, 2, 3, and 4 and sometimes by WN viruses; and systemic undifferentiated illness associated with Banzi, Wesselsbron, and Zika.

Mortality varies, depending, among other circumstances, on the organ affected. It is greatest with certain encephalitides (JE, RSSE, MVE, and SLE), yellow fever, KFD and OHF hemorrhagic fevers, and in infants and young children that develop dengue shock syndrome; it is almost nil with uncomplicated dengue and systemic infections.

Laboratory Diagnosis

Clinical and epidemiological data may suggest a specific diagnosis, but this can only be attained by laboratory procedures. Isolation and identification of the virus and/or detection of antibody development in the patient are the bases for diagnosis.

Table 1
TOGAVIRUS: FLAVIVIRUS

Virus type	Clinical manifestations	Source of specimen	Laboratory diagnosis, Isolation technique: Cell culture	Laboratory diagnosis, Isolation technique: Animal	Laboratory diagnosis, Isolation technique: Fluorescent antibody and/or electron microscopy	Serologic diagnosis[a]	Ref.
Central European tick-borne encephalitis	Meningoencephalitis, encephalitis; low mortality	Brain, CSF, blood	Chick embryo	Mouse, newborn and weanlings; chick embryo	No reports	N, CF, HI, FAT, agar gel precipitation	15
Dengue 1	Febrile illness with rash; hemorrhagic fever; dengue shock syndrome with considerable mortality	Blood	Primate cells, LLC-MK_2; mosquito cells	Newborn mouse, mosquito	In inoculated mosquitoes	N, CF, HI, FAT	16
Dengue 2	Febrile illness with rash; hemorrhagic fever; dengue shock syndrome with considerable mortality	Blood	Primate cells, LLC-MK_2; mosquito cells	Newborn mouse, mosquito	In inoculated mosquitoes	N, CF, HI, FAT	16
Dengue 3	Febrile illness with rash; hemorrhagic fever; dengue shock syndrome with considerable mortality	Blood	Primate cells, LLC-MK_2; mosquito cells	Newborn mouse, mosquito	In inoculated mosquitoes	N, CF, HI, FAT	17
Dengue 4	Febrile illness with rash; hemorrhagic fever; dengue shock syndrome with considerable mortality	Blood, liver	Primate cells, LLC-MK_2; mosquito cells	Newborn mouse, mosquito	In inoculated mosquitoes	N, CF, HI, FAT	17
Ilheus	Encephalitis, systemic	Blood	Hamster kidney	Mouse, newborn and weanling	No reports	N, CF, HI	18
Japanese encephalitis	Encephalitis, high mortality	Blood, brain, CSF, viscera	Chick embryo	Mouse, newborn and weanling; chick embryo	No reports	N, CF, HI	19

Table 1 (continued)
TOGAVIRUS: FLAVIVIRUS

			Laboratory diagnosis, Isolation technique				
Virus type	Clinical manifestations	Source of specimen	Cell culture	Animal	Fluorescent antibody and/or electron microscopy	Serologic diagnosis[a]	Ref.
Kyasanur forest disease	Hemorrhagic fever	Blood, brain, CSF, viscera	HeLa, primate, other mammal, chick embryo	Mouse, newborn and weanling; hamster	No reports	N, CF, HI, agar gel precipitation	20
Louping ill	Encephalitis	Blood, CSF	Possible; no reports	Mouse, newborn and weanling; sheep	No reports	N, CF, HI, agar gel precipitation	21
Murray Valley encephalitis	Encephalitis	Brain	Possible; no reports	Mouse, newborn and weanling; chick embryo	No reports	N, CF, HI	22
Omsk hemorrhagic fever	Hemorrhagic fever	Blood, CSF, heart	Chick embryo	Mouse, newborn and adult; chick embryo; guinea pig; *Microtus*	No reports	N, CF, HI, agar gel precipitation	23
Russian spring-summer encephalitis	Encephalitis	Brain, blood, CSF, viscera	Chick embryo; others not reported	Mouse, newborn and weanling; chick embryo; guinea pig	No reports	N, CF, HI, agar gel precipitation	24
St. Louis encephalitis	Encephalitis	Brain, blood	Possible; no reports	Mouse, newborn and weanling	No reports	N, CF, HI FAT	25
West Nile	Febrile with rash; also encephalitis	Blood, CSF	Possible; no reports	Mouse, newborn and weanling	No reports	N, CF, HI	26
Yellow Fever	Hemorrhagic fever with severe hepatitis	Blood, liver	Primates, hamster, pig	Mouse, newborn and weanling; monkeys	No reports	N, CF, HI, FAT	27

Note: Human cases, natural and/or laboratory acquired, have been reported in small numbers with Apoi, Banzi, Bussuquara, Kunjin, Negishi (fatal), Powassan (fatal), Rio Bravo, Sepik, Spondweni, Wesselsbron, and Zika. Rocio, a newly discovered flavivirus has caused a large epidemic in Brazil.[28]

[a] N, neutralization; CF, complement fixation; HI, hemagglutination inhibition; FAT, fluorescent antibody test.

Specimen Collection

The specimen of choice for virus isolation is blood serum taken early in the first phase; unfortunately, such a specimen is often unavailable due to lack of foresight. Most patients are seen by a physician during the second phase when signs and symptoms of localization point to an encephalitis or other severe form of illness; at this time, virus is no longer circulating at an appreciable titer. Isolations have also been made from cerebrospinal fluid (CSF) in cases of encephalitis or meningoencephalitis. In fatal cases, virus is isolated from various organs, especially brain and liver and less frequently, spleen and lymph nodes. A method used extensively in the early days of yellow fever surveys was to take, in fatal cases, a piece of liver by means of a viscerotome and make a histological diagnosis. This procedure, while not strictly specific, was amazingly accurate in experienced hands.

In all instances in which a flavivirus infection is suspected, a second specimen of blood serum should be taken during the late phase of the disease or early convalescence (from 10 to 20 days after onset) to test for antibodies. The late or early convalescent sample tested simultaneously with the early or acute one shows whether or not antibodies have developed during the course of the infection.

Virus Isolation and Identification

Ideally, a sequential series of steps should be taken for isolation and identification of a flavivirus. In actual practice , some of the steps are eliminated or taken out of order. The ideal stepwise procedure is as follows:

1. The clinical or autopsy specimen is inoculated into a susceptible host system — either newborn mice or cell cultures. Among the latter are extensively used VERO or BHK-21 cells. If a dengue virus is suspected, LLC-MK_2 cells used under an overlay are preferable. A recent development for isolation of dengue virus consists of inoculation of the mosquito *Aedes albopictus* or other species. The susceptibility of this method is far greater than any method previously employed for the dengue viruses. Inoculation of arthropod cells in culture, especially *A. albopictus* cells, is an alternative procedure. With either the whole insect or insect cells in culture, a successful isolation requires demonstration of the virus either serologically or by passage through another host that will develop disease or show cytopathic effect (CPE).
2. If an infectious agent is isolated, it must be established as a virus by elimination of other possible microorganisms. Filtration through readily available filters of average pore diameter (APD) 220 nm is an easy and generally effective procedure.
3. If the filtrate is passed serially, indigenous viruses in the laboratory animals or cell cultures used must be excluded. The risk is to use mice from a colony infected with murine viruses such as mouse polioencephalitis, lymphocytic choriomeningitis (LCM), hepatitis , or other agents. Indigenous or passenger viruses may also occur in cell cultures.[8] An easy procedure is to test immune reference sera for murine viruses against an antigen prepared with the isolate.
4. Because arboviruses belonging to other antigenic groups can be isolated in circumstances similar to those in which flaviviruses prevail, a practical step at this point is to determine the group affiliation of the isolate. This is done by testing an antigen, usually a CF antigen, prepared with the isolate against grouping polyvalent arbovirus sera or ascitic fluids. Such reagents are available for antigenic groups A, B, C, Bunyamwera, Bwamba, vesicular stomatitis, and other groups.[9] Placement of the isolate in a group facilitates its identification.
5. If the isolate is a flavivirus, it is next typed by testing against specific sera by CF. The operation is simplified if the unknown is first placed in a subgroup or complex

through the use of the adequate serum (dengue complex, tick-borne complex, or SLE-JE-WN complex). Once the identification is done by an in vitro test, the final identification requires that a neutralization test be performed either by the i.c. route in mice or, preferably, by plaque reduction neutralization.

Characterization in Cell Cultures

The morphology and morphogenesis of a virus in cell cultures does not give a specific diagnosis. At best, electron microscopy will give an indication that a virus is a flavivirus. Lesions in cultured cells, range of susceptible cells and other pathological characteristics, and host range susceptibility are not pathognomonic.

Serological Tests

The serological tests commonly used for identification of flaviviruses are neutralization, either in cells in culture or in mice; HI; and CF. Flourescent antibody technique (FAT) is increasingly being used, especially since incident light microscopes have become generally available. Other tests such as agar gel diffusion and precipitation (AGDP), indirect hemagglutination, ferritin precipitation, and radioimmunoassay are less frequently used as routine diagnostic procedures. Serological tests are employed for identification of virus isolates by using known immune reagents and for diagnosis of past infections through the detection of antibodies in sera by employing known antigens.

Identification of an isolate is accomplished by following the stepwise procedure outlined above. A complete collection of immune sera or ascitic fluids is required. For technical reasons, these reagents are prepared in mice. The ascitic fluids are obtained by injecting immunized mice with either Freund's complete adjuvant or certain lines of mouse sarcoma cells. The homologous titer of the immune reagent and the titers of the reagent against other viruses of the subgroup or complex are determined. In general, cross-reactivity among viruses of a subgroup or complex is more marked with the flaviviruses than it is with the alphaviruses, especially with HI tests.

Serological tests are also used for presumptive specific diagnosis of infection in recent patients from whom no virus has been isolated during the acute phase or from persons participating in a seroepidemiological survey. In the latter case, it is often very difficult to determine the virus responsible for the antibodies detected. In certain areas of the world, several flaviviruses may be prevalent; thus superinfections (i.e., sequential infections) may occur. A person sequentially infected will develop widely reacting antibodies not only against the infecting viruses but also against other of or even outside the complex [10,11] In such a situation, HI results will give the most pronounced cross-reactions, followed by the CF test; the neutralization test, either by the i.c. route in mice, plaque reduction in cell monolayers under an overlay, or CPE neutralization in fluid cultures, will give the highest degree of specificity, although the specificity is not complete.

The CF test is widely employed and is especially useful as an aid in the diagnosis of a current illness. The antibody titer rises more slowly than in the HI or neutralization tests and ordinarily decreases rapidly, usually within a few months. The HI test and especially the neutralization test remain positive for much longer periods of time, the latter presumably for a lifetime; therefore, these tests are useful for seroepidemiological surveys.

Diagnostic serological tests with the togavirus, both alphavirus and flavivirus, require availability of the specific reagents needed as well as a certain expertise. A number of State and Federal health laboratories are now fully competent to conduct such tests.

FAT is currently used as a rapid, sensitive, and accurate method for the diagnosis of dengue viruses. The material suspected of containing virus is parenterally inoculated into susceptible mosquitoes. After a few days, an impression smear is made with the head or salivary glands of the mosquito, which is then stained with a fluorescein-conjugated dengue-immune serum.[12] Identification of St. Louis virus by means of FAT has also been used routinely.[13]

Pathogenicity in Experimental Animals

Flaviviruses are lethally pathogenic for newborn Swiss mice after i.c. inoculation. The only exceptions appear to be some strains of dengue viruses, especially strains of dengue 3 and 4 virus, which are isolated relatively easily by mosquito inoculation or plaque formation on LLC-MK_2 cells. By serial passages, most of these strains can be adapted to newborn mice. Weanling and young adult mice are leathally infected following i.c. inoculation by the great majority of flavivirus, especially after they have been passaged in newborn mice. Other small laboratory animals including guinea pigs, hamsters, and rabbits may be susceptible to some flaviviruses, but not to others; no host is as uniformly susceptible as the newborn mouse.

A variety of cell cultures (VERO, BHK-21, HeLa, and LLC-MK_2) support replication of flaviviruses with CPE and plaque formation. CPE under fluid medium is often not complete; therefore, problems in the interpretation of results may occasionally arise. However, a few passages in the cell culture usually improve the result. Of particular interest is the use of LLC-MK_2 monolayers under an overlay for work with dengue viruses. The cells are excellent for isolation and for plaque reduction tests for type identification. As with alphaviruses, *Aedes albopictus* and *A. aegypti* cells in cultures support replication of mosquito-borne flaviviruses, but with no visible CPE. The virus can be evidenced either by transfer to a system that shows lesions as a result of infection or by serological tests, especially CF tests.[14]

GENERAL REFERENCES

Casals, J., The arthropod-borne group of animal viruses, *Trans. N.Y. Acad. Sci.* (Ser. 2), 19, 219,1957.

Casals, J. , Arbovirus infections, in *Serological Epidemiology*, Paul, J. R. and White, C., Eds., Academic Press, New York, 1973, 99

Berge, T. O., Ed. International Catalogue of Arboviruses, 2nd ed., Publ. No. (CDC) 75-8301, Department of Health, Education, and Welfare, Washington, D. C., 1975.

Fenner, F., The classification and nomenclature of viruses, *J. Gen. Virol.*, 31, 463, 1976.

Porterfield, J.S., Casals, J., Chumakov, M. P., Gaidamovich, S. Ya., Hannoun, C., Holmes, I. H., Horzinek, M. C., Mussgay, M., Oker-Blom, N., and Russell, P. K., Togaviridae, *Intervirology*, 9, 129, 1978.

REFERENCES

1. **Fenner, F.**, The classification and nomenclature of viruses, *J. Gen. Virol.*, 31, 463, 1976.
2. **Casals J.**, Viruses: the versatile parasite: the arthropod-borne group of animal viruses, *Trans. N Y. Acad. Sci.* Ser. 2, 19, 219, 1957
3. **Igarashi, A., Harrap, K. A., Casals, J., and Stollar, V.**, Morphological, biochemical and serological studies on a viral agent (CFA) which replicates in and causes fusion of *Aedes albopictus* (Singh) cells, *Virology*, 74, 174, 1976.

4. **Trousdale, M., Trent, D., and Shelokov, A.**,Studies on simian hemorrhagic fever (SHF), in preparation,
4a. **Murphy, F. A. , Harrison, A. K., Gary, G. W., Jr., Whitfield, S. G., and Forrester, F. T.**, St. Louis encephalitis virus infection of mice. Electron microscopic studies of central nervous system, *Lab. Invest.*, 19, 652, 1968.
5. **Abdelwahab, K. S. E., Almeida, J. D., Doane, F. W., and McLean, D. M.**, Powassan virus: morphology and cytopathology,*Can. Med. Assoc. J.*, 90, 1068, 1964.
6. **Porterfield, J. A., Casals, J., Chumakov, M. P., Gaidamovich, S. Ya., Hannoun, C., Holmes, I. H., Horzinek, M. C. , Oker-Blom, N., and Russell, P. K.**, Togarviridae, *Intervirology*, 9, 129, 1978.
7. **Berge, T. O., Ed.**, International Catalogue of Arboviruses, 2nd ed., Pub. No. (CDC) 75-8301, Department of Health, Education, and Welfare, Atlanta, 1975.
8. **Cunningham, A., Buckley, S. M., Casals, J., and Webb, S. R.**, Isolation of chikungunya virus contaminating an *Aedes albopictus* cell lines, *J. Gen. Virol.*, 27, 97, 1975.
9. **Cunningham, S. and Nutter, J. E., Eds.**, National Institute of Allergy and Infectious Disease Catalog of Research reagents 1975-77, Pub. No. (NIH) 75-899, Department of Health, Education, and Welfare, Washington, D. C., 1975.
10. **Casals, J.**, Arbovirus infections, in *Serological Epidemiology*, Paul, J. R. and White, C., Eds., Academic Press, New York, 1973, 99
11. **Buckley, S. M., Davis, J. L., III, Madalengoitia, J., Flores, W., and Casals, J.**, Arbovirus neutralization tests with Peruvian sera in Vero cell cultures, *Bull. Who*, 46, 451, 1972.
12. **Rosen, L. and Gubler, D.**, The use of mosquitoes to detect and propagate dengue viruses, *Am. J. Trop. Med. Hyg.*, 23, 1153, 1974.
13. **Emmons, R. W., Dodero, D. V., Devlin, V., and Lennette, E. H.**, Serologic diagnosis of Colorado tick fever. A comparison of complement-fixation, immunofluorescence, and plaque reduction methods, *Am. J. Trop. Med. Hyg.*, 18, 796, 1969.
14. **Ajello, C., Gresikova, M., Buckley, S. M., and Casals, J.**, Detection of West Nile complement-fixing antigen in *Aedes albopictus* cell cultures, *Acta Virol.* (Eng. Ed.) 19, 441, 1975.
15. **Gallia, F., Rampas, J., and Hollender, L.**, Laboratory infection with encephalitis virus, *Cas. Lek. Cesk.*, 88, 224, 1949.
16. **Sabin, A. B. and Schlesinger, R. W.**, Production of immunity to dengue with virus modified by propagation in mice, *Science*, 101, 640, 1945.
17. **Hammon, W. McD., Rudnick, A., and Sather, G. E.** , Viruses associated with epidemic hemorrhagic fevers of the Philippines and Thailand, *Science*, 131, 1102, 1960.
18. **Laemmert, H. W., Jr. and Hughes, T. P.**, The virus of Ilheus encephalitis. Isolation, serological specificity and transmision, *J. Immunol.*, 55, 61, 1947.
19. **Taniguchi, T., Hosokawa, M., and Kuga, S.**, A virus isolated in 1935 epidemic of summer encephalitis of Japan, *Jpn. J. Exp. Med.*, 14, 185, 1936.
20. **Work, T. H., Trapido, H., Murthy, D. P. N., Rao, R. L., Bhatt, P. N., and Kulkarni, K. G.** Kyasanur Forest Disease. III. A preliminary report on the nature of the infection and clinical manifestations in human beings, *Indian J. Med. Sci.* 11, 619, 1957.
21. **Pool, W. A. , Brownlee, A., and Wilson, R. D.**, The etiology of "louping ill," *J. Comp. Pathol Ther.*, 43, 253, 1930.
22. **French, E. L.**, Murray Valley encephalitis: isolation and characterization of the aetiological agent, *Med. J. Aust.*, 1, 100, 1952.
23. **Chumakov, M. P.**, Results of the Institue of Neurology in the study of Omsk hemorrhagic fever, *Vest. Akad. Med. Nauk SSR*, 2, 19, 1948.
24. **Levkovich, E. N., Shubladze, A. K., Chumakov, M. P., and Soloviev, V. D.**, Etiology of spring-summer epidemic encephalitis, *Arkh. Biol. Nauk*, 52, 162, 1938.
25. **Webster, L. T. and Fite, G. L.**, A virus encountered in the study of material from cases of encephalitis in the St. Louis and Kansas City epidemics of 1933, *Science*, 78, 463, 1933.
26. **Smithburn, K. C., Hughes, T. P., Burke, A. W., and Paul, J. H** , A neurotropic virus isolated from the blood of a native of Uganda, *Am. J. Trop. Med.*, 20, 471, 1940.
27. **Stokes, A., Bauer, J. H., and Hudson, N. P.**, Experimental transmission of yellow fever to laboratory animals, *Am. J. Trop. Med.*, 8, 103, 1928.
28. **Karabatsos, N., Ed.** Suppl. to Int. Cat. of Arboviruses, *Am. J. Trop. Med. Hyg.*, 27 (Suppl. 2) 418, 1978.

RUBIVIRIDAE: RUBELLA VIRUS

Paul D. Parkman, Hope E. Hopps, and Harry M. Meyer, Jr.

Introduction

For more than 200 years after its recognition, rubella (German measles, three-day measles) was regarded as one of the milder rash diseases common in children. The viral nature of the causative agent was established in 1938 by inoculation of filtered throat washings into volunteers.[1] In 1941, Gregg[2] observed the association of rubella in pregnant mothers with the occurrence of congenital defects in their infants. This finding stimulated an increased interest in the disease and attempts to recover the causative virus. However, it was not until 1962 that two groups of investigators reported isolation of the virus.[3,4] Through the ability to work with this virus in the laboratory, rapid advances have occurred in our knowledge of its characteristics, the disease it produces and its epidemiology and pathogenesis, and means for preventing the disease through use of "attenuated" (modified) live virus vaccines.

Rubella virus is an RNA- and lipid-containing enveloped virus of a single serotype. Although descriptions of rubella virus morphology have been variable, it is now generally accepted that it resembles the group A arthropod-borne viruses, and it has been classified[5] as a member of the togavirus group (rubivirus). No serologic cross-reactions have been found with other togaviruses,[6] and its transmission by arthropods has not been shown.[7] Rubella retains the unique attribute of being the first disease for which immunization of one generation is being encouraged with the primary goal of preventing disease in the next.

Morphology

The descriptions in the literature of the electron microscopic appearance of rubella virus vary considerably. Early studies were hampered by low yields of virus and the fact that the virus appears to be fragile and its morphology easily distorted. Assessments of the diameter of virus particles in the extracellular fluids of cell cultures have varied over the range of 40 to 280 nm and the internal component from 40 to 130 nm. The surface of particles was described as smooth, granular, and ragged or as having a variety of kinds of projections.[8-17] However, the consensus of recent studies using sevieral methods, all involving use of cell cultures producing high yields of infectious virus,[11,12,14,15,17] have placed the particle diameter in the range of 40 to 80 nm with a central electron-dense core. These authors agree that the particles tend to be spherical in shape; no helical ribonucleoprotein-like structures have been seen. Surface structures of rubella virus have been described as fringe-like[14] or spicular;[13] in the best preserved preparations they appear as well-defined spike-like projections.[12,17]

The morphology of rubella virus development during intracellular replication presents similar problems in interpretation. However, the best documented observations suggest that virus production and assembly occurs in the cytoplasm in the absence of significant changes in nuclear morphology. Virus development appears to start as a condensation of electron-dense material near a cytoplasmic membrane, followed by encasement by the membrane of 35-nm cores which become pinched off (Figure 1), releasing the completed particle.[13,16,18-21]

Physical and Chemical Properties

Rubella virus is a relatively fragile RNA-containing virus. Its infectivity is readily destroyed by ultraviolent (UV) and gamma irradiation, treatment with organic solvents, trypsin, formaldehyde, acid or alkaline pH, and heating (Table 1). It is readily

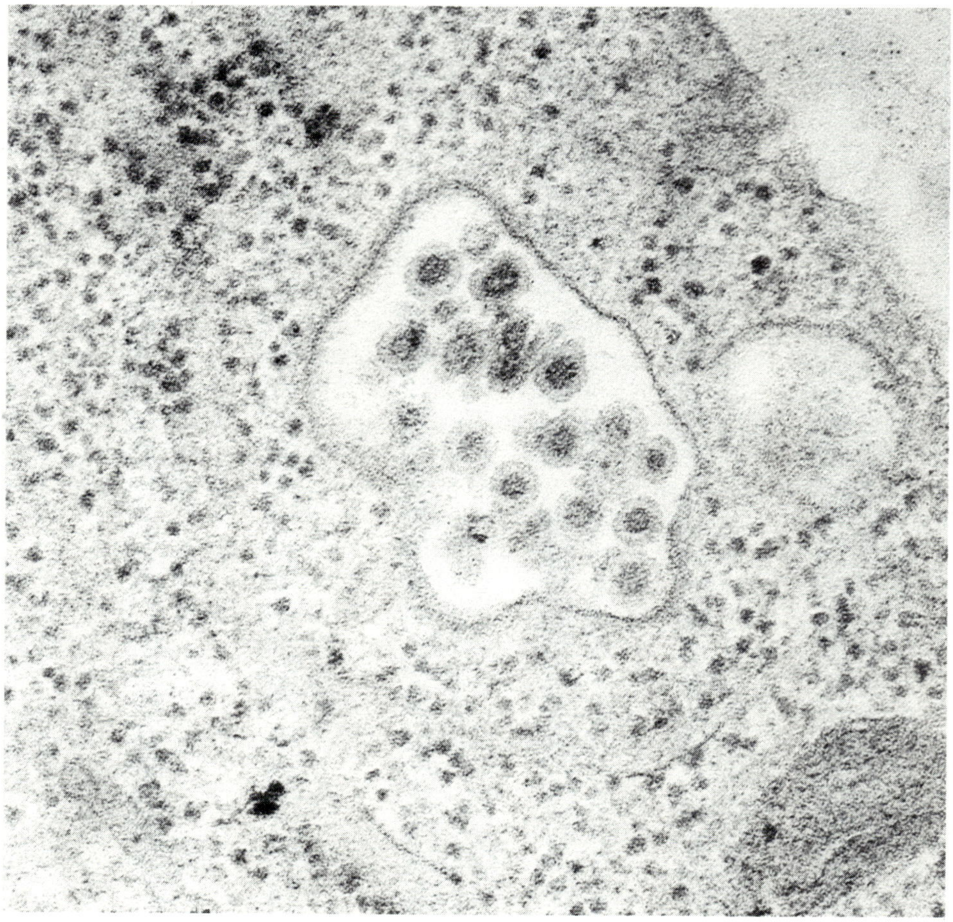

FIGURE 1. Electron photomicrograph of rubella virus particles in a cytoplasmic vacuole of an infected Vero cell. Magnification × 124,000.

filterable through membranes of 220 to 260 nm average pore diameter, but is retained by those of 50 to 70 nm in size. Its growth in cell cultures is not diminished by inhibitors of DNA synthesis. The antiviral agent 1-adamantanamine (amantidine) inhibits viral replication in cell cultures. Its effect is apparent during the early phase of the virus replicative cycle, perhaps involving virus penetration of the cell. Amantidine has not been shown to have a demonstrable effect on rubella in vivo.

Attempts have been made to define rubella virus polypeptide components by polyacrylamide gel electrophoresis techniques. Depending on the method used, the molecular weight and the number of major and minor polypeptides observed vary, probably as a result of differences in virus preparation methods and electrophoretic conditions.[17,44,45] Weak neuraminidase activity was found in one study when mucin was used as a substrate;[17] this finding has not yet been confirmed.

Growth of Rubella Virus

The fact that rubella virus does not produce profound cytopathic changes or extraordinary amount of virus in cell cultures commonly used for virus isolation accounts for the relatively late initial successful virus isolations by Weller and Neva[3] and Parkman et al.[4] Rubella virus has been found to be detectable in cell cultures by the interference it produces with the growth of other viruses, the production of cytopathic effects (CPE), and specific immunofluorescence. Generally, the interference phenomenon is pronounced and readily interpreted; the CPE is often subtle and is very dependent on the maintenance of special cell lines and special efforts to adapt the virus by serial passage in the cells selected specifically for this purpose. The CPE seen in the rabbit kidney cell line RK_{13} is shown in Figure 2. Interference reliably develops in African green monkey kidney cultures (GMK). Because these cells may contain extraneous simian agents, however, many workers have chosen to use a cell line (such as Vero) which appears to be of approximately equal sensitivity for initial isolation and subpassage fluids from these cultures to GMK, which are then tested for interference. The interference phenomenon becomes pronounced within 7 to 10 days of inoculation.[22] Rubella-infected cultures resist superinfection virtually completely, while uninfected controls are completely destroyed by the CPE of the challenge virus (Figure 3). Such interference with the growth of other viruses (including many enterovirus types, myxoviruses, other togaviruses and vesicular stomatitis virus) has been shown.[22] Interference is incomplete or absent with herpesviruses and adenoviruses.

Cell Cultures Used for Detection and Propagation of Rubella Virus

Rubella virus mulitplies in many cell types, including primary cells and cell lines. Cultures of human, simian, and rabbit origin have been found useful for detection of virus in clinical specimens and for general virologic work. Vero and BHK-21 cells have served a special role in the production of rubella preparations with high infectivity titers. Avian (duck embryo), rabbit, and canine kidney primary cultures and a human cell line (WI-38) have been used for preparation of live virus vaccines licensed in the United States for commercial distribution. (See Table 2 for a listing of cell cultures, their susceptibility to rubella virus infection, and appropriate references).

Virus Antigens

The virus has several components which react with antibodies developing in sera from persons or animals convalescent from infection. These components, usually referred to as the antigens of the virus, vary from the intact infectious virus itself to subviral components. Information about the source and behavior of these virus components, their use in studying rubella virus, and their use for diagnostic purposes are presented in Table 3 (appropriate references are included in the table). The hemagglutinating (HA) and complement-fixing antigens are the most common reagents used in diagnostic laboratories. A word of caution about rubella fluorescence may be appropriate: this phenomenon is difficult to reproduce consistently and is fraught with the potential for false positive results. Thus, laboratory results based entirely on fluorescent antibody (FA) techniques should be viewed with some concern. They may be correct, but careful correlation of the FA results with one of the standard virus isolation and identification methods based on interference or CPE is a highly important initial step in developing a reliable fluorescence assay.

Rubella Virus Infection of Experimental Animals

Rubella virus produces subclinical infections in several common laboratory animal species including rabbits,[108,110] guinea pigs,[108] hamsters,[108,109] mice,[108,110] and rats.[111] Each of these animals experience infection, as evidenced by the development of antibodies in response to the administration of relatively small amounts of virus.

In the species which have been tested (including guinea pigs, rabbits, and rats), virus is not regularly detectable in the upper respiratory tract and viremia is rarely shown. The ferret experiences an asymptomatic infection similar to subclinical rubella in man;

animals shed virus from the respiratory tract and develop antibodies during the 2 weeks after inoculation by intranasal or parenteral routes. Several of these species have been studied in attemts to develop a convenient animal model for congenital rubella.[107,108,111] In general, these studies have not provided a model which adequately reflects the human disease, and the published work presents a number of inconsistencies from one study to the next. Thus, no universally accepted small animal model has been found.

Monkeys, including *Cercopithecus* species[118] and *Macacca mulatta* (rhesus),[119-121] have also been shown to experience subclinical infection. The rhesus has been studied most extensively. In these animals the pattern of viremia, virus shedding from the pharynx, and development of antibodies parrallels that seen in the human infection. Infections are transmissible from monkey to monkey. In infected animals the virus is widely distributed in many organs and tissues including the respiratory and intestinal tracts, lymph nodes and spleen, liver, skeletal muscle, and kidney. [121] Studies involving pregnant rhesus monkeys inoculated during the first third of Gestation has indicated that virus is frequently transmitted to the placenta and amnion of the developing embryo.[122,123] Virologic evidence for transmission of virus to the fetus itself is uncommon.[121-124] Both the rhesus monkey and the rabbit have been used in marker testing to differentiate virulent and attenuated rubella virus.[76,123,125]

Specific rubella antiserum for use in identifying rubella virus isolates can be prepared in any of a number of small laboratory animals. The rabbit is convenient for this purpose because such an animal develops high levels of rubella antibody after repeated immunization and is of sufficient size to enable production of antiserum pools of reasonable volume.[85] Rubella virus has been tested for oncogenic effects in newborn hamsters but no tumorigenic potential was apparent.[109,126]

Clinical Features

Rubella is a common rash-associated disease occuring worldwide.[127] It is significant because of its role in producing congenital defects. Reports of the clinical disease originally appeared in the German medical literature; thus, it was commonly referred to in early English papers as "German measles."[128] By the late 1800s it was recognized that a case of rubella precluded recurrence of the disease, but did not prevent subsequent attacks of measles or scarlet fever. Two general forms of rubella virus infection may occur, based on the mode of infection: (1) postnatally acquired, in which the virus is transmitted from person to person, probably by the respiratory route; and (2) prenatally acquired, in which the developing fetus becomes infected via the maternal viremia occuring during concurrent primary maternal infection.

The clinical features of postnatally acquired rubella virus infection are summarized in Table 4. Such infections may range from subclinical to clinically apparent disease of sufficient severity to result in an incorrect clinical diagnosis of rubeola.[128,130] Rash appears 17 ± 3 days after exposure. Typical cases include a rash and generalized lymphadenopathy with the nodes in the posterior cervical area particularly prominent; low-grade fever may occur. The rash may span the spectrum of severity from a transient blush, which probably frequently goes undetected, to a measles-like rash lasting several days. It usually appears first on the face and neck and quickly spreads to the trunk and arms and then to the legs. The rash on the upper portion of the body commonly fades by the time it becomes fully developed on the lower extremities. An enanthem frequently occurs, but it does not have the same characteristic appearance or pathognomonic significance of Koplik's spots in rubeola. Unlike rubeola, prominent respiratory symptoms are unusual. The clinical illness appears to be somewhat more pronounced in adults, especially in adult women. The adult may have headache, malaise, myalgia, painful teeth or gums, peripheral neuritis with paresthesias, and prolonged episodes of polyarthralgia with or without overt signs of arthritis.

In the U. S., rubella is the only virus infection commonly associated with joint symptoms. Age- and sex-related differences are especially pronounced in the incidence and

Table 1
RUBELLA VIRUS CHARACTERISTICS

Characteristic	Result	Method	Ref.
Size and morphology	Generally spherical, 40—80 nm in diameter with central electron-dense nucleoid; surface covered with projections; released into intracellular vesicles and from cell membrane	Electron microscopy of virus agglutinated with rubella antiserum or partially purified; infectivity retained by 50-nm filter membrane; most of virus retained by 100-nm membrane; readily passes 220-nm and larger pore diameter filters	3, 4, 11—26
Nucleic acid	RNA, single stranded	Not affected by inhibitors of DNA synthesis; infectious RNA has been extracted from partially purified virus	22, 27—31
Effect of chemical and physical agents			
Ethyl ether	Inactivated	Equal volume of virus suspension and ethyl ether, 18 hr at 4°C	3, 4, 22
Chloroform	Inactivated	Equal volumes shaken at 4° C	4, 22
Sodium Deoxycholate	Inactivated	Deoxycholate, 0.013 to 2.0%; room temperature for 60 min	3, 32
Formalin, 1:4,000	Inactivated	10^3 to 10^4 $\log_{10}$ ID_{50}; inactivated within 1 hr at 37°C	22
1-Adamantanamine	Inhibits growth in cell cultures	Inhibition in concentrations from 10—25 μg/mℓ; no evidence for substantial effectiveness shown in experimental animals or man	33—39
pH	Infectivity stable pH 6.8 to 8.1; loss in infectivity at pH 5.9, with complete inactivation at pH 2.0	pH adjusted using phosphate buffers, NaOH, and HCl	22, 40, 41
Temperature	Inactivated within 30 min at 56°C; 0.3 to 0.4 $\log_{10}$ ID_{50}/hr loss at 37°C; stable for longer than 7 days at 4°C and for years at −60°C	Virus suspended in maintenance medium containing 2% fetal bovine serum at indicated temperature	3, 4, 22, 42
Lyophilization	Freeze-drying in protein-containing medium done without significant titer loss (<0.5 $\log_{10}ID_{50}$); stable at 25°C for 1 month or longer	Lyophilized to <2% moisture in sealed glass ampules	22
Irradiation			
Ultraviolet	Rapidly inactivated by ultraviolet light	Exposure to mercury arc lamp at 7.5-in. distance for intervals of between 1 and 30 min	42
Gamma	Inactivated exponentially	Virus exposed in frozen state to ^{137}Ce or ^{60}Co	43

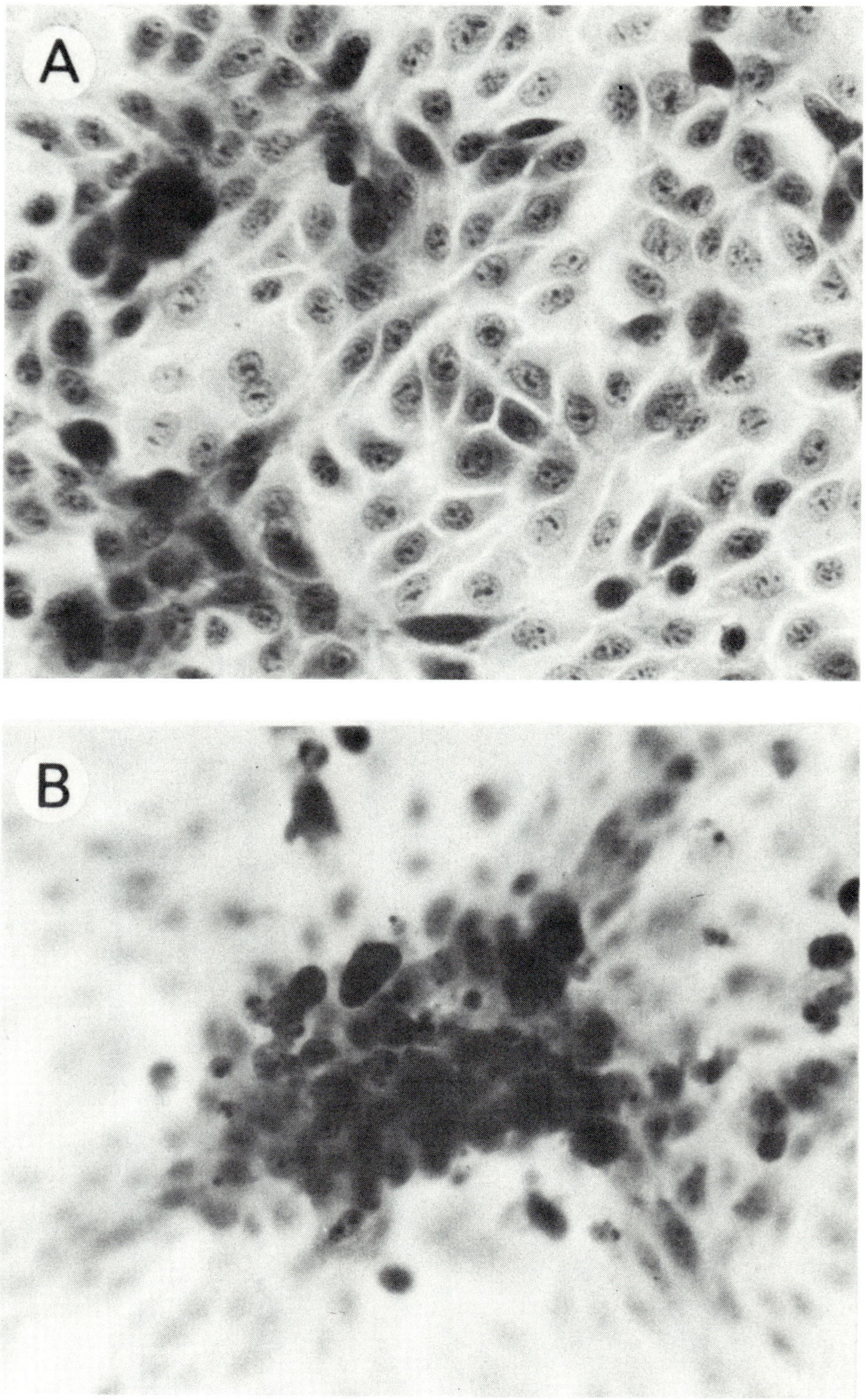

FIGURE 2. Detection of rubella virus by cytopathic effect (CPE) in RK_{13} cultures (stained preparations). (A) Uninoculated RK_{13} continuous rabbit kidney control culture. (B) Early focal "piled up" area of rubella CPE. (C) Moderately advanced CPE with scattered, rounded-up cells and loss of cells from the glass. (D) Advanced CPE. (Magnification × 300.) (Reproduced with permission from the *American Journal of Clinical Pathology,* Volume 57, p. 807, 1972.)

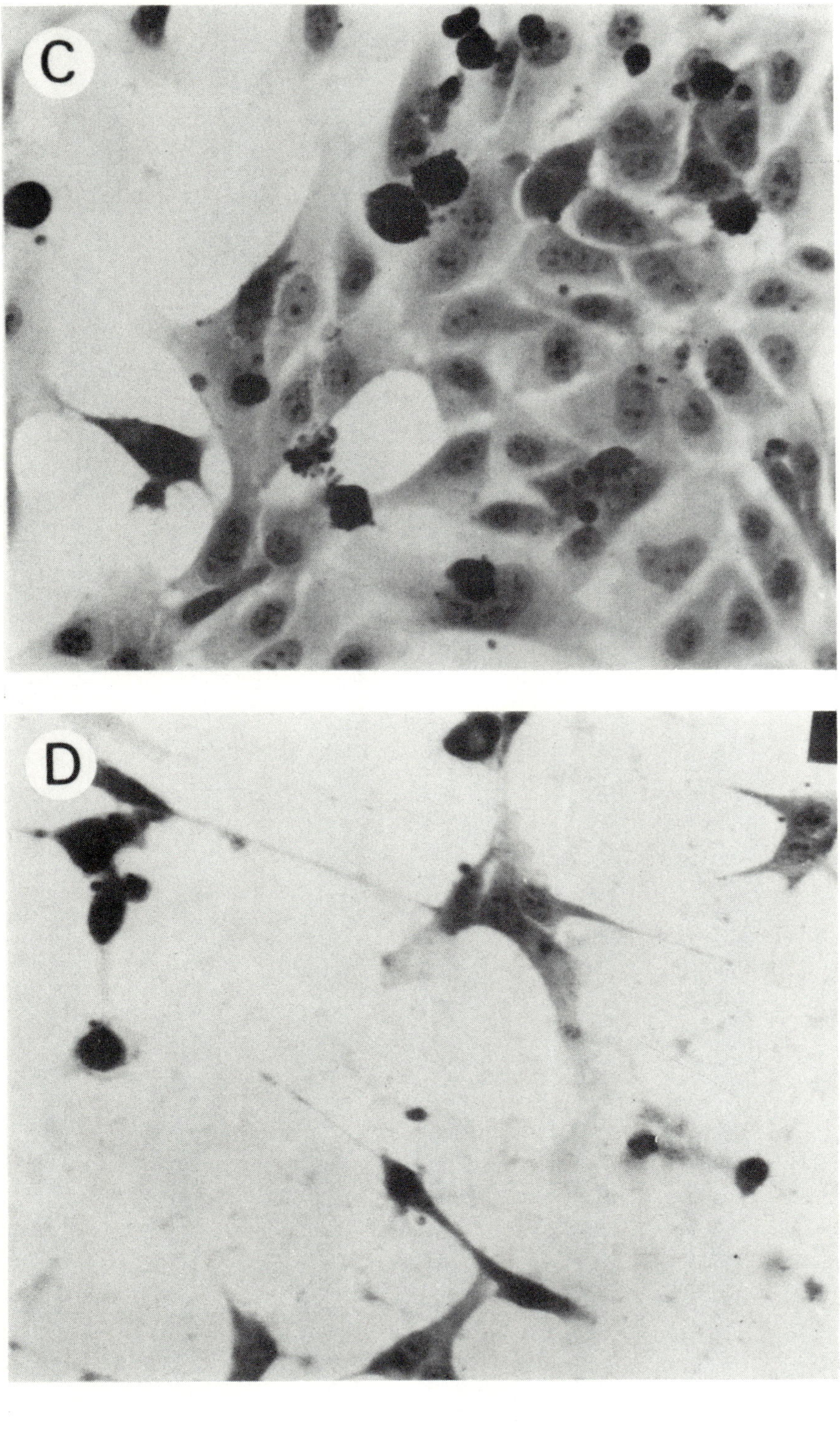

FIGURE 2 (continued)

Table 2
CELL CULTURE SYSTEMS FOR DETECTION AND PROPAGATION OF RUBELLA VIRUS

Cell culture	Virus detected by[a]: Interference	CPE	Plaques	Comments; special uses	Refs.
Simian					
Primary					
Cercopithecus monkey kidney	+ +	0	0	Primary isolation, virus attenuation	4, 22
Erythrocebus patas monkey kidney	+ +	0	NT	Primary isolation	46
Cell lines					
Cercopithecus monkey kidney					
BS-C-1	+	0	0		22, 47
Vero	0	+ or 0	+ or 0	Primary isolation with subpassage to primary *Cercopithecus* kidney for detection by interference; production of high virus yields, thus useful for HA and CF antigen production	48—50
AH-1	NT	+ +	+		51
Rhesus monkey kidney					
LLC-MK2	+	0	0	Chronically infected LLC-MK2 infected line has been used for fluorescence testing	22, 52, 53
MA-104	+	0	0	Cell line derived from monkey embryo	22, 54
Rhesus lung FRhL-1	NT	±	NT	Possibly suitable for vaccine production	55
Human					
Primary					
Amnion	+	+ or 0	NT	Primary isolation; requires incubation for long periods (CPE onset at 17—34 days) and excellent culture morphology	3, 56
Embryo kidney	+ or 0	0	NT	Interference varies with cell lot	22
Thyroid	±	+ or 0	NT	Interference partial; CPE variable depending on morphology of cell cultures	57
Cell lines					
HEp-2	+ or 0	0	NT	Variable results by different investigators; may vary with cell line used	3, 22, 57
Embryonic				Generally support growth of virus without CPE or pronounced interference	58, 59
WI-38 (lung)	±	0	NT	Partial interference by rubella with herpes and poliovirus type 1; cells used for virus attenuation and vaccine production	60—62
MRC-5 (lung)	NT	0	NT	Cultures potentially useful for vaccine production	63

Table 2 (continued)
CELL CULTURE SYSTEMS FOR DETECTION AND PROPAGATION OF RUBELLA VIRUS

Cell culture	Virus detected by[a] Interference	CPE	Plaques	Comments; special uses	Refs.
Rabbit					
Primary					
Rabbit kidney	+ +	+ or 0	0	Subtle CPE produced involving epithelioid culture elements only; used for vaccine production	22, 64, 65
Rabbit embryo kidney	NT	+	NT		66
Cell lines					
Rabbit kidney					
GL-RK_{13}	+	+ +	+	Most widely used cell type for detection of CPE and plaques	57, 67
LLC-RK_1		+			68
T-17	NT	+ +	+		69
Rabbit cornea					
SIRC	NT	+ +	+		70—72
Avian					
Primary					
Duck embryo	NT	0	NT	Cells used for vaccine production	73, 74
Chick embryo	+	0	NT	Wild virus does not propagate in these cells; attenuated high-passage viruses adapted with difficulty	75
Quail embryo				Used for vaccine production in Japan (MEQ_{11} vaccine)	74, 76
Miscellaneous					
Primary					
Canine kidney	NT	0	NT	Cells derived from a closed colony of beagles used for vaccine production in past	77
Guinea pig kidney	+	0	NT	Interference shown with vesicular stomatitis virus; used for virus attentuation in Japan (TO-336 vaccine)	76
Bovine embryo skin muscle	+	0	NT		78, 79
Cell lines					
Hamster kidney					
BHK-21	0	+ or 0	+ or 0	Production of high virus yields; useful for production of CF and HA antigens	80—84

[a] + +, marked effect; +, moderate effect; ±, weakly positive effect; + or 0, inconsistent effect between cell batches or cell lines; 0, no effect; NT, not tested.

Table 3
RUBELLA VIRUS BIOLOGICAL CHARACTERISTICS

Antigenic Makeup

Antigen	Isolation methods	Means of production	Use	Refs.
Infectious virus	Crude cell culture harvests; at 15,000 × *G*, 90% of virus sedimented in 6 min	Cell culture harvests	Detection of virus in clinical specimens and in cell culture systems used for virus propagation	3, 4
			Neutralization of virus infectivity as measured by inhibition of cytopathic effect (CPE), plaque reduction; originally this test was used for serodiagnosis of acute infection and prior rubella infection, but now is largely supplanted by hemagglutination inhibition (HAI) test	49, 50-52, 57, 71, 72, 78, 80, 85, 86-88
			Used for detection of antibodies in fluorescent antibody assay	89, 90
Hemagglutinin (HA)	Associated with infectious virus, in Sephadex® G-200 column eluates, sucrose density gradients	Harvests of cell cultures yielding high titers of infectious virus (Vero, BHK-21); produced by use of HA inhibitor-free serum or in alkaline-extracted cells (both procedures eliminate inhibitors of HA)	HAI test; most commonly used for serodiagnosis, seroepidemiologic studies; identification of virus isolates	84, 91-96
Complement fixation (CF)	"Cell pack" preparations	Infected cells disrupted and concentrated by low-speed centrifugation	Specific application in rubella serodiagnosis	42, 49, 83, 97-103

Table 3 (continued)
RUBELLA VIRUS BIOLOGICAL CHARACTERISTICS

Antigenic Makeup

Antigen	Isolation methods	Means of production	Use	Refs.
"Large particle" or "sedimentable"	Associated with infectious virus, HA activity in Sephadex G-200 columns, sucrose density gradients; especially large amounts in early harvests (<4 days postinoculation); retained by 50-nm filter	Sedimented cells from monolayer or suspension cultures ("cell packs") or alkaline extraction of cell cultures	Basic virologic studies of rubella virus	
"Small particle" or "soluble"	Smaller and lighter density than infectious virus and hemagglutinin; appears later in course of infection (7+ days); filterable through 10-nm average pore diameter filters	As above or by prolonged ether treatment of "large particle" CF antigen	Basic virologic studies of rubella virus	
Precipitins Theta Iota	Subviral in size	Supernatants of cell cultures or alkaline extraction of cells	Basic studies of virus and serologic response	104-106

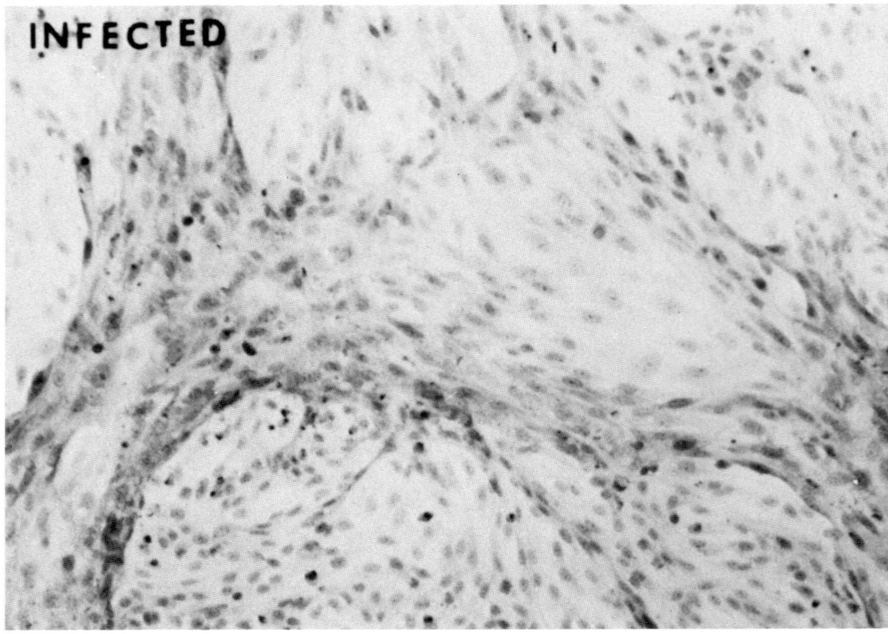

A

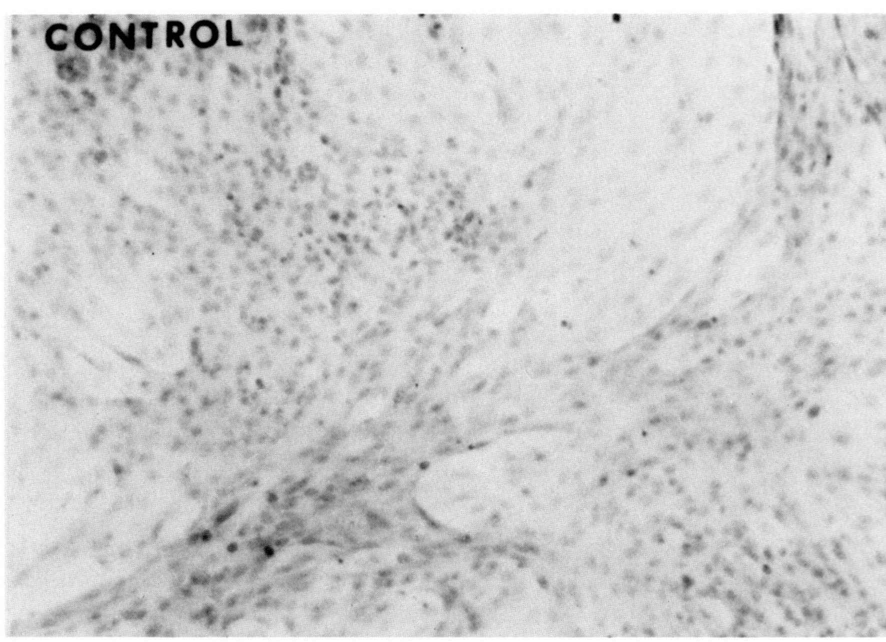

B

FIGURE 3. Demonstration of rubella virus in primary African green monkey kidney cell cultures using the interference detection system. (A) Cell culture inoculated 7 days earlier with throat washings from a rubella patient. No detectable change has occurred. (B) Uninoculated control culture. (C) Rubella-infected culture and (D) uninoculated control culture, each shown 2 days after addition of 10,000 $TCID_{50}$ of echovirus type 11. The control cells are extensively destroyed by the enterovirus while the rubella-infected cells remain normal in appearance. (Magnification × 30.) (Reproduced with permission from the *American Journal of Clinical Pathology,* Volume 57, p. 806, 1972.)

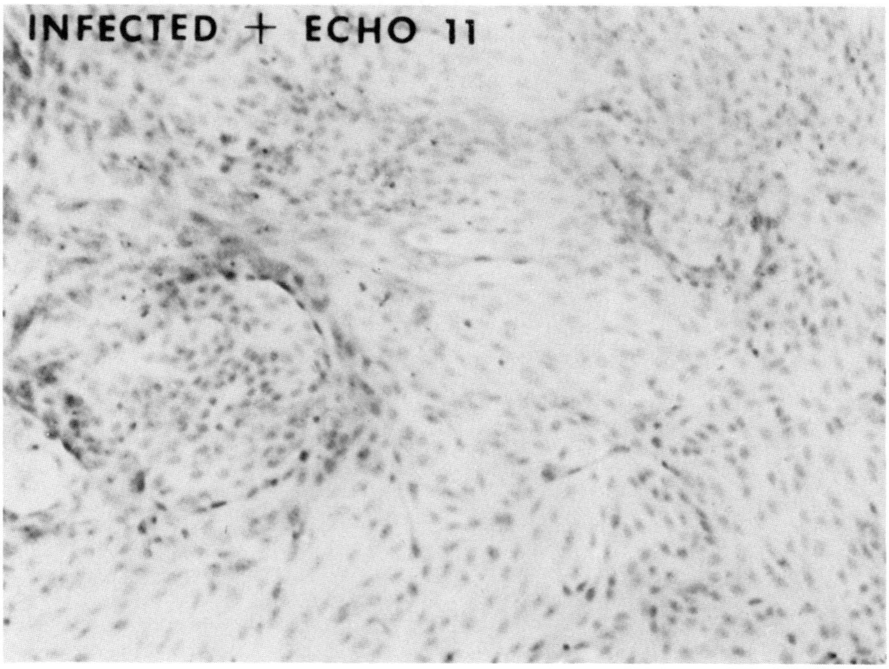

FIGURE 3C

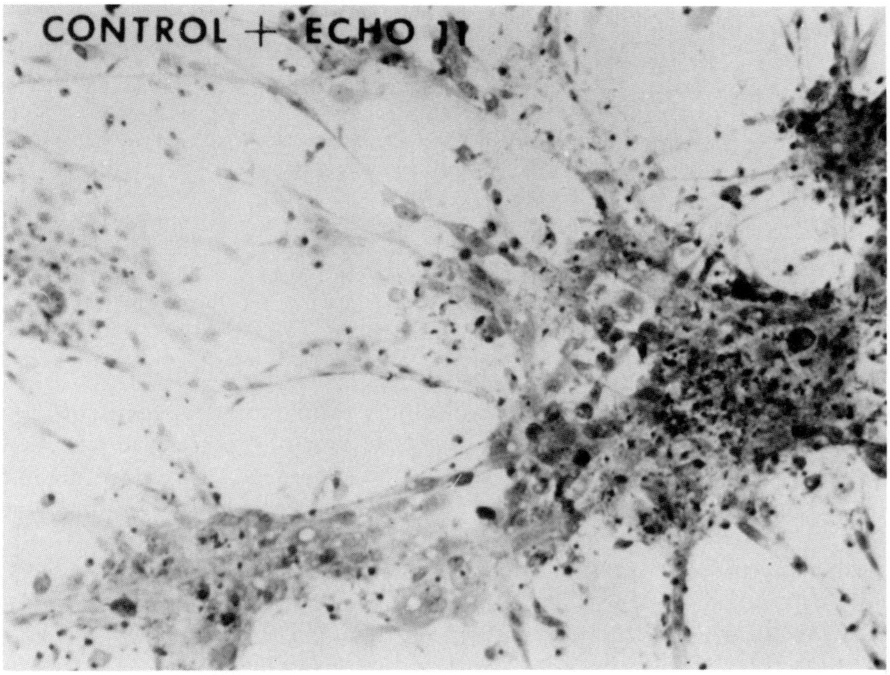

FIGURE 3D

Table 4
A SUMMARY OF CLINICAL FINDINGS IN RUBELLA ACQUIRED BY CONTACT

Category of clinical disease	Signs and symptoms	Duration	Frequency
Subclinical	None apparent or unnoticed	—	Estimates have varied from 0 to 86% of cases; depends in large measure on intensity of surveillance
Clinically evident	Lymph node enlargement, especially postauricular, suboccipital, and cervical	5—7 days before onset of rash; may persist for weeks after rash onset	Virtually 100%
	Mild conjunctivitis and coryza	1—5 days before rash to 2 days after rash onset	Most commonly observed in adolescents and adults
	Fever (ordinarily <102°F)	Day 1 of rash	Approximately 50%
	Rash, maculopapular, erythematous, beginning on face and neck, spreading downward to arms, trunk, and lower extremities	Usually 3 days; ranges from few hours to 5 days	In virtually 100% of patients observed twice daily by experienced medical personnel; lesser percentages as degree of surveillance decreases
Complications	Arthritis occurs quite frequently, especially in metacarpophalangeal and interphalangeal joints, ankles, knees, and elbows; peripheral neuritis	Appears 1—5 days after onset of rash, ordinarily lasts less than 2 weeks, rarely lasts for 1 year or longer	In 25% of children under age 11 years, 52% of persons 11 years of age or older; among adults, women more commonly have symptoms
	Encephalitis occurs rarely; 20% mortality; no demyelination	Appears during first few days after onset; survivors usually recover without sequellae	In 1 of every 5,000 or more cases; adults involved more often than children
	Purpura, with or without thrombocytopenia	Onset several days to weeks after rubella	1 of every 3,000 cases
	Platelets depressed in most cases of rubella	2 weeks to several months	Children more often involved than adults; eventual complete recovery expected

severity of arthritic symptoms. Joint symptoms occur more frequently after puberty and increase thereafter with age. They appear to be more common in females.[131] Approximately 25% of children may be expected to show such symptoms. In adults over age 20, more than 50% will experience joint-related symptoms. The joint symptoms most often appear several days after the onset of rash, although in some patients symptoms have appeared before rash onset or as late as several weeks afterward. The distribution of joint involvement follows that often seen in rheumatoid arthritis; however, rubella arthritis has not shown a tendency to become chronic or to produce deformities.[132] In most instances, the joint symptoms following rubella last less than 2 weeks. Occasionally, especially in older women, symptoms may recur over periods of a year or more.

Relatively rare complications include encephalitis and thrombocytopenic purpura. Encephalitis is seen more often in adults than in children.[133] Signs and symptoms commonly appear several days after the onset of the exanthem. Virus has not been re-

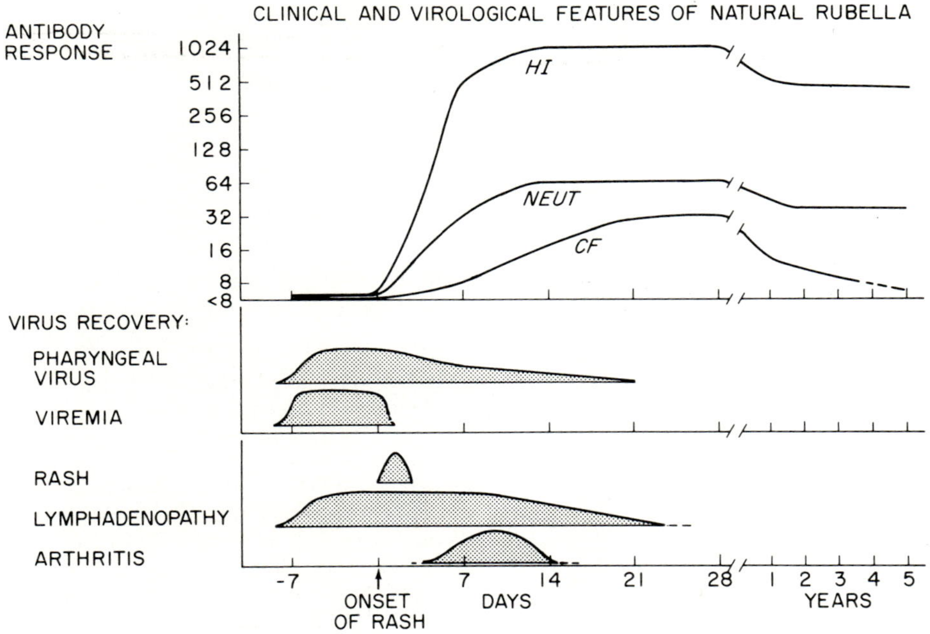

FIGURE 4. Correlation of virus shedding and serologic response with clinical findings in postnatally acquired rubella. (Reproduced with permission from the *American Journal of Clinical Pathology,* Volume 57, p. 809, 1972).

covered from the cerebrospinal fluid or central nervous system tissues; the pathogenesis of this complication is not well understood.[134] Thrombocytopenic purpura may also occur during the same time period and may manifest itself by purpura and hemorrhage from the nose, gums, and gastrointestinal and urinary tracts.[135] The prognosis for complete recovery is good. Treatment of acute rubella and its complications is symptomatic.

A summary of the timing of the virologic events in postnatal rubella is shown in Figure 4. Virus reaches sufficient concentration to be detected in blood and the respiratory tract 1 to 1½ weeks before the onset of rash.[128,136,138] The highest concentrations of virus in the respiratory secretions are attained during the period from 3 days before to 3 days after rash; this corresponds with the maximum period of infectivity as defined by epidemiologic studies. Virus can also be isolated from the rash itself[139] and from enlarged lymph nodes.[57] Viremia disappears shortly after the onset of rash, concomitant with the appearance of neutralizing and hemagglutination-inhibiting (HAI) antibodies. These antibodies reach maximum titers within 1 to 4 weeks of rash and persist in substantial titer for years.[140] Complement-fixing (CF) antibodies appear 7 to 14 days after rash; their persistence is variable. Virus is shed in the pharynx for periods of up to 28 days after rash onset. Both viremia and virus shedding from the respiratory tract are present in subclinical infections. The persistence of virus in tissues is unknown; however, in experimental infections in monkeys, detectable infectious virus disappears at about the same time as does nasopharyngeal excretion.

Individuals with antibodies ordinarily do not develop symptoms, but may develop an anamnestic antibody response following exposure to the disease.[138] In a study of a small number of pregnant women known to have had antibodies prior to exposure and shown to have anamnestic antibody rises, none had infection transmitted to the fetus.[141]

Congenital Rubella

If maternal infection occurs during the early months of pregnancy, virus invasion

Table 5
A SUMMARY OF POSSIBLE PREGNANCY OUTCOMES AFTER RUBELLA DURING THE FIRST TRIMESTER OF PREGNANCY

Transmission of maternal virus to products of conception	Infection	Estimated frequency (%)	Pathogenesis	Disease produced
Does not occur	None	Probably 10 or less	Seeding of placenta with virus does not occur	None
Does occur	Placental and/or fetal infection			
	Subclinical	55—70	Minimal damage to placenta and fetus; virus may persist for weeks or months; some have virus shedding at birth	None apparent
	Clinical	20—35	Major damage to placenta and/or the fetus; virus persists in most from time of infection to neonatal period (80%) and early infancy (10—20%)	Abortion, stillbirth, congenital defects; if infection is in 1st month, defects in about 50%, 2nd month, defects in about 25%; 3rd month, defects in about 10%[a]

[a] The risk of fetal infection and disease is highest in the early weeks of pregnancy; some defects, especially deafness and milder degrees of motor and mental retardation, have been reported after infections occurring early in the second trimester.

of the placenta and the fetus frequently occurs; once fetal infection is established, virtually any organ system can be damaged and congenital defects result (Table 5). The several combinations of defects which commonly occur often produce a clinically recognizable pattern termed the "rubella syndrome." The risk of severe defects is highest in fetuses infected in the early weeks of gestation. Under these circumstances the syndrome may involve the eye, inner ear, heart, and brain and may produce derangements in other organ systems, resulting in bone lesions detectable on radiologic examination, hemolytic anemia, purpura, hepatitis, low birth weight, and poor growth and development. The effects of the infection may be of sufficient severity to lead to spontaneous abortion or stillbirth.[142,143] In the sections that follow and in Table 6, the major effects on the several organ systems are briefly described.

Hearing Disorders

Deafness is the most common consequence of congenital rubella virus infection.[146] It can result from infection occurring at any time during the first 4 months of gestation.[147] The hearing loss is of the nerve-deafness type and may affect one or both ears. Severe deafness may be detectable early in life; however, as lesser degrees of hearing loss may occur, it is important that careful audiometric testing be performed at intervals during the first several years in order to detect and institute early corrective measures.[148,149] Early treatment of such defects can reduce the deleterious effects of deafness on speech and psychologic development.

Visual Disorders

Ocular defects are common in congenitally affected infants. In fact, the first abnormality to be associated with the rubella syndrome was congenital cataract.[144,145] The cataract may affect one or both eyes and may range from grossly apparent severe opacification to minimal changes requiring careful ophthalmoscopic techniques for detection. Congenital glaucoma occurs less commonly; the cornea is enlarged and appears hazy. The anterior chamber of the eye is enlarged, with increased intraocular

Table 6
A SUMMARY OF CLINICAL FINDINGS IN RUBELLA-AFFECTED OFFSPRING OF MOTHERS INFECTED DURING PREGNANCY[a]

Effects	Approximate incidence (%)	Timing of occurrence; duration
Transient		
Thrombocytopenia with purpura	28—45	Most severe during first week of life; rarely lasts until second or third month
Hepatitis with jaundice	14—19	Onset during first days of life, improves slowly after period of several weeks to several months
Pneumonitis	27—30	Interstitial pneumonitis probably due to rubella virus; occurs in first months of life
Hemolytic anemia	17	First several weeks of life; spontaneous recovery by 6 months of age
Bone lesions	20	X-rays show radiolucent and dense streaking of metaphyses of the long bones and ribs; usually resolves by 2 months of age, rarely may last 8—12 months
Growth retardation		
Intrauterine	60—71	Weight gain, head size, development of dentition, height are delayed; effect most severe in those with major congenital defects and those who experience continuing active rubella virus infection during first year of life
Neonatal	57	
Permanent		
Deafness	52—82	Unilateral or bilateral; commonly detectable only when child old enough for audiologic testing; most common defect occurring without other associated defects
Eye defects	30—60	May be apparent at birth or may develop and increase in extent during first months of life
Cataracts	19	
Glaucoma	4	
Retinopathy only	48	Total percentage higher; not functionally significant
Cardiac and vascular defects, distribution of abnormalities in those with defects	52—59	
Patent ductus arteriosus	49—78	Present at birth, may require corrective surgery during first year of life if cardiac failure develops; if not, defer until age 2.
Right pulmonary artery stenosis	70	
Left pulmonary artery stenosis	56	
Valvular pulmonic stenosis	40—45	
Aortic valvular stenosis	14	
Ventricular septal defect	1—10	
Aberrant right subclavian	11	

Table 6 (continued)
A SUMMARY OF CLINICAL FINDINGS IN RUBELLA-AFFECTED OFFSPRING OF MOTHERS INFECTED DURING PREGNANCY[a]

Effects	Approximate incidence (%)	Timing of occurrence; duration
Neurological defects	81	
Chronic encephalitis	33—41	Chronic, presistent rubella virus infection of CNS with increased CSF cells, protein; lethargy, irritability, full fontanelle; symptoms of acute process usually disappear by 12—18 months
Psychomotor retardation	40—56	Becomes apparent over first several years of life; delayed developmental milestones
Of these: mild	49	
moderate	24	
severe	27	
With Cerebral Palsy	27	
Abnormal dermatoglyphic pattern	47	Present at birth
Mortality	10—20	Usually occurs during first 6 months of life due to cardiac failure, sepsis, general debility
Long-term		
Rubella-specific immunologic defects		
Neonatal	5	Elevated IgM, normal, high or low IgG antibodies; usually all become normal by age 1 year
During later infancy and childhood	20	Some infants fail to develop rubella antibodies; 20% of those with antibodies may lose detectable titers by age 5 years; such children may not develop antibodies when vaccinated; several documented instances of apparent and inapparent rubella infection have been reported;[148,157,158] the exact mechanism of these unusual immunologic phenoma is unknown
Chronic progressive panencephelitis of late onset	Rare; exact incidence unknown	Slowly progressive motor and mental deterioration with elevated protein and gamma globulin levels and rubella antibody levels in CSF and serum

[a] Note that the percentages given represent the approximate incidence of defects in children diagnosed as being rubella affected. Estimates of the incidence in children of mothers exposed to rubella would be substantially less, as it is estimated that only 20 to 35% of such children will have clinically apparent disease.

pressure. High degrees of myopia are sometimes seen in children with congenital rubella. Prompt and early detection of this visual defect is important, as it will allow early use of corrective lenses. The retina is commonly involved, with the appearance of discrete patchy black pigmentation and depigmentation. This retinal abnormality is not associated with clinically apparent visual loss.

Heart and Vascular Defects

The most common cardiac defect associated with the rubella syndrome is patency of the ductus arteriosus with or without stenosis of the pulmonary valve or the pulmonary artery and its branches. Numerous other cardiac and vascular defects have been described in association with the syndrome. The more common of these are aortic stenosis, malformation of the aortic arch including coarctation of the aorta, and defects of the interventricular septum.[146]

Central Nervous System (CNS) Disorders

Disorders of the CNS are common in infants with congenital rubella.[150-152] Symptoms include lethargy, irritability, disturbances of muscle tone, and difficulty in swallowing. The fontanelle frequently shows a fullness consistent with increased cerebrospinal fluid (CSF) pressure. The CSF commonly shows moderate increases in leukocytes and protein content. Evidence for CNS disease may continue during the early months of life. By age 18 months mental retardation may be evident and may be accompanied by cerebral palsy, hyperactivity, restlessness, seizures, and other abnormal motor and psychologic signs and symptoms.

Other Organ Systems

Neonatal thrombocytopenic purpura is a common consequence of intrauterine rubella.[153,154] The platelet count may be depressed during the first few weeks of life, with gradual return to normal levels, usually over a period of several weeks. It is rarely life threatening; however, its appearance in association with severe congenital defects is a poor prognostic sign.

Enlargement of the liver and spleen may occur. Hepatitis with elevated serum bilirubin levels presents problems in management during the early weeks of life.[154] The occurrence of hemolytic anemia and pneumonitis have been noted.[155] The rubella syndrome has also been associated with a variety of other abnormalities of lesser importance in clinical management, including abnormal dermatoglyphic patterns.[156]

The persistence of antibodies in children who have had congenital rubella infection differs from that seen in those who have had postnatal infections. Among a large group of children with documented intrauterine rubella virus infection, the decline in antibody in the early years of life was significantly more rapid than that seen in children with postnatally acquired infection.[157,158] In the congenitally infected group, 18.5% of the children who had reached their fifth birthday were seronegative. Moreover, when 19 of these seronegative children were given rubella vaccine, 17 failed to develop antibody, and the 2 children who did respond developed only low levels of antibody. The immunologic phenomena responsible for these findings are unclear. However, they suggest that the lack of response to immunization may be of assistance in the retrospective diagnosis of congenital rubella in children with the stigmata of the disease who do not have detectable antibodies. Whether or not such children would be uniformly susceptible to reinfection and, if so, what percentage would experience clinical rubella is unknown at present.

Four patients with congenital rubella have been described who, after a period of 11 to 14 years of relatively normal health, showed the onset of progressive neurological symptoms characterized by a deterioration of mental and motor function.[159,160] These clinical findings were associated with elevated levels of IgG and high rubella antibody

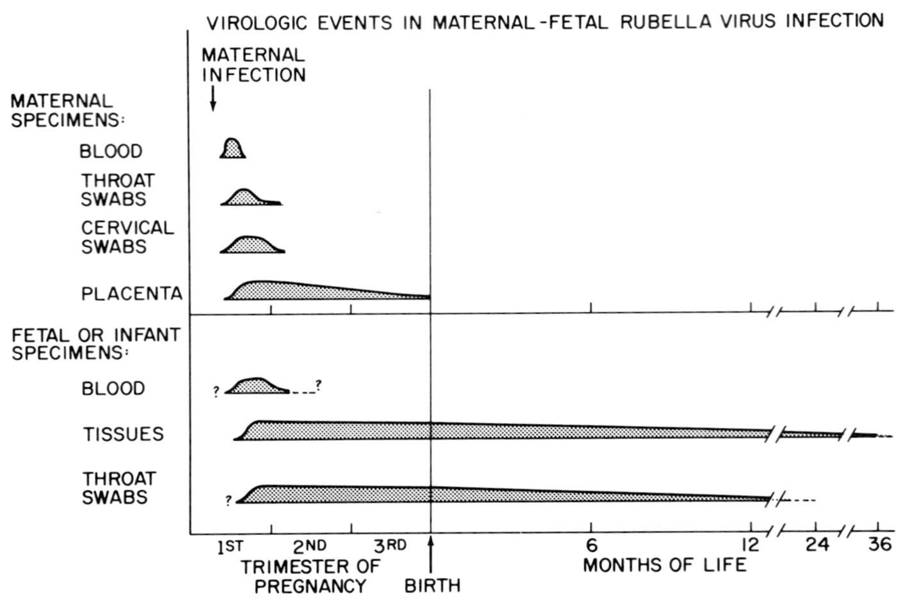

FIGURE 5. Virologic events in maternal-fetal rubella virus infection. (Reproduced with permission from the *American Journal of Clinical Pathology,* Volume 57, p. 808, 1972.)

titers in the CSF. Pathologically, the brains of these patients showed an inflammatory panencephalitis with areas of calcification. In one, rubella virus was recovered from brain biopsy material. It is likely that this affliction, like juvenile paresis following maternal syphilis during pregnancy, is a relatively rare complication of congenital infection.[161]

Through work in the laboratory made possible by the isolation of rubella virus, the pathogenesis of congenital rubella has been considerably clarified. The virus has been recovered from virtually all organs and tissues of congenitally infected infants. It exerts its effects on the fetus through a variety of mechanisms including causing cell abnormality or death (e.g., cells in the developing lens); slowing the growth rate of cells, an effect which may vary with the type of cells involved; decreasing mitotic activity and producing chromosomal abnormalities in fetal cells;[162,163] and inducing vascular damage in chorionic and embryonic tissues, leading to malformation through interference with blood supply.

The mechanism by which virus persists in the affected infant despite the presence of both passive maternal and actively produced antibody has been postulated to be related to a transient defect in delayed hypersensitivity or the persistence of clones of infected cells that continue to produce virus as long as they survive.[164-166] Infants congenitally infected with rubella virus may shed virus in the upper respiratory secretions and urine and are commonly infectious for persons caring for them.[167-169] The percent of infants from whom virus was recovered at various time periods after birth was as follows: 0 to 1 months, 84%; 1 to 4 months, 62%; 5 to 8 months, 33%; 9 to 12 months, 11%; and 13 to 20 months, 3%. No virus was detected in secretions of children over age 19 months.[146] Virus may be present in the lens for periods up to the third year. Recovery[170] of virus from a patient with chronic encephalitis[160] suggests that the virus may, probably in rare circumstances, be present in some form for even longer periods of time. A summary of the pathogenesis of fetal rubella in relation to maternal infection is shown in Figure 5.

Pathologically, the lesions seen in rubella-infected embryos and fetuses generally consist of cellular damage with little inflammatory or fibrotic response. Pneumonitis,

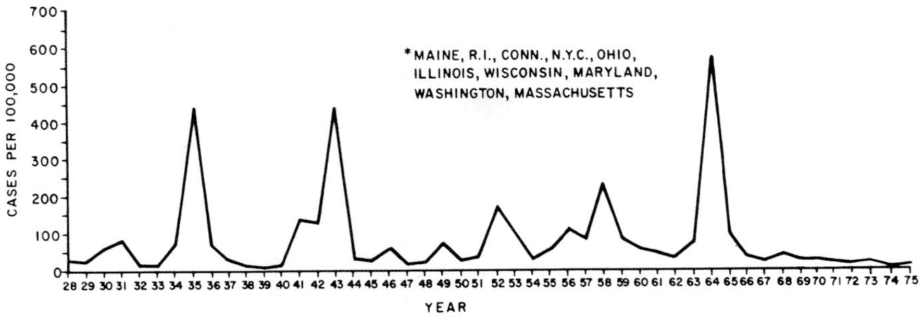

FIGURE 6. Rubella incidence in ten selected areas (*) of the United States, 1928—1975. (From Rubella Surveillance, July 1973—December 1975, Center for Disease Control, Atlanta, August 1976.)

meningitis, and iridocyclitis represent lesions in which inflammatory changes may be more pronounced. Disturbed growth of connective tissues with nodular sclerosis of the cardiac valves and vascular sclerosis of the aorta, pulmonary arteries, and, less commonly, other arteries occurs. Occasional areas of focal necrosis of cells are seen in blood vessel endothelium, myocardial cells, skeletal muscle cells, lens fiber cells, the developing cochlea and stria vascularis of the inner ear, and the teeth.[171-174]

Rubella Epidemiology

Because rubella is a poorly reported illness, the figures concerning its incidence represent estimates. However, if one makes the assumption that, in general, the percent of cases reported remains relatively constant and couples this information with that obtained from serological surveys, a clear picture of the epidemiology emerges.[126,175] Sporadic outbreaks occur each year. Major epidemics involving the entire country or large geographic areas have occurred in the past approximately every 6 to 9 years (Figure 6). A significant proportion of the population escapes rubella infection in childhood. Approximately 15% of women of childbearing age residing in the continental U.S. lack serologic evidence of earlier rubella infection. The seasonal occurrence of rubella shows a pattern similar to that of other illnesses transmitted via the respiratory tract, with a peak incidence occurring in the later winter and early spring months. The disease is most infectious during a period of a few days immediately before and after the onset of rash, corresponding to the period of maximum virus shedding.

Since the extensive use of rubella vaccines in the U.S. began, there has been a decline in the number of reported cases of rubella and the occurrence of the rubella syndrome (Figure 6, Table 7). There has not been a major epidemic of rubella in more than a decade.[131,175,176]

Laboratory Diagnosis

Clinical Laboratory Tests

The common clinical laboratory tests are of little assistance in the diagnosis of rubella. Infections acquired by contact may exhibit a mild leukopenia with a relative lymphocytosis and the appearance of plasma cells in peripheral blood smears. The platelets may be slightly reduced. In rubella arthritis, when effusions occur joint aspirates show a pale yellow viscid fluid with a protein content of between 1.9 and 3.4 g%. Cell counts per cubic millimeter have ranged between 14,500 and 27,000. The differential counts have been variable; some patients have predominantly polymorphonuclear cells; others have primarily macrophages, lymphocytes, and synovial cells.[177,178]

In congenital rubella the laboratory findings are related to the organs involved, e.g.,

Table 7
NATIONAL REGISTRY OF CONGENITAL RUBELLA SYNDROME

Year	No. cases by year of birth
1969	78
1970	92
1971	52
1972	40
1973	38
1974	25
1975	33
1976	16[a]

[a] Reporting for recent years is incomplete, as some cases are not diagnosed until later childhood.

Established by the Center for Disease Control, Atlanta, 1969. Data provided by the Center for Disease Control.

thrombocytopenia or anemia with a bone marrow showing hyperplastic erythroid elements. The serum bilirubin may be elevated in the presence of hepatitis. In patients with congenital rubella there are almost always immunoglobulin abnormalities.[164,165] The IgM levels are commonly elevated during the first year, while IgG antibodies may be high, normal, or (less commonly) low. These abnormalities usually disappear by 12 months of age. Elevations in CSF cell counts and protein levels occur during the neonatal period, but ordinarily return to normal by age 3 months. In infants with chronic encephalitis, however, continuing CSF abnormalities may be evidenced for months. Radiologic abnormalities may be present, with linear areas of radiolucency and increased bone density in the metaphyses. These changes are especially prominent in the long bones of the extremities, and they heal radiographically by 2 to 3 months of age.

Virus Laboratory Tests

The diagnosis of rubella can be definitively established only by specific virologic or serologic test procedures. A brief outline of the procedures involved in virus isolation is shown in Table 8. In rubella virus acquired by contact, secretions collected by swabbing of the nasopharynx and oropharynx are the richest source of virus.[55,140] Isolates are less often made from heparinized whole blood, urine, or rectal swabs. Swab specimens should be immersed in a suitable transport medium containing a small amount of protein as a stabilizer and should be protected from inactivation by heat and light. Blood and urine samples should be collected in sterile containers. Such specimens should be held at refrigerator temperature pending inoculation of cell cultures if this can be accomplished within 48 hr of collection. If longer time periods are involved or if the materials are to be shipped, freezing in appropriately insulated containers in dry ice is necessary. The optimum time periods during the course of the disease for collection of such specimens are indicated in Figure 4.

Virus isolation is a relatively complex and costly procedure requiring a well-equipped laboratory and an experienced virologist. For these reasons, diagnosis is more commonly attempted through the use of serologic procedures. Most state health departments and many commercial testing laboratories are capable of performing rubella serologic tests. For this purpose paired sera are collected, the first soon after exposure

Table 8
RUBELLA VIRUS ISOLATION PROCEDURES

- Rubella clinical specimen (throat swab, urine, blood, stool or rectal swab, tissues obtained surgically or at autopsy)
  - Store at 4°C; protect from drying, heat, light
  - Process
    - Add antibiotics including fungistatic agents as needed to prevent overgrowth of fungi and bacteria
    - Swab specimens rinsed in balanced salt solution
    - Triturate tissues; prepare 10% suspensions
  - Inoculate cell cultures, e.g., primary African green monkey kidney (GMK), BS-C-1, Vero or others
    - Observe for
      - Interference (GMK, BS-C-1)
      - Specific fluorescence with rubella antisera (any of the cultures)
  - Subpassage in cell cultures
    - Observe for
      - Interference (GMK, BS-C-1)
      - Fluorescence with specific rubella as above, or CPE (RK_{13}, SIRC, Vero); may require repeated subpassages for a clear-cut CPE)
    - Identify as rubella virus by
      - Neutralization with specific antiserum
      - Fluorescence with specific rubella antiserum
      - Preparation of antigens (BHK-21, Vero)
        - HA: identify by HAI with specific antiserum
        - CF: identify by CF with specific antiserum

(e.g., in a pregnant woman exposed to rubella) or as soon as possible after onset of clinical illness and the second 6 to 7 weeks after exposure or 3 to 4 weeks after the onset of illness. Dilutions of serum are then tested for antibody, and the highest serum dilution producing a significant effect in the test system employed is taken as the antibody titer. Both sera should be assayed in the same test, as variation in biological tests is too great to allow meaningful interpretations among tests performed at different times. Ordinarily, a fourfold or greater rise in antibody is considered diagnostic (e.g., a titer of 1:8 in the acute specimen or 1:32 or more in the convalescent). The hemagglutination inhibition (HAI) test is the simplest and least expensive test and is therefore the most commonly used.[84,179]

However, some laboratories may perform complement fixation (CF) or fluorescent antibody (FA) tests. The CF test may be of particular value if the initial serum sample is obtained late (e.g., a week or more after rash onset), as these antibodies appear more slowly (Figure 4). For the determination of immunity status, the HAI test is most commonly used because HAI antibodies, as do neutralizing and fluorescent antibodies, appear promptly after disease onset and persist for many years with little change in level. The CF antibodies may disappear within months or years and thus are not well suited for this purpose. In primary infection the early antibody response involves IgM antibodies; IgG antibodies appear shortly thereafter, increase in level, and in approximately four weeks replace IgM as the predominant antibody type. Thus, by assaying for rubella-specific IgM, it may be possible to obtain evidence for recent rubella on the basis of a single serum specimen.[179] Such tests involving specialized immunological techniques are ordinarily conducted only by research laboratories.[180-182]

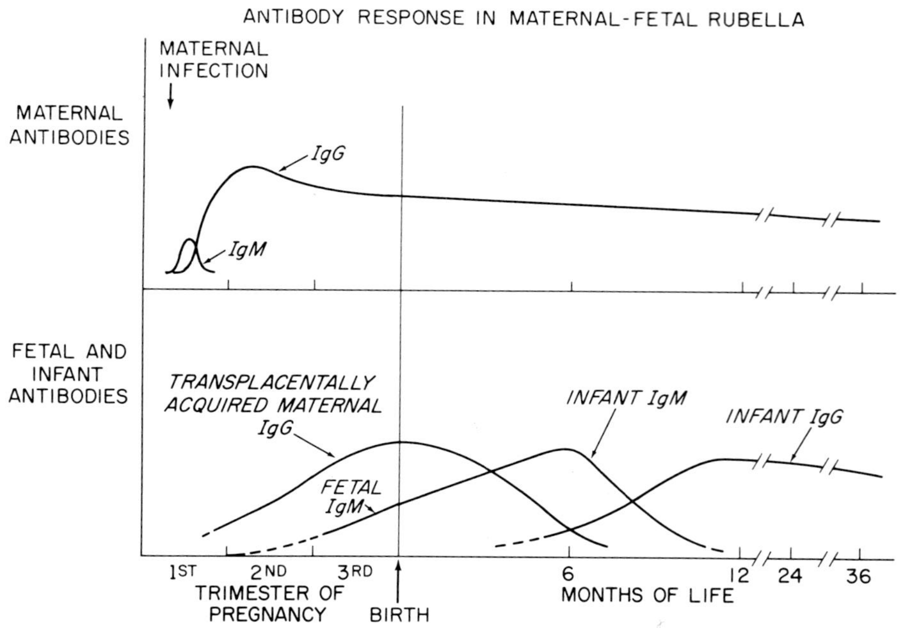

FIGURE 7. Characteristics of the antibody response in maternal-fetal rubella. (Reproduced with permission from the *American Journal of Clinical Pathology,* Volume 57, p. 808, 1972).

In congenital rubella, virus isolation assumes greater importance. Both placental and fetal tissues may contain virus, and the affected infant often sheds virus in the respiratory secretions, urine, and stool (Figure 6). Tissues of infants contain virus. Surgically removed cataractous lenses have been positive the longest; virus has been shown in children several years of age.[170] Clinical specimens should be handled in the same manner as described above. Tissues should be collected in sterile containers, refrigerated or frozen, and shipped to the virus isolation laboratory. The neonate with congenital rubella has circulating antibodies; these represent transplacentally acquired maternal IgG antibodies and actively produced fetal IgM antibodies. During the first several months of life, the passively acquired IgG antibodies decline and the actively produced IgM antibodies become predominant. IgG antibody production by the infant appears during the early months and late in the first year replaces IgM as the predominant antibody type.[183] Thus, the presence of significant amounts of rubella-specific IgM antibody during the first year of life is evidence for congenital rubella infection (Figure 7).

Treatment and Prophylaxis

Therapy

No specific chemotherapeutic measures are available, and treatment is symptomatic. Analgesics are sometimes required in the management of rubella arthritis. Steroids are commonly used in the treatment of rubella-induced thrombocytopenic purpura. Steroids have also been used in treating rubella encephalitis, although there is no convincing evidence for their effectiveness in this condition.

In congenital rubella, the therapy is designed to alleviate the specific defects. Special attention should be paid to correction of auditory and visual deficits, as early treatment of these problems may have a beneficial effect in preventing speech disorders and psychological problems.

Table 9
RUBELLA VIRUS VACCINES LICENSED FOR COMMERCIAL DISTRIBUTION IN THE UNITED STAES

Virus strain	Manufacturer	No. passages in cell culture	Cell culture used for production	Cell culture derivation
HPV-77[a] DE5	Merck Sharp & Dohme	77 in primary GMK[b] + 5 in duck embryo (DE)	Primary duck embryo	Closed flock
HPV-77[c] DK12	Philips-Roxane (also distributed by Parke, Davis & Co.)	77 in primary GMK + 12 in primary canine kidney (DK)	Primary canine kidney	Closed beagle colony
Cendehill	Recherche et Industrie Therapeutique S.A. (distributed in U.S. by Smith, Kline & French)	3 in GMK + 53 in primary rabbit kidney cells (RK)	Primary rabbit kidney	Closed rabbit colony
RA-27/3	Wellcome Foundation	29 in WI-38 cells	WI-38 human diploid	Continuously propagated cell line derived from human fetal lung fibroblasts

[a] HPV, high-passage virus; a derivative of the M-33 strain of rubella virus subcultured 77 times in primary African green monkey kidney cell cultures.

[b] GMK, primary African green monkey kidney cell cultures.

[c] Not currently marketed; this vaccine was associated with a higher incidence of joint manifestations and was withdrawn from the market in 1973.

Gamma Globulin

Gamma globulin has been used in attempts to prevent rubella in the exposed pregnant woman. The administration of large doses (20 ml of immune serum globulin, human) may modify the clinical symptoms[184] and has been shown to effect some diminution in the duration of viremia in experimentally infected volunteers.[129] However, globulin is definitely not highly effective, as children with the congenital rubella syndrome have been born to mothers given appropriately timed large doses of globulin.[185] If this prophylactic measure has an effect in reducing the numbers of infected fetuses is unknown. Whether or not globulin is used, careful serologic follow-up should be conducted to determine whether infection has occurred following exposure. Specimens should be collected 3 weeks after exposure and at weekly intervals thereafter for 3 weeks. Because there is some evidence that globulin administration may delay the appearance of natural antibodies, the specimen collected at 6 weeks postexposure should not be neglected.[179] These sera and the serum obtained soon after exposure should be tested simultaneously in order to establish the diagnosis.

Vaccines

Live rubella virus vaccines first became available in 1969. As shown in Table 9, four types of vaccine have been licensed for commercial production in the U.S.A.[62,77,186,187] The vaccine viruses have been modified by serial subpassage in cell cultures and, as indicated in Table 10, can be distinguished from natural rubella viruses by a variety of in vitro and in vivo laboratory tests.[76,188] The vaccines produce a modified rubella infection in susceptible vaccinees, with the production of antibodies. This infection usually occurs in the absence of symptoms; however, adults (especially women) may experience transient rash and lymph node enlargement, and, as in the natural disease, may have symptoms referable to the joints and peripheral nerves. Joint complaints usually occur approximately 3 to 5 weeks postimmunization, but may sometimes be

Table 10
"MARKER" CHARACTERISTICS OF ATTENUATED RUBELLA VIRUS

	Virus characteristics	
Test	Natural "wild type" virus	Attenuated vaccine virus
In vitro		
CPE in RK_{13} cells	Slight at 35°C	Prompt, complete at 35°C
Plaques in RK_{13} cells	Small, slow to develop	Large, develop quickly
Induction of Interferon	None or low level	High level
In vivo		
Experimental infection in the Rhesus monkey		
Infectivity by intranasal route of inoculation	High; one tissue culture infectious dose equals one monkey infectious dose	Low; approximately 100 tissue culture infectious doses required to infect
Virus in respiratory secretions	Present in 90% or more; sustained over period of 1 week	Rare; when observed usually a single positive specimen
Viremia	Detected in 50% or more	Not seen
Communicability to cage contacts	Often occurs	Not seen
Serologic response		
Neutralizing	High titers	Low titers
HAI	High titers	Low titers
CF	High titers	Low titers or absent
Distribution of virus in infected animal	Reticuloendothelial and many other tissues positive	Only reticuloendothelial tissues positive
Experimental infection of rabbits and guinea pigs	Produces brisk antibody response	Produces little or no antibody

delayed until as long as 8 weeks.[176] The incidence of these reactions ranges from less than 10% in children to as high as 25 to 60% in adult females. As in the natural disease, these reactions are usually mild, of brief duration, and have not shown any tendency to progress to chronic deforming arthritis. Otherwise, the vaccines have been essentially free of consequential side reactions. Vaccinees briefly shed small amounts of the attenuated virus in the pharynx during the period 10 to 20 days after vaccination; despite this, carefully controlled trials under field conditions have not shown evidence of transmission of vaccine virus to contacts.[176,189]

Vaccine-induced antibodies are of lower titer than those following the natural disease; however, the levels have been shown to persist for a period of years.[190] Vaccine-induced antibodies have been shown to protect against the natural disease.[138] The strategy for rubella prevention in the U. S. calls for vaccination of children 15 months of age and older by private physicians and in public health programs. The immunization of women in the childbearing years who lack serologic evidence of infection is also encouraged. Special efforts must be taken here to exclude from receiving vaccine women who may be pregnant, as vaccine virus has been recovered from fetuses of women undergoing abortions because of inadvertant vaccination during pregnancy.[191-193] While there is no documented instance of the attenuated virus having caused the rubella syndrome, the possibility that it may do so still exists. Thus, it is important to assure that women who elect to receive vaccine not be pregnant at the time of immunization or for a period of at least 2 months thereafter, a length of time sufficient to allow the vaccine virus infection to subside and active immunity to develop.

Rubella vaccines are currently most commonly used in combination products containing either live attenuated measles vaccine or both attenuated measles and mumps viruses.[194] The long interval of time since the last major epidemic in 1964 suggests that if the extensive use of vaccines continues, it may be possible to prevent such occurrences in future years.

GENERAL REFERENCES

Wesselhoeft, C., Medical progress, Rubella (German measles), *N. Engl. J. Med.*, 236, 943, 1947.

Ingalls, T. H., Plotkin, S. A., Meyer, H. M., Jr., and Parkman, P. D., Rubella: epidemiology, virology and immunology, *Am. J. Med. Sci.*, 253, 349, 1967.

Krugman, S., Ed., Rubella Symposium, *Am. J. Dis. Child.*, 110, 345, 1965.

Schiff, G. M. and Sever, J. L., Rubella: recent laboratory and clinical advances, *Prog. Med. Virol.*, 8, 30, 1966.

Regamey, R. H., de Parkieri, A., Hennessen, W., Ikic, D., and Perkins, F. T., Eds., *Proc. 23rd Symp. Microbiological Standardization: Rubella Vaccines, London, November 1968*, S. Karger, Basel, 1969.

Parkman, P. D. and Meyer, H. M., Jr., Prospects for a rubella virus vaccine, *Prog. Med. Virol.*, 11, 80, 1969.

Krugman, S., International Conference on Rubella Immunization, *Am. J. Dis. Child.*, 118, 1, 1969.

Rubella and its control: Proceedings of an International Symposium held at St. Vincent's Hospital, Dublin on 18 November 1971, *Postgrad. Med. J. Suppl.*, 48(3), 3, 1972.

Rawls, W. E., Virology and epidemiology of rubella virus, *Int. Ophthalmol. Clin.*, 12, 21, 1972.

Rudolph, A. J. and Desmond, M. M., Clinical manifestations of the congenital rubella syndrome, *Int. Ophthalmol. Clin.*, 12, 3, 1972.

REFERENCES

1. **Hiro, V. and Tasaka, S.**, Die rotheln sind eine Viruskankheit, *Monatsschr. Kinderheilkd.*, 76, 328, 1938.
2. **Gregg, N. M.**, Congenital cataract following German measles in the mother, *Trans. Ophthalmol. Soc. Aust.*, 3, 35, 1941.
3. **Weller, T. H. and Neva, F. A.**, Propagation in tissue culture of cytopathic agents from patients with rubella-like illness, *Proc. Soc. Exp. Biol. Med.*, 111, 215, 1962.
4. **Parkman, P. D., Buescher, E. L., and Artenstein, M. S.**, Recovery of rubella virus from army recruits, *Proc. Soc. Exp. Biol. Med.*, 111, 225, 1962.
5. **Fenner, F.**, Classification and nomenclature of viruses, second report of the International Committee on taxonomy of viruses, *Intervirology*, 7, 44, 1976.
6. **Mettler, N. E., Petrelli, R. L., and Casals, J.**, Absence of antigenic cross-reactions between rubella virus and arboviruses, *Virology*, 36, 503, 1968.
7. **Tesh, R. B. and Rosen, L.**, Failure of rubella virus to replicate in mosquitoes, *Intervirology*, 5, 216, 1975.
8. **Norrby, E., Magnusson, P., Friding, B., and Gard, S.**, A note on the morphology of rubella virus, *Arch. Gesamte Virusforsch.*, 13, 421, 1963.
9. **Magnusson, P. and Norrby E. C. J.**, On the morphology of rubella virus, *Arch. Gesamte Virusforsch.*, 16, 412, 1965.
10. **Phillips, C. A., Behbehani, A. M., Johnson, L. W., and Melnick, J. L.**, Isolation of rubella virus: an epidemic characterized by rash and arthritis, *JAMA*, 191, 615, 1965.
11. **Best, J. M., Banatvala, J. E., Almeida, J. D., and Waterson, A. P.**, Morphological characteristics of rubella virus, *Lancet*, 2, 237, 1967.
12. **Holmes, I. H. and Warburton, M. R.**, Is rubella an arbovirus? *Lancet*, 2, 1233, 1967.
13. **Holmes, I. H., Wark, M. C., Jack, I., and Grutzner, J.**, Identification of two possible types of virus particle in rubella-infected cells, *J. Gen. Virol.*, 2, 37, 1968.
14. **Smith, K. O. and Hobbins, T. E.**, Physical characteristics of rubella virus, *J. Immunol.*, 102, 1016, 1969.

15. **Almeida, J. D. and Laurence, G. D.,** Heated and unheated antiserums on rubella virus: morphological effect, *Am. J. Dis. Child.,* 118, 101, 1969.
16. **McCombs, R. M., Braunschweig, J. P., and Rawls, W. E.,** Morphology of rubella virus, *Exp. Mol. Pathol.,* 9, 27, 1968.
17. **Bardeletti, G., Kessler, N., and Aymard-Henry, M.,** Morphology, biochemical analysis and neuraminidase activity of rubella virus, *Arch. Virol.,* 49, 175, 1975.
18. **Murphy, F. A., Halonen, P. E., and Harrison, A. K.,** Electron microscopy of the development of rubella virus in BHK-21 cells, *J. Virol.,* 2, 1223, 1968.
19. **Holmes, I. H., Wark, M. C., and Warburton, M. F.,** Is rubella an arbovirus? II. Ultrastructural morphology and development, *Virology,* 37, 15, 1969.
20. **Edwards, M. R., Cohen, S. M., Bruno, M., and Diebel, R.,** Micromorphological aspects of the development of rubella virus in BHK-21 cells, *J. Virol.,* 3, 439, 1969.
21. **vonBonsdorff, C. H. and Vaheri, A.,** Growth of rubella virus in BHK-21 cells: electron microscopy of morphogenesis, *J. Gen. Virol.,* 5, 47, 1969.
22. **Parkman, P. D., Buescher, E. L., Artenstein, M. S., McCown, J. M., Mundon, F. K., and Druzd, A. D.,** Studies of rubella. I. Properties of the virus, *J. Immunol.,* 93, 595, 1964.
23. **Plotkin, S. A., Dudgeon, J. A., and Ramsay, A. M.,** Laboratory studies on rubella and the rubella syndrome, *Br. Med. J.,* 2, 1296, 1963.
24. **Schiff, G. M. and Sever, J. L.,** Rubella: recent laboratory and clinical advances, *Prog. Med. Virol,* 8, 30, 1966.
25. **Russell, B., Selzer, G., and Goetze, H.,** The particle size of rubella virus, *J. Gen. Virol.,* 1, 305, 1967.
26. **Ver, B. A., Melnick, J. L., and Wallis, C.,** Efficient filtration and sizing of viruses with membrane filters, *J. Virol.,* 2, 21, 1968.
27. **Cusamano, C. L., Schiff, G. M., Sever, J. L., and Huebner, R. J.,** Rubella virus: nucleic acid studies using 5-iodo-2′-deoxyuridine, *J. Pediatr.,* 65, 138, 1964.
28. **Maassab, H. F. and Cochran, K. W.,** Influence of 5-fluorodeoxyruridine on growth characteristics of rubella virus, *Proc. Soc. Exp. Biol. Med.,* 117, 410, 1964.
29. **Hovi, T. and Vaheri, A.,** Rubella virus-specific ribonucleic acids in infected BHK-21 cells, *J. Gen. Virol.,* 6, 77, 1970.
30. **Hovi, T. and Vaheri, A.,** Infectivity and some physicochemical characteristics of rubella virus ribonucleic acid, *Virology,* 42, 1, 1970.
31. **Sedwick, W. D. and Sokol, F.,** Nucleic acid of rubella virus and its replication in hamster kidney cells, *J. Virol.,* 5, 478, 1970.
32. **Auletta, A. E. and Marlowe, M. L.,** Effect of sodium deoxycholate on rubella virus, *Appl. Microbiol.,* 16, 1624, 1968.
33. **Maassab, H. F. and Cochran, K. W.,** Rubella virus: inhibition *in vitro* by amantadine hydrochloride, *Science,* 145, 1443, 1964.
34. **Plotkin, S. A.,** Inhibition of rubella virus by amantadine, *Arch. Gesamte Virusforsch.,* 16, 438, 1965.
35. **Oxford, J. S. and Schild, G. C.,** *In vitro* inhibition of rubella virus by 1-adamantanamine hydrochloride, *Arch. Gasamte Virusforsch.,* 17, 313, 1965.
36. **Oxford, J. S. and Schild, G. S.,** Inhibition of the growth of influenza and rubella viruses by amines and ammonium salts, *Br. J. Exp. Pathol.,* 48, 235, 1967.
37. **Oxford, J. S. and Schild, G. C.,** The evaluation of antiviral compounds for rubella virus using organ cultures, *Arch. Gesamte Virusforsch.,* 22, 349, 1967.
38. **Stephenson, J. A., Artenstein, M. S., Parkman, P. D., Buescher, E. L., and Druzd, A. D.,** Effect of amantadine hydrochloride on rubella virus infection in the rhesus monkey, *Antimicrob. Agents Chemother.,* p. 548, 1965.
39. **Plotkin, S. A., Klaus, R. M., and Whitely, J. P.,** Hypogammaglobulinemia in an infant with congenital rubella syndrome: failure of 1-adamantanamine to stop virus excretion, *J. Pediatr.,* 69, 1085, 1966.
40. **Chagnon, A. and Laflamme, P.,** Effect of acidity on rubella virus, *Can. J. Microbiol.,* 10, 501, 1964.
41. **Parkman, P. D.,** Biological characteristics of rubella virus, *Arch. Gesamte Virusforsch.,* 16, 401, 1965.
42. **Fabiyi, A. K., Sever, J. L., Ratner, N., and Caplan, B.,** Rubella virus: growth characteristics and stability of infectious virus and complement-fixing antigen, *Proc. Soc. Exp. Biol. Med.,* 122, 392, 1966.
43. **Kenney, M. T., and Albright, K. L., Emery, J. B., and Bittle, J. L.,** Inactivation of rubella virus by gamma radiation, *J. Virol.,* 4, 807, 1969.
44. **Vaheri, A. and Hovi, T.,** Structural proteins and subunits of rubella virus, *J. Virol.,* 9, 10, 1972.
45. **Liebhaber, H. and Gross, P. A.,** The structural proteins of rubella virus, *Virology,* 47, 684, 1972.
46. **Best, J. M. and Banatvala, J. E.,** A comparison of RK13, vervet monkey kidney and patas monkey kidney cell cultures for the isolation of rubella virus, *J. Hyg.,* 65, 263, 1967.

47. **Schmidt, N. J., Lennette, E. H., Woodie, J. D., and Ho, H. H.**, Identification of rubella virus isolates by immunofluorescent staining, and a comparison of the sensitivity of three cell culture systems for recovery of virus, *J. Lab. Clin. Med.*, 68, 502, 1966.
48. **Desmyter, J., Melnick, J. L., and Rawls, W. E.**, Defectiveness of interferon production and of rubella virus interference in a line of green monkey kidney cells (Vero), *J. Virol.*, 2, 955, 1968.
49. **Liebhaber, H., Riordan, J. T., and Horstmann, D. M.**, Replication of rubella virus in a continuous line of African green monkey kidney cells (Vero), *Proc. Soc. Exp. Biol. Med.*, 125, 636, 1967.
50. **Rhim, J. S. and Schell, K.**, Cytopathic and plaque assay of rubella virus in a line of African green monkey kidney cells (Vero), *Proc. Soc. Exp. Biol., Med.*, 125, 602, 1967.
51. **Günalp, A.**, Growth and cytopathic effect of rubella virus in a line of green monkey kidney cells, *Proc. Soc. Exp. Biol. Med.*, 118, 85, 1965.
52. **Veronelli, J. A., Maassab, H. F., and Hennessy, A. V.**, Isolation in tissue culture of an interfering agent from patients with rubella, *Proc. Soc. Exp. Biol. Med.*, 111, 472, 1962.
53. **Maassab, H. F. and Veronelli, J. A.**, Characteristics of serially propagated monkey kidney cell cultures with persistent rubella infection, *J. Bacteriol.*, 91, 436, 1966.
54. **Parkman, P. D., Hopps, H. E., and Meyer, H. M., Jr.**, Virus isolation procedures, in *Rubella: 1st Annu. Symp. Eastern Pennsylvania Branch, American Society for Microbiology*. Friedman, H. and Prier, J. E., Eds., Charles C Thomas, Springfield, Ill., 1973, 33.
55. **Wallace, R. E., Vasington, P. J., Petricciani, J. C., Hopps, H. E., and Lorenz, D. E.**, Diploid cell lines from subhuman primates as substrates for virus vaccine production, *Prog. Immunobiol. Stand.*, 5, 181, 1972.
56. **Burnett, J. W. and Alford, C. A., Jr.**, Comparative effectiveness of two culture techniques for isolation of rubella virus, *Proc. Soc. Exp. Biol. Med.*, 120, 569, 1965.
57. **McCarthy, K. G., Taylor-Robinson, C. H., and Pillinger, S. E.**, Isolation of rubella virus from cases in Britain, *Lancet*, 2, 593, 1963.
58. **Hoskins, J. M. and Plotkin, S. A.**, Behavior of rubella virus in human diploid cell strains. I. Growth of virus, *Arch. Gesamte Virusforsch.*, 21, 283, 1967.
59. **Hoskins, J. M. and Plotkin, S. A.**, Behaviour of rubella virus in human diploid cell strains. II. Studies of infected cells, *Arch. Gesamte Virusforsch.*, 21, 296, 1967.
60. **Plotkin, S. A., Boue, A., and Boue, J. G.**, The in vitro growth of rubella virus in human embryo cells, *Am. J. Epidemiol.*, 81, 71, 1965.
61. **Plotkin, S. A.**, Development of RA27/3 attenuated rubella virus grown in WI-38 cells, in *Proc. 23rd Int. Symp. Microbiological Standardization: Rubella Vaccines, London, November 1968*, Regamey R. H., de Parkieri, A., Hennessen, W., Ikic, D., and Perkins, F. T., Eds., S. Karger, Basel, 1969, 249.
62. **Plotkin, S. A., Farquhar, J. D., Katz, M., and Buser, F.**, Attenuation of RA 27-3 rubella virus in WI-38 human diploid cells, *Am. J. Dis. Child.*, 118, 178, 1969.
63. **Jacobs, J. P.**, Some comparative characteristics of WI-38 and MRC-5 cells and their suitability for the production of viral vaccines, in *Proc. Symp. Human Diploid Cells*, Ikic, D., Ed., Yugoslavian Academy of Science and Arts, Zagreb, 1970, 43.
64. **McCarthy, K. and Taylor-Robinson, C. H.**, Growth and cytopathic effects of rubella virus in primary rabbit tissue culture, *Arch. Gesamte Virusforsch.*, 16, 415, 1965.
65. **Huygelen, C., Sigel, M. M., Zygraich, N., Petermans, J. H., Colinet, G., Leyten, R., Raupp, W. G., Pinto, C. A., Garg, S. G., Boyle, J. J., and Haff, R. F.**, Safety testing of rubella virus vaccine (Cendehill strain), preparation in primary rabbit kidney cells, *Am. J. Dis. Child.*, 118, 362, 1969.
66. **Reddick, A. and Roesel, C. E.**, Rubella virus: growth and cytopathic effect in primary cultures of cells of rabbit embryos, *Science*, 151, 1405, 1966.
67. **Taylor-Robinson, C. H., McCarthy, K., Grylls, S. G., and O'Ryan, E. M.**, Plaque formation by rubella virus, *Lancet*, 1, 1364, 1964.
68. **Hull, R. N. and Butorac, G.**, The utility of rabbit kidney cell strain LLC-RK1 to rubella virus studies, *Am. J. Epidemiol.*, 83, 509, 1966.
69. **Van der Veen, J.**, personal communication.
70. **Leerhoy, J.**, Cytopathic effect of rubella virus in a rabbit-cornea cell line, *Science*, 149, 633, 1965.
71. **Phillips, C. A., Melnick, J. L., and Burkhardt, M.**, Isolation, propagation and neutralization of rubella virus in cultures of rabbit cornea (SIRC) cells, *Proc. Soc. Exp. Biol. Med.*, 122, 783, 1966.
72. **Rhim, J. S., Schell, K., and Huebner, R. J.**, Plaque assays of rubella virus in cultures of rabbit cornea (SIRC) cells, *Proc. Soc. Exp. Biol. Med.*, 125, 1271, 1967.
73. **Buynak, E. B., Larson, V. M., McAleer, W. J., Mascoli, C. C., and Hilleman, M. R.**, Preparation and testing of duck embryo cell culture rubella vaccine, *Am. J. Dis. Child.*, 118, 347, 1969.
74. **Fuccillo, D. A., Gitnick, G. L., Traub, R., Wong, K., Sever, J. L., and Huebner, R. J.**, Cultivation of rubella virus in avian tissue cultures and embryonated eggs, *Proc. Soc. Exp. Biol. Med.*, 125, 1015, 1967.

75. **Parkman, P. D., Meyer, H. M., Jr., and Hopps, H. E.,** Production and laboratory testing of experimental live rubella virus vaccine, in *Proc. 23rd Int. Symp. Microbiological Standardization: Rubella Vaccines, London, November 1968,* Regamy, R. H., de Parkieri, A., Hennessen, W., Ikic, D., and Perkins, F. T., Eds., S. Karger, Basel, 1969, 261.
76. **Shishido, A. and Ohtawara, M.,** Development of attenuated rubella virus vaccines in Japan, *Jpn. J. Med. Sci. Biol.,* 29, 227, 1976.
77. **Musser, S. J. and Hilsabeck, L. J.,** Production of rubella virus vaccines live, attenuated in canine renal cell cultures, *Am. J. Dis. Child.,* 118, 355, 1969.
78. **Neva, F. A. and Weller, T. H.,** Rubella interferon and factors influencing the indirect neutralization test for rubella antibody, *J. Immunol.,* 93, 466, 1964.
79. **Weller, T. H. and Neva, F. A.,** Biological characteristics of rubella virus as assayed in a human amnion culture system, *Arch. Gesamte Virusforsch.,* 16, 393, 1965.
80. **Vaheri, A., Sedwick, W. D., Plotkin, S. A., and Maes, R.,** Cytopathic effect of rubella virus in BHK-21 cells and growth to high titers in suspension culture, *Virology,* 27, 239, 1965.
81. **Maes, R., Vaheri, A., Sedwick, D., and Plotkin, S.,** Synthesis of virus and macromolecules by rubella-infected cells, *Nature,* 210, 384, 1966.
82. **Vaheri, A., Sedwick, W. D., and Plotkin, S. A.,** Growth of rubella virus in BHK-21 cells. I. Production, assay, and adaptation of virus, *Proc. Soc. Exp. Biol. Med.,* 125, 1086, 1967.
83. **Schell, K., Wong, K. T., Turner, H. C., and Huebner, R. J.,** Production of rubella complement-fixing antigen in BHK-21 cells, *Proc. Soc. Exp. Biol. Med.,* 123, 832, 1966.
84. **Stewart, G. L., Parkman, P. D., Hopps, H. E., Douglas, R. D., Hamilton, J. P., and Meyer, H. M., Jr.,** Rubella-virus hemagglutination-inhibition test, *N. Engl. J. Med.,* 276, 554, 1967.
85. **Parkman, P. D., Mundon, F. K., McCown, J. M., and Buescher, E. L.,** Studies of rubella. II. Neutralization of the virus, *J. Immunol.,* 93, 608, 1964.
86. **Plotkin, S. A.,** Plaquing of rubella virus in RK 13 cells, *Arch. Gesamte Virusforsch.,* 16, 423, 1965.
87. **Marcus, P. I. and Carver, D. H.,** Hemadsorption-negative plaque test: new assay for rubella virus revealing a unique interference, *Science,* 149, 983, 1965.
88. **Rawls, W. E., Desmyter, J., and Melnick, J. L.,** Rubella virus neutralization by plaque reduction, *Proc. Soc. Exp. Biol. Med.,* 124, 167, 1967.
89. **Brown, G. C., Maassab, H. F., Veronelli, J. A., Francis, T., Jr.,** Rubella antibodies in human serum: detection by the indirect fluorescent-antibody techniques, *Science,* 145, 943, 1964.
90. **Woods, W. A., Johnson, R. T., Hostetler, D. D., Lepow, M., and Robbins, F. C.,** Immunofluorescent studies on rubella-infected tissue cultures and human tissues, *J. Immunol.,* 96, 253, 1966.
91. **Feldman, H. A.,** Removal by heparin-$MnCl_2$ of nonspecific rubella hemagglutinin serum inhibitor., *Proc. Soc. Exp. Biol. Med.,* 127, 570, 1968.
92. **Liebhaber, H.,** Measurement of rubella antibody by hemagglutination inhibition. I. Variables affecting rubella hemagglutination, *J. Immunol.,* 104, 818, 1970.
93. **Liebhaber, H.,** Measurement of rubella antibody by hemagglutination inhibition. II. Characteristics of an improved HAI test employing a new method for the removal of non-immunoglobulin HA inhibitors from serum, *J. Immunol.,* 104, 826, 1970.
94. **Halonen, P. E., Ryan, J. M., and Stewart, J. A.,** Rubella hemagglutinin prepared with alkaline extraction of virus grown in suspension culture of BHK-21 cells, *Proc. Soc. Exp. Biol. Med.,* 125, 162, 1967.
95. **Furukawa, T., Plotkin, S. A., Sedwick, W. D., and Profeta, M. L.,** Studies on hemagglutination by rubella virus, *Proc. Soc. Exp. Biol. Med.,* 126, 745, 1967.
96. **Palmer, D. F., Herrmann, K. L., and Lincoln, R. E.,** A procedural Guide to the Performance of the Standardized Rubella Hemagglutination-inhibition Test, Immunology Ser. No. 2, Center for Disease Control, Atlanta, 1970, 1.
97. **Sever, J. L., Huebner, R. J., Castellano, G. A., Sarma, P. S., Fabiyi, A., Schiff, G. M., and Cusumano, C. L.,** Rubella complement fixation test, *Science,* 148, 385, 1965.
98. **Stern, H.,** Rubella virus complement-fixation test, *Nature,* 208, 200, 1965.
99. **Veronelli, J. A. and Eckert, E. A.,** A new technique for preparation of rubella complement fixing antigen, *Proc. Soc. Exp. Biol. Med.,* 121, 1223, 1966.
100. **Halonen, P. E., Casey, H. L., Stewart, J. A., and Hall, A. D.,** Rubella complement-fixing antigen prepared by alkaline extraction of virus grown in suspension culture of BHK-21 cells, *Proc. Soc. Exp. Biol. Med,* 125, 167, 1967.
101. **Schmidt, N. J. and Lennette, E. H.,** The complement-fixing antigen of rubella virus, *Proc. Soc. Exp. Biol. Med.,* 121, 243, 1966.
102. **Schmidt, N. J., Lennette, E. H., Gee, P. S., and Dennis, J.,** Physical and immunologic properties of rubella antigens, *J. Immunol.,* 100, 851, 1968.
103. **Schmidt, N. J. and Styk, B.,** Immunodiffusion reactions with rubella antigens, *J. Immunol.,* 101, 210, 1968.

104. **Le Bouvier, G. L.**, Precipitinogens of rubella virus-infected cells, *Proc. Soc. Exp. Biol. Med.*, 130, 51, 1969.

105. **Le Bouvier, G. L.**, Rubella precipitins, in *Proc. 23rd Int. Symp. Microbiological Standardization: Rubella Vaccines, London, November 1968*, Regamey, R. H., de Parkieri, A., Hennessen, W., Ikic, D., and Perkins, F. T., Eds., S. Karger, Basel, 1969, 133.

106. **Le Bouvier, G. L.**, Physicochemical characteristics of rubella antigens theta and iota, *Nature*, 221, 78, 1969.

107. **Belcourt, R. J. P., Wong, F. C., and Walcroft, M. J.**, Growth of rubella virus in rabbit foetal tissues and cell cultures, *Can. J. Public Health*, 56, 253, 1965.

108. **Oxford, J. S.**, The growth of rubella virus in small laboratory animals, *J. Immunol.*, 98, 697, 1967.

109. **Oxford, J. S. and Potter, C. W.**, Persistent rubella virus infection in laboratory animals, *Arch. Gesamte Virusforsch.*, 34, 75, 1971.

110. **Carver, D. H., Seto, D. S. Y., Marcus, P. I., and Rodrigues, L.**, Rubella virus replication in the brains of suckling mice, *J. Virol.*, 1, 1089, 1967.

111. **Cotlier, E. and Bohigian, G.**, Further studies on pregnant rats infected with rubella virus and their offsprings in *Proc. 23rd Int. Symp. Microbiological Standardization: Rubella Vaccines, London, November 1968*, Regamey, R. H., de Parkieri, A., Hennessen, W., Ikic, D., and Perkins, F. T., Eds., S. Karger, Basel, 1969, 205.

112. **Schiff, G. M., Sever, J. L., and Huebner, R. J.**, Experimental rubella infection in ferrets, *Clin. Res.*, 12, (Abstr.), 245, 1964.

113. **Rorke, L. B., Fabiyi, A., Elizan, T. S., and Sever, J. L.**, Experimental cerebrovascular lesions in congenital and neonatal rubella-virus infections of ferrets, *Lancet*, 2, 153, 1968.

114. **Kono, R., Hayakawa, Y., Hibi, M., and Ishii, K.**, Experimental vertical transmission of rubella virus in rabbits, *Lancet* 1, 343, 1969.

115. **London, W. T., Fucillo, D. A., Anderson, B., and Sever, J. L.**, Concentration of rubella virus antigen in chondrocytes of congenitally infected rabbits, *Nature*, 226, 172, 1970.

116. **Elizan, T. S., Fabiyi, A., and Sever, J. L.**, Experimental teratogenesis in ferrets using rubella virus, *J. Mt. Sinai Hosp. N.Y.*, 36, 103, 1969.

117. **Elizan, T. S., Fabiyi, A., and Sever, J. L.**, Study of rubella virus as a teratogen in experimental animals: a short review, *J. Mt. Sinai Hosp. N.Y.*, 36, 108, 1969.

118. **Sigurdardottir, B., Givan, K. F., Rozee, K. R., and Rhodes, A. J.**, Association of virus with cases of rubella studied in Toronto: propagation of the agent and transmission to monkeys, *Can. Med. Assoc. J.*, 88, 128, 1963.

119. **Heggie, A. D. and Robbins, F. C.**, Experimental rubella in rhesus monkeys, *Proc. Soc. Exp. Biol. Med.*, 114, 750, 1963.

120. **Parkman, P. D., Phillips, P. E., Kirschstein, R. L., and Meyer, H. M., Jr.**, Experimental rubella virus infection in the rhesus monkey, *J. Immunol.*, 95, 743, 1965.

121. **Parkman, P. D., Meyer, H. M., Jr., Hopps, H. E., and Kirschstein, R. L.**, Laboratory studies with an attenuated rubella virus, in *1st Int. Conf. Vaccines Against Viral and Rickettsial Diseases of Man*, Sci. Publ. No. 147, Pan American Health Organization and World Health Organization, Washington, D. C., May 1967, 381.

122. **Parkman, P. D., Phillips, P. E., and Meyer, H. M., Jr.**, Experimental rubella virus infection in pregnant monkeys, *Am. J. Dis. Child.*, 110, 390, 1965.

123. **Parkman, P. D. and Meyer, H. M., Jr.**, Prospects for a rubella virus vaccine, *Prog. Med. Virol.*, 11, 80, 1969.

124. **Delahunt, C. S. and Rieser, N.**, Rubella induced embryopathies in monkeys, *Am. J. Obstet. Gynecol.*, 99, 580, 1967.

125. **Huygelen, C. and Peetermans, J.**, Attenuation of rubella virus by serial passage in primary rabbit kidney cell cultures. II. Experiments in animals, *Arch. Gesamte Virusforsch.*, 21, 357, 1967.

126. **Girardi, A. J., Larson, V. M., and Hilleman, M. R.**, Further tests in hamsters for oncogenic quality of ordinary viruses and mycoplasma with correlative review, *Proc. Soc. Exp. Biol. Med.*, 118, 173, 1965.

127. **Cockburn, W. C.**, World aspects of the epidemiology of rubella, *Am. J. Dis. Child.*, 118, 112, 1969.

128. **Veale, H.**, History of an epidemic of rötheln with observations on its pathology, *Edinburgh Med. J.*, 12, 404, 1866.

129. **Green, R. H., Balsamo, M. R., Giles, J. P., Krugman, S., and Mirick, G. S.**, Studies of the natural history and prevention of rubella, *Am. J. Dis. Child.*, 110, 348, 1965.

130. **Heggie, A. D. and Robbins, F. C.**, Natural rubella acquired after birth, clinical features and complications, *Am. J. Dis. Child.*, 118, 12, 1969.

131. **Judelsohn, R. G. and Wyll, S. A.**, Rubella in Bermuda; termination of epidemic by mass vaccination, *JAMA*, 223, 401, 1973.

132. **Kantor, T. G. and Tanner, M.**, Rubella arthritis and rheumatoid arthritis, *Arthritis Rheum.*, 5, 378, 1962.

133. **Miller, H. G., Stanton, J. B., and Gibbons, J. L.,** Para-infectious encephalomyelitis and related syndromes, a critical review of the neurological complications of certain specific fevers *Q. J. Med.*, 25, 427, 1956.
134. **Sherman, F. E., Michaels, R. H., and Kenny, F. M.,** Acute encephalopathy (encephalitis) complicating rubella; report of cases with virologic studies, *JAMA*, 192, 675, 1965.
135. **Wallace, S. J.,** Thrombocytopenic purpura after rubella, *Lancet*, 1, 139, 1963.
136. **Meyer, H. M., Jr. and Parkman, P. D.,** Rubella: the immunology of rubella, *Am. J. Med. Sci.*, 253, 349, 1967.
137. **Sever, J. L., Brody, J. A., Schiff, G. M., McAlister, R., and Cutting R.,** Rubella epidemic on St. Paul Island in the Pribilofs, 1963. II. Clinical and laboratory findings for the intensive study population, *JAMA*, 191, 624, 1965.
138. **Davis, W. J., Larson, H. E., Simsarian, J. P., Parkman, P. D., and Meyer, H. M., Jr.,** A study of rubella immunity and resistance to infection, *JAMA*, 215, 600, 1971.
139. **Heggie, A. D.,** Pathogenesis of the rubella exanthem: detection of rubella virus from the skin, *N. Engl. J. Med.*, 285, 664, 1971.
140. **Parkman, P. D., Hopps, H. E., and Meyer, J. M., Jr.,** Rubella virus, isolation characterization and laboratory dignosis, *Am. J. Dis. Child.*, 118, 68, 1969.
141. **Boue, A., Nicolas, A., and Montagnon, B.,** Reinfection with rubella in pregnant women, *Lancet*, 1, 1251, 1971.
142. **Dudgeon, J. A.,** Congenital rubella, *J. Pediatr.*, 87, 1078, 1975.
143. **Sheridan, M. D.,** Final report of a prospective study of children whose mothers had rubella in early pregnancy, *Br. Med. J.*, 2, 536, 1964.
144. **Cooper, L. Z., Ziring, P. R., Ockerse, A. B., Fedun, B. A., Kiely, B., and Krugman, S.,** Rubella: clinical manifestations and management, *Am. J. Dis. Child.*, 118, 18, 1969.
145. **Hardy, J. B., McCracken, G. H., Gilkeson, M. R., and Sever, J. L.,** Adverse fetal outcome following maternal rubella after the first trimester of pregnancy, *JAMA*, 207, 2414, 1969.
146. **Jackson, A. D. M. and Fisch, L.,** Deafness following maternal rubella; results of prospective investigation, *Lancet*, 2, 1241, 1958.
147. **Karmody, C. S.,** Subclinical maternal rubella and congenital deafness, *N. Engl. J. Med.*, 278, 809, 1968.
148. **Cooper, L. Z. and Krugman, S.,** Clinical manifestations of postnatal and congenital rubella, *Arch. Ophthalmol.*, 77, 434, 1967.
149. **Boniuk, M. and Zimmerman, L. E.,** Ocular pathology in the rubella syndrome, *Arch. Ophthalmol.*, 77, 455, 1967.
150. **Rorke, L. B. and Spiro, A. J.,** Cerebral lesions in congenital rubella syndrome, *J. Pediatr.*, 70, 243, 1967.
151. **Desmond, M. M., Wilson, G. S., Melnick, J. L., Singer, D. B., Zion, T. E., Rudolph, A. J., Pineda, R. G., Ziai, M., and Blattner, R. J.,** Congenital rubella encephalitis, course and early sequelae, *J. Pediatr.*, 71, 311, 1967.
152. **Zausmer, E.,** Congenital rubella: pathogenesis of motor deficits, *Pediatrics*, 47, 16, 1971.
153. **Banatvala, J. E., Horstmann, D. M., Payne, M. C., and Gluck, L.,** Rubella syndrome and thrombocytopenic purpura in newborn infants, *N. Engl. J. Med.*, 273, 474, 1965.
154. **Cooper, L. Z., Green, R. H., Krugman, S., Giles, J. P., and Mirick, G. S.,** Neonatal thrombocytopenic purpura and other manifestations of rubella contracted in utero, *Am. J. Dis. Child.*, 110, 416, 1965.
155. **Korones, S. B., Ainger, L. E., Monif, G. R. G., Roane, J., Sever, J. L., and Fuste, F.,** Congenital rubella syndrome: myocardial damage and other new clinical aspects, study of 22 infants, *Am. J. Dis. Child.*, 110, 434, 1965.
156. **Purvis-Smith, S. G. and Menser, M. A.,** Genetic and environmental influences on digital dermatoglyphics in congenital rubella, *Pediatr. Res.*, 7, 215, 1973.
157. **Hardy, J. B., Sever, J. L., and Gilkeson, M. R.,** Declining antibody titers in children with congenital rubella, *J. Pediatr.*, 75, 213, 1969.
158. **Cooper, L. Z., Florman, A. L., Ziring, P. R., and Krugman, S.,** Loss of rubella hemagglutination inhibition antibody in congenital rubella. Failure of seronegative children with congenital rubella to respond to HPV-77 rubella vaccine. *Am. J. Dis. Child.*, 122, 397, 1971.
159. **Townsend, J. J., Baringer, J. R., Wolinsky, J. S., Malamud, N., Melnick, J. L., Panitch, H. S., Scott, R. A. T., Oshiro, L. S., and Cremer, N. E.,** Progressive rubella panencephalitis, late onset after congenital rubella *N. Engl. J. Med.*, 292, 990, 1975.
160. **Weil, M. L., Itabashi, H. H., Cremer, N. E., Oshiro, L. S., Lennette, E. H., and Carnay, L.,** Chronic progressive panencephalitis due to rubella virus simulating subacute sclerosing panencephalitis, *N. Engl. J. Med.*, 292, 994, 1975.
161. **Johnson, R. T.,** Editorial: progressive rubella encephalitis, *N. Engl. J. Med.*, 292, 1023, 1975.

162. **Chang, T. H., Moorhead, P. S., Boué, J. G., Plotkin, S. A., and Hoskins, J. M.,** Chromosome studies of human cells infected *in utero* and *in vitro* with rubella virus, *Proc. Soc. Exp. Biol. Med.*, 122, 236, 1966.
163. **Nusbacher, J., Hirschhorn, K., and Cooper, L. Z.,** Chromosomal abnormalities in congenital rubella, *N. Engl. J. Med.*, 276, 1409, 1967.
164. **Michaels, R. H.,** Immunologic aspects of congenital rubella, *Pediatrics*, 43, 339, 1969.
165. **Schimke, R. N., Bolano, C., and Kirkpatrick, C. H.,** Immunologic deficiency in the congenital rubella syndrome, *Am. J. Dis. Child.*, 118, 626, 1969.
166. **Simons, M. J.,** Congenital rubella: an immunological paradox? *Lancet*, 2, 1275, 1968.
167. **Rudolph, A. J., Yow, M. D., Phillips, C. A., Desmond, M. M., Blattner, R. J., and Melnick, J. L.,** Transplacental rubella infection in newly born infants, *JAMA*, 191, 843, 1965.
168. **Hardy, J., Monif, G., Medearis, D., and Sever, J. L.,** Postnatal transmission of rubella virus to nurses, *JAMA*, 191, 1034, 1965.
169. **Schiff, G. M., and Dine, M. S.,** Transmission of rubella from newborns; a controlled study among young adult women and report of an unusual case, *Am. J. Dis. Child.*, 110, 447, 1965.
170. **Menser, M. A., Harley, J. D., Hertzberg, R., Dorman, D. C., and Murphy, A. M.,** Persistence of virus in lens for three years after prenatal rubella, *Lancet*, 2, 387, 1967.
171. **Gray, J. E.,** Rubella in pregnancy: a report on six embryos, *Br. Med. J.*, 1, 1388, 1960.
172. **Tondury, G. and Smith, D. W.,** Fetal rubella pathology, *J. Pediatr.*, 68, 867, 1966.
173. **Singer, D. B., Rudolph, A. J., Rosenberg, H. S., Rawls, W. E., and Boniuk, M.,** Pathology of the congenital rubella syndrome, *J. Pediatr.*, 71, 665, 1967.
174. **Driscoll, S. G.,** Histopathology of gestational rubella, *Am. J. Dis. Child.*, 118, 49, 1969.
175. **Modlin, J. F., Brandling-Bennett, A. D., Witte, J. J., Campbell, C. C., and Meyers, J. D.,** A review of five years' experience with rubella vaccine in the United States, *Pediatrics*, 55, 20, 1975.
176. **Meyer, H. M., Jr. and Parkman, P. D.,** Rubella vaccination: a review of practical experience, *JAMA*, 215, 613, 1971.
177. **Chambers, R. J. and Bywaters, E. G. L.,** Rubella synovitis, *Ann. Rheum. Dis.*, 22, 263, 1963.
178. **Yanez, J. E., Thompson, G. R., Mikkelsen, W. M., and Bartholomew, L. E.,** Rubella arthritis, *Ann. Intern. Med.*, 64, 722, 1966.
179. **Banatvala, J. E.,** Maternal rubella and its virologic diagnosis, *Postgrad. Med. J.*, 48, (Suppl.), 11, 1972.
180. **Vesikari, T. and Vaheri, A.,** Rubella: a method for rapid diagnosis of a recent infection by demonstration of IgM antibodies, *Br. Med. J.*, 1, 221, 1968.
181. **Baublis, J. V. and Brown, G. C.,** Specific response of the immunoglobulins to rubella infection, *Proc. Soc. Exp. Biol. Med.*, 128, 206, 1968.
182. **Gupta, J. D., Peterson, V., Stout, M., and Murphy, A. M.,** Single-sample diagnosis of recent rubella by fractionation of antibody on a Sephadex G-200 column, *J. Clin. Pathol.*, 24, 547, 1971.
183. **Cooper, L. Z., Chandor, S. B., Ockerse, A. B., Feinstein, P., and Krugman, S.,** Immunologic consequences of congenital rubella, *Pediatr. Res.*, 1, (Abstr.), 217, 1967.
184. **Brody, J. A., Sever, J. L., and Schiff, G. M.,** Prevention of rubella by gamma globulin during an epidemic in Barrow, Alaska, in 1964, *N. Engl. J. Med.*, 272, 127, 1965.
185. **Avery, G. B., Monif, G. G. R., Sever, J. L., and Leikin, S. L.,** Rubella syndrome after inapparent maternal illness, *Am. J. Dis. Child.*, 110, 444, 1965.
186. **Hilleman, M. R., Buynak, E. B., Whitman, J. E., Weibel, R. W., and Stokes, J., Jr.,** Live attenuated rubella virus vaccines: experiences with duck embryo cell preparations, *Am. J. Dis. Child.*, 118, 166, 1969.
187. **Prinzie, A., Huygelen, C., Gold, J., Farquhar, J., and McKee, J.,** Experimental live attenuated rubella virus vaccine: clinical evaluation of Cendehill strain, *Am. J. Dis. Child.*, 118, 172, 1969.
188. **Hopps, H. E., Parkman, P. D., and Meyer, H. M., Jr.,** Laboratory testing in rubella vaccine control, *Am. J. Dis. Child.*, 118, 338, 1969.
189. **Meyer, H. M., Jr., Parkman, P. D., and Panos, T. C.,** Attenuated rubella virus. II. Production of an experimental live-virus vaccine and clinical trial, *N. Engl. J. Med.*, 275, 575, 1966.
190. **Herrmann, K. L., Halstead, S. B., Brandling-Bennett, D. A., Witte, J. J., Wiebenga, N. H., and Eddins, D. L.,** Rubella immunization: persistence of antibody four years after a large scale field trial, *JAMA*, 235, 2201, 1976.
191. **Larson, H. E., Parkman, P. D., Davis, W. J., Hopps, H. E., and Meyer, H. M., Jr.,** Inadvertent rubella virus vaccination during pregnancy, *N. Engl. J. Med.*, 284, 870, 1971.
192. **Fleet, W. F., Jr., Benz, E. W., Jr., Karzon, D. T., Lefkowitz, L. B., and Herrmann, K. L.,** Fetal consequences of maternal rubella immunization, *JAMA*, 227, 621, 1974.
193. **Modlin, J. F., Herrmann, K. L., Brandling-Bennett, A. D., Eddins, D. L., and Hayden, G. H.,** Risk of congenital abnormality after inadvertent rubella vaccination of pregnant women, *N. Engl. J. Med.*, 294, 972, 1976.
194. **Buynak, E. B., Weibel, R. E., Whitman, J. E., Stokes, J., Jr., and Hilleman, M. R.,** Combined live measles, mumps and rubella virus vaccines, *JAMA*, 207, 2259, 1969.

REOVIRIDAE: REOVIRUSES

Rex S. Spendlove

Reoviridae

Viruses of the family Reoviridae are nonenveloped and have icosahedral symmetry. Their genomes consist of several (generally 10 to 12) segments of double-stranded RNA encapsidated within a single virus particle.[1] Two genera in the family have been described: *Reovirus,* with a host range of vertebrates, and *Orbivirus,* with a host range of vertebrates and insects.[1,2]

In addition to *Reovirus* and *Orbivirus,* other double-stranded RNA viruses which have not yet been assigned to taxonomic groups include

1. The reo-like gastroenteritis viruses (rotaviruses)
2. Plant viruses represented by wound tumor and rice dwarf viruses
3. Insect viruses that include the cytoplasmic polyhedrosis viruses that have at least 80 different species of insects as hosts[3]
4. Mycoviruses that infect fungi
5. The ϕ6 bacteriophage of *Pseudomonas phaseocola*[4]
6. Virus-like particles from killer strains of *Saccharomyces cerevisiae*[5]
7. Infectious pancreatic necrosis virus of fish[6]

Reoviruses

Introduction

The icosahedral reovirion has a double-layered capsid that contains a genome composed of ten segments of double-stranded RNA. Stanley et al.[7,8] were the first to isolate a reovirus. The virus was obtained from the stools of an aboriginal child and was given the name hepato-encephalomyelitis virus. In 1954, Ramos-Alvarez and Sabin[9] recovered several reoviruses from the stools of healthy children. These isolates were classified as echovirus type 10. Some strains of reoviruses that have been isolated are shown in Table 1.

Sabin[11] pointed out that isolates of the echovirus type 10 had a number of characteristics that differed from those of the picornaviruses, and he proposed the name reovirus. The name is an acronym that stresses the respiratory, enteric, and orphan (REO) characteristics of the virus.

Morphology

Reoviruses have two coats that are icosahedral in structure with five-, three-, and two-fold symmetry; however, the arrangement of the capsomers in the two shells has not been definitely determined yet. Earlier, workers [22-27] described a $P = 1$ virion that possessed four capsomers per edge, i.e., a virion with a triangulation number (T) of 9 and a capsomer number of 92. The capsomers were reported to be elongated and hollow.

Vasquez and Tournier[28] reinterpreted the earlier observations and suggested a model in which the capsomers were not separate pentagonal or hexagonal columns as previously described, but morphological units made up of interconnected rings sharing structural components. Amano et al.[29] and Vasquez[30] further substantiated the shared structural components theory through studies in which they observed disrupted reovirions.

Luftig et al.[31] presented results that cast doubt on reovirus models that postulated 92 holes. They observed a total of 20 capsomers per complete periphery (Figure 1) and

Table 1
REOVIRUS STRAINS[10]

Virus strain	Isolated from/in	Refs.
Serotype 1		
Lang (Echovirus 10) prototype strain	Stools of healthy child/in monkey kidney cells	9
Cocrel	Healthy child/in monkey kidney cells	11
CV3 Pan 4	Brain of chimpanzee with kuru disease/in human embryonic kidney cells	12
Perez	Human stool/in human embryonic cells	13
716	Rectal swab of child with mild respiratory illness and diarrhea/in monkey kidney cells	14
Serotype 2		
Amy	Child with steatorrheic enteritis/in monkey kidney cells	15
Barton	Human stool specimen/in human embryonic cells	16
CV4-Pan 3	Cerebellum of chimpanzee with kuru/in human embryonic kidney cells	12
D5-Jones prototype strain	Rectal swab of child with diarrhea/in monkey kidney cells	17,18
988	Child/in monkey kidney cells	19
Serotype 3		
Abney	Anal swab of child with upper respiratory infection/in rhesus monkey kidney cells	20
Carter	Human stool/in human embryonic cells	13,16
Dearing prototype strain	Child with diarrhea/in monkey kidney cells	18,21
Hepatoencephalomyelitis virus (HEV)	Stools of child with bronchiectasis, bronchopneumonia, alopecia, and conjunctivitis/in suckling mice	7,8

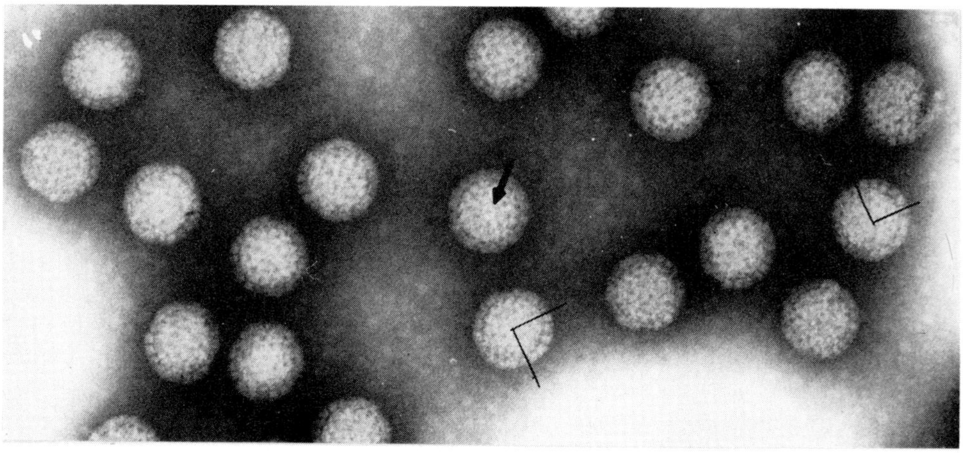

FIGURE 1. Intact reovirus (Dearing strain) stained with 4% PTA × 135,000. (From Luftig, R. B., Kilham, S., Hay, A. J., Zweerink, H. J., and Joklik, W. K., *Virology*, 48, 170, 1972. With permission.)

FIGURE 2. A 122-hole (P = 3, T = 12), 42 capsomer (P = 1, T = 4) shared subunit model. (The kit from which the model was constructed was provided through the courtesy of Jensen Research Laboratories, Provo, Utah.)

suggested that the capsomer arrangement would best conform to the P = 3 class of icosahedral deltahedra with 240 subunits arranged around 122 holes (Figure 2).

The outer coat polypeptides of the reovirion can be removed sequentially by controlled treatment with chymotrypsin or several other proteolytic enzymes.[32-37] Partial removal of the outer shell produces subviral particles, while more complete digestion yields core particles. The digestion products (Figure 3) contain 12 projections or spikes located on five-fold vertices.[31]

Physicochemical Properties

The outer reovirus capsid is composed of three polypeptides, while the inner shell contains four. The central virion cavity contains ten segments of double-stranded RNA and a mixture of oligoribonucleotides (short, single-stranded RNA molecules) from 2 to 20 nucleotides long. Several enzyme activities are associated with the virion and with subviral particles. Some properties of the reovirus are presented in Tables 2 and 3.

Virus Replication*

The mechanisms reoviruses use for attaching to and penetrating host cells in productive infections have not been fully determined. Several methods of penetration (Figure

* A number of recent reviews[39,44-47] can be consulted for a more detailed discussion of different aspects of reovirus replication.

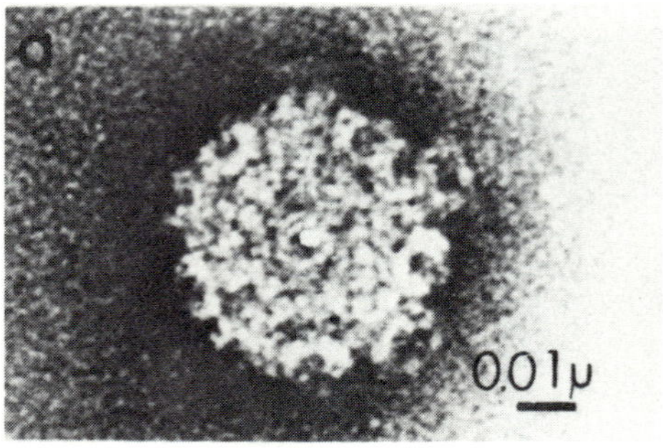

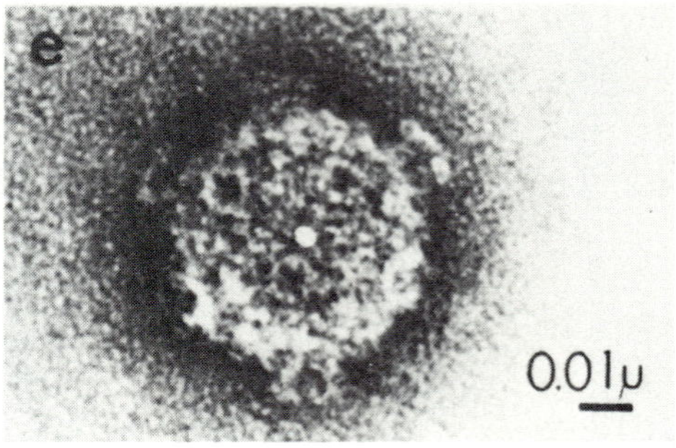

FIGURE 3. Two reovirus core particles the central axes of which were either on a presumptive (a) fivefold or (b) threefold vertex. (From Luftig, R. B., Kilham, S., Hay, A. J., Zweerink, H. J., and Joklik, W. K., *Virology*, 48, 170, 1972. With permission.)

Table 2
REOVIRUS CAPSID POLYPEPTIDES[38]

Location	Designation	Mol wt (daltons)	Percent in virion	No. of molecules/virion (Ref. 39)
Core	$\gamma 1$	155,000	16, 15 (Ref. 39)	113, 110
	$\gamma 2$	140,000	10, 11 (Ref. 39)	80, 90
	$\sigma 2$	38,000	7	202, 200
Outer shell	$\mu 1^a$	80,000	2	23, 20
	$\mu 2^b$	72,000	36	550
	$\sigma 1$	42,000	1	31, 30
	$\sigma 3$	34,000	28	890, 900

[a] Joklik[39] considers $\mu 1$ to be a core polypeptide.
[b] Of the 550 $\mu 2$ polypeptides per virion, 10 to 20 are glycosylated.[40]

4) have been proposed: direct penetration (step 1),[48] uncoating or partial uncoating of the viruses prior to their being taken into the cell (step 1a)[35,49] and pinocytosis (viropexis) (step 1b)[50,51] Step 1c shows a virion passing through the membrane of a pinocytotic vesicle. This would be analogous to direct penetration as pictured in step 1. Another possibility is that subviral particles pass through lysosomal membranes (see step preceding 1e).

Most of the investigations of reovirus penetration and uncoating have been directed toward showing that virions are taken into cells by viropexis (step 1b) and that the pincytotic vesicles fuse with lysosomal enzymes (step 1d), after which the viral outer coat is removed by lysosomal enzymes (step 2c) [50,51] (Reoviruses are not completely uncoated in the infectious process[52-54]). Although little evidence supports direct penetration and partial uncoating in the cytoplasm, there is an equal lack of strong evidence to show that the virus which initiates infection is partially uncoated in lysosomes.

When all of the polypeptide $\sigma 3$ and an 8000-dalton fragment of polypeptide $\mu 2$ (see Table 2) are removed from the outer capsid of the reovirion, the transcriptase in the core is activated[32] and the viral genome is transcribed (step 3). It is thought that the single-stranded mRNA, which is transcribed from each of the ten genome segments, passes through holes in the spikes of the core particle (step 3).[55] It is not clear whether all ten segments are transcribed into early mRNA or whether the L_1, M_3, S_3, and S_4 genomic segments (see Table 3) are selectively transcribed at this time.[34,39,56-64] Some investigators have postulated that transcription may take place in lysosomes.[45]

Table 3
GENOME SEGMENTS OF REOVIRUS

Segment number	Size designation[a]	Mol wt (× 10^6 daltons)		
1	L	2.5[41]	2.5[42]	2.55[43]
2	L	2.4	2.37	2.45
3	L	2.3	2.3	2.45
4	M	1.6	1.58	1.65
5	M	1.6	1.51	1.52
6	M	1.4	1.45	1.47
7	S	0.9	0.82	0.82
8	S	0.76	0.73	0.74
9	S	0.64	0.67	0.63
10	S	0.61	0.62	0.63
11		—	—	—
Total		14.7	14.55	14.92

[a] L, large; M, medium; S, small

The nascent mRNA is methylated and capped,[65-71] and translated into viral protein (step 4). Soon thereafter, new viral protein and single-stranded viral RNA are found associated together (steps 5 and 6). Some recognition mechanism must function in bringing together the ten segments of single-stranded RNA (synthesized respectively on the ten parental genome segments); however, no experimental evidence supports the diagram shown (the product of steps 5 and 6). It is not known whether protein or nucleic acid provides the recognition mechanism for sequestering the ten RNA segments.[39]

Viral proteins are sequentially added to the single-stranded RNA complex (steps 7 and 8) to form a core particle. In an early virion precursor complex (represented by step 7), the replicase (the SS → DS RNA polymerase)[72] forms a complementary strand on each of the ten single strands of RNA enclosed by viral protein. This method of replication permits the parental genome to be conserved. Sometime after the double-stranded RNA is formed and the virus is nearing completion, oligoadenylic acid is synthesized within the viral particles.[73,74]

Core particles can be further coated to form a virion (step 9) or used in the synthesis of new mRNA (step 10). In the latter case, a repeatable cycle is completed. Sometime during the infectious cycle, the parental virus core can be recoated (step 11).[54]

Early during infection, viral antigens (factories) are present in localized areas (upper cell, Figure 5A). The antigens subsequently spread and form a filamentous network throughout the cytoplasm (lower cell, Figure 5A). The virus has an affinity for microtubules[75-80] of both the mitotic spindle and interphase cells. Colchicine disrupts microtubules and causes any viral antigen in interphase cells to form globular masses. As a result of the affinity of reovirus antigen for host cell microtubules, the viral factories are transported (by the microtubules) to a perinuclear position (step 12, Figure 5B). Some of the virions are released from host cells as they mature (step 13).[81] In the late stages of an infection, viral antigen diffuses away from the nucleus, and it is released from the cells in blebs that pinch off from the cytoplasm (step 14, Figure 5C). This ballooning is probably not a method of viral release until the cell dies and the membranes lose their integrity.

A number of studies on one-step reovirus growth cycles have been conducted.[14,21,75,81,82] With Dearing strain virus infections of L cells, maximum yields of virus are produced after approximately 14 hr of infection, with little release of the cell-associated virus.

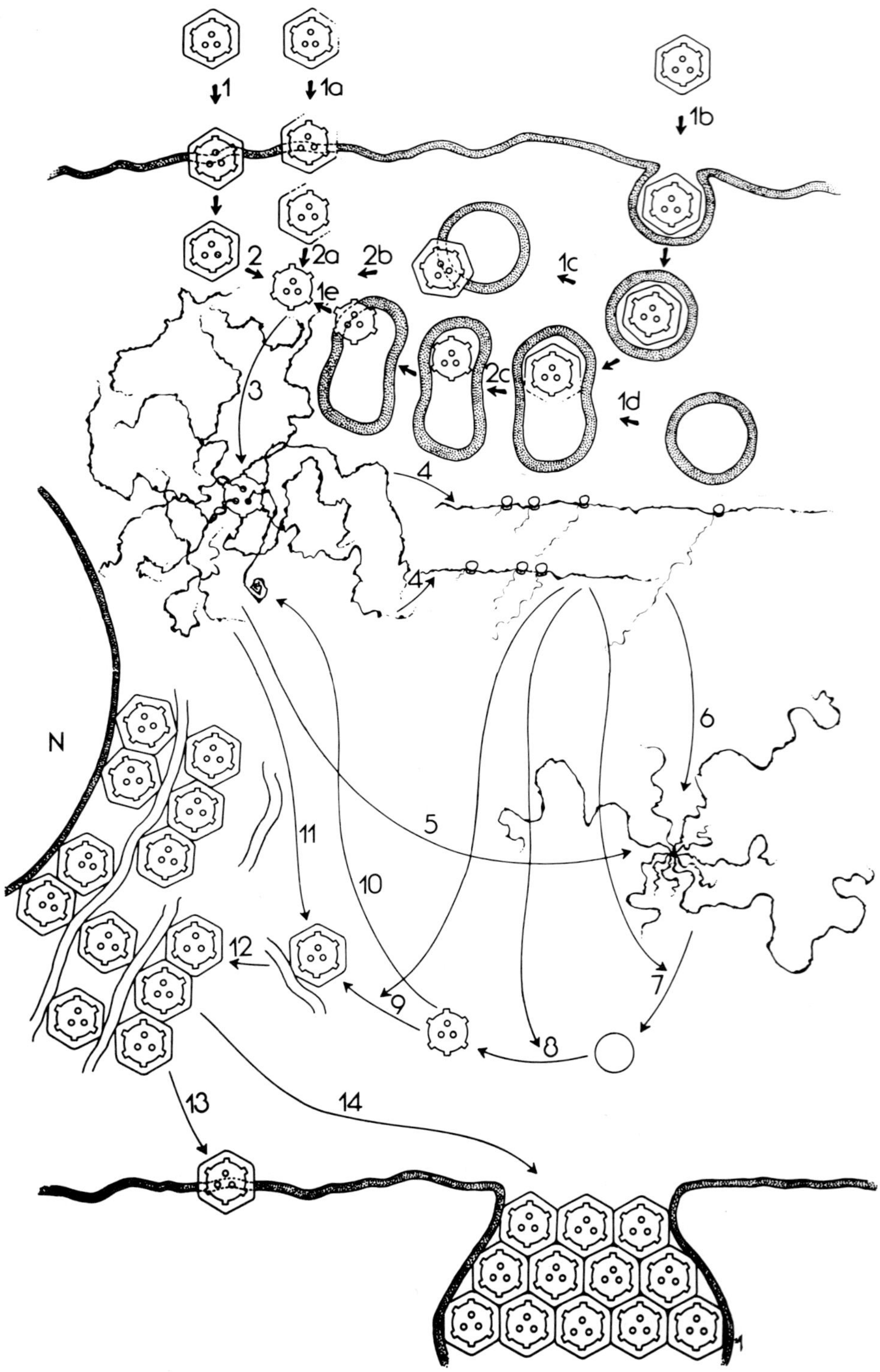

FIGURE 4. Schematic diagram showing the steps of reovirus replication.

In one investigation using the Lang strain of type 1 reovirus in human amnion (FL) cells,[81] the cell-free and cell-associated viruses were assayed before and after chymotrypsin treatment (Figure 6). Chymotrypsin removes the outer viral coat and converts potentially infectious virions to infectious subviral particles.[33] The study[81] showed that

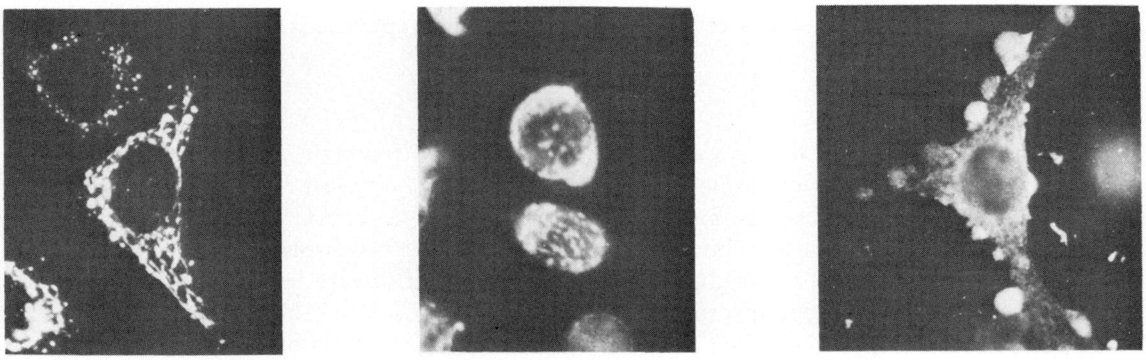

A B C

FIGURE 5. Photomicrographs of HeLa cells infected with type 1 reovirus and stained with fluorescent antiviral antibody. The four stages of viral development are illustrated: (A) Stage 1, (upper left) antigen in particulate form; Stage 2, (center) network-like structure throughout the cytoplasm; (B) Stage 3, antigen in perinuclear position; (C) Stage 4, antigen release in "bubbling" cellular processes. (From Spendlove, R. S., Lennette, E. H., Knight, C. O., and Chin, J. N., *J. Immunol.*, 90, 548, 1963. With permission.)

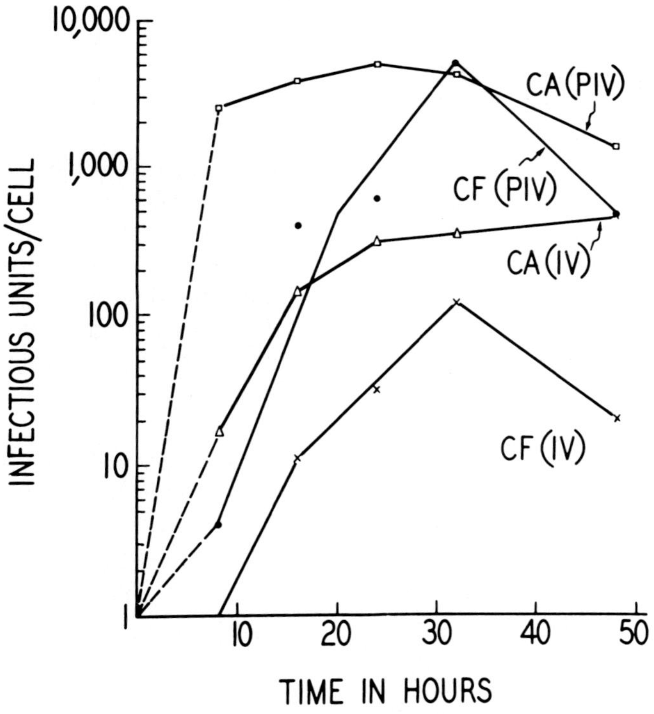

FIGURE 6. Production of infectious virus (IV) and potentially infectious virus (PIV) in reovirus infected FL cells. The broken lines representing the first 8 hr of infection indicate that the length of the latent period is unknown. CA, cell-associated virus; CF, cell-free virus. (From Spendlove, R. S., Lennette, E. H., Knight, C. O., and Chin, J. N., *J. Immunol.*, 90, 548, 1963. With permission.)

infectivity titers of virus produced during infection were more than ten times higher if the viruses were treated with enzyme before assay; and that contrary to previous reports, most of the Lang strain reovirus in infected FL cells incubated at 35°C was released from the cells. To detect the cell-free virus, the preparations had to be enzyme treated before assay.

Clinical Features

Although reoviruses have a world-wide distribution with a high incidence of infection, they are probably not important human pathogens, i.e., most reovirus infections are either subclinical or very mild.[83] However, type three reovirus appears to have caused at least three deaths in humans.[84-86] Reoviruses have been isolated from patients with fever, exanthems, upper and lower respiratory tract disease, gastroenteritis (including steatorrhea), hepatitis, meningitis, and encephalitis, [84,85,87-90]; however, an etiologic relationship has not been proven. Reports of reovirus infections in human have been reviewed by Rosen,[83,91] and Jackson and Muldoon.[90]

When volunteers were challenged with a nasal secretion containing a type 1 reovirus from an adult with a common cold[92] or with monkey kidney cell-passed virus,[93] asymptomatic infections were common. Symptoms were observed more often in the volunteers who did not have detectable reovirus antibodies.

Laboratory Diagnosis

Choice of specimens for laboratory diagnosis of reovirus is varied. Reoviruses are most often isolated from stool specimens and rectal swabs; however, they have also been recovered from throat swabs, nasal secretions, urine, blood, cerebrospinal fluid, and various organs at autopsy.[83] It is best to collect the specimens early in the disease and on several successive days.[83] Collection and handling of specimens for laboratory diagnosis have been described by Lennette et al.[94] (Figures 7 and 8). One or two specimens (2 to 4 g) collected 24 to 48 hr apart are frozen in a screw-capped jar. Rectal swabs containing fecal material are rinsed in 3 ml of tissue culture medium containing antibiotics; the medium is expressed by pressing the swab on the side of the container. After the swab is removed, the medium is frozen. The blood is collected aseptically, neither anticoagulants nor preservatives are added; the blood is not frozen before it is processed, and the serum is collected. Acute-phase blood specimens for serological tests should be collected no later than 5 to 7 days after onset of the illness, while convalescent-phase specimens should be collected 14 to 21 days or longer after onset.[94]

Serologic Tests

The hemagglutination inhibition (HI) test is by far the most widely used test in diagnostic serology of reovirus infections. It is simple to perform and is more sensitive than the neutralization and complement fixation (CF) tests. Reoviruses are assigned among three serotypes using the HI test, and the type 2 viruses can be further separated into four subtypes.[95] CF antibodies are group specific and cannot be used to type the reoviruses.

Reoviruses of the three serotypes produce good hemagglutination titers when grown in primary rhesus monkey kidney cells [96,97] However, these cells are expensive and have limited availability. Other cells that are satisfactory for propagating reoviruses have been recommended[13,98,99]

Several techniques by which to improve hemagglutinin production have been reported. Halonen,[100] when cultivating type 1 (Lang strain) reovirus, found that the hemagglutinin titers continued to increase 3 or 4 days after infectivity titers leveled off. Zalan and Labzoffsky[101] reported that reovirus type 3 (Abney strain) infectivity and hemagglutinins were produced synchronously, but the infectious particles were released much more rapidly. At least ten cycles of freezing and thawing and an additional 3-day incubation period were required to release the hemagglutinins from infected primary African green monkey kidney cells. The tendency of the type 3 hemagglutinin to

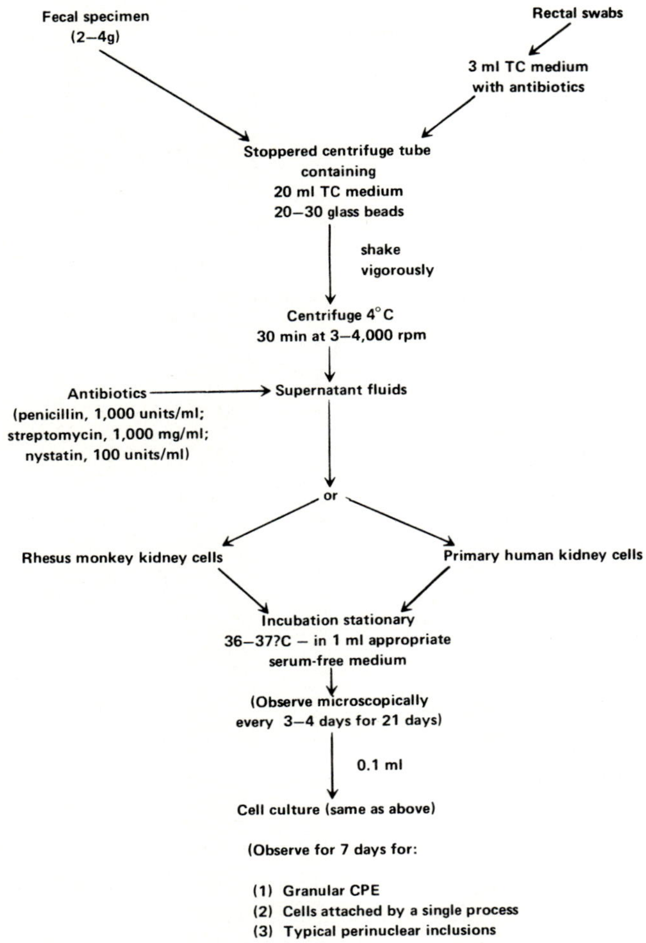

FIGURE 7. Flow chart for fecal specimens and rectal swabs.[94] Learner et al.[95] prefer roller tube cultures to stationary incubation.

remain cell associated was cited as a possible explanation for the inconsistent hemagglutination results with type 3 reoviruses using human red cells. The results of Zalan and Labzoffsky[101] also suggested that the type 3 infectious particles and human red cell hemagglutinins exist as separate entities.

Lerner[102] sedimented debris of cells infected with type 2 reovirus, incubated it for 2 hr at 4°C with 100 μg/ml of deoxycholate, and produced hemagglutination titers of 820,000 to 1,640,000/ml. Halonen and Pyhtila[103] found that the first and second fluorocarbon treatments of reovirus preparations significantly increased hemagglutination titers. Two fluorocarbon treatments removed 85% of the total protein from unconcentrated reovirus preparations. The possible relationship between inhibitor removal and the increase in hemagglutination activity was not discussed.

Erythrocytes of the four human ABO groups are agglutinated by all three reovirus serotypes at temperatures ranging from 4 to 37°C. However, hemagglutination titers are slightly higher, when the group A factor is present.[104] For practical purposes, group O erythrocytes are usually used.

Not all strains of serotype 3 agglutinate human red cells, but they do agglutinate bovine red cells at 4°C.[105,106] Type 1 and 2 strains do not agglutinate bovine red cells. The receptors on bovine erythrocytes are removed by neuraminidase, and the human red cell receptors are not.[105]

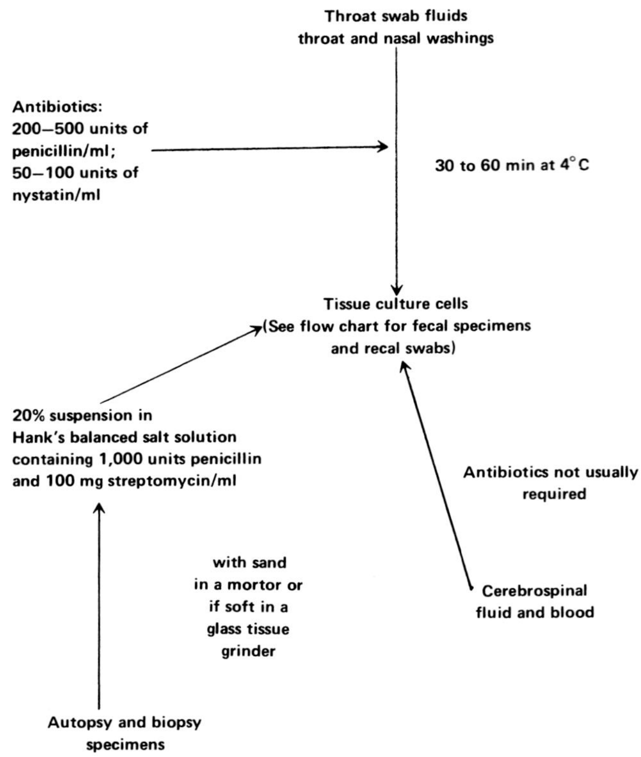

FIGURE 8. Flow chart for throat swab fluids and throat and nasal washings.[94]

Nonspecific inhibitors of reovirus hemagglutination are found in sera and in serum-free infectious culture fluids. The serum inhibitor(s) appears to be a heat-stable phospholipid-protein complex that is inactivated by a combination of chymotrypsin and phospholipase C.[107] The inhibitors found in serum-free tissue culture fluids possess characteristics of lipoprotein.[108] Bentonite and Kaolin will adsorb serum inhibitors from most specimens, but they may also remove specific antibody. Other methods for removing serum inhibitors[109-111] have not been used widely.

Pathogenicity in Experimental Animals

Probably all mammals are susceptible to reovirus infections. Results of investigations of animals infected with reoviruses have recently been reviewed by Stanley,[112] Kilham and Margolis,[113] and Margolis and Kilham.[114] The most complete studies of experimental infections have been conducted using rodents. Oronasal infection of neonatal mice with type 3 strains of reovirus has resulted in steatorrhea, an oily hair effect, tremors, retardation of growth, incoordination, and paresis prior to death.

Recently, intracerebral inoculation of suckling rats with reovirus types 1, 2, or 3 was found to induce formation of distinctive intranuclear bodies corresponding most closely to those designated as Cowdry type B inclusions.[128] Therefore, the reovirus-infected rat provides a model for the study of intranuclear body formation. This model may be especially important because the number of nuclear bodies and complexity of forms are strikingly increased in cells stimulated to active protein synthesis attending hypertrophy and hyperplasia, neoplasia, secretory activity, antibody formation, and viral infection.[128]

Raine and Fields,[129,130,133] Fields,[131] and Raine[132] have used suckling rats infected with temperature-sensitive mutants of reovirus type 3 as a model system for studying persistent, degenerative, central nervous system diseases. Mutants were found that had the same degree of neurotropism, yet produced less destructive lesions. In many cases, the mutants produced changes in the central nervous system that differed from those induced by wild-type virus. The similarities between infections involving wild-type reoviruses and their temperature-sensitive mutants and infections involving measles and subacute sclerosing panencephalitis viruses were emphasized. These studies show the potential hazard of naturally occurring spontaneous mutants and attenuated mutants used in vaccines.[131] The reports on studies of an oncolytic reovirus of mice[134-140] are also relevant.

Viremia has been reported to result in a pantropic spread of infection to the liver, central nervous system, pancreas, heart, lung, spleen, thymus, salivary gland, kidney, connective tissue, and skeletal muscle.[115-123] The acute disease of reovirus types 1 and 2 is similar to the type 3 disease except that cardiac and pulmonary lesions are more prominent; the thymus is less severely affected;[122, 124, 125] and the oily hair effect, liver lesions, and steatorrhea are rarely produced.[122]

The disease may persist in survivors until they die. These persistent infections have great value as models of human disease.[112,114,126,127] Depending on the reovirus serotype used, the clinical-pathological syndromes of survivors include hydrocephalus, runting, hepatitis, lymphoma, and amyloidosis. In addition, reovirus-infected mice provide models for viral-induced bacterial pneumonias, chronic obstructive jaundice,[112] and transplacental infections.[114]

GENERAL REFERENCES

Wood, H. A., Viruses with double-stranded RNA genomes, *J. Gen. Virol.*, 20, 61, 1973.

Jackson, G. G. and Muldoon, R. L., Viruses causing common respiratory infection in man. IV. Reoviruses and adenoviruses, *J. Infect. Dis.*, 128, 811, 1973.

Joklik, W. K., Reproduction of reoviridae, in *Comprehensive Virology,* Vol. 2 Fraenkel-Conrat, H. and Wagner, R. R., Eds., Plenum Press, New York, 1974, 231.

Shatkin, A. J., Animal RNA viruses: genome structure and function, *Annu. Rev. Biochem.*, 43, 643, 1974.

Rosen, L., Reoviruses, in *Manual of Clinical Microbiology,* 2nd ed., Lennette, E. H., Spaulding, E. H., and Truant, J. P., Eds., American Society for Microbiology, Washington, D. C., 1974, 735.

Stanley, N.F., The reovirus murine models, *Prog. Med. Virol.*, 18, 257, 1974.

Kilham, L. and Margolis, G., Problems of human concern arising from animal models of intrauterine and neonatal infections due to viruses: a review. I. Introduction and virologic studies, *Prog. Med. Virol.*, 20, 113, 1975.

Margolis, G. and Kilham, L., Problems of human concern arising from animal models of intrauterine and neonatal infections due to viruses. A review. II. Pathologic studies, *Prog. Med. Virol.*, 20, 144, 1975.

Silverstein, S. C., Christman, J. K., and Acs, G., The reovirus replicative cycle, in *Annu. Rev. Biochem.*, 45, 375, 1976.

Ramig, R. F. and Fields, B. N., Reoviruses, in *Molecular Biology of Animal Viruses,* Nayak, D., Ed., Marcel Dekker, New York, 1977, 383.

REFERENCES

1. **Fenner, F., Pereira, H. G., Porterfield, J. S., Joklik, W. K., and Downie, A. W.,** Family and generic names for viruses approved by the International Committee on Taxonomy of Viruses, June, 1974 *Intervirology,* 3, 193, 1974.
2. **Fenner, F.,** The classification and nomenclature of viruses, summary of results of meetings of the International Committee on Taxonomy of Viruses in Madrid, September 1975, *Virology,* 71, 371, 1976.

3. **Smith, K. M.,** The cytoplasmic virus diseases, in *Insect Pathology,* Vol. 1, Steinhaus, E. A., Ed., Academic Press, New York, 1963, 457.
4. **Coplin, D. L., Van Etten, J. L., Koski, R. K., and Vidaver, A. K.,** Intermediates in the biosynthesis of double-stranded ribonucleic acids of bacteriophage φ6, *Proc. Natl. Acad. Sci. U.S.A.,* 72, 849, 1975.
5. **Adler, J., Wood, H. A., and Bozarth, R. F.,** Virus-like particles from killer, neutral, and sensitive strains of *Saccharomyces cerevisiae, J. Virol.,* 17, 472, 1976.
6. **McDonald, R. D. and Yamamoto, T.,** The structure of infectious pancreatic necrosis virus RNA, *J. Gen. Virol.,* 34, 235, 1977.
7. **Stanley, N. F., Dorman, D. C., and Ponsford, J.,** Studies on the pathogenesis of a hitherto undescribed virus (*hepato-encephalomyelitis)* producing unusual symptoms in suckling mice, *Aust. J. Exp. Biol. Med. Sci.,* 31, 147, 1953.
8. **Stanley, N. F., Dorman, D. C., and Ponsford, J.,** Studies on the hepato-encephalomyelitis virus (HEV), *Aust. J. Exp. Biol. Med. Sci.,* 32, 543, 1954.
9. **Ramos-Alvarez, M. and Sabin, A. B.,** Characteristics of poliomyelitis and other enteric viruses recovered in tissue culture from healthy American children, *Proc. Soc. Exp. Biol. Med.,* 87, 655, 1954.
10. **Shatkin, A. J.,** Reoviruses in *Strains of Human Viruses,* Majer, J. and Plotkin, S. A., Eds., S. Karger, New York, 1972, 183.
11. **Sabin, A. B.,** Reoviruses: a new group of respiratory and enteric viruses formerly classified as ECHO-type 10 is described, *Science,* 130, 1387, 1959.
12. **Rogers, N. G., Basnight, M., Gibbs, C. J., Jr., and Gajdusek, D. C.,** Latent viruses in chimpanzees with experimental kuru, *Nature,* 216, 446, 1967.
13. **McClain, M. E., Spendlove, R. S., and Lennette, E. H.,** Infectivity assay of reoviruses: comparison of immunoflourescent cell count and plaque methods, *J. Immunol.,* 98, 1301, 1967.
14. **Rhim, J. S., Smith, K. O., and Melnick, J. L.,** Complete and coreless forms of reovirus (ECHO 10). Ratio of number of virus particles to infective units in the one-step growth cycle, *Virology,* 15, 428, 1961.
15. **Sabin, A. B.,** The significance of viruses recovered from the intestinal tracts of healthy infants and children, *Ann. N. Y. Acad. Sci.,* 66, 226, 1956.
16. **Spendlove, R. S. and Schaffer, F. L.,** Enzymatic enhancement of infectivity of reovirus, *J. Bacteriol.,* 89, 597, 1965.
17. **Ramos-Alvarez, M.,** Cytopathogenic enteric viruses associated with undifferentiated diarrheal syndromes in early childhood, *Ann. N. Y. Acad. Sci.,* 67, 326, 1957.
18. **Ramos-Alvarez, M. and Sabin, A. B.,** Enteropathogenic viruses and bacteria, role in summer diarrheal diseases of infancy and early childhood, *JAMA* 167, 147, 1958.
19. **Lerner, A. M., Cherry, J. D., Klein, J. O., and Finland, M.** Infections with reoviruses, *N. Engl. J. Med.,* 267, 947, 1962.
20. **Rosen, L., Hovis, J. F., Mastrota, F. M., Bell, J. A., and Huebner, R. J.,** Observations on a newly recognized virus (Abney) of the reovirus family, *Am. J. Hyg.,* 71, 258, 1960.
21. **Gomatos, P. J., Tamm, I., Dales, S , and Franklin, R. M.,** Reovirus type 3: physical characteristics and interaction with L cells, *Virology,* 17, 441, 1962.
22. **Jordan, L. E. and Mayor, H. D.,**The fine structure of reovirus, a new member of the icosahedral series, *Virology,* 17, 597, 1962.
23. **Vasquez, C. and Tournier, P.,** The morphology of reovirus, *Virology,* 17, 503, 1962.
24. **Mayor, H. D., Jamison, R. M., Jordan, L. E., and Mitchell, M. V.,** Reoviruses. II: structure and composition of the virion, *J. Bacteriol.,* 89, 1548, 1965.
25. **Mayor, H. D. and Jordan, L. E.,** Studies on reovirus. I. Morphologic observations on the development of reovirus in tissue culture, *Exp. Mol. Pathol.,* 4, 40, 1965.
26. **Loh, P. C., Hohl, H. R., and Soergel, M.,** Fine structure of reovirus type 2, *J. Bacteriol.,* 89, 1140, 1965.
27. **Muller, G., Schneider, C. C., and Peters, D.,** Zur Feinstruktur des Reovirus (Typ 3), *Arch. Ges. Virusforsch.,* 19, 110, 1966.
28. **Vasquez, C. and Tournier, P.,** New interpretation of the reovirus structure, *Virology,* 24, 128, 1964.
29. **Amano, Y., Katagiri, S., Ishida, N., and Watanabe, Y.,** Spontaneous degradation of reovirus capsid into subunits, *J. Virol.,* 8, 805, 1971.
30. **Vasquez, C.,** Morphology of reovirus structural components, Letter *J Microsc.,* (Paris) 14, 219, 1972.
31. **Luftig, R. B., Kilham, S. S., Hay, A. J., Zweerink, H. J., and Joklik, W. K.,** An ultrastructural study of virions and cores of reovirus type 3, *Virology,* 48, 170, 1972.
32. **Joklik, W. K.,** Studies on the effect of chymotrypsin on reovirions, *Virology,* 49, 700, 1972.
33. **Spendlove, R. S., McClain, M. E., and Lennette, E. H.,** Enhancement of reovirus infectivity by extracellular removal or alteration of the virus capsid by proteolytic enzymes, *J. Gen. Virol.,* 8, 83, 1970.

34. **Shatkin, A. J. and LaFiandra, A. J.**, Transcription by infectious subviral particles of reovirus, *J. Virol.*, 10, 698, 1972.
35. **Borsa, J., Sargent, M. D., Long, D. G., and Chapman, J. D.**, Extraordinary effects of specific monovalent cations on activation of reovirus transcriptase by chymotrypsin in vitro, *J Virol.*, 11, 207, 1973.
36. **Borsa, J., Copps, T. P., Sargent, M. D., Long, D. G., and Chapman, J. D.**, New intermediate subviral particles in the in vitro uncoating of reovirus virions by chymotrypsin, *J. Virol.*, 11, 552, 19753.
37. **Borsa, J., Sargent, M. D., Copps, T. P., Long, D. G., and Chapman, J. D.**, Specific monovalent cation effects on modification of reovirus infectivity by chymotrypsin digestion in vitro, *J. Virol.*, 11, 1017, 1973.
38. **Smith, R. E., Zweerink, H. J., and Joklik, W. K.**, Polypeptide components of virions, top component and cores of reovirus type 3, *Virology*, 39, 791, 1969.
39. **Joklik, W. K.**, Reproduction of Reoviridae, in *Comprehensive Virology*, Vol. 2, Fraenkel-Conrat, H. and Wagner, R. R., Eds., Plenum Press, New York, 1974, 231.
40. **Krystal, G., Perrault, J., and Graham, A. F.**, Evidence for a glycoprotein in reovirus, *Virology*, 72, 308, 1976.
41. **Shatkin, A. J., Sipe, J. D., and Loh, P.**, Separation of ten reovirus genome segments by polyacrylamide gel electrophoresis, *J. Virol.*, 2, 986, 1968.
42. **Wood, H. A. and Streissle, G.**, Wound tumor virus: purification and fractionation of the double-stranded ribonucleic acid, *Virology*, 40, 329, 1970.
43. **Fujii-Kawata, I., Miura, K. I., and Fuke, M.**, Segments of genome of viruses containing double-stranded ribonucleic acid, *J. Mol. Biol.*, 51, 247, 1970.
44. **Shatkin, A. J.**, Replication of reovirus, *Adv. Virus. Res.*, 14, 63, 1969.
45. **Silverstein, S. C., Christman, J. K., and Acs, G.**, The reovirus replicative cycle, *Annu. Rev. Biochem.*, 45, 375, 1976.
46. **Shatkin, A. J.**, Viruses with segmented ribonucleic acid genomes: multiplication of influenza versus reovirus, *Bacteriol. Rev.*, 35, 250, 1971.
47. **Graham, A. F. and Millward, S.**, Replication of Reovirus, in *Nucleic Acid-Protein Interactions and Nucleic Acid Synthesis in Viral Infections*, Vol. 2, Ribbons, D. W., Woesner, J. F. and Schultz, J., Eds., Elsevier, Amsterdam, 1971, 333.
48. **Morgan, C.**, Entry of reovirus, *Bacteriol. Proc.*, 203, 1970.
49. **Spendlove, R. S.**, Unique reovirus characteristics, *Prog. Med. Virol.*, 12, 161, 1970.
50. **Dales, S., Gomatos, P. J., and Hsu, K. C.**, The uptake and development of reovirus in stain L cells followed with labeled viral ribonucleic acid and ferritin-antibody conjugates, *Virology*, 25, 193, 1965.
51. **Silverstein, S. C. and Dales, S.**, The penetration of reovirus RNA and initiation of its genetic function in L-strain fibroblasts, *J. Cell Biol.*, 36, 197, 1968.
52. **Silverstein, S. C., Astell, C., Levin, D. H., Schonberg, M., and Acs, G.**, The mechanisms of reovirus uncoating and gene activation in vivo, *Virology*, 47, 797, 1972.
53. **Schonberg, M., Silverstein, S. C., Levin, D. H., and Acs, G.**, Asynchronous synthesis of the complementary strands of the reovirus genome, *Proc. Natl. Acad. Sci. U.S.A.*, 68, 505, 1971.
54. **Chang, C. T. and Zweerink, H. J.**, Fate of parental reovirus in infected cell, *Virology*, 46, 544, 1971.
55. **Bartlett, N. M., Gillies, S. C., Bullivant, S., and Bellamy, A. R.**, Electron microscopy study of reovirus reaction cores, *J. Virol.*, 14, 315, 1974.
56. **Silverstein, S. C., Schonberg, M., Levin, D. H., and Acs, G.**, The reovirus replicative cycle: conservation of parental RNA and protein, *Proc. Natl. Acad. Sci. U.S.A.*, 67, 275, 1970.
57. **Spandidos, D. A., Krystal, G., and Graham, A. F.**, Regulated transcription of the genomes of defective virions and temperature-sensitive mutants of reovirus, *J. Virol.*, 18, 7, 1976.
58. **Millward, S. and Nonoyama, M.**, Segmented structure of the reovirus genome, *Cold Spring Harbor Symp. Quant. Biol.*, 35, 773, 1970.
59. **Watanabe, Y., Millward, S., and Graham, A. F.**, Regulation of transcription of the reovirus genome, *J. Mol. Biol.*, 36, 107, 1968.
60. **Nonoyama, M., Millward, S., and Graham, A. F.**, Control of transcription of the reovirus genome, *Nucleic Acids Res.*, 1, 373, 1974.
61. **Lau, R. Y., Van Alstyne, D., Berckmans, R., and Graham, A. F.**, Synthesis of reovirus-specific polypeptides in cells pretreated with cycloheximide, *J Virol.*, 16, 470, 1975.
62. **Ito, Y. and Joklik, W. K.**, Temperature-sensitive mutants of reovirus. I. Patterns of gene expression by mutants of groups C, D, and E, *Virology*, 50, 189, 1972.
63. **Cross, R. K. and Fields, B. N.**, Temperature-sensitive mutants of reovirus type 3: studies on the synthesis of viral RNA, *Virology*, 50, 799, 1972.
64. **Skehel, J. J. and Joklik, W. K.**, Studies on the in vitro transcription of reovirus RNA catalyzed by reovirus cores, *Virology*, 39, 822, 1969.

65. **Levin, D. H., Acs, G., and Silverstein, S. C.,** Chain initiation by reovirus RNA transcriptase in vitro, *Nature,* 227, 603, 1970.
66. **Borsa, J., Grover, J., and Chapman, J. D.,** Presence of nucleoside triphosphate phosphohydrolase activity in purified virions of reovirus, *J. Virol.,* 6, 295, 1970.
67. **Banerjee, A. K., Ward, R., and Shatkin, A. J.,** Initiation of reovirus mRNA synthesis in vitro, *Nature,* 230, 169, 1971.
68. **Both, G. W., Furuichi, Y., Muthukrishnan, S., and Shatkin, A. J.,** Ribosome binding to reovirus mRNA in protein synthesis requires 5′ terminal 7-methylguanosine, *Cell,* 6, 185, 1975.
69. **Shatkin, A. J.,** Methylated messenger RNA synthesis in vitro by purified reovirus, *Proc. Natl. Acad. Sci. U.S.A.,* 71, 3204, 1974.
70. **Furuichi, Y., Morgan, M., Muthukrishnan, S., and Shatkin, A. J.,** Reovirus messenger RNA contains a methylated, blocked 5′-terminal structure: m^7 G(5′)ppp(5′)G^mpCp-, *Proc. Natl. Acad. Sci. U.S.A.,* 72, 362, 1975.
71. **Faust, M., Hastings, K. E. M., and Millward, S.,** m^7 $G^{5'}$ $ppp^{5'}$ CmpCpUp at the 5′ terminus of reovirus messenger RNA, *Nucleic Acids Res.,* 2, 1329, 1975.
72. **Zweerink, H. J., Ito, Y., and Matsuhisa, T.,** Synthesis of reovirus double-stranded RNA within virionlike particles, *Virology,* 50, 349, 1972.
73. **Silverstein, S. C., Astell, C., Christman, J., Klett, H., and Acs, G.,** Synthesis of reovirus oligo adenylic acid in vivo and in vitro, *J Virol.,* 13, 740, 1974.
74. **Stoltzfus, C. M., Morgan, M., Banerjee, A. K., and Shatkin, A. J.,** Poly (A) polymerase activity in reovirus, *J. Virol.,* 13, 1338, 1974.
75. **Spendlove, R. S., Lennette, E. H., Knight, C. O., and Chin, J. N.,** Development of vital antigen and infectious virus in HeLa cells infected with reovirus, *J. Immunol.,* 90, 548, 1963.
76. **Spendlove, R. S., Lennette, E. H., and John, A. C.,** The role of the mitotic apparatus in the intracellular location of reovirus antigen, *J. Immunol.,* 90, 554, 1963.
77. **Spendlove, R. S., Lennette, E. H., Chin, J. N., and Knight, C. O.,** Effect of antimitotic agents on intracellular reovirus antigen, *Cancer Res.,* 24, 1826, 1964.
78. **Kohler, M. R. and Spendlove, R. S.,** Reovirus infected cells for studying microtubules and spindle poisons, *Cytobios,* 9, 131, 1974.
79. **Dales, S.,** Association between the spindle apparatus and reovirus, *Proc. Natl. Acad. Sci. U.S.A.,* 50, 268, 1963.
80. **Dales, S., Gomatos, P. J., and Hsu, K. C.,** The uptake and development of reovirus in strain L cells followed with labeled viral ribonucleic acid and ferritin-antibody conjugates, *Virology,* 25, 193, 1965.
81. **Spendlove, R. S., Lennette, E. H., Knight, C. O., and Chin, J. H.,** Production in FL cells of infectious and potentially infectious reovirus, *J. Bacteriol.,* 92, 1036, 1966.
82. **Rhim, J. S., Jordan, L. E., and Mayor, H. D.,** Cytochemical, fluorescent-antibody, and electron microscopic studies on the growth of reovirus (ECHO 10) in tissue culture, *Virology,* 17, 342, 1962.
83. **Rosen, L.,** Reoviruses, in *Manual of Clinical Microbiology,* 2nd ed., Lennette, E. H., Spaudling, E. H., and Truant, J. P., Eds., American Society for Microbiology, Washington, D. C., 1974, 735.
84. **Krainer, L. and Aronson, B. E.,** Disseminated encephalomyelitis in the human with recovery of hepatoencephalitis virus (HEV): pathologic and virologic report, (Abstr.) *J. Neurophathol. Exp. Neurol.,* 18, 339, 1959.
85. **Joske, R. A., Keall, D. D., Leak, P. J., Stanley, N. F., and Walters, M. N. I.,** Hepatitis-encephalitis in humans with reovirus infection, *Arch. Intern. Med.,* 113, 811, 1964.
86. **Tillotson, J. R. and Lerner, A. M.,** Isolation of reovirus type 3 from post-mortem tissues of a child with a nonbacterial pneumonia, *Clin. Res.,* 14, 344, 1966.
87. **DeLavergne, E., Olive, D., and LeMoyne, M. T.,** Les reovirus, *Presse Med.,* 73, 951, 1965.
88. **Tillotson, J. R. and Lerner, A. M.,** Reovirus type 3 associated with fatal pneumonia, *N. Engl. J. Med.,* 276, 1060, 1967.
89. **Edmonson, J. H., Millian, S. J., Goodenow, M., and Lee, S. L.,** Persistent viruria due to reovirus in a patient treated for Hodgkin's disease in a protected environment, *J. Infect. Dis.,* 121, 438, 1970.
90. **Jackson, G. G. and Muldoon, R. L.,** Viruses causing common respiratory infection in man. IV. Reoviruses and adenoviruses, *J. Infect. Dis.,* 128, 811, 1973.
91. **Rosen, L.,** Reoviruses, *Virol. Monogr.,* 1, 73, 1968.
92. **Jackson, G. G., Muldoon, R. L., Johnson, G. C., and Dowling, H. F.,** Studies in volunteer groups: contributions of volunteers to studies on the common cold, *Am. Rev. Respir. Dis.,* 88(Suppl.), 120, 1963.
93. **Rosen, L., Evans, H. E., and Spickard, A.,** Reovirus infections in human volunteers, *Am. J. Hyg.,* 77, 29, 1963.
94. **Lennette, E. H., Melnick, J. L., and Maggofin, R. L.,** Clinical virology: introduction to methods, in *Manual of Clinical Microbiology,* 2nd edition, Lennette, E. H., Spaulding, E. H., and Truant, J. P., Eds., American Society For Microbiology, Washington, D. C., 1974, 667.

95. **Hartley, J. W., Rowe, W. P., and Austin, J. B.**, Subtype differentiation of reovirus type 2 strains by hemagglutination-inhibition with mouse antisera, *Virology,* 16, 94, 1962.
96. **Lerner, A. M., Cherry, J. D., and Finland, M.**, Enhancement of cytopathic effects of reoviruses in rolled cultures of Rhesus kidney, *Proc. Soc. Exp. Biol. Med.,* 110, 727, 1962.
97. **Rhim, J. S. and Melnick, J. L.**, Quantitative studies of reovirus (ECHO-10) in monkey kidney cell cultures, *Tex. Rep. Biol. Med.,* 19, 851, 1961.
98. **Rhim, J. S., Kato, J. I., and Pelon, W.**, Hemagglutination by reoviruses propagated in various cell lines, *Proc. Soc. Exp. Biol. Med.,* 118, 453, 1965.
99. **Hsiung, G. D.**, Some distinctive biological characteristics of ECHO-10 virus, *Proc. Soc. Exp. Biol. Med.,* 99, 387, 1958.
100. **Halonen, P.**, Growth, stability and hemagglutination of a reovirus, *Ann. Med. Exp. Fenn.,* 39, 132, 1961.
101. **Zalan, E. and Labzoffsky, N. A.**, The hemagglutinin of reovirus type 3, *Arch. Ges. Virusforsch.,* 21, 25, 1967.
102. **Lerner, A. M.**, Concentration and purification of viruses with special reference to reoviruses, *Bacteriol. Rev.,* 28, 391, 1964.
103. **Halonen, P. and Pyhtila, M.**, Purification of reovirus and measles virus hemagglutinin by fluorocarbon, on calcium phosphate and by differential gradient centrifugation, *Ann. Med. Exp. Fenn.,* 40, 365, 1962.
104. **Brubaker, M. M., West, B., and Ellis, R. J.**, Human blood group influence on reovirus hemagglutination titers, *Proc. Soc. Exp. Biol. Med.,* 115, 1118, 1964.
105. **Gomatos, P. J. and Tamm, I.**, Reactive sites of reovirus type 3 and their interaction with receptor substances, *Virology,* 17, 455, 1962.
106. **Eggers, H. J., Gomatos, P. J., and Tamm, I.**, Agglutination of bovine erythrocytes: a general characteristic of reovirus type 3, *Proc. Soc. Exp. Biol. Med.,* 110, 879, 1962.
107. **Schmidt, N. J., Dennis, J., Hoffman, M. N., and Lennette, E. H.**, Inhibitors of echovirus and reovirus hemagglutination. II. Serum and phospholipid inhibitors, *J. Immunol.,* 93, 377, 1964.
108. **Schmidt, N. J., Dennis, J., Hoffman, M. N., and Lennette, E. H.**, Inhibitors of echovirus and reovirus hemagglutination. I. Inhibitors in tissue culture fluids, *J. Immunol.,* 93, 367, 1964.
109. **Mann, J. J., Rosen, R. D., Lehrich, J. R., and Kasel, J. A.**, The effect of kaolin on immunoglobulins: an improved technique to remove the nonspecific serum inhibitor of reovirus hemagglutination, *J. Immunol.,* 98, 1136, 1967.
110. **Schmidt, N. J., Dennis, J., and Lennette, E. H.**, Studies on filtrates from cultures of a psychrophylic *Pseudomonas* sp. which inactivate nonspecific serum inhibitors for certain hemagglutinating viruses, *J. Immunol.,* 93, 140, 1964.
111. **Styk, B., Schmidt, N J., and Dennis, J.**, The use of rivanol treatment for removal from sera of nonspecific inhibitors of enterovirus and reovirus hemagglutination, *Am. J. Epidemiol.,* 88, 398, 1968.
112. **Stanley, N. F.**, The reovirus murine models, *Prog. Med. Virol.,* 18, 256, 1974.
113. **Kilham, L. and Margolis, G.**, Problems of human concern arising from animal models of intrauterine and neonatal infections due to viruses: a review. I. Introduction and virologic studies, *Prog. Med. Virol.,* 20, 113, 1975.
114. **Margolis, G. and Kilham, L.**, Problems of human concern arising from animal models of intrauterine and neonatal infections due to viruses. A review. II. Pathologic studies, *Prog. Med. Virol.,* 20, 144, 1975.
115. **Hassan, S. A., Rabin, E. R., and Melnick, J. L.**, Reovirus myocarditis in mice: an electron microscopic, immunofluorescent, and virus assay study, *Exp. Mol. Pathol.,* 4, 66, 1965.
116. **Joske, R. A., Leak, P., Stanley, N. F., and Walters, M. N.-I.**, Hepatic changes following reovirus III infection in infant mice, *Gut,* 5, 490, 1964.
117. **Papadimitriou, J. M.**, Electron micrographic features of acute murine reovirus hepatitis, *Am. J. Pathol.,* 47, 565, 1965.
118. **Popadimitriou, J. M.**, An electron microscopic study of murine reovirus 3 encephalitis, *Am. J. Pathol.,* 50, 59, 1967.
119. **Papadimitriou, J. M.**, Renal damage in murine infection with reovirus 3: an ultrastructural study, *Pathology,* 1, 35, 1969.
120. **Papadimitriou, J. M. and Walters, M. N.-I.**, Studies on the exocrine pancreas. II. Ultrastructural investigation of reovirus pancreatitis, *Am. J. Pathol.,* 51, 387, 1967.
121. **Stanley, N. F. and Leak, P J.**, Murine infection with reovirus type 3 and the runting syndrome, *Nature,* 199, 1309, 1963.
122. **Walters, M. N.-I., Joske, R. A., Leak, P. J., and Stanley, N. F.**, Murine infection with reovirus: I. Pathology of the acute phase, *Br. J. ExP. Pathol.,* 44, 427, 1963.
123. **Walters, M. N.-I., Leak, P. J., Joske, R. A., Stanley, N. F., and Perret, D. H.**, Murine infection with reovirus. III. Pathology of infection with types 1 and 2, *Br. J. Exp. Pathol.,* 46, 200, 1965.

124. **Phillips, P. A., Keast, D., Walters, M. N.-I., and Stanley, N. F.,** Murine lymphoma induced by reovirus 3, *Pathology,* 3, 133, 1971.
125. **Phillips, P. A., Walters, M. N.-I., Stanley, N. F., and Keast D.,** Amyloidosis, immunocompetence, and the chronic murine diseases induced by reovirus types 1 and 2, *Pathology,* 3, 267, 1971.
126. **Stanley, N. F. and Joske, R. A.,** Animal model: chronic murine hepatitis induced by reovirus type 3, *Am. J. Pathol.,* 80, 181, 1975.
127. **Stanley, N. F. and Joske, R. A.,** Animal model: chronic biliary obstruction caused by reovirus type 3, *Am. J. Pathol.,* 80, 185, 1975.
128. **Margolis, G., Kilham, L., and Baringer, J. R.,** Identity of cowdry type B inclusions and nuclear bodies: observations in reovirus encephalitis, *Exp. Mol. Pathol.,* 23, 228, 1975.
129. **Raine, C. S. and Fields, B. N.,** Neurotropic viruses and the developing brain, *N. Y. State J. Med.,* 73, 1169, 1973.
130. **Raine, C. S. and Fields, B. N.,** Reovirus type III encephalitis: a virologic and ultrastructure study, *J. Neuropathol. Exp. Neurol.,* 32, 19, 1973.
131. **Fields, B. N.,** Genetic manipulation of reovirus-a model for modification of disease?, *New Eng. J. Med.,* 287, 1026, 1972.
132. **Raine, C. S.,** Differential effects of viruses and their mutants upon central nervous system development, in *The Neurosciences,* Volume 3, Schmitt, F. O. Worden, F., Eds., MIT Press, Cambridge, Mass., 1974, 1043.
133. **Raine, C. S. and Fields, B. N.,** Neurotropic virus-host relationship alterations due to variation in viral genome as studied by electron microscopy, *Am. J. Pathol.,* 75, 119, 1974.
134. **Nelson, J. B. and Tarnowski, G. S.,** An oncolytic virus recovered from Swiss mice during passage of an ascites tumor, *Nature,* 188, 866, 1960.
135. **Bennette, J. G.,** Isolation of a non-pathogenic tumor-destroying virus from mouse ascites, *Nature,* 187, 72, 1960.
136. **Hartley, J. W., Rowe, W. P., and Huebner, R. J.,** Recovery of reoviruses from wild and laboratory mice, *Proc. Soc. Exp. Biol. Med.,* 108, 390, 1961.
137. **Nelson, J. B.,** Response of mice to reovirus type 3 in presence and absence of ascites tumor cells, *Proc. Soc. Exp. Biol. Med.,* 116, 1086, 1964.
138. **Bennette, J. G., Bush, P. V., and Steele, R. D.,** Characteristics of a newborn runt disease induced by neonatal infection with an oncolytic strain of reovirus type 3 (REO3MH), I. Pathological investigations in rats and mice, *Br. J. Exp. Pathol.,* 48, 251, 1967.
139. **Bennette, J. G., Bush, P. V., and Steele, R. D.,** Characteristics of a newborn runt disease induced by neonatal infection with an oncolytic strain of reovirus type 3 (REO3MH). II. Immunological aspects of the disease in mice, *Br. J. Exp. Pathol.,* 48, 267, 1967.
140. **Keast, D. and Chesterman, F. C.,** Reovirus pathogenesis and the identity of REO3MH, *Br. J. Exp. Pathol.,* 53, 90, 1972.

REOVIRIDAE: ROTAVIRUSES

Rex S. Spendlove

Introduction

The reo-like gastroenteritis viruses (rotaviruses) have a double-layered capsid and 11 segments of double-stranded RNA. The viruses that have been tested are serologically related to each other,[1,2,3] but are not serologically related to viruses of the *Reovirus* and *Orbivirus* genera.[2]

Rotaviruses cause acute diarrhea in infants, calves, mice, piglets, rabbits, deer, dogs, lambs and other animals. The individual viruses are referred to by the following names: infantile gastroenteritis virus (IGV), reo-like virus, human reovirus-like agent (HRVLA), reovirus-like (RVL) and orbivirus-like agents, neonatal calf diarrhea virus or Nebraska calf diarrhea virus (NCDV), neonatal calf diarrhea reovirus-like (NCDR) agent, and epidemic diarrhea or epizootic diarrhea of infant mice (EDIM) virus. The simian virus SA 11 and the related offal (O) agent isolated from offal from butchered sheep and cattle in South Africa[4] are rotaviruses; however, they are not known to be associated with gastroenteritis.

Two categorizing names have been proposed for these reo-like viruses. Flewett et al.[1] recommended "rotavirus" (from the Latin "rota," "a wheel"). Davidson et al.[5] suggested the name "duovirus" to describe the characteristic double-shelled capsid structure and the growth of the viruses in the duodenum of infected animals. The rotavirus designation is most widely accepted and will be used in this communication.

Rotaviruses are very difficult to propagate in cell cultures. Consequently, detection of human rotaviruses was not reported until 1973.[6] Although virus isolates have been reported from human beings,[3,7-17] it has usually been necessary to use fluorescent antibody to detect the infected cells in inoculated cultures. Rotavirus strains are shown in Table 1.

Morphology

Flewett et al.[1] described the rotavirion (Figure 1) as a wheel-like structure with a large hub-like core from which short cylindrical capsomers radiate like spokes of a wheel. An outer layer of capsomers attaches to the ends of the inner capsomers to form a smooth-edged layer that has the appearance of a rim. This wheel-like appearance led them to recommend the name "rotavirus." Viral particles without the outer capsid layer are often observed. These particles have rough outer edges. The virion diameter has been reported to be 60 to 75 nm,[1,7,24-28] viral particles without the outer capsid are 50 to 60 nm,[24,26-28] and the electron-dense core is 33 to 40 nm in diameter.[1,7,24,26,27]

In addition to two capsid layers, some investigators[4,6,29] have reported occasionally seeing virions that are enclosed in an envelope obtained from host membranes. This envelope is thought not to be required for infectivity, nor to be part of the virion.

The most complete study of rotavirus morphology has been conducted by Martin et al.[24] They studied infantile gastroenteritis virus particles that lacked the outer capsid layer and reported that the virus had 32 capsomers, i.e., it is a P = 3, T = 3 icosahedron. This structure is identical to the 92-hole shared-subunit model proposed by some investigators for the reoviruses.[30-32]

Evidently, the number and arrangement of the structural subunits of reoviruses and rotaviruses are the same.[33] The unresolved question is whether they have 18 or 20 peripheral capsomers. Peripheral capsomer counts have been used by different investigators to postulate capsomer numbers of 32[24,33] and 42,[34] respectively.

Table 1
ROTAVIRUS ISOLATES

Virus name	Strain designation	Isolated from	Refs.
Neonatal calf diarrhea (NCD) virus	Cody	Diarrheic calf	18
Reovirus-like agent	Lincoln	Diarrheic calf	18
Calf rotavirus	None	Diarrheic calf	19,20
SA 11	H 96	Monkey rectal swab	21, 22
O agent	None	Intestinal washings of sheep and cattle	22
Human reovirus-like agent	Fh and D	Diarrheic child	2, 23

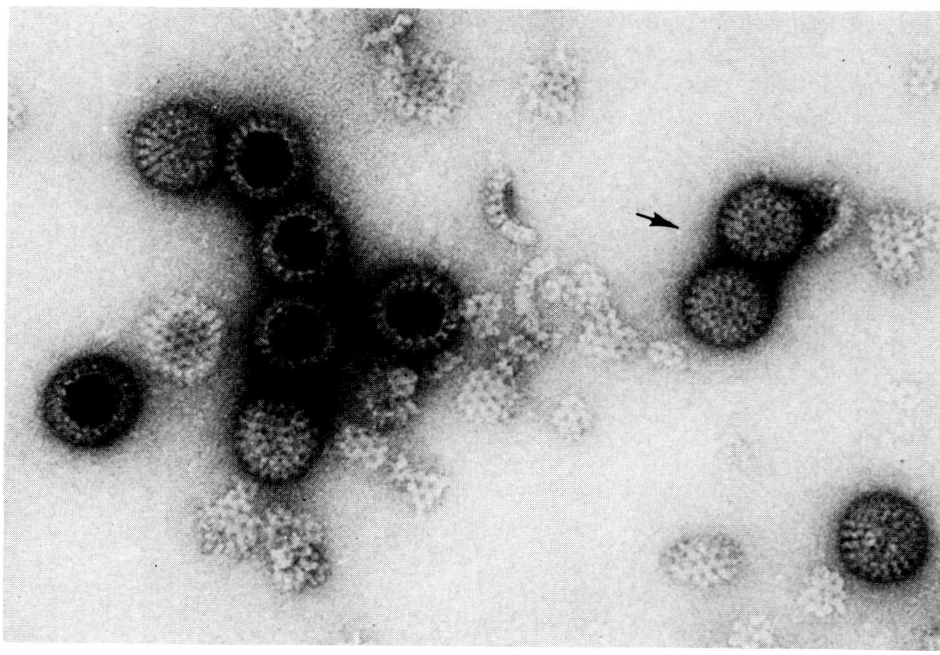

FIGURE 1. Survey micrograph of infantile gastroenteritis virus particles present in lower band of density gradients. The arrow indicates particles with ring-shaped capsomers. In particles at lower right and upper left, the capsomers have a spoke-like arrangement. The shared subunits can be observed in the disrupted capsids. All particles lack the outer "rim" capsid. Magnification × 234,360. (From Martin, M. L., Palmer, E. L., and Middleton, P. J., *Virology*, 68, 146, 1975. With permission.)

Physicochemical Properties

The rotavirion has 5 to 10 polypeptides and 11 segments of double-stranded RNA, with a total molecular weight of approximately 10.7×10^6 daltons (see Tables 2 and 3). Calf rotavirus is stable at pH 3 and resistant to lipid solvents, but it is inactivated when heated to 50°C for 1 hr in the presence of 1.0 *M* $MgCl_2$.[41] Virus in feces held at room temperature retains full infectivity for 7 months.[42]

Because rotaviruses are difficult to propagate to high concentrations in cell cultures and to assay reproducibly, relatively few of their physicochemical characteristics have been determined. To circumvent this problem, some investigators[24,26-29,35,36,38-40] have obtained viruses from stool specimens for their studies.

Table 2
REPORTED MOLECULAR WEIGHTS OF ROTAVIRUS CAPSID POLYPEPTIDES

	Virus (daltons)					
Polypeptide	Lamb[35]	Calf[36]	Human[36]	Calf[37]	Calf[25]	Calf[26]
1	128,000	131,000	127,000	128,000	135,000	125,000[a]
2	108,000	103,000	103,000	99,000	122,000	103,000
3	99,000	97,000	97,000	83,000	96,000	98,000[b]
4	96,000	92,000	88,000	41,000	36,000	63,000
5	72,000[c]	58,000	58,000	34,000	30,000	44,000
6	50,000	32,000	32,000	15,000		
7	48,000	22,000	26,000			
8	45,000[c]	16,000	21,000			
9	37,000[c]	14,000				
10	30,000[c]					
Total	713,000	565,000	552,000	400,000	419,000	433,000

[a] Only detected occasionally.[26]
[b] Appears to be an outer coat polypeptide.[26]
[c] Appear to be outer polypeptides.[35]

Table 3
GENOME SEGMENTS OF ROTAVIRUSES

		Molecular weights (× 10^6 daltons)						
Segment number	Size designation[a]	Calf virus[37]	Calf virus[25]	Calf virus[36]	Calf virus[39]	Human virus[38]	Pig virus[40]	Lamb virus[35]
1	L	2.18	2.2	2.2	2.2	2.04	2.1	2.1
2	L	1.73	1.9	1.6	1.8	1.58	1.7	1.7
3	L	1.64	1.9	1.6	1.73	1.58	1.6	1.6
4	L	1.48	1.7	1.5	1.60	1.40	1.5	1.5
5	M	0.94	0.9	0.96	0.94	0.81	0.9	0.9
6	M	0.77	0.8	0.75	0.78	0.75	0.8	0.8
7	S	0.50	0.5	0.51	0.53	0.50	0.5	0.5
8	S	0.50	0.5	0.51	0.50	0.50	0.5	0.5
9	S	0.50	0.5	0.51	0.50	0.50	0.5	0.5
10	S	0.29	0.3	0.32	0.26	0.28	0.3	0.3
11	S	0.22	0.2	0.24	0.20	0.23	0.3	0.3
Total		10.62	11.12	10.7	11.04	10.17	10.7	10.7

[a] L, large; M, medium; S, small.

Replication

In the one reported investigation of rotavirus growth, neonatal calf diarrhea virus was studied in primary bovine embryonic kidney cell cultures.[43,44] Virus was detected as early as 2 hr after infection, and its concentration increased until 18 hr postinfection, at which time the yield of cell-free virus approximately equaled the yield of cell-associated virus. When viral antigen production was determined using the fluorescent antibody technique, antigen was first detected 4 hr after inoculation. The infected cells contained discrete perinuclear cytoplasmic fluorescent granules. By 8 hr postinoculation, diffusely stained antigen was present throughout the entire cytoplasm. After 24 hr, some stained cells detached from the monolayer.

Clinical Features

Shepherd et al.[45] believe that the clinical features of rotavirus gastroenteritis may be distinct enough to indicate which infants should be screened for infection. The clinical manifestations include loose, yellow-green, offensive stools without blood or mucus.[45] Raised temperature, vomiting, and abdominal cramps are common symptoms,[46] and severe dehydration with an electrolyte imbalance may occur.[45] The prolonged high temperature observed in infants may be due to viremia.[47] Disaccharidase levels are depressed.[48] The absence of fecal leukocytes indicates that the virus does not invoke an inflammatory response.[46]

Symptoms regularly last 5 to 8 days[45,46,49] and occasionally persist for up to 28 days.[42] If treated, the disease is usually uncomplicated and recovery is complete. The diarrhea is aggravated by feeding and relieved by fasting.[48] Symptoms can range from subclinical to severe, and in some cases dehydration can result in death.[5,50] Some children show no symptoms even when seroconversion occurs.[7,42,45,51] The disease appears to have its most severe effects in children from 6 months to 2 years of age,[52] possibly because many younger infants derive some passive immunity from breast-feeding. Whorwell et al.[53] have presented evidence that suggests rotavirus as a cause of Crohn's disease.

Laboratory Diagnosis

Epidemiology

Although epidemic gastroenteritis ranks second in incidence only to viral respiratory infection during winter months,[54] the major etiologic agent of human gastroenteritis was not detected until 1973. The first observation came when duodenal mucosa biopsy material was examined by electron microscopy[55] and reo-like viruses were found.

The subsequent use of electron microscopy to detect rotaviruses directly in human stool specimens[7,56-59] stimulated numerous investigators throughout the world to look for rotaviruses. Davidson et al.,[5] in summarizing some of these reports, found that 396 (48%) of 827 young children with gastroenteritis excreted rotaviruses. The virus was observed in only 2 of 357 children who did not have diarrhea. A recent report of an extensive 18-month investigation[60] stated that approximately half of the children surveyed who had gastroenteritis were infected with rotavirus. The peak incidence of rotavirus infection (78%) was observed in December and January. The authors[60] concluded that rotavirus is the major etiologic agent of diarrhea in infants and young children during the cooler months of the year in temperate climates.

The epidemiologic picture has not yet been clarified; however, the infection probably spreads from human to human via stools[48] and is most common in the winter months.[5,50,52,60,61] The age groups reported to have the highest incidence of infection vary. Chrystie et al.[62] found the average age to be 5 days, with the disease more frequent in bottle-fed infants. Hamilton[48] and Shepherd et al.[45] reported the disease to be more prevalent in youngsters under 2 years of age, with males infected more frequently than females. Other workers have observed the highest rate of infectivity to be in 6- to 12-month-old infants.[5] Another investigation produced results that suggested rotavirus as the cause of a widespread epidemic of gastroenteritis in school children aged 6 to 12 years.[63]

Although 80% of children tested have acquired antibody by the time they are 3 years of age,[64] adults can occasionally be infected. Two cases of child-mother transmission have been reported,[65,66] and the disease has been transmitted to a seronegative volunteer by administering rotavirus through a tube into the second part of the duodenum.[50] Von Bonsdorff et al.[67] found the incidence of gastroenteritis in adults to be comparable to that found in children, and Meurman and Laine[68] reported a limited epidemic at a military base. Most individuals who have not had previous contact with rotavirus are apparently susceptible to infection. It is not uncommon to acquire a rotavirus infection in a hospital.[48] Children younger than 12 months of age admitted to wards close to a gastroenteritis ward are especially susceptible.[5,42]

Incubation has been reported to require 12 to 48 hr,[69] 48 to 78 hr,[45] and 4 to 6 days or less.[50] This variation may be similar to that observed in calf gastroenteritis, in which case the incubation period increases with age[70] and decreases as the amount of virus to which the individual is exposed increases.[71]

Choice of Specimen

The specimens that have been used in diagnosing human rotavirus infections are stools, paired sera, duodenal mucosa biopsy, intestinal juice (duodenal aspirates), and mucosa from the upper jejunum. Stool specimens should be collected on admission to the hospital or early in the illness. Kapikian et al.[2] collected specimens within 4 days, while Bryden et al.[52] did not include patients in their reports unless feces had been collected within 7 days of the onset of the disease. Many different methods are used in preparing stool specimens for use in serologic tests. For example, most investigators clarify and concentrate fecal suspensions by differential centrifugation.[27,49,50,56] Other investigators use the suspensions directly without clarification,[72] filter,[2] homogenize with fluorocarbon,[27,57] precipitate with polyethylene glycol,[57] or centrifuge the suspensions through gradients.[27,57]

Paired sera are usually obtained during the acute phase of the disease and again 3 to 6 weeks later.[2] Duodenal mucosa biopsy material can be obtained by the method of Barnes and Townley.[73] The duodenal aspirates are collected using a French Argyle feeding tube under fluoroscopic control.[50] The upper jejunum mucosa is obtained at necropsy.[50] Portions of the mucosa can be washed and scraped with a scalpel in order to harvest cells for electron microscopy and fluorescent antibody staining.[50]

Direct Detection of Fecal Rotaviruses by Electron Microscopy

Flewett[74] has recently described two procedures used in the direct examination of fecal specimens by electron microscopy. A 20% fecal suspension is centrifuged to remove bacteria, after which the viruses in the supernatant fluids are sedimented by high-speed centrifugation. The pellet is resuspended in distilled water and examined by electron microscopy after negative staining. The lowest limit of detection by this procedure is approximately 10^5 virus particles per gram of feces. If the number of viruses in a stool specimen is very high, a water suspension of the stool can be partially dried on a grid, washed, and negatively stained.[74]

Distinguishing among Rotaviruses from Different Animal Species

Several groups of workers[39,40,75] have studied the electrophoretic patterns of the RNA genome segments of rotaviruses as evidenced in acrylamide gels. The RNAs of human, monkey (SA 11), calf, and O agent rotaviruses all give different gel patterns;[75] calf and pig RNAs are indistinguishable;[40] the wild-type calf rotavirus RNA differs from that of cell culture-adapted calf rotavirus.[39]

Neutralization tests[3,16] and hemagglutination inhibition tests[76] can be used to detect type-specific rotavirus antigens, while complement fixation, gel diffusion, immunoelectron microscopy, and immunofluorescence detect group-specific rotavirus antigens.[3]

Detection of Rotaviruses in Cell Cultures

Recently developed procedures increase the efficiency of isolating rotaviruses in cell cultures. Banatvala et al.[9] inoculated coverslip cultures of cells with stool filtrates and then centrifuged the inoculated coverslip preparations for 2 hr at 3000 × G. After 24 hr of incubation at 37°C, the infected cells were stained with fluorescent antibody. Virus was detected in six of seven cultures inoculated with stool filtrates that were known to contain rotavirus. The effectiveness of centrifugation in improving the efficiency of detection of fecal rotaviruses in cell cultures has been confirmed by other investigators.[13,15-17]

McNulty et al.[14] infected cell cultures by diluting freshly trypsinized cells at least tenfold with fecal filtrates. The monolayers that formed from diluted cells had more infected cells when stained with fluorescent antibody than did cells infected as established monolayers. It was assumed that actively growing cells were more susceptible to the virus. Woode et al.[3] were able to demonstrate replication of the murine rotavirus (epizootic diarrhea of infant mice virus) in mouse kidney cells by inoculating cell suspensions. The virus did not replicate using conventional tissue culture procedures. Matsuno et al.[77] demonstrated that the Lincoln strain of the bovine rotavirus forms plaques when diethylaminoethyl-dextran and trypsin are incorporated into the agar overlay medium.

Serologic Tests

Since 1974, a number of serologic tests have been described for use in detecting human rotaviruses in stool specimens and in determining antibody titers in sera. If used properly, all of the tests discussed below will give satisfactory results. The test procedures reported in the literature have not been standardized, and studies comparing various tests have, in most cases, been superficial. In determining which test(s) is appropriate for a particular laboratory, the relevant factors are the available equipment, the number of specimens to be tested, and past experience with the various tests.

Immune electron microscopy has been the most widely used procedure for detecting rotaviruses in stools,[2,58,59,78] and it has also been useful for assaying antiviral antibody.[2] (The rotavirus-antibody complex as it appears in electron micrographs is shown in Figure 2.) An electron microscopic detection of rotaviruses in stool specimens indicates a very recent infection,[52] as the virus is probably not a common latent virus,[6] and there are no reports of chronic infections.

The complement fixation (CF) test was first used by Kapikian et al.[64] in diagnosing human rotavirus infections. In some investigations, the CF test has proved almost as sensitive as electron microscopy in detecting rotavirus in stools.[79] In CF tests to detect antibodies in human serum, Kapikian et al.[64] used neonatal calf diarrhea virus grown in cell cultures and virus purified from stool specimens[2] as antigens. In the latter case, the antigens may be anticomplementary[80] and are sometimes difficult to acquire in adequate quantities. Gust et al.[81] designed a special bed for patients so that more stools could be collected to provide virus for the CF test.

Several tests using electrophoresis have been described. Spence et al.[79] found that with counterimmunoelectrophoresis, only fecal suspensions with a titer by CF of 1:40 or higher gave positive results. Middleton et al.[82] used counterimmunoelectroosmophoresis to detect rotavirus in stools and antibody in sera. When unconcentrated stools from patients were examined by electroosmophoresis, the test detected 35 of 39 (92%) that were positive for rotavirus by electron microscopy. Tufvesson and Johnsson[80] found a 100% correlation between detection by electron microscopy and immunoelectroosmophoresis, if reaction gels were stained before examination.

Fluorescent antibody can be used to detect rotavirus-infected cells sloughed from the intestinal tract[83] or viral antigen in biopsy material,[47,84] determine growth of fecal virus in cell cultures,[3,7-17] assay rotavirus antibody by indirect staining[8,60,64,85,86] or in neutralization tests,[1,3,16,87] detect virus in stools using a fluorescent virus precipitin test,[78,88,89] and assay rotavirus using the immunofluorescent cell counting procedure.[90]

Recently, Yolken et al.[89] pointed out that the fluorescent virus precipitin test was preferable to several other tests used to detect rotaviruses in stool specimens. The disadvantages cited for the other tests were: (1) electron microscopy is not readily available to most laboratories and cannot be used with large numbers of specimens, (2) the

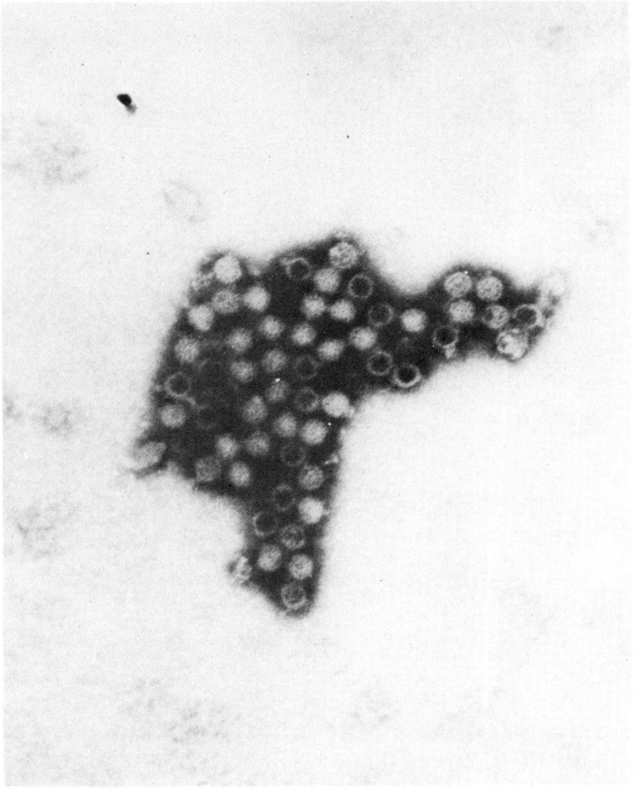

FIGURE 2. Electron micrograph of rotavirus antibody complex. (From Foster, L. G., Peterson, M. W., and Spendlove, R. S., *Proc. Soc. Exp. Biol. Med.,* 150, 155, 1975. With permission.)

material in many stools is anticomplementary in the complement fixation test, (3) counterimmunoelectrophoresis often fails to detect small amounts of rotavirus antigen unless special concentration or staining procedures are used, and (4) fluorescent antibody staining of infected cells in smears made from stool specimens gives a large number of false positive results.

Two solid-phase radioimmunoassay procedures have been described for rotavirus detection.[91,72] These tests will be very beneficial to those in laboratories equipped for radioimmunoassays.

An enzyme-labeled antibody procedure has been reported to be effective for diagnosing rotavirus infections.[92,93] Methods using enzyme labels will undoubtedly play a major role in rotavirus detection tests.

Spence et al.[76] prepared a hemagglutinin from African green monkey kidney cells infected with bovine rotavirus. The hemagglutinin that was associated with the virus particle, agglutinated human group O cells over a wide pH range and was inhibited by antiviral antibody. A flow chart showing some of the tests that are being used in diagnosing rotavirus infections is presented in Figure 3.

Pathogenicity in Experimental Animals

Monkeys,[86,94] piglets,[23,95-97] and calves[98,99] are susceptible to the human rotavirus. In a monkey study,[86] the small intestine of affected animals was edematous with cor-

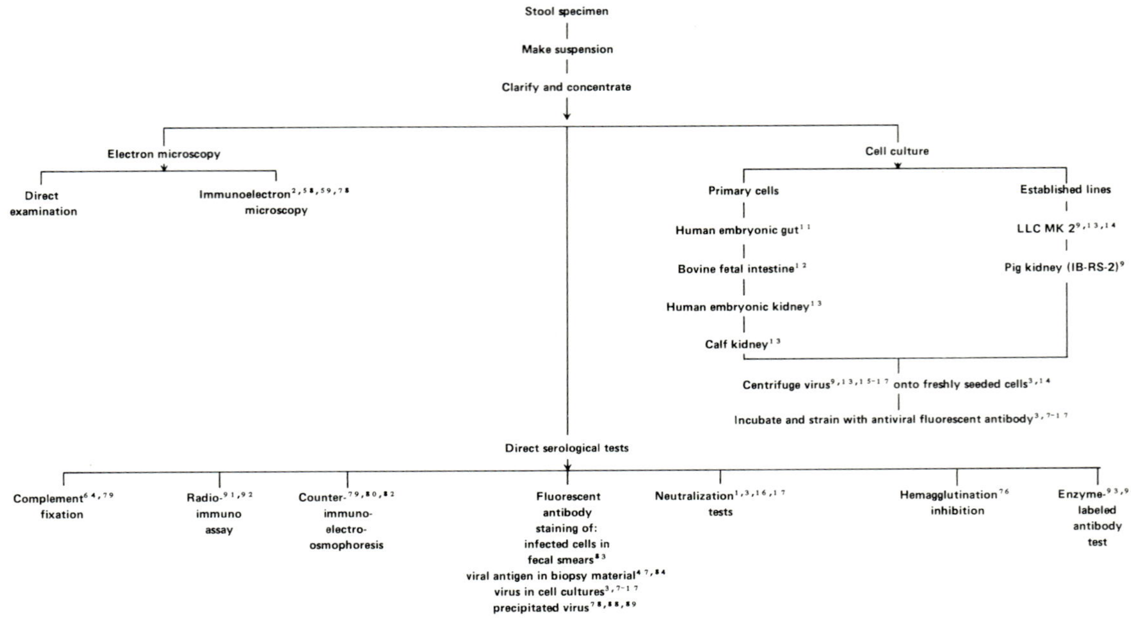

FIGURE 3. Flow chart for virus isolation and identification.

rugation of the mucosa and the distal two thirds was fluid filled. The serosal vessels of the large intestines, the three mesenteric lymph nodes, and the adrenals were enlarged. Fluorescent antiviral antibody staining of frozen sections of tissue from the gastrointestinal tract revealed coarsely granular cytoplasmic fluorescence in mid-distal portions of the small intestine. Viral antigen appeared to be associated with mucosal and submucosal cells of the villi. When conventionally raised or germ-free piglets were infected with the human rotavirus,[23,95,96] the severity of illness varied with the different passages. Studies on tissues taken at necropsy were not reported.

Mebus et al.[99] studied human rotavirus infections in gnotobiotic calves. The lesions in the calves were similar to those described in children. The infection rapidly progressed distally from the proximal part of the small intestine. According to them, "The sequence of events in the small intestine was: (i) infection of the adsorptive villous epithelial cells, (ii) replacement of the tall columnar villous epithelial cells with cuboidal and squamous cells, shortening of the villi, and enlargement of reticular cells and lymphocytic infiltration into the villous lamina propria; and (iii) repair." The small intestinal villous epithelial changes are shown in Figures 4 to 9.

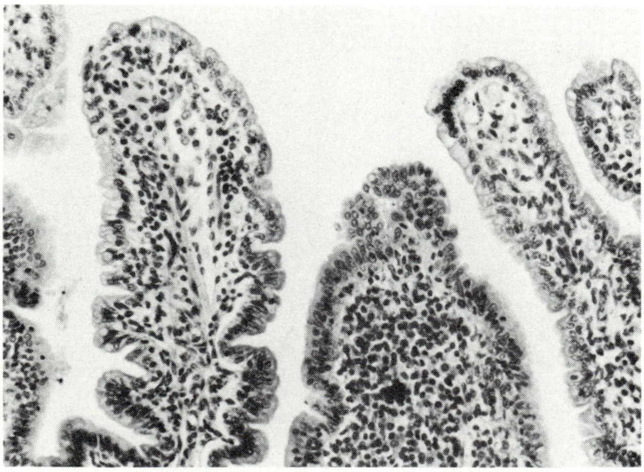

FIGURE 4. Distal part of the small intestine of a calf 3 hr after onset of diarrhea caused by a human rotavirus. Epithelial cells on the lower part of the villi have a columnar appearance and those toward the tip are columnar with apical vacuolation. (From Mebus, C. A., Wyatt, R. G., and Kapikian, A. Z., *Vet. Pathol.,* 14, 273, 1977. With permission.)

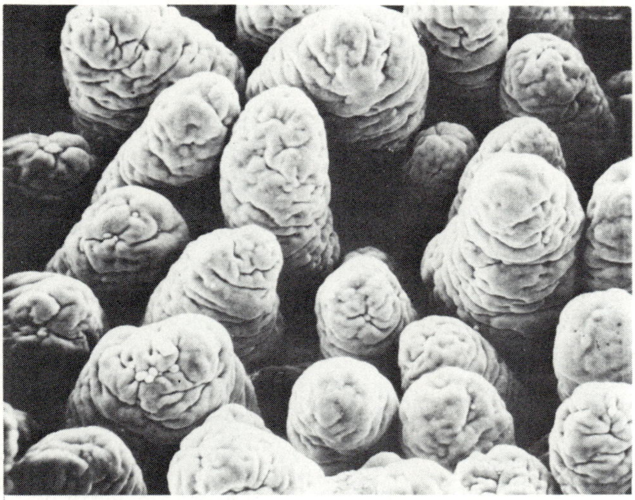

FIGURE 5. Scanning electron micrograph of villi in the distal part of the small intestine of a calf 7 hr after onset of human rotavirus diarrhea. The villi are shortened. Villous epithelial cells are of various sizes giving the surface a roughened appearance. Tips of 4 villi are denuded. (From Mebus, C. A., Wyatt, R. G., and Kapikian, A. Z., *Vet. Pathol.,* 14, 273, 1977. With permission.)

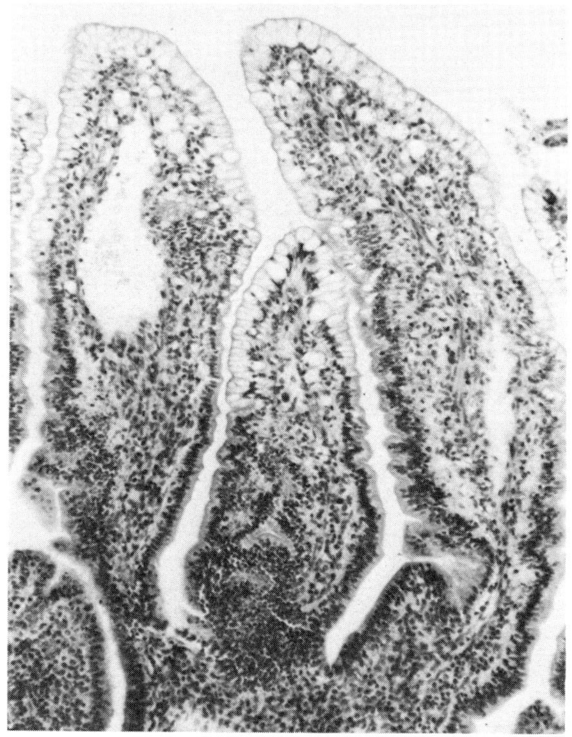

FIGURE 6. Higher magnification of one of the denuded villous tips shown in Figure 5. Epithelial cells at the edge of the exposed lamina propria have a squamous appearance. (From Mebus, C. A., Wyatt, R. G., and Kapikian, A. Z., *Vet. Pathol.*, 14, 273, 1977. With permission.)

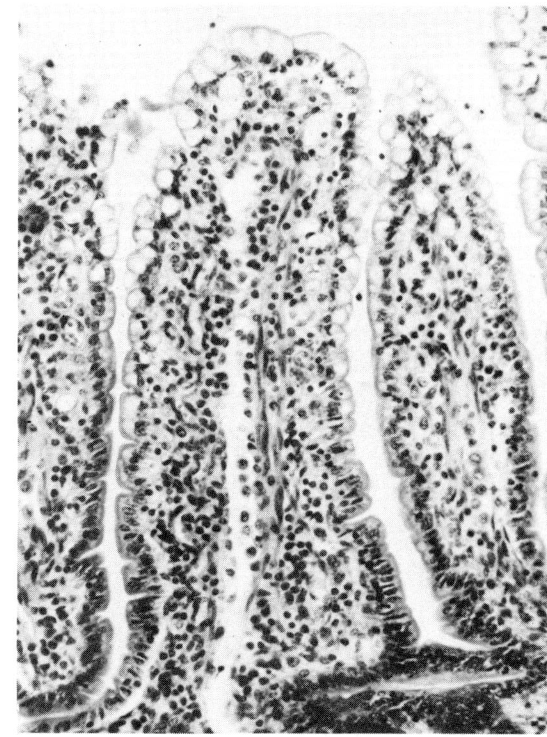

FIGURE 7. Distal part of the small intestine 7 hr after onset of diarrhea. The villi are shortened. Villous epithelial cells have a low columnar appearance. The lamina propria has a mild lymphocytic infiltrate. Hematoxylin-eosin. (From Mebus, C. A., Wyatt, R. G., and Kapikian, A. Z., *Vet. Pathol.*, 14, 273, 1977. With permission.)

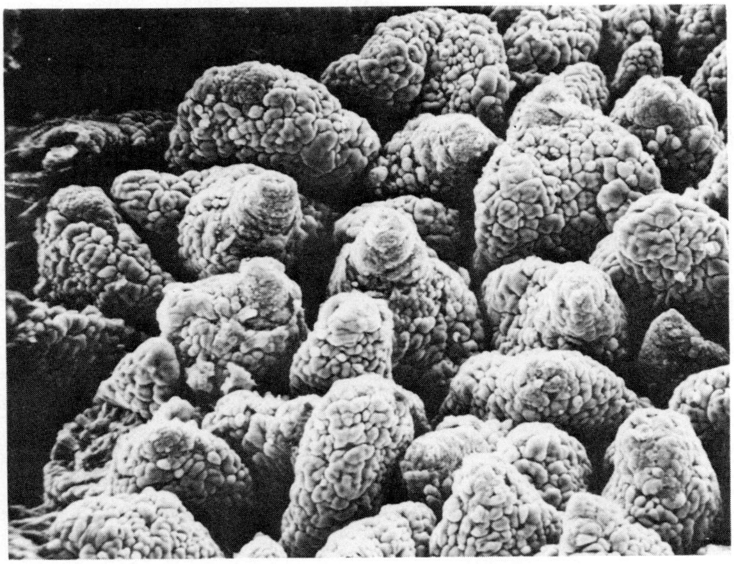

FIGURE 8. Scanning electron micrograph of the distal part of the small intestine 48 hr after onset of diarrhea. The villous surfaces have a smooth appearance and resemble those in the normal calf. (From Mebus, C. A., Wyatt, R. G., and Kapikian, A. Z., *Vet. Pathol.*, 14, 273, 1977. With permission.)

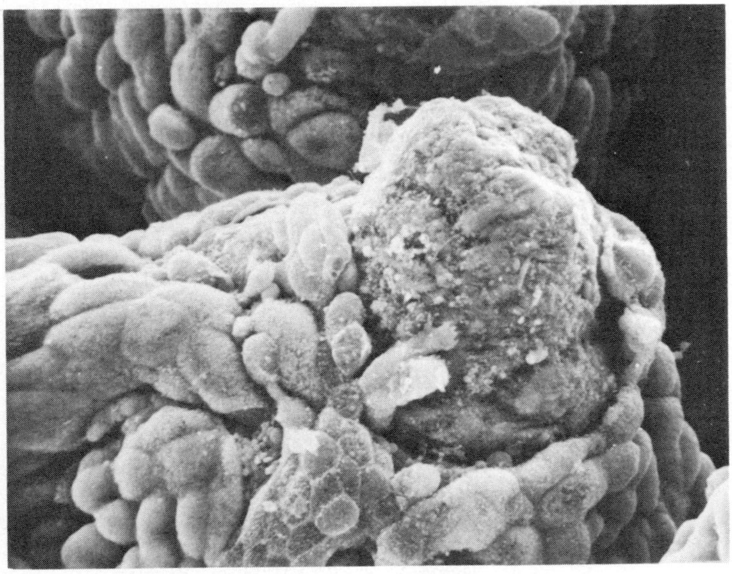

FIGURE 9. Distal part of the small intestine 48 hr after onset of diarrhea. Villi over Peyer's patch have tall columnar vacuolated epithelial cells. Hematoxylin-eosin. (From Mebus, C. A., Wyatt, R. G., and Kapikian, A. Z., *Vet. Pathol.*, 14, 273, 1977. With permission.)

GENERAL REFERENCES

DuPont, H. L., Portnoy, B. L., and Conklin, R. H., Viral agents and diarrheal illness, *Annu. Rev. Med.,* 28, 167, 1977.

Mebus, C. A., Reovirus-like calf enteritis, *Am. J. Dig. Dis.,* 21, 592, 1976.

Wyatt, R. G., Kalica, A. R., Mebus, C. A., Kim, H. W., London, W. T., Chanock, R. M., and Kapikian, A. Z., Reovirus-like agents (rotaviruses) associated with diarrheal illness in animals and man. *Perspect. Virol.,* 10, in press.

Kapikian, A. Z., Kim, H. W., Wyatt, R. G., Cline, W. L., Parrott, R. H., Chanock, R. M., Arrobio, J. O., Brandt, C. D., Rodriguez, W. J., Kalica, A. R., and Van Kirk, D. H., Recent advances in the aetiology of viral gastroenteritis, *Ciba Found. Symp.,* 42, 273, 1976.

REFERENCES

1. **Flewett, T. H., Bryden, A. S., Davies, H., Woode, G. N., Bridger, J. C., and Derrick, J. M.,** Relation between viruses from acute gastroenteritis of children and newborn calves, *Lancet,* 2, 61, 1974.
2. **Kapikian, A. Z., Kim, H. W., Wyatt, R. G., Rodriguez, W. J., Ross, S., Cline, W. L., Parrott, R. H., and Chanock, R. M.,** Reoviruslike agent in stools: associated with infantile diarrhea and development of serologic tests, *Science,* 185, 1049, 1974.
3. **Woode, G. N., Bridger, J. C., Jones, J. M., Flewett, T. H., Bryden, A. S., Davies, H. A., and White, G. B. B.,** Morphological and antigenic relationships between viruses (rotaviruses) from acute gastroenteritis of children, calves, piglets, mice, and foals, *Infect. Immun.,* 14, 804, 1976.
4. **Els, H. J. and Lecatsas, G.,** Morphological studies on simian virus S. A. 11 and the 'related' O agent, *J. Gen. Virol.,* 17, 129, 1972.
5. **Davidson, G. P., Townley, R. R. W., Ruck, B. J., Bishop, R. F., and Holmes, I. H.,** Importance of a new virus in acute sporadic enteritis in children, *Lancet,* 1, 242, 1975.
6. **Bishop, R. F., Davidson, G. P., Holmes, I. H., and Ruck, B. J.,** Virus particles in epithelial cells of duodenal mucosa from children with acute non-bacterial gastroenteritis, *Lancet,* 2, 1281, 1973.
7. **Flewett, T. H., Davies, H., Bryden, A. S., and Robertson, M. J.,** Diagnostic electron microscopy of faeces. 2. Acute gastroenteritis associated with reovirus-like particles, *J. Clin. Pathol.,* 27, 608, 1974.
8. **Wyatt, R. G., Kapikian, A. Z., Thornhill, T. S., Sereno, M. M., Kim, H. W., and Chanock, R. M.,** In vitro cultivation in human fetal intestinal organ culture of a reovirus-like agent associated with nonbacterial gastroenteritis in infants and children, *J. Infect. Dis.,* 130, 523, 1974.
9. **Banatvala, J. E., Totterdell, B., Chrystie, I. L., and Woode, G. N.,** In-vitro detection of human rotaviruses, Letter, *Lancet,* 2, 821, 1975.
10. **Purdham, D. R., Purdham, P. A., Evans, N., and McNeish, A. S.,** Isolation of human rotavirus using human embryonic gut monolayers, Letter, *Lancet,* 2, 977, 1975.
11. **Wyatt, R. G., Gill, V. W., Sereno, M. M., Kalica, A. R., VanKirk, D. H., Chanock, R. M., and Kapikian, A. Z.,** Probable in vitro cultivation of human reovirus-like agent of infantile diarrhea, *Lancet,* 1, 98, 1976.
12. **Albrey, M. B. and Murphy, A. M.,** Rotavirus growth in bovine monolayers, Letter, *Lancet,* 2, 753, 1976.
13. **Bryden, A. S., Davies, H. A., Thouless, M. E., and Flewett, T. H.,** Diagnosis of rotavirus infection by cell culture, *J. Med. Microbiol.,* 10, 121, 1977.
14. **McNulty, M. S., Allan, G. M., Pearson, G. R., McFerran, J. B., Curran, W. L., and McCracken, R. M.,** Reovirus-like agent (rotavirus) from lambs, *Infect. Immun.,* 14, 1332, 1976.
15. **Chasey, D. and Lucas, M.,** Detection of rotavirus in experimentally infected piglets, *Res. Vet. Sci.,* 22, 124, 1977.
16. **Thouless, M. E., Bryden, A. S., Flewett, T. H., Woode, G. N., Bridger, J. C., Snodgrass, D. R., and Herring, J. A.,** Serological relationships between rotaviruses from different species as studied by complement fixation and neutralization, *Arch. Virol.,* 53, 287, 1977.
17. **De Silva, L. M. and Marshall, J.,** Identification of rotaviruses, Letter, *Lancet,* 1, 609, 1977.
18. **Mebus, C. A., Kono, M., Underdahl, N. R., and Twiehaus, M. J.,** Cell culture propagation of neonatal calf diarrhea (scours) virus, *Can. Vet. J.,* 12, 69, 1971.
19. **Woode, G. N., Bridger, J. C., Hall, G., and Dennis, J. M.,** The isolation of a reovirus-like agent associated with diarrhea in colostrum-deprived calves in Great Britain, *Res. Vet. Sci.,* 16, 102, 1974.
20. **McNulty, M. S., Allan, G. M., and McFerran, J. B.,** Isolation of cytopathic calf rotavirus, *Res. Vet. Sci.,* 21, 114, 1976.

21. **Malherbe, H. H. and Strickland-Cholmley, M.**, Simian virus SA11 and the related O agent, *Arch. Ges. Virusforsch.*, 22, 235, 1967.
22. **Malherbe, H. H., Strickland-Cholmley, M., and Geyer, S. M.**, Viruses in abattoir effleunts, in *Transmission of Viruses by the Water Route*, Berg, G., Ed., John Wiley & Sons, New York, 1967, 347.
23. **Torres-Medina, A., Wyatt, R. G., Mebus, C. A., Underdahl, N. R., and Kapikian, A. Z.**, Diarrhea caused in gnotobiotic piglets by the reovirus-like agent of human infantile gastroenteritis, *J. Infect. Dis.*, 133, 22, 1976.
24. **Martin, M. L., Palmer, E. L., and Middleton, P. J.**, Ultrastructure of infantile gastroenteritis virus, *Virology*, 68, 146, 1975.
25. **Newman, J. F. E., Brown, F., Bridger, J. C., and Woode, G. N.**, Characterization of a rotavirus, *Nature*, 258, 631, 1975.
26. **Bridger, J. C. and Woode, G. N.**, Characterization of two particle types of calf rotavirus, *J. Gen. Virol.*, 31, 245, 1976.
27. **Welch, A. B.**, Purification, morphology and partial characterization of a reovirus-like agent associated with neonatal calf diarrhea, *Can. J. Comp. Med.*, 35, 195, 1971.
28. **Holmes, I. H., Ruck, B. J., Bishop, R. F., and Davidson, G. P.**, Infantile enteritis viruses: morphogenesis and morphology, *J. Virol.*, 16, 937, 1975.
29. **Adams, W. R. and Kraft, L. M.**, Electron-microscopic study of the intestinal epithelium of mice infected with the agent of epizootic diarrhea of infant mice (EDIM virus), *Am. J. Pathol.*, 51, 39, 1967.
30. **Amano, Y., Katagiri, S., Ishida, N., and Watanabe, Y.**, Spontaneous degradation of reovirus capsid into subunits, *J. Virol.*, 8, 805, 1971.
31. **Vasquez, C. and Tournier, P.**, New interpretation of the reovirus structure, *Virology*, 24, 128, 1964.
32. **Muller, G., Schneider, C. C., and Peters, D.**, Zur Feinstruktur des Reovirus (Typ 3), *Arch. Gesamte Virusforsch.*, 19, 111, 1966.
33. **Palmer, E. L. and Martin, M. L.**, The fine structure of the capsid of reovirus type 3, *Virology*, 76, 109, 1977.
34. **Luftig, R. B., Kilham, S. S., Hay, A. J., Zweerink, H. J., and Joklik, W. K.**, An ultrastructural study of virions and cores of reovirus type 3, *Virology*, 48, 170, 1972.
35. **Todd, D. and McNulty, M. S.**, Biochemical studies on a reovirus-like agent (rotavirus) of lambs, *J. Virol.*, 21, 1215, 1977.
36. **Rodger, S. M., Schnagl, R. D., and Holmes, I. H.**, Biochemical and biophysical characteristics of diarrhea viruses of human and calf origin, *J. Virol.*, 16, 1229, 1975.
37. **Barnett, B. B.**, Partial Characterization of the Genome of Neonatal Calf Diarrhea Virus Reoviruslike Agent, Ph.D. thesis, Utah State University, Logan, 1975.
38. **Schnagl, R. D. and Holmes, I. H.**, Characteristics of the genome of human infantile enteritis virus (rotavirus), *J. Virol.*, 19, 267, 1976.
39. **Verly, E. and Cohen, J.**, Demonstration of size variation of RNA segments between different isolates of calf (rotavirus), *J. Gen. Virol.*, 35, 583, 1977.
40. **Todd, D. and McNulty, M. S.**, Characterization of pig rotavirus RNA. *J. Gen. Virol.*, 33, 147, 1976.
41. **Welch, A. B. and Thompson, T. L.**, Physicochemical characterization of a neonatal calf diarrhea virus, *Can. J. Comp. Med.*, 37, 295, 1973.
42. **Flewett, T. H., Bryden, A. S., Davies, H., and Morris, C. A.**, Epidemic viral enteritis in a long-stay children's ward, *Lancet*, 1, 4, 1975.
43. **Welch, A. B.**, Biological Studies in Cell Culture and Physicochemical Characterization of a Neonatal Calf Diarrhea Virus, M.S. thesis, University of Nebraska, Lincoln, 1972.
44. **Welch, A. B. and Twiehaus, M. J.**, Cell culture studies of a neonatal calf diarrhea virus, *Can. J. Comp. Med.*, 37, 287, 1973.
45. **Shepherd, R. W., Turslow, S., Walker-Smith, J. A., Bird, R., Cutting, W., Darnell, R., and Barker, C. M.**, Infantile gastroenteritis: a clinical study of reovirus-like agent infection, *Lancet*, 2, 1082, 1975.
46. **Ryder, R. W., Sack, D. A., Kapikian, A. Z., McLaughlin, J. C., Chakraborty, J., Mizanur Rahman, A. S. M., Merson, M. H., and Wells, J. G.**, Enterotoxigenic *Escherichia coli* and reovirus-like agent in rural Bangladesh, *Lancet*, 1, 659, 1976.
47. **Davidson, G. P., Goller, I., Bishop, R. F., Townley, R. R. W., Holmes, I. H., and Ruck, B. J.**, Immunofluorescence in duodenal mucosa of children with acute enteritis due to a new virus, *J. Clin. Pathol.*, 28, 263, 1975.
48. **Hamilton, J. R., Gall, D. G., Kerzner, B., Butler, D. G., and Middleton, P. J.**, Recent developments in viral gastroenteritis, *Pediatr. Clin. North Am.*, 22, 747, 1975.
49. **Schoub, B. D., Koornhof, H. J., Lecatsas, G., Prozesky, O. W., Frieman, I., Hartman, E., and Kassel, H.**, Viruses in acute summer gastroenteritis in black infants, *Lancet*, 1, 1093, 1975.
50. **Middleton, P. J., Szymanski, M. T., Abbott, G. D., Bortolussi, R., and Hamilton, J. R.**, Orbivirus acute gastroenteritis of infancy, *Lancet*, 1, 1241, 1974.

51. **Bishop, R. F., Hewstone, A. S., Davidson, G. P., Townley, R. R. W., Holmes, I. H., and Ruck, B. J.**, An epidemic of diarrhoea in human neonates involving a reovirus-like agent and 'enteropathogenic' serotypes of *Escherichia coli*, *J. Clin. Pathol.*, 29, 46, 1976.
52. **Bryden, A. S., Davies, H. A., Hadley, R. E., Flewett, T. H., Morris, C. A., and Oliver, P.**, Rotavirus enteritis in the west midlands during 1974, *Lancet*, 2, 241, 1975.
53. **Whorwell, P. J., Phillips, C. A., Beeken, W. L., Little, P. K., and Roessner, K. D.**, Isolation of reovirus-like agents from patients with Crohn's disease, *Lancet*, 1, 1169, 1977.
54. **Webb, C. H. and Wallace, W. M.**, Diagnosis and treatment: epidemic gastroenteritis, presumably viral, *Pediatrics*, 38, 494, 1966.
55. **Bishop, R. F., Davidson, G. P., Holmes, I. H., and Ruck, B. J.**, Evidence for viral gastroenteritis, *N. Engl. J. Med.*, 289, 1096, 1973.
56. **Flewett, T. H., Bryden, A. S., and Davies, H.**, Electron microscopy of faeces in acute gastroenteritis, *J. Clin. Pathol.*, 27, 512, 1973.
57. **Bishop, R. F., Davidson, G. P., Holmes, I. H., and Ruck, B. J.**, Detection of a new virus by electron microscopy of faecal extracts from children with acute gastroenteritis, *Lancet*, 1, 149, 1974.
58. **Albrey, M. B. and Murphy, A. M.**, Rotaviruses and acute gastroenteritis of infants and children, *Med. J. Aust.*, 1, 82, 1976.
59. **Flewett, T. H., Bryden, A. S., and Davies, H.**, Diagnostic Electron Microscopy of faeces. I. The viral flora of the faeces as seen by electron microscopy, *J. Clin. Pathol.*, 27, 603, 1974.
60. **Kapikian, A. Z., Kim, H. W., Wyatt, R. G., Cline, W. L., Arrobio, J. O., Brandt, C. D., Rodriguez, W. J., Sack, D. A., Chanock, R. M., and Parrott, R. H.**, Human reovirus-like agent as the major pathogen associated with "winter" gastroenteritis in hospitalized infants and young children, *N. Engl. J. Med.*, 294, 965, 1976.
61. **Tufvesson, B. and Johnsson, T.**, Occurrence of reo-like viruses in young children with acute gastroenteritis, Diagnosis established by electron microscopy and complement fixation using the reo-like virus as antigen, *Acta Pathol. Microbiol. Scand. Sect. B*, 84, 22, 1976.
62. **Chrystie, I. L., Totterdell, B., Baker, M. J., Scopes, J. W., and Banatvala, J. E.**, Rotavirus infections in a maternity unit, *Lancet*, 2, 79, 1975.
63. **Hara, M., Mukoyama, J., Tsuruhara, T., Saito, Y., and Tagaya, I.**, Duo-virus in school children with gastroenteritis, *Lancet*, 1, 311, 1976.
64. **Kapikian, A. Z., Cline, W. L., Mebus, C. A., Wyatt, R. G., James, H. D., Jr., Kalica, A. R., VanKirk, D., Chanock, R. M., and Kim, H. W.**, New complement-fixation test for the human reovirus-like agent of infantile gastroenteritis, Nebraska calf diarrhea virus used as antigen, *Lancet*, 1, 1056, 1975.
65. **Gomez-Baretto, J., Palmer, E., Hatch, M., and Nahmias, A. J.**, Occurrence of "viral particles" in diarrhea, Letter, *N. Engl. J. Med.*, 292, 1297, 1975.
66. **Zissis, G., Lambert, J. P., Fonteyne, J., and de Kegel, D.**, Child-mother transmission of rotavirus, Letter, *Lancet*, 1, 96, 1976.
67. **Von Bonsdorff, C.-H., Hovi, T., Måkelå, P., Hovi, L., and Tevalvoto-Aarnio, M.**, Rotavirus associated with acute gastroenteritis in adults, *Lancet*, 2, 423, 1976.
68. **Meurman, O. H. and Laine, M. J.**, Rotavirus epidemic in adults, Letter, *N. Engl. J. Med.*, 296, 1298, 1977.
69. **Cameron, D. J. S., Bishop, R. F., Davidson, G. P., Townley, R. R. W., Holmes, I. H., and Ruck, B. J.**, Rotavirus infections in obstetric hosptials, Letter, *Lancet*, 2, 124, 1975.
70. **Woode, G. N., Jones, J., and Bridger, J.**, Levels of colostral antibodies against neonatal calf diarrhoea virus, *Vet. Rec.*, 97, 148, 1975.
71. **Mebus, C. A.**, Viral calf enteritis, *J. Dairy Sci.*, 59, 1175, 1976.
72. **Kalica, A. R., Purcell, R. H., Sereno, M. M., Wyatt, R. G., Kim, H. W., Chanock, R. M., and Kapikian, A. Z.**, A microtiter solid phase radioimmunoassay for detection of the human reovirus-like agent in stools, *J. Immunol.*, 118, 1275, 1977.
73. **Barnes, G. L. and Townley, R. R. W.**, Duodenal mucosal damage in 31 infants with gastroenteritis, *Arch. Dis. Child.*, 48, 343, 1973.
74. **Flewett, T. H.**, Diagnosis of enteritis virus, *Proc. R. Soc. Med., (London)*, 69, 693, 1976.
75. **Kalica, A. R., Garon, C. F., Wyatt, R. G., Mebus, C. A., Van Kirk, D. H., Chanock, R. M., and Kapikian, A. Z.**, Differention of human and calf reovirus-like agents associated with diarrhea using polyacrylamide gel electrophoresis of RNA, *Virology*, 74, 86, 1976.
76. **Spence, L., Fauvel, M., Petro, R., and Bloch, S.**, Haemagglutinin from rotavirus, Letter, *Lancet*, 2, 1023, 1976.
77. **Matsuno, S., Inouye, S., and Kono, R.**, Plaque assay of neonatal calf diarrhea virus and the neutralizing antibody in human sera, *J. Clin. Microbiol.*, 5, 1, 1977.
78. **Peterson, M. W., Spendlove, R. S., and Smart, R. A.**, Detection of neonatal calf diarrhea virus, infant reovirus-like diarrhea virus, and a coronavirus using the fluorescent virus precipitin test, *J. Clin. Microbiol.*, 3, 376, 1976.
79. **Spence, L., Fauvel, M., Bouchard, S., Babiuk, L., and Saunders, J. R.**,Test for reovirus-like agent, *Lancet*, 2, 322, 1975.

80. **Tufvesson, B. and Johnsson, T.**, Immunoelectroosmophoresis for detection of reo-like virus: methodology and comparison with electron microscopy, *Acta. Pathol. Microbiol. Scand. Sect. B*, 84, 225, 1976.
81. **Gust, I. D., Seger, V. J., Sandland, A. M., and Pringle, R. C.**, Acute nonbacterial gastroenteritis: collection of large quantities of stools from young children, Letter, *Lancet*, 2, 933, 1975.
82. **Middleton, P.J., Petric, M., Hewitt, C. M., Szymanski, M. T., and Tam, J. S.**, Counter-immuno-electro-osmophoresis for the detection of infantile gastroenteritis virus (orbi-group) antigen and antibody, *J. Clin. Pathol.*, 29, 191, 1976.
83. **Mebus, C. A., Rhodes, M. B., and Stair, E. L.**, Laboratory techniques for demonstrating Nebraska calf diarrhea virus, *Proc. 75th Annu. Meet., U.S. Animal Health Association*, Spencer Printing, Richmond, 1971, 599.
84. **Mebus, C. A., Stair, E. L., Underdahl, N. R., and Twiehaus, M. J.**, Pathology of neonatal calf diarrhea induced by reo-like virus, *Vet. Pathol.*, 8, 490, 1971.
85. **Echeverria, P., Blacklow, N. R., and Smith, D. H.**, Role of heat-labile toxigenic *Escherichia coli* and reovirus-like agent in diarrhoea in Boston children, *Lancet*, 2, 1113, 1975.
86. **Wyatt, R. G., Sly, D. L., London, W. T., Palmer, A. E., Kalica, A. R., VanKirk, D. H., Chanock, R. M., and Kapikian, A. Z.**, Induction of diarrhea in colostrum-deprived newborn rhesus monkeys with the human reovirus-like agent of infantile gastroenteritis, *Arch. Virol.*, 50, 17, 1976.
87. **Mebus, C. A., White, R. G., Bass, E. P., and Twiehaus, M. J.**, Immunity to neonatal calf diarrhea virus, *J. Am. Vet. Med. Assoc.*, 163, 880, 1973.
88. **Foster, L. G., Peterson, M. W., and Spendlove, R. S.**, Fluorescent virus precipitin test, *Proc. Soc. Exp. Biol. Med.*, 150, 155, 1975.
89. **Yolken, R. H., Wyatt, R. G., Kalica, A. R., Kim, H. W., Brandt, C. D., Parrott, R. H., Kapikian, A. Z., and Chanock, R. M.**, Use of a free viral immunofluorescence assay to detect human reovirus-like agent in human stools, *Infect. Immun.*, 16, 467, 1977.
90. **Barnett, B. B., Spendlove, R. S., Peterson, M. W., Hsu, L. Y., LaSalle, V. A., and Egbert, L. N.**, Immunofluorescent cell assay of neonatal calf diarrhea virus, *Can. J. Comp. Med.*, 39, 462, 1975.
91. **Middleton, P. J., Holdaway, M. D., Petric, M., Szymanski, M. T., and Tam, J. S.**, Solid-phase radioimmunoassay for the detection of rotavirus, *Infect. Immun.*, 16, 439, 1977.
92. **Scherrer, R. and Bernard, S.**, Application of enzyme-linked immunosorbent assay (ELISA) to the detection of calf rotavirus and rotavirus antibodies, *Ann. Microbiol.*, (Inst. Pasteur), 128(A), 499, 1977.
93. **Yolken, R. H., Kim, H. W., Clem, T., Wyatt, R. G., Kalica, A. R., Chanock, R. M., and Kapikian, A. Z.**, Enzyme linked immunosorbent assay (ELISA) for detection of human reovirus-like agent of infantile gastroenteritis, *Lancet*, 2, 263, 1977.
94. **Mitchell, J. D., Lambeth, L. A., Sosula, L., Murphy, A., and Albrey, M.**, Transmission of rotavirus gastroenteritis from children to a monkey, *Gut*, 18, 156, 1977.
95. **Middleton, P. J., Petric, M., and Syzmanski, M. T.**, Propagation of infantile gastroenteritis virus (Orbi-group) in conventional and germfree piglets, *Infect. Immun.*, 12, 1276, 1975.
96. **Bridger, J. C., Woode, G. N., Jones, J. M., Flewett, T. H., Bryden, A. S., and Davies, H.**, Transmission of human rotaviruses to gnotobiotic piglets, *J. Med. Microbiol.*, 8, 565, 1975.
97. **Torres-Medina, A., Wyatt, R. G., Mebus, C. A., Underdahl, N. R., and Kapikian, A. Z.**, Patterns of shedding of human reovirus-like agent in gnotobiotic newborn piglets with experimentally-induced diarrhea, *Intervirology*, 7, 250, 1976.
98. **Mebus, C. A., Wyatt, R. G., and Kapikian, A. Z.**, Intestinal lesions induced in gnotobiotic calves by the virus of human infantile gastroenteritis, *Vet. Pathol.*, 14, 273, 1977.
99. **Mebus, C. A., Wyatt, R. G., Sharpee, R. L., Sereno, M. M., Kalica, A. R., Kapikian, A. Z., and Twiehaus, M. J.**, Diarrhea in gnotobiotic calves caused by the reovirus-like agent of human infantile gastroenteritis, *Infect. Immun.*, 14, 471, 1976.

REOVIRIDAE: ORBIVIRUSES

I. HOLMES

Introduction

The orbiviruses of the family Reoviridae constitute a large group of double-stranded RNA arboviruses transmitted to vertebrate animals by ticks, mosquitoes, *Culicoides* midges, or phlebotomine sandflies. Most orbiviruses are not believed to be pathogenic for man, but a few, including Colorado tick fever, Kemerovo, and Changuinola, cause a febrile illness in man. In addition, important veterinary diseases such as bluetongue of sheep and African horse sickness are caused by orbiviruses.

Morphology

The inner capsid of orbivirus particles displays clearly defined, ring-shaped capsomers when negatively stained, but in most particles this symmetry is partially obscured by the amorphous outer capsid layer. This was first discovered during studies on purification and analysis of the type species, bluetongue virus,[1,2] but is probably true of all orbiviruses[3] (Figure 1). The dimensions vary somewhat from one orbivirus to another, but the inner capsid diameter is generally within the range 60 to 68 nm and the outer diameter 65 to 80 nm.[3,4]

The capsid symmetry is clearly icosahedral, and it is generally agreed that the inner capsid consists of 32 capsomers.[4,5] It is these relatively large (10 to 15 nm) doughnut or ring-shaped capsomers which suggested the generic name, from the Latin *Orbis* = ring or circle.[4] In crude virus suspensions, flattened tubules about the same width as virus particles may be found. Cross striations on these tubules indicate a repeating subunit structure and their dimensions (similar to that of capsomers) suggest that they may be composed of capsid subunits.

When thin sections of orbivirus-infected cells are studied, many resemblances between orbiviruses and reoviruses are immediately apparent. Virus particles appear in association with moderately electron-dense fibrillar and/or granular matrices in the cytoplasm.[4,6-18,20-23] Particles may be found scattered throughout the matrix or mainly at the periphery (Figures 2, 3a). The densely stained core (35 to 45 nm) is surrounded by a well-defined capsid zone; the total diameter is usually 65 to 70 nm.[4] Thin-walled tubules, whose diameter varies somewhat, but is usually comparable to that of virus particles, may be seen in large numbers in the cytoplasm near the "virus factories" or matrices. These correspond to the flattened tubules seen in negatively stained preparations, and are characteristic of bluetongue and epizootic hemorrhagic disease viruses, and probably all the Kemerovo group viruses (Figure 3b).[4,9,16,21]

The viral matrices are free in the cytoplasm, i.e., they are not membrane bound, but in the case of some orbiviruses, particles accumulate within cisternae of the rough endoplasmic reticulum. A proportion of these particles bud through the membrane of the endoplasmic reticulum and thus acquire an envelope or "pseudoenvelope". The latter term has been suggested because there is no evidence that such envelopes are essential for infectivity or contain virus-coded proteins. Both unenveloped and enveloped particles (diameter 90 to 100 nm) accumulate within the cisternae in cells infected with Colorado tick fever and Orungo viruses (Figure 2).[9,23] Other orbiviruses, e.g., in the bluetongue, African horse sickness, Eubenangee, and Warrego groups, acquire their pseudoenvelopes by budding from the cell surface.[8,12,15,17,20,22]

A variety of fibrillar structures are found in cells infected with certain orbiviruses, either within the viral matrices or "factories"[4,13,14,22] (Figure 3a) or adjacent to them

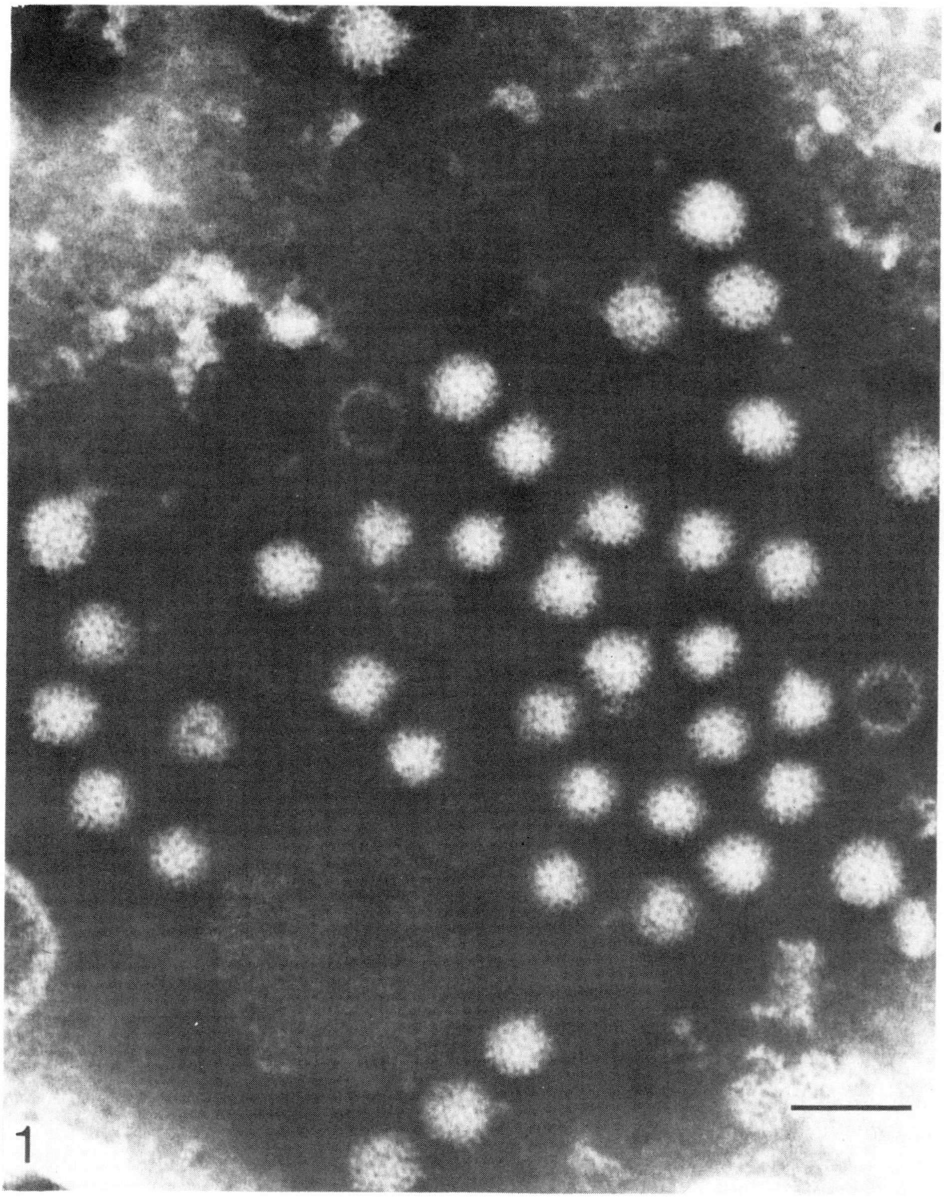

FIGURE 1. Particles of Eubenangee virus from suckling mouse brain, negatively stained with potassium phosphotungstate. Purification may remove surface polypeptides from orbivirus particles, producing "cleaner" looking particles but reducing infectivity. Marker = 100 nm. (Magnification × 180,000.)

(Fig. 2).[10,11,13,14,17] These have not been fully characterized, but they appear to consist of protein[14] antigenically related to the virus.[19] Virus-related fibrils can even accumulate within the nuclei of infected cells.[10,11,17] In the case of Colorado tick fever virus, particles and fibrils can be demonstrated within mature erythrocytes, which are presumably infected at an earlier stage of differentiation.[19]

Physicochemical Properties

Particles of bluetongue virus, the best characterized orbivirus, have a sedimentation coefficient of 550 S and band at densities 1.36 and 1.38 g/cm^3 in cesium chloride

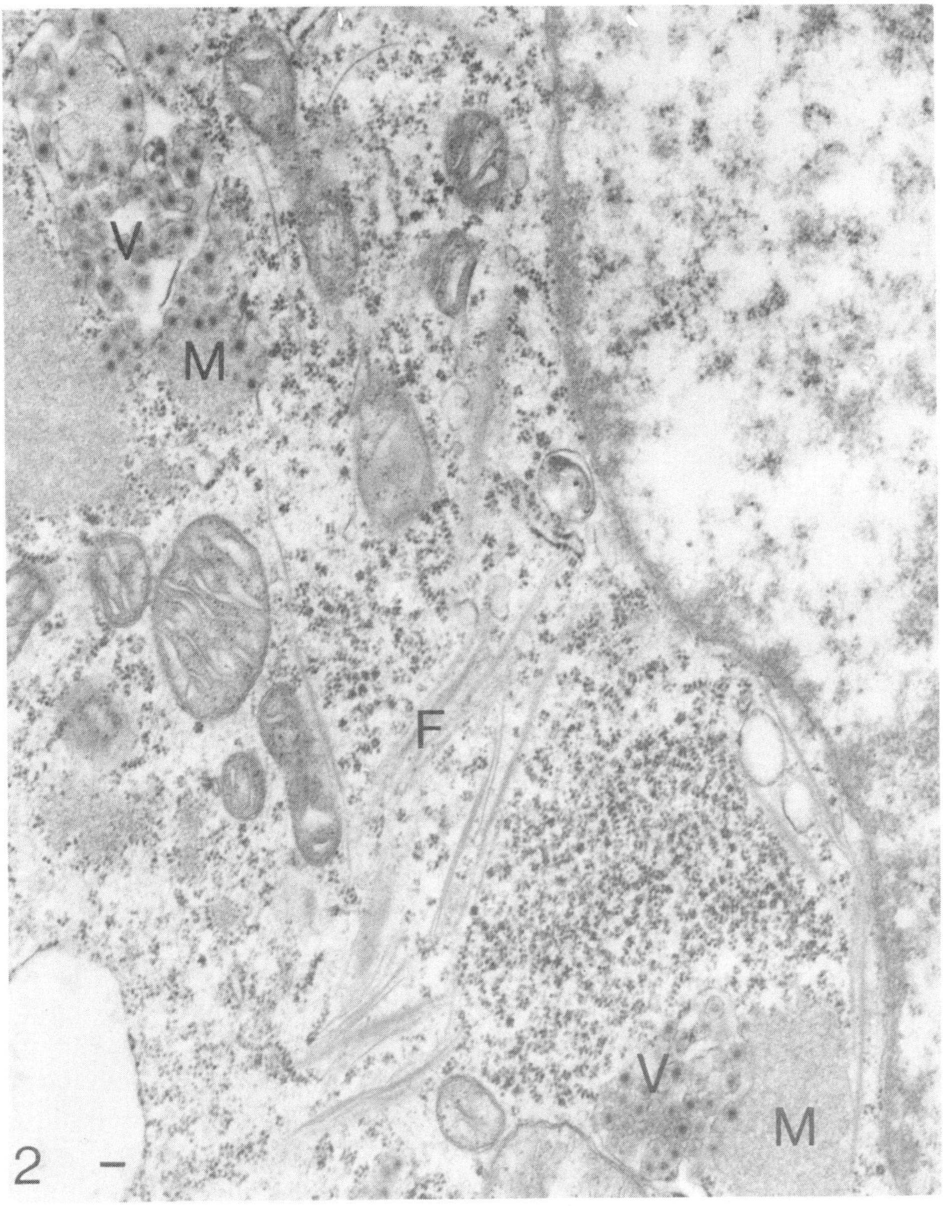

FIGURE 2. Thin section of BHK-21 cell infected with Colorado tick fever virus. Note the intracytoplasmic matrices (M) and the accumulation of virus particles (V) within cisternae. Most intracisternal particles are surrounded by a pseudoenvelope. Virus-specific fibrils (F) are also present. (Magnification ×36,000.) (Courtesy of F. A. Murphy.)

gradients.[1,2] The lighter particles contain seven polypeptides, whereas the denser have lost two (that comprise the fuzzy outer capsid) and have a lower infectivity/particle.

Compared to togaviruses, orbiviruses are resistant to inactivation by ether, chloroform, or sodium deoxycholate, although a small degree of inactivation does occur.[24] Orbiviruses are also acid sensitive, being inactivated within 3 hr at 4°C at pH 3.0, which readily distinguishes them from reoviruses.[24]

Virus Replication

The orbiviruses which have been analyzed contain a genome composed of ten seg-

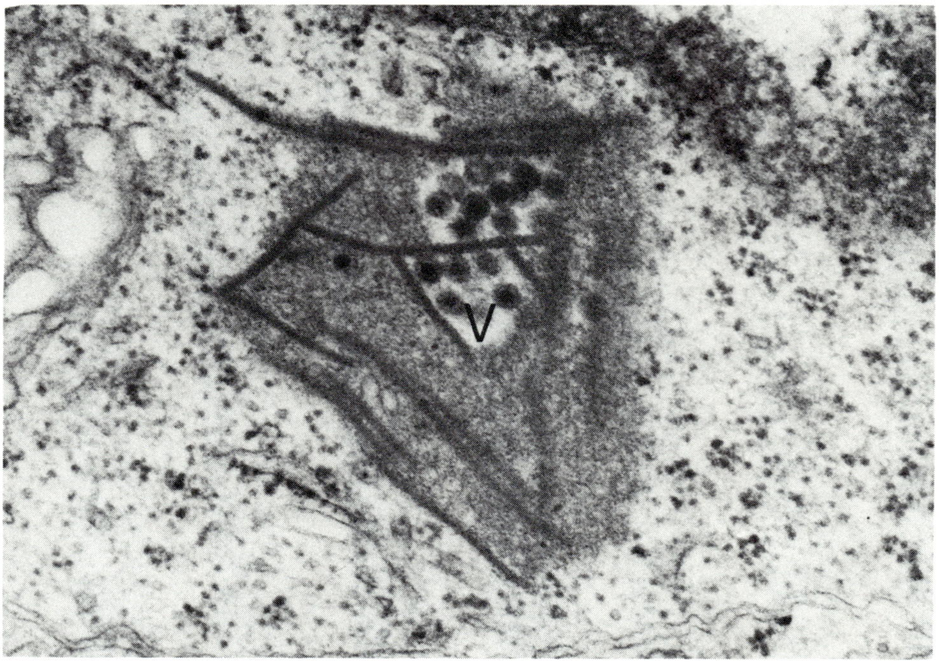

A

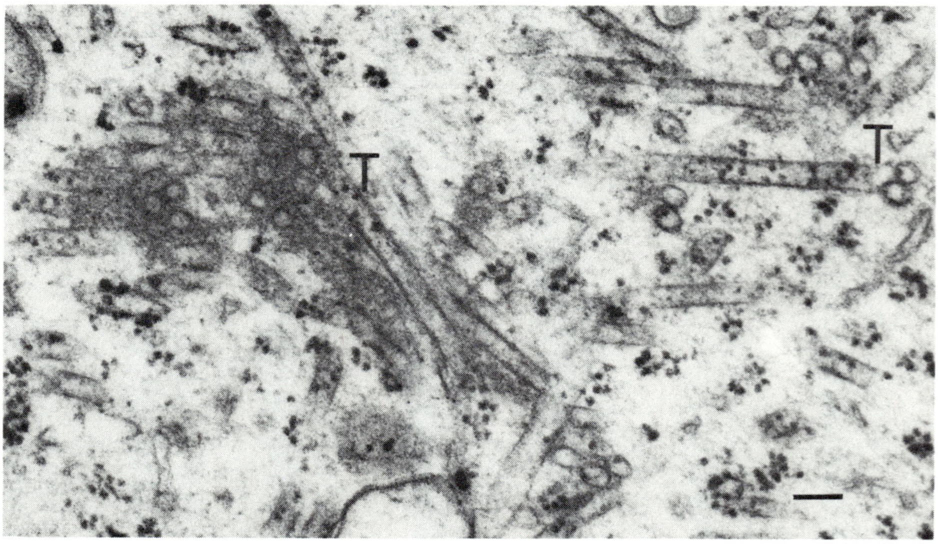

B

FIGURE 3. (A and B) Infant mouse neurone infected with Nugget virus of the Kemerovo serogroup. The linear structures within the matrix "virus factory" (V) consist of closely packed fibrils or sheets of protein, and are characteristic of Kemerovo group viruses. The tubules (T) of capsomeric material are seen both in longitudinal and cross sections. Such tubules are found in cells infected by various orbiviruses. (Magnification × 75,000.)

ments of double-stranded RNA, ranging in molecular weight from 0.3 to 2.7 × 10^6 daltons, and a total molecular weight about 12 million daltons.[1,2,25] When particles enter cells after adsorption, or are converted to the dense (1.38g/cm^3) form by loss of two outer polypeptides in vitro, RNA polymerases in the particle cores transcribe messenger RNA molecules corresponding in length to the genome segments.[1,2]

In the case of bluetongue virus infection, cellular protein and DNA syntheses are rapidly inhibited.[26] Viral RNA and protein synthesis occur in regions of "viroplasm" in the cytoplasm of infected cells (Figures 2 and 3a), and virus particles assemble in and around these virus "factories".

In vivo, Colorado tick fever virus can apparently multiply within precursors of erythrocytes and thus both particles and antigenically related fibrils can be visualized within mature erythrocytes.[19] Intraerythrocytic localization of virions is presumably the explanation of prolonged, cell-associated viremia which can persist in the presence of antibody in infections due to Colorado tick fever and other orbiviruses.

Virus Strains

Serologically, ten subgroups of orbiviruses have been defined,[23,24] but only the monotypic Colorado tick fever and some members of the Kemerovo serogroup will be listed here (Table 1). Human infections with members of other subgroups are on record in the International Catalogue of Arboviruses,[27] but are less common.

Clinical Features

Orbiviruses are best known for the diseases they cause in domestic animals, but Colorado tick fever virus is an important human pathogen within its geographical area. This naturally coincides with the distribution of its vector, the wood tick, *Dermacentor andersoni,* i.e., the Rocky Mountain area of the U.S. and Canada.

Symptoms appear 3 to 6 days after a tick bite, usually suddenly with fever, chilly sensations, headache, deep ocular pain, and muscle pains, followed by anorexia and nausea. After 2 to 3 days, fever and other symptoms subside for a similar period, then typically recur, so the disease has a diphasic course.[31] Leukopenia is notable by the third day and cell counts decrease further until the fifth to sixth day, with an increased proportion of immature granulocytes. Fever can recur more than once, but usually the patient is symptom free by the ninth to tenth day, although weakness may persist during convalescence.[32]

Serious, but not usually fatal, complications can occur in children up to the age of 10 years: the frequency is approximately 15% in this group. Various degrees of en-

Table 1
SOME ORBIVIRUSES CAUSING HUMAN DISEASE

				Laboratory diagnosis			
Virus Strain	Diseases associated	Geographic distribution	Vector	Source of specimen	Isolation in animal	Serologic diagnosis	Ref.
Colorado tick fever virus	Colorado tick fever	Northwestern North America	Wood tick *(Dermacentor)*	Erythrocytes	SM[a]	IF[b], NT[c]	28
Kemerovo	Fever	Western Siberia	Ixodid tick	CSF or serum	SM or chick embryo	NT, CF[d]	29
Tribec	Fever (?)	Czechoslovakia	Ixodid tick				30

[a] SM = 1- 3-day old (suckling) mouse, intracerebral inoculation.
[b] IF = immunofluorescence, see text.
[c] NT = neutralization tests, in cell culture.
[d] CF = complement fixation tests.

cephalitis result in hyperesthesia, a stiff neck, and drowsiness. In the most severe cases, the initial intense frontal headache is followed either by coma or by hyperirritability, convulsions, and delirium.[33] The other serious complication is bleeding, which can occur from the nose, mouth, and other mucosae, and can be accompanied by purpura or a petechial skin rash.[31]

The disease produced by Kemerovo or Tribecyviruses may be a mild fever or aseptic meningitis. It was recognized in the course of studies on the potentially more dangerous tick-borne encephalitis, or Russian spring — summer encephalitis, which is caused by a flavivirus. These Kemerovo group viruses are not only found in the same areas as the flavivirus, i.e., from Siberia to Czechoslovakia, but they even share the same vectors, species of Ixodid ticks.[29,30]

Laboratory Diagnosis

The occurrence and persistence of Colorado tick fever virus within erythrocytes of the infected host is unusual, but assists virus isolation. Successful virus isolations from erythrocytes have been obtained from blood samples taken as early as 1 day and as late as 120 days after development of symptoms.[34] This means that infectious virus can be extracted from erythrocytes even after clinical recovery from the disease, and after development of a normal convalescent antibody response.

Both virus isolation and serological diagnosis can be attempted if acute phase blood samples are obtained. After clotting, serum is retained for subsequent antibody tests, and the clot can be homogenized and frozen and thawed as the virus source. With samples taken more than 2 weeks after onset of symptoms, trypsin treatment or thorough washing of the erythrocytes is necessary before lysis to remove antibody. Erythrocyte lysates should be inoculated intracerebrally (0.015 mℓ per mouse) into litters of 1- to 4-day-old mice. Mice are the most sensitive detection system, although the virus grows well in cultures of BHK-21 or KB cell lines.[35] Mice are observed and if encephalitic symptoms (hyperirritability, incoordination, and limb tremors) develop within about a week, brains should be harvested and homogenized in buffered saline to provide a stock virus suspension. This stock can be inoculated into cultures of KB or BHK-21 cells and the identity of the virus confirmed by either neutralization or immunofluorescence tests with reference antisera.[36,37]

Rapid diagnosis by direct immunofluorescent staining of fixed smears of the patients erythrocytes is possible, although interpretation of the results may require some experience.[37] Granular fluorescence identifies the small percentage of erythrocytes containing viral antigens.

When paired sera are available for serologic diagnosis, both complement fixation and neutralization tests have been used, but indirect immunofluorescence tests employing BHK-21 cells infected with stock virus (e.g., Florio strain) are recommended.[36,37] Antibody is detectable by immunofluorescence or neutralization 2 to 4 weeks after onset of symptoms, and by complement fixation after 4 to 6 weeks. Neutralizing antibody persists for long periods after natural infection, or following experimental immunization with a formalinized vaccine.[38]

For Kemerovo virus isolation, acute phase serum or cerebrospinal fluid samples should be inoculated intracerebrally into suckling mice or into cultures of chick embryo fibroblasts. The virus kills suckling mice in 2 to 5 days, and produces extensive cytopathic effects in chick embryo fibroblasts within 48 to 72 hr.[29] This latter feature helped to distinguish Kemerovo virus from the tick-borne encephalitis (flavi-)virus. Strain identification is carried out by complement fixation or neutralization tests with reference sera.[39,40] While Kemerovo virus certainly causes clinical disease in man, serological surveys suggest that Tribec virus rarely infects man and nothing is known of the symptoms it produces.[30]

GENERAL REFERENCES

Borden, E. C., Shope, R. E., and Murphy, F. A., Physicochemical and morphological relationships of some arthropod-borne viruses to bluetongue virus — a new taxonomic group. Physicochemical and serological studies, *J. Gen. Virol.*, 13, 261, 1971.

Murphy, F. A., Borden, E. C., Shope, R. E., and Harrison, A., Physicochemical and morphological relationships of some arthropod-borne viruses to bluetongue virus — a new taxonomic group. Electron microscopic studies, *J. Gen. Virol.*, 13, 273, 1971.

Holmes, I. H., Arboviruses other than togaviruses, in *Ultrastructure of Animal Viruses and Bacteriophages: An Atlas*, Dalton, A. J. and Haguenau, F., Eds., Academic Press, New York, 1973, 197.

Berge, T. O., *International Catalogue of Arboviruses, including certain other viruses of vertebrates*, Center for Disease Control, Atlanta, 1975.

REFERENCES

1. **Verwoerd, D. W., Els, H. J., DeVilliers, E.-M., and Huismans, H.**, Structure of the bluetongue virus capsid, *J. Virol.*, 10, 783, 1972.
2. **Martin,S. A. and Zweerink, H. J.**, Isolation and characterization of two types of bluetongue virus particles, *Virology*, 50, 495, 1972.
3. **Lecatsas, G. and Gorman, B. M.**, Visualization of the extracapsid coat in certain bluetongue-type viruses, *Onderstepoort J. Vet. Res.*, 39, 193, 1972.
4. **Murphy, F. A., Bordon, E. C., Shope, R. E., and Harrison, A.**, Physicochemical and morphological relationships of some arthropod-borne viruses to bluetongue virus — a new taxonomic group. Electron microscopic studies, *J. Gen. Virol.*, 13, 273, 1971.
5. **Els, H. L. and Verwoerd, D. V.**, Morphology of bluetongue virus, *Virology*, 38, 213, 1969.
6. **Shestopalova, N. M., Reingold, V. N., Tikhomirova, T. I., Karpovich, L. G., and Chumakov, M. P.**, Electron microscope study of chick embryo cell culture infected with Kemerovo virus, *Acta Virol.* (Praha), 8, 88, 1964.
7. **Bowne, J. G. and Jones, R. H.**, Observations on bluetongue virus in salivary glands of an insect vector, *Culicoides variipennis*, *Virology*, 30, 127, 1966.
8. **Lecatsas, G. and Erasmus, B. J.**, Electron microscopic study of the formation of African horse sickness virus, *Arch. Gesamte Virusforsch.*, 22, 442, 1967.
9. **Lecatsas, G.**, Electron microscopic study of the formation of bluetongue virus, *Onderstepoort J. Vet. Res.*, 35, 139, 1968.
10. **Murphy, F. A., Coleman, P. H., Harrison, A. K., and Gary, G. W., Jr.**, Colorado tick fever virus: an electron microscopic study, *Virology*, 35, 28, 1968.
11. **Oshiro, L. S. and Emmons, R. W.**, Electron microscopic observations of Colorado tick fever virus in BHK21 and KB cells, *J. Gen. Virol.*, 3, 279, 1968.
12. **Schnagl, R. D., Holmes, I. H., and Doherty, R. L.**, An electron microscope study of Eubenangee, an Australian arbovirus, *Virology*, 38, 347, 1969.
13. **Breese, S. S., Jr., Ozawa, Y., and Dardiri, A. H.**, Electron microscopic characterization of African horse-sickness virus, *J. Am. Vet. Med. Assoc.*, 155, 391, 1969.
14. **Breese, S. S., Jr. and Ozawa, Y.**, Intracellular inclusions resulting from infection with African horsesickness virus, *J. Virol.*, 4, 109, 1969.
15. **Bowne, J. G. and Ritchie, A. E.**, Some morphological features of bluetongue virus, *Virology*, 40, 903, 1970.
16. **Tsai, K. S. and Karstad, L.**, Epizootic haemorrhagic disease virus of deer; an electron microscopic study, *Can. J. Microbiol.*, 16, 427, 1970.
17. **Schnagl, R. D. and Holmes, I. H.**, A study of Australian arboviruses resembling bluetongue virus, *Aust. J. Biol. Sci.*, 24, 1151, 1971.
18. **Gorman, B. M. and Lecatsas, G.**, Formation of Wallal virus in cell culture, *Onderstepoort J. Vet. Res.*, 39, 229, 1972.
19. **Emmons, R. W., Oshiro, L. S., Johnson, H. N., and Lennette, E. H.**, Intra-erythrocytic location of Colorado tick fever virus, *J. Gen. Virol.*, 17, 185, 1972.
20. **Ito, Y., Tanaka, Y., Inaba, Y., and Omori, T.**, Electron microscopy of Ibaraki virus, *Arch. Gesamte Virusforsch.*, 40, 29, 1973.
21. **Lecatsas, G., Erasmus, B. J., and Els, H. J.**, Electron microscopic studies on equine encephalosis virus, *Onderstepoort J. Vet. Res.*, 40, 53, 1973.

22. **Schnagl, R. D. and Holmes, I. H.,** Electron microscopy of Japanaut and Tilligerry viruses: two proposed members of the Orbivirus group, *Aust. J. Biol. Sci.,* 28, 425, 1975.
23. **Tomori, O., Fabiyi, A., and Murphy, F.,** Characterization of Orungo virus, an orbivirus from Uganda and Nigeria, *Arch. Virol.,* 51, 285, 1976.
24. **Borden, E. C., Shope, R. E., and Murphy, F. A.,** Physicochemical and morphological relationships of some arthropod-borne viruses to bluetongue virus — a new taxonomic group. Physicochemical and serological studies, *J. Gen. Virol.,* 13, 261, 1971.
25. **Schnagl, R. D. and Holmes, I. H.,** Polyacrylamide gel electrophoresis of the genomes of two orbiviruses: D'Aguilar and Eubenangee, *Intervirology,* 5, 300, 1975.
26. **Huismans, H.,** Host cell protein synthesis after infection with bluetongue virus and reovirus, *Virology,* 46, 500, 1971.
27. **Berge, T. O.,** *International Catalogue of Arboviruses, including certain other viruses of vertebrates,* Center for Disease Control, Atlanta, 1975.
28. **Florio, L., Stewart, M. O., and Mugrage, E. R.,** The experimental transmission of Colorado tick fever, *J. Exp. Med.,* 80, 165, 1944.
29. **Chumakov, M. P., Karpovich, L. G., Sarmanova, E. S., Sergeeva, G. I., Bychkova, M. B., Tapupere, V. O., Libikova, H., Mayer, V., Rehacek, J., Kozuch, O., and Ernek, E.,** Report on the isolation from *Ixodes persulcatus* ticks and from patients in western Siberia of a virus differing from the agent of tick-borne encephalitis, *Acta Virol.* (Praha), 7, 82, 1963.
30. **Grešiková, M., Nosek, J., Kozuch, O., Ernek, E., and Lichard, M.,** Study of the ecology of Tribec virus, *Acta Virol.* (Praha), 9, 83, 1965.
31. **Florio, L., Stewart, M. O., and Mugrage, E. R.,** The etiology of Colorado tick fever, *J. Exp. Med.,* 83, 1, 1946.
32. **Casals, J. and Clarke, D. H.,** Arboviruses other than Groups A and B, in *Viral and Rickettsial Infections of Man,* Horsfall, F. L. and Tamm, I., Eds., Lippincott, Philadelphia, 1965, 659.
33. **Eklund, C. M.,** Sequelae to the arthropod-borne encephalitis. VI. Colorado tick fever, *Neurology,* 8, 889, 1958.
34. **Hughes, L. E., Casper, E. A., and Clifford, C. M.,** Persistence of Colorado tick fever virus in red blood cells, *Am. J. Trop. Med. Hyg.,* 23, 530, 1974.
35. **Emmons, R. W.,** Colorado tick fever along the Pacific slope of North America, *Jpn. J. Med. Sci. Biol.,* Suppl. 20, 166, 1967.
36. **Gerloff, R. K. and Eklund, C. M.,** A tissue culture neutralization test for Colorado tick fever antibody and use of the test for serological surveys, *J. Infect. Dis.,* 104, 174, 1959.
37. **Emmons, R. W., Dondero, D. V., Devlin, V., and Lennette, E. H.,** Serologic diagnosis of Colorado tick fever: a comparison of complement fixation, immunofluorescence and plaque reduction methods, *Am. J. Trop. Med. Hyg.,* 18, 796, 1969.
38. **Thomas, L. A., Philip, R. N., Patzer, E., and Casper, E.,** Long duration of neutralizing antibody response after immunization of man with a formalinized Colorado tick fever vaccine, *Am. J. Trop. Med. Hyg.,* 16, 60, 1967.
39. **Libíková, H. and Casals, J.,** Serological characterization of eurasian Kemerovo group viruses. I. Cross complement fixation tests, *Acta Virol.* (Praha), 15, 65, 1971.
40. **Libíková, H. and Buckley, S. M.,** Serological characterization of eurasian Kemerovo group viruses II. Cross neutralization tests, *Acta Virol.* (Praha), 15, 79, 1971.

BUNYAVIRIDAE

I. Holmes and R. E. Shope

Introduction

The bunyaviruses are enveloped, spherical particles which contain single-stranded, three-segmented RNA and three structural proteins. Bunyaviruses make up one of four major families of arboviruses. Bunyamwera virus is the type virus.[1] Representatives of the family Bunyaviridae occur throughout the world. Many are not pathogenic for man, but those that are include California encephalitis,[2] Crimean hemorrhagic fever-Congo,[3] Oropouche,[4] Rift Valley fever,[5] and Sandfly fever viruses.[6] All are serious human pathogens. In addition, some members of the Bunyaviridae family, such as Rift Valley fever and Nairobi sheep disease, cause major illness in domestic animals. Each virus is maintained in a cycle involving ticks, mosquitoes, or phlebotomines, and one or more wild vertebrates. Each has a characteristic geographic distribution which may assist in differential diagnosis and selection of antigens for serologic tests.

Morphology

Bunyavirus particles are enveloped, rather pleomorphic, and usually 90 to 100 nm in diameter. When negatively stained, particles of Bunyamwera virus and of other viruses of the Bunyavirus genus have a fuzzy outline, the glycoprotein "spikes" in the envelope being ill-defined and approximately 5 to 10 nm in length[7,8] (Figure 1A). The family Bunyaviridae probably also includes at least one other group or genus, whose best studied representative is Uukuniemi (UUK) virus. Although the basic particle structure is similar, the peplomers of UUK virus are clearly defined rings 10 to 12 nm in diameter.[9]

The particles contain filamentous, probably helical ribonucleoprotein strands, about 9 nm wide, which occur in the form of closed circles in three size classes. Such strands can be seen in virus preparations disintegrating as a result of storage at 4°[7] but the circular nature of the ribonucleoprotein has only been demonstrated by disruption of the particles with detergents such as Triton X-100[10,11] or Nonidet P40.[12]

In thin sections of infected cells, Bunyavirus particles are recognized by their size (90 to 100 nm diameter) and localization in Golgi vesicles (Figure 1B) and, in the case of the more prolific members of the group, in smooth membrane vesicles found throughout the cytoplasm[13-16] (Figure 2). Vesicles apparently extrude their contents to the cell surface, so virus particles are released without cell lysis.[16]

Physicochemical properties

Bunyavirus particles have been reported to have buoyant densities in sucrose in the range of 1.17 to 1.19 g/cm^3 and a sedimentation coefficient about 450S.[8,9,17,18] Infectivity is most stable at pH 7 to 9, but markedly unstable at pH <6.[8] Infectivity is destroyed by lipid solvents and ionic or nonionic detergents.

Structural components

Only a small proportion of the many bunyaviruses have been chemically analysed. The genome consists of three segments of single-stranded RNA.[9,11,18] Molecular weight estimates are shown in Table 1.

These viral RNA molecules may or may not exist as closed loops,[11,20] but their protein complexes, i.e., the viral ribonucleoproteins, certainly occur as twisted circular structures.[10,11,12]

Bunyavirus particles contain three structural proteins; two glycoproteins associated

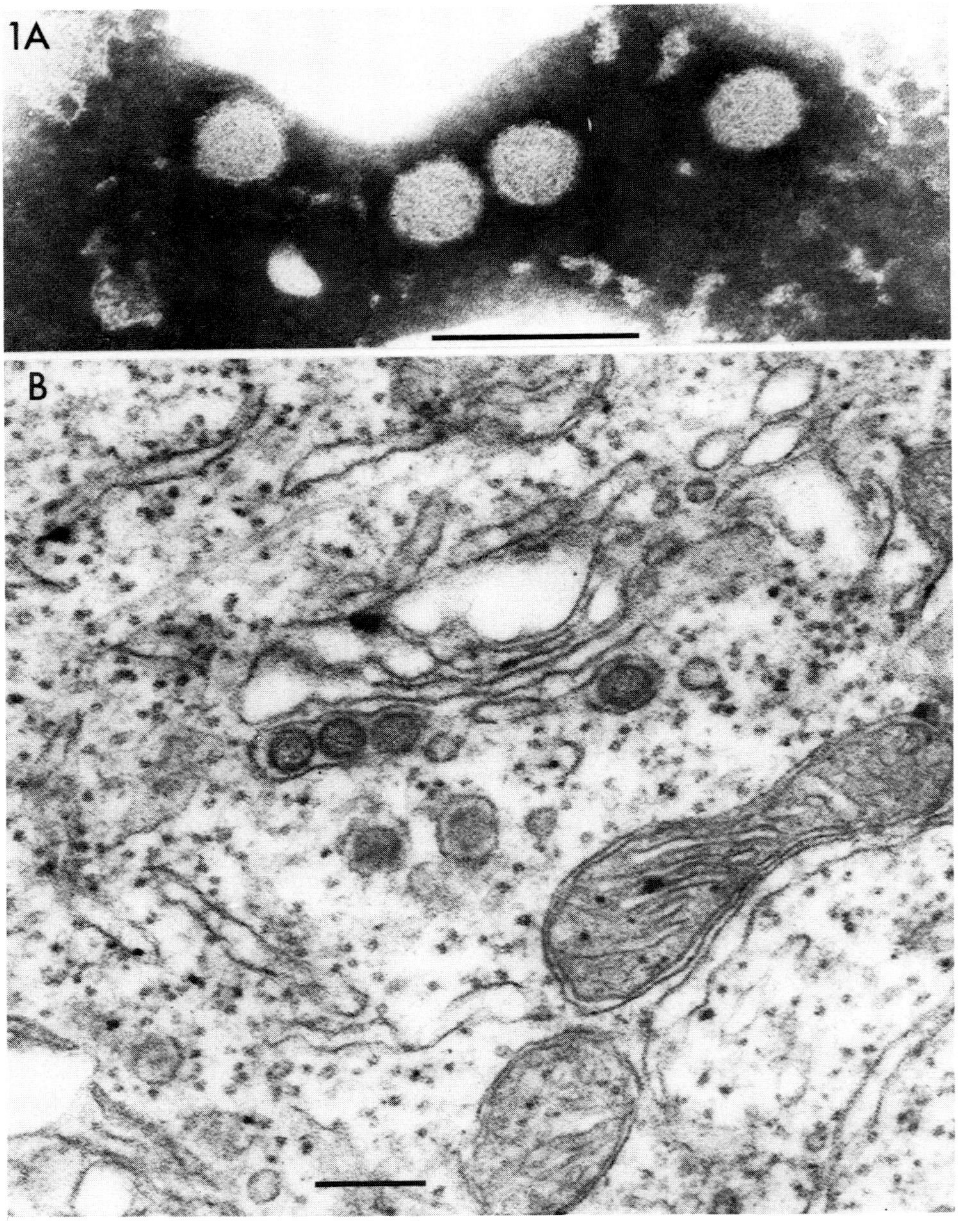

FIGURE 1. (A.) Bunyamwera virus particles, negatively stained with potassium phosphotungstate, X150000 (Magnification marker 200 nm). (B.) Thin section of a neurone of a suckling mouse, fixed 66 hr after intracerebral inoculation of Oropouche virus (Simbu group). The virus particles are in Golgi saccules or smooth membrane lined vesicles. (Marker = 200 nm) (Magnification × 80,000.)

with the envelope and a nucleocapsid protein of lower molecular weight. Purification of bunyaviruses is not easy, and some molecular weight estimates are being revised, so Table 2 incorporates only a recent selection of estimates.

It is not known whether one or both of the envelope glycoproteins are responsible for hemagglutinating activity. (Many, but not all bunyaviruses agglutinate erythrocytes of geese or 1-day-old chicks; the optimum pH varies with the virus but usually lies between pH 5.8 and 6.8.)

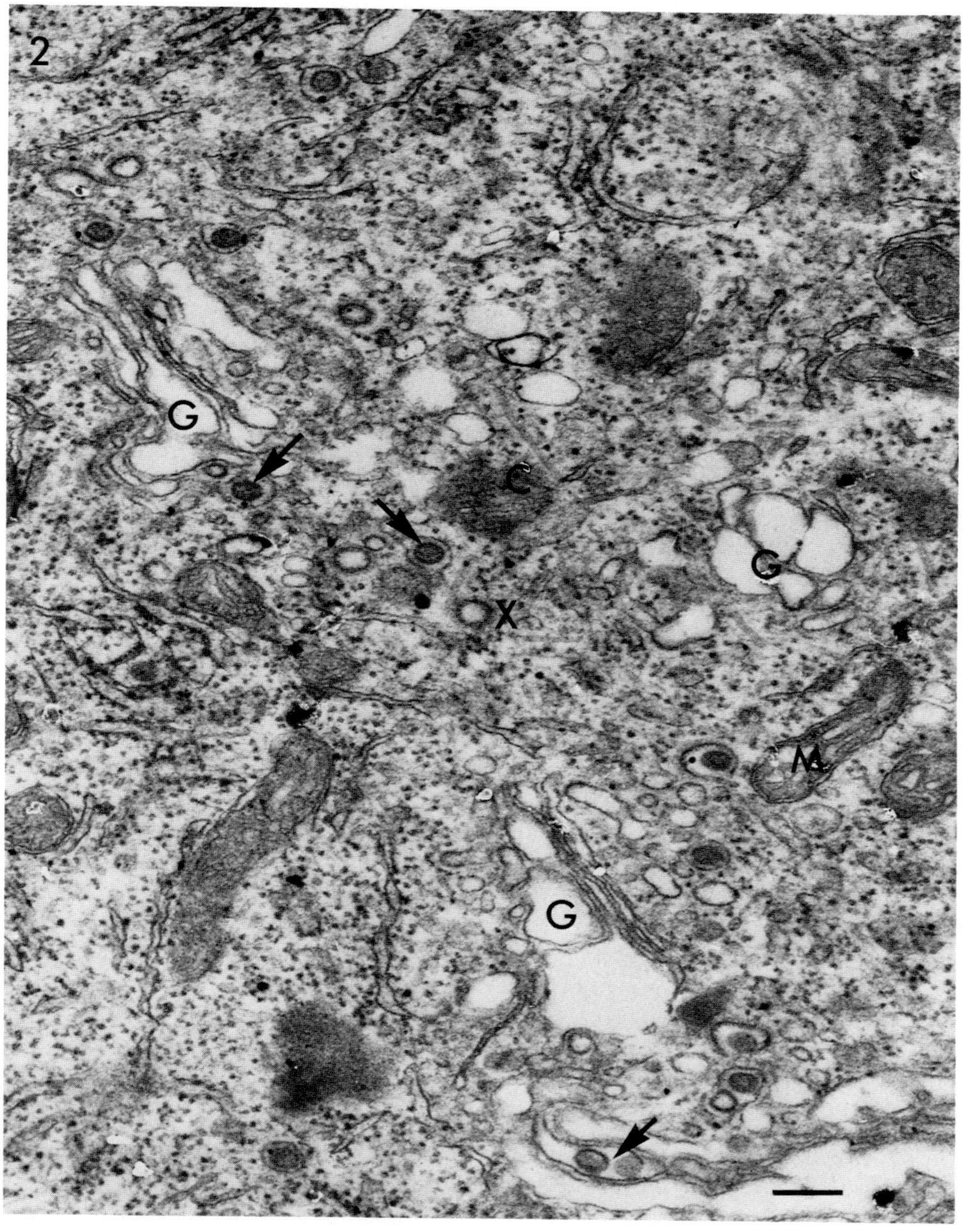

FIGURE 2. General view of the Golgi zone of a mouse neurone infected with Oropouche virus. Features include centriole (*C*), Golgi apparatus (*G*), mitochondria (*M*) and virus particles (some arrowed). Virus particles are frequently hard to find, and must be distinguished from coated vesicles (*X*) which are common in the Golgi zone even of uninfected cells. (Marker = 200 nm) (Magnification × 50,000.)

An RNA-dependent RNA-polymerase activity is associated with purified bunyavirions,[9,23] and it may be detectable as a minor, high molecular weight polypeptide.[22]

Virus replication

Bunyaviruses replicate in arthropod as well as in mammalian cells, since they are arboviruses and thus alternate between arthropods and animals. In addition transovarial transmission of certain bunyaviruses in mosquitoes or ticks has been demonstrated.

Although some reports have claimed that isolated bunyavirus RNA is infectious,

Table 1
MOLECULAR WEIGHTS OF GENOME RNA SEGMENTS OF BUNYAVIRUSES

	Molecular Weights of Segments[a]			
Virus	Small	Medium	Large	Ref.
Lumbo	3.0	2.0	0.5	18
La Crosse	2.9	1.8	0.4	11
Uukuniemi	4.1	1.9	0.8	19

[a] In millions of daltons.

Table 2
MOLECULAR WEIGHTS OF BUNYAVIRUS POLYPEPTIDES

Virus	Mol. Wt. ($\times 10^3$ daltons)			Ref.
Bunyamwera	gp 116	gp 37	p 27	8
Bunyamwera	gp 125	gp 30	p 22	21
La Crosse	gp 120	gp 34	p 23	22

both the segmented nature of the genome and the finding of RNA polymerase in the virions make this unlikely. Although it has not been formally proven that all the RNA segments are essential for infectivity, bunyaviruses are probably "negative-strand" RNA viruses and their replication is expected to proceed via the production of M-RNA complementary to the genome RNA.

Very little has been published on biochemical events occurring in bunyavirus-infected cells. There is a rapid shutdown of cellular RNA and protein synthesis in Bunyamwera virus-infected cells.[24] Synthesis of the viral structural proteins can be measured but no nonstructural virus proteins have been identified. In laboratory studies of Bunyamwera virus, infection of cells at multiplicities greater than about 1 pfu/cell results in a great reduction in yield of infectious virus, i.e., the shutdown appears to affect viral macromolecular syntheses as well as those of the host.[24] This unusual phenomenon also occurs with other members of the group, for example, La Crosse virus, in which case maximal yields are obtained by infecting cells at a multiplicity of 0.001 pfu/cell[22] and may be a general feature of the Bunyavirus genus.

On the other hand, UUK virus does not shutdown host cell macromolecular syntheses.[25] It has been reported that passage of Rift Valley fever virus at high multiplicity in vivo (in suckling mouse brain) results in the production of "incomplete" virus, or in current terminology, defective-interfering particles, and this also may prove to be a general property of bunyaviruses.[26-28]

Virus strains and antigenic properties

The genus *Bunyavirus* contains at least 84 viruses as well as the type species, *Bunyamwera* virus.[29] These are all antigenically related and can be subdivided serologically into 10 groups. In addition, the family Bunyaviridae contains another 56 "possible member" viruses which are likely to comprise at least two more genera, represented possibly by one of the Phlebotomus (sandfly) fever viruses and Uukuniemi virus. Each bunyavirus has a restricted geographic distribution and not all cause disease in man; those that cause disease in man are listed in Table 3. For a complete listing, the *International Catalogue of Arboviruses*[30] should be consulted.

Clinical features

Altogether 34 of the bunyaviruses have been reported to infect man, but the great majority cause "undifferentiated" fevers. Of the strains listed in Table 3, Bunyamwera and Oropouche viruses are typical of this class. Symptoms include headache, fever, sometimes backache and muscle pains, and photophobia. Onset is often sudden, and the duration of disease may range from 1 day to 5 to 7 days. Oropouche is notable in that it has caused three epidemics in Para, Brazil[31] whereas with Bunyamwera and most other bunyaviruses, only isolated cases have been reported.

Table 3
BUNYAVIRUSES CAUSING HUMAN DISEASE

Virus strain	Serological group	Geographical distribution	Vector	Clinical manifestations	Laboratory diagnosis: Source of specimen	Laboratory diagnosis: Isolation in animal[a]	Laboratory diagnosis: Serologic diagnosis[b]
Apeu	Group C	South America	Mosquito	Fever	Serum	SM	HI,NT,CF
Caraparu		South America	Mosquito	Fever	Serum	SM	HI,NT,CF
Itaqui		South America	Mosquito	Fever	Serum	SM	HI,NT,CF
Madrid		South America	Mosquito	Fever	Serum	SM	HI,NT,CF
Marituba		South America	Mosquito	Fever	Serum	SM	HI,NT,CF
Murutucu		South America	Mosquito	Fever	Serum	SM	HI,NT,CF
Oriboca		South America	Mosquito	Fever	Serum	SM	HI,NT,CF
Ossa		South America	Mosquito	Fever	Serum	SM	HI,NT,CF
Restan		South America	Mosquito	Fever	Serum	SM	HI,NT,CF
Bunyamwera	Bunyamwera group	Africa	Mosquito	Fever	Serum	SM	HI,NT,CF
Germiston		Africa	Mosquito	Fever	Serum	SM	HI,NT,CF
Guaroa		South America, Panama	Mosquito	Fever	Serum	SM	HI,NT,CF
Ilesha		Africa	Unknown	Fever	Serum	SM	HI,NT,CF
Wyeomyia		South America, Panama	Mosquito	Fever	Serum	SM	HI,NT,CF
Bwamba	Bwamba group	Africa	Mosquito	Fever	Serum	SM	HI,NT,CF
California	California group	U.S.A., Canada	Mosquito	Encephalitis	(brain)	SM	HI,NT,CF
LaCrosse		U.S.A., Canada	Mosquito	Encephalitis	(brain)	SM	HI,NT,CF
Tahyna		Europe	Mosquito	Fever	Serum	SM	HI,NT,CF
Catu	Guama group	South America	Mosquito	Fever	Serum	SM	HI,NT,CF
Guama		South America	Mosquito	Fever	Serum	SM	HI,NT,CF
Candiru	Sandfly fever group (Phlebotomus fever)	South America	Unknown	Fever	Serum	SM	HI,NT,CF
Chagres		Central America	Phlebotomine	Fever	Serum	SM	HI,NT,CF
Naples type		Europe, Africa, Asia	Phlebotomine	Fever	Serum	SM	HI,NT,CF
Punta Toro		Panama	Phlebotomine	Fever	Serum	SM	HI,NT,CF
Sicilian type		Europe, Africa, Asia	Phlebotomine	Fever	Serum	SM	HI,NT,CF

Table 3 (continued)
BUNYAVIRUSES CAUSING HUMAN DISEASE

Virus strain	Serological group	Geographical distribution	Vector	Clinical manifestations	Laboratory diagnosis: Source of specimen	Laboratory diagnosis: Isolation in animal[a]	Laboratory diagnosis: Serologic diagnosis[b]
Shuni	Simbu group	Africa	Mosquito	Fever	Serum	SM	HI,NT,CF
Oropouche		South America	Mosquito	Fever	Serum	SM	HI,NT,CF
Crimean hemorrhagic-Congo	Congo group	Europe, Africa	Tick	Hemorrhagic fever	Serum	SM	CF,FA
Dugbe	Ungrouped	Africa	Tick	Fever	Serum	SM	HI,NT,CF
Nairobi sheep disease		Africa, India	Tick	Fever	Serum	SM	NT,CF
Rift Valley		Africa	Mosquito	Hemorrhagic fever, encephalitis	Serum	SM	HI,NT,CF
Tataguine		Africa	Mosquito	Fever	Serum	SM	NT,CF
Thogoto		Africa, Europe	Tick	Meningitis	Serum, CSF	SM	HI,NT,CF

[a] SM 1 to 3 day old (suckling) mouse, intracerebral inoculation

[b] HI = Hemagglutination inhibition test, NT = neutralization test, CF = complement fixation test, FA = fluorescent antibody test

(From Burge, T. O., Ed., *International Catalogue of Arboviruses, Including Certain Other Viruses of Vertebrates.* With permission.)

California group virus infections may be subclinical, or can range in severity from mild fever, through influenza-like illnesses to severe encephalitis.[32] La Crosse virus appears to be the most pathogenic strain for humans.[33] In the more severe cases the common symptoms are fever, headache, nausea, neck rigidity, and usually convulsions and lethargy or coma.[32] Fever persists 7 to 10 days. Minor sequelae have been reported but major sequelae are uncommon.[32]

Crimean hemorrhagic fever is a serious disease transmitted by Hyalomma ticks and is found in various regions of the U.S.S.R. and adjacent countries. The sudden onset is signaled by chills, fever, headache, pains and malaise. Duration of the fever is 7 to 9 days, and hemorrhagic signs appear at about the 4th day in the form of a skin rash or purpura, or internal hemorrhages.[34]

Sandfly fever is a nonfatal but incapacitating disease, transmitted by the small biting fly *Phlebotomus papatasi.* It has long been known in countries in the Mediterranean area and Middle East but came under intensive study as a result of outbreaks in troops during the Second World War. Abdominal symptoms or malaise sometimes occur during the 3 to 4 day incubation period, then there is a sudden onset with headache, photophobia and retro-orbital pains, and pains in the back and joints. The main feature is fever which lasts about 3 days, but there is no rash (which distinguishes the disease from dengue fever) although there may be a pronounced flushing of the face and neck, resembling sunburn. Fatigue or weakness is common during convalescence.[34] Serologically related viruses capable of causing mild fevers in man are found in Panama and Brazil.[35]

Rift Valley fever is caused by a mosquito-borne virus which primarily affects sheep and cattle. It often produces a fatal hepatitis in lambs and calves, and also causes a severe but not fatal disease in humans. After a 4 to 6 day incubation period, a high fever develops with headache, muscular pains, vertigo, nausea and vomiting. The fever lasts 2 to 3 days, and a second peak may occur after an additional 1 to 2 days. Photophobia is common and ocular complications (retinitis) can occur later, but complete recovery is usual.[34] Encephalitis and fatal hemorrhagic disease with hepatitis are recorded.[36]

Laboratory diagnosis

Bunyaviruses cause viremia in man and can be isolated from the serum and/or blood usually for 3 days from onset of fever, and sometimes longer (Table 3). The encephalitis of California group viruses[32] occurs after the viremic period; in these cases the virus is presumably recoverable from brain, but since the patient rarely dies, isolation from brain is not done in practice.

The baby mouse is the host of choice for isolation of bunyaviruses.[37] Mice are inoculated intracerebrally and observed for sickness or death over a 2-week period. Alternatively cell cultures such as Vero and BHK-21 are usually susceptible to bunyaviruses. If Crimean hemorrhagic fever virus is suspected, the CER cell[38] should be used in addition to baby mice.

The complement-fixation (CF) test is uniformly applicable to bunyavirus diagnosis. Antigens are prepared by sucrose-acetone extraction of baby mouse brain and/or liver.[37] A fourfold or greater rise in titer between acute and convalescent sera is diagnostic. Not all patients develop CF antibody however, and in some patients the CF titer rises only late in illness.

Hemagglutinating antigens from mouse brain, liver, serum, or from tissue culture fluids are available from virtually all bunyaviruses.[37] Such antigens are technically difficult to produce, however, with some of the important human pathogens such as LaCrosse and Crimean hemorrhagic fever viruses. Where applicable, mouse serum antigens are superior to brain or liver because of increased sensitivity of the HI test. Mouse brain and liver antigens which have been extracted by the sucrose-acetone technique are improved in titer by sonication.[39] Bunyavirus antigens from mouse brain, liver, or serum should be used in the presence of 0.4 M or higher molarity NaCl for optimum titers.[40]

The neutralization test in mice is expensive but practical under field conditions.[37] The plaque reduction neutralization test, while not in general use for bunyavirus diagnosis because it is more expensive than the CF or HI test, and because it is less adaptable to conditions in field laboratories in developing nations, is nevertheless quite specific and reliable. BHK-21 and Vero cells are most widely used.[33] Other techniques such as radioimmunoassay, fluorescent antibody, and gel diffusion precipitin tests are not in general use, either because they require sophisticated technology or because they do not offer substantial advantage in cost, sensitivity, or specificity over CF, HI, or neutralization tests.

The choice of antigens for serologic diagnosis should be guided by geographic location of the patient (or likely site of exposure to arthropods) since bunyaviruses have distinct focal distributions (Table 3).

GENERAL REFERENCES

Murphy, F. A., Harrison, A. K., and Whitfield, S. G., Bunyaviridae: morphologic and morphogenetic similarities of Bunyamwera serologic supergroup viruses and several other arthropod-borne viruses, *Intervirology,* 1, 297, 1973.

Henderson, B. E. and Coleman, P. H., The growing importance of California arboviruses in the etiology of human disease, *Prog. Med. Virol.,* 13, 403, 1971.

Holmes, I. H., Arboviruses other than togaviruses, in *Ultrastructure of Animal Viruses and Bacteriophages: An Atlas,* Dalton, A. J. and Haguenau, F., Eds., Academic Press, New York, 1973, 197.

Berge, T. O., *International Catalogue of Arboviruses, Including Certain Other Viruses of Vertebrates,* Center for Disease Control, Atlanta, 1975.

Watts, D. M. and Eldridge, B. F., Transovarial transmission of arboviruses by mosquitoes: a review, *Med. Biol.,* 53, 271, 1975.

Porterfield, J. S., Casals, J., Chumakov, M. P., Gaidamovich, S. Ya., Hannoun, C., Holmes, I. H., Horzinek, M. C., Mussgay, M., Oker-Blom, N., and Russell, P. K., Bunyaviruses and Bunyaviridae, *Intervirology,* 6, 13, 1975/76.

REFERENCES

1. **Smithburn, K. C., Haddow, A. J., and Mahaffy, A. F.,** A neurotropic virus isolated from Aedes mosquitoes caught in the Semliki Forest, *Am. J. Trop. Med. Hyg.,* 26, 189, 1946.
2. **Thompson, W. H., Kalfayan, B., and Anslow, R. O.,** Isolation of California encephalitis group virus from a fatal human illness, *Am. J. Epidemiol.,* 81, 245, 1965.
3. **Chumakov, M. P.,** Studies of virus haemorrhagic fevers, *J. Hyg. Epidemiol. Microbiol. Immunol.,* 7, 125, 1963.
4. **Anderson, C. R., Spence, L., Downs, W. G., and Aitken, T. H. G.,** Oropouche virus: a new human disease agent from Trinidad, West Indies, *Am. J. Trop. Med. Hyg.,* 10, 574, 1961.
5. **Daubney, R., Hudson, J. R., and Garnham, P. C.,** Enzootic hepatitis or Rift Valley fever, an undescribed virus disease of sheep cattle and man from East Africa, *J. Pathol. Bacteriol.,* 34, 545, 1931.
6. **Sabin, A. B.,** Experimental studies on Phlebotomus (Pappataci, sandfly) fever during World War II, *Arch. Gesamte Virusforsch.,* 4, 367, 1951.
7. **von Bonsdorff, C.-H., Saikku, P., and Oker-Blom, N.,** The inner structure of Uukuniemi and two Bunyamwera supergroup arborviruses, *Virology,* 39, 342, 1969.
8. **Lazdins, I. and Holmes, I. H.,** Purification and the identification of polypeptides of Bunyamwera virus, (in preparation).

9. **Pettersson, R. F.**, The structure of Uukuniemi virus, a proposed member of the bunyaviruses, *Med. Biol.*, 53, 418, 1975.
10. **Pettersson, R. F. and von Bonsdorff, C.-H.**, Ribonucleoproteins of Uukuniemi virus are circular, *J. Virol.*, 15, 386, 1975.
11. **Obijeski, J. F., Bishop, D. H. L., Palmer, E. L., and Murphy, F. A.**, Segmented genome and nucleocapsid of La Crosse virus, *J. Virol.*, 20, 664, 1976.
12. **Samso, A., Bouloy, M., and Hannoun, C.**, Présence de ribonucléoproteines circulaires dans le virus Lumbo (Bunyavirus), *C. R. Acad. Sci. Ser. D*, 280, 779, 1975.
13. **Murphy, F. A., Harrison, A. K., and Tzianabos, T.**, Electron microscopic observations of mouse brain infected with Bunyamwera group arboviruses, *J. Virol.*, 2, 1315, 1968.
14. **Murphy, F. A., Whitfield, S. G., Coleman, P. H., Calisher, C. H., Rabin, E. R., Jenson, A. B., Melnick, J. L., Edwards, M. R., and Whitney, E.**, California group arboviruses: electron microscopic studies, *Exp. Mol. Pathol.*, 9, 44, 1968.
15. **Holmes, I. H.**, Morphological similarity of Bunyamwera supergroup viruses, *Virology*, 43, 708, 1971.
16. **Murphy, F. A., Harrison, A. K., and Whitfield, S. G.**, Bunyaviridae: morphologic and morphogenetic similarities of Bunyamwera serologic supergroup viruses and several other arthropod-borne viruses, *Intervirology*, 1, 297, 1973.
17. **Rosato, R. R., Dalrymple, J. M., Brandt, W. E., Cardiff, R. D., and Russell, P. K.**, Biophysical separation of major arbovirus serogroups, *Acta Virol.* (Praha), 18, 25, 1974.
18. **Bouloy, M., Krams-Ozden, S., Horodniceanu, F., and Hannoun, C.**, Three-segment RNA genome of Lumbo virus (Bunyavirus), *Intervirology*, 2, 173, 1973/74.
19. **Pettersson, R. and Kääriäinen, L.**, The ribonucleic acids of Uukuniemi virus, a non-cubical tick-borne arbovirus, *Virology*, 56, 608, 1973.
20. **Samso, A., Bouloy, M., and Hannoun, C.**, Mise en evidence de molecules d'acide ribonucleique circulaire dans le virus Lumbo (Bunyavirus), *C. R. Acad. Sci. Ser. D*, 282, 1653, 1976.
21. **Vorndam, A. V. and Trent, D. W.**, Structural proteins of the Bunyaviridae: I. The California encephalitis group, *J. Virol.*, 1977.
22. **Obijeski, J. F., Bishop, D. H. L., Murphy, F. A., and Palmer, E. L.**, Structural proteins of La Crosse virus, *J. Virol.*, 19, 985, 1976.
23. **Bouloy, M., Colbere, F., Krams-Ozden, S., Vialat, P., Garapin, A.-C., and Hannoun, C.**, Activite RNA polymeraseque associe a un bunyavirus (Lumbo), *C. R. Acad. Sci. Ser. D*, 280, 213, 1975.
24. **Lazdins, I. and Holmes, I. H.**, Protein synthesis in Bunyamwera virus-infected cells, (in preparation).
25. **Pettersson, R. F.**, Effect of Uukuniemi virus infection on host cell macromolecular synthesis, *Med. Biol.*, 52, 90, 1975.
26. **Mims, C. A.**, Rift Valley fever virus in mice *IV*. Incomplete virus: its production and properties, *Br. J. Exp. Pathol.*, 37, 129, 1956.
27. **Pettersson, R. F.**, Uukuniemi virus, a non-cubical tick-borne arbovirus, Ph.D. thesis, University of Helsinki, Finland, 1974.
28. **David-West, T. S. and Porterfield, J. S.**, Dugbe virus: a tick-borne arbovirus from Nigeria, *J. Gen. Virol.*, 23, 297, 1974.
29. **Fenner, F.**, Classification and nomenclature of viruses, *Intervirology*, 7, 37, 1976.
30. **Burge, T. O., Ed.**, *International Catalogue of Arboviruses, Including Certain Other Viruses of Vertebrates*, Center for Disease Control, Atlanta, 1975.
31. **Pinheiro, F., Pinheiro, M., Bensabath, G., Causey, O. R., and Shope, R. E.**, Epidemia de virus Oropouche em Belem, *Rev. Saude Publica*, 12, 15, 1962.
32. **Henderson, B. E. and Coleman, P. H.**, The growing importance of California arboviruses in the etiology of human disease, *Prog. Med. Virol.*, 13, 404, 1971.
33. **Lindsey, H. S., Calisher, C. H., and Mathews, J. H.**, Serum dilution neutralization test for California group virus identification and serology, *J. Clin. Microbiol.*, 4, 503, 1976.
34. **Casals, J. and Clarke, D. H.**, Arboviruses other than Groups A and B, in *Viral and Rickettsial Infections of Man*, Horsfall, F. L. and Tamm, I., Eds., Lippincott, Philadelphia, 1965, 659.
35. **McLean, D. M.**, Phlebotomus (Sandfly) fever, in *Textbook of Virology*, Rhodes, A. J. and Van Rooyen, C. E., Eds., Williams and Wilkins Co., Baltimore, 1968, 762.
36. **Van Velden, D. J. J., Meyer, J. D., Olivier, J., Gear, J. H. S., and McIntosh, B.**, Rift Valley fever affecting humans in South Africa. A clinicopathological study, *S. Af. Med. J.*, 51, 867, 1977.
37. **Hammon, W. McD. and Sather, G. E.**, Arboviruses, in *Diagnostic Procedures for Viral and Rickettsial Infections*, 4th ed., Lennette, E. H. and Schmidt, N. J., Eds., American Public Health Association, New York, 1969, 227.
38. **Smith, A. L., Tignor, G. H., Mifune, K., and Motohashi, T.**, Isolation and assay of rabies serogroup viruses in CER cells, *Intervirology*, 8, 92, 1977.
39. **Ardoin, P., Clarke, D. H., and Hannoun, C.**, The preparation of arbovirus hemagglutinins by sonication and trypsin treatment, *Am. J. Trop. Med. Hyg.*, 18, 592, 1969.
40. **Beaty, B. J., Shope, R. E., and Clarke, D. H.**, Salt-dependent hemagglutination with Bunyaviridae antigens, *J. Clin. Microbiol.*, 5, 548, 1977.

RHABDOVIRIDAE: RABIES AND RABIES-RELATED VIRUSES

Robert E. Shope

Introduction

Rabies virus causes an acute central nervous system (CNS) disease of man and animals characterized by fever, convulsions, salivation, excitation, anxiety, difficulty in swallowing, and death. Reservoir animals such as dogs, cats, bats, and wild carnivores transmit rabies virus to man. The disease occurs in most parts of the world and has been recognized in dogs and man for at least 4000 years.[1] The illness has no cure, but can be prevented by vaccination and immune serum treatment given immediately after exposure. In rare cases, man has recovered from or survived clinical central nervous system disease.[2,3] Naturally occurring virus is called "street" rabies virus and has a relatively long incubation period in laboratory animals. After laboratory passage, the incubation period shortens and becomes constant. Such laboratory strains are called "fixed."

There are five African viruses serologically related to rabies.[4] Two of these, Duvenhage[5] and Mokola,[6] cause fatal human CNS disease, and Lagos bat virus causes a rabies-like illness in dogs and monkeys inoculated in the laboratory.[7]

Morphology and Morphogenesis

Murphy[9] has reviewed the morphology of rabies virus, and his description is used freely in the account. Rabies virus has rhabdovirus morphology with cylindrical particles which measure 180 × 75 nm excluding surface projections.[8]

When negatively stained, the virus particle has surface projections measuring 6 to 8 nm (Figure 1). These spikes represent the glycoprotein of the virion and are responsible functionally for stimulation of neutralizing antibody and hemagglutination[10] and probably for attachment to cells. The membrane envelope of the virion is a lipid bilayer which surrounds the helical ribonucleoprotein capsid (Figure 2). The helix may be seen as cross-striations with an interval of 4.5 nm. The planar end of the virion characteristically exhibits a bleb of extra membrane (Figure 1), and a hollow axial channel is often visible within the virus particle.

In thin sections of infected animal brains the virus particles are seen budding primarily from plasma membranes during infection with fixed strains. Budding from the endoplasmic reticulum is more common with street strains and often occurs in association with massed nucleocapsid material, which forms a matrix (Figure 3). It is this matrix or inclusion which is recognized in light microscopy as the Negri body.

On rare occasions, unusual forms of the particles are seen as pairs or triplets attached to each other by the planar ends. Defective particles are also recognized in fixed-virus infections initiated by high-multiplicity inoculum.[11] The defective particles are truncated and are therefore called "T" particles. T particles are identical in morphology to fully infective B particles, except for their shorter lengths of 70 to 100 nm.

Three other members of the rabies serogroup, Mokola, Lagos bat, and Duvenhage, are not morphologically distinguishable from rabies virus. Rabies serorelatives (kotonkan and Obodhiang viruses, isolated from *Culicoides* midges and mosquitoes, respectively) differ in morphology from rabies virus in that a variable proportion of their particles are cone shaped rather than bullet shaped.[12] The significance of the cone shape is not known.

Physicochemical Properties

Rabies virus contains single-stranded RNA with a total molecular weight of approximately 4.6×10^6 daltons.[13] The RNA is noninfectious,[14] nonsegmented, and sediments

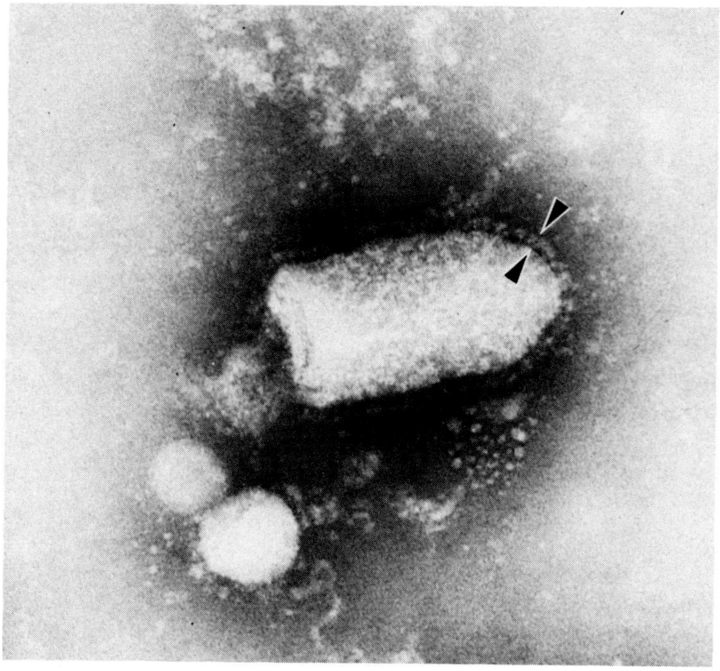

FIGURE 1. Negatively stained rabies virus particle illustrating symmetrically spaced surface projections (arrows) and the bleb of extra membrane characteristically seen at the planar end of the particle. Magnification × 220,000. (Micrograph courtesy of F. A. Murphy.)

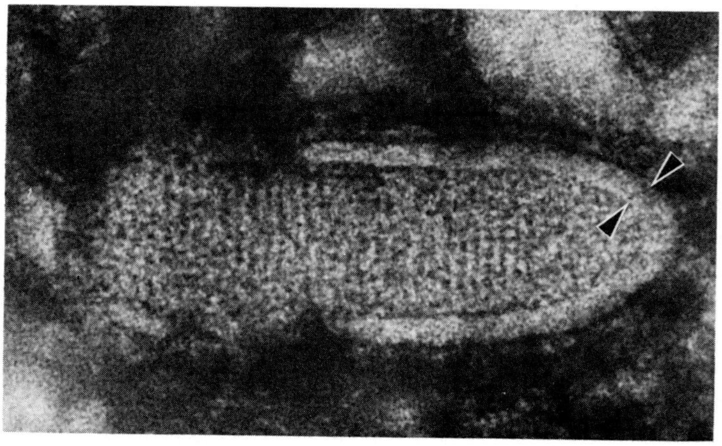

FIGURE 2. Negatively stained rabies virus particles (challenge virus standard {CVS} strain) demonstrating the lipid bilayer membrane envelope (arrows) which surrounds the helical nucleocapsid. Magnification × 400,000. (From Murphy, F. A., in *The Natural History of Rabies*, Vol. 1, Baer, G. M., Ed., Academic Press, New York, 1975, 33. With permission.)

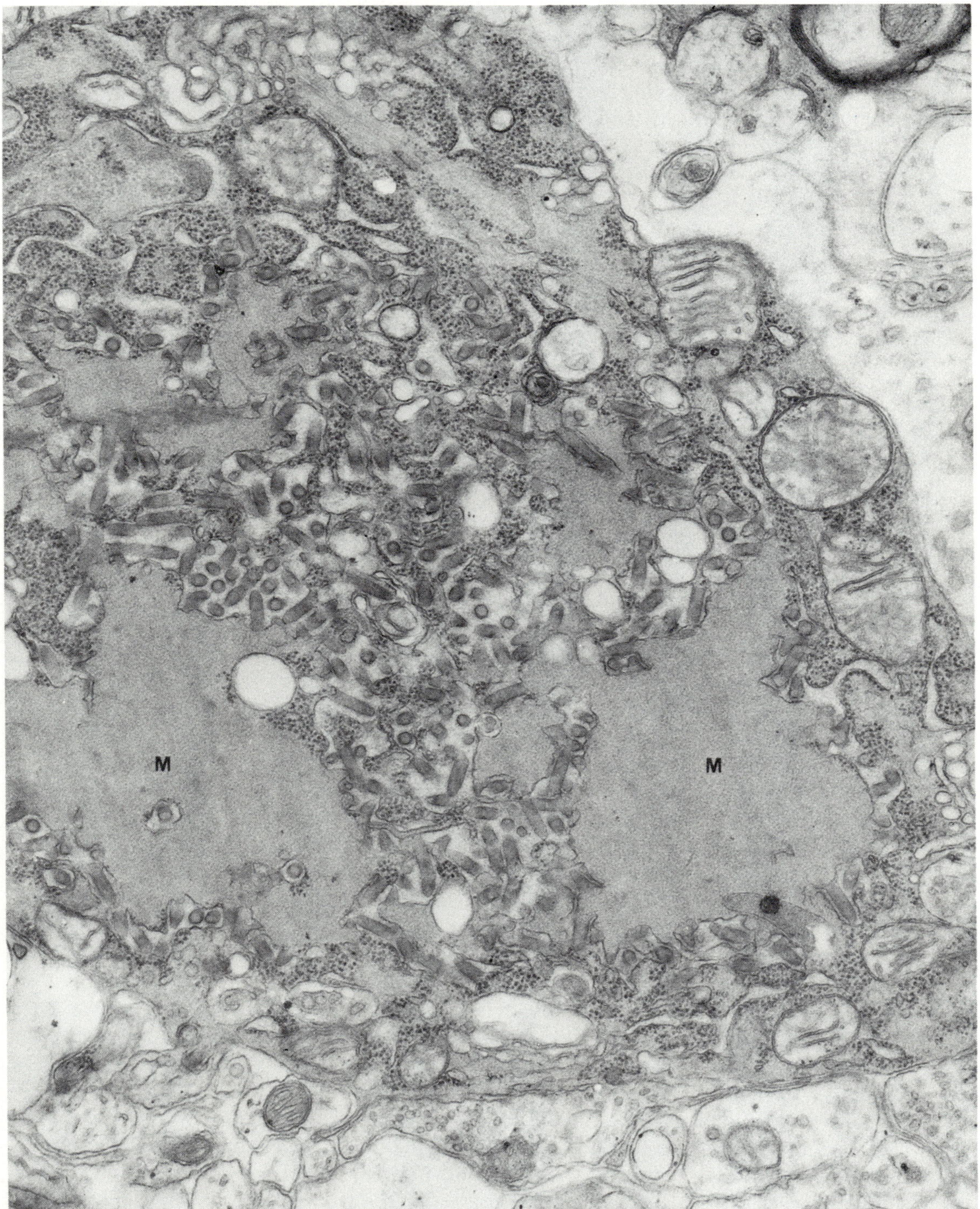

FIGURE 3. Rabies virus (human isolate, first passage) infection of mouse neuron. Virus particles are budding from intracytoplasmic membrane in association with matrix (M). Magnification × 35,000.(From Murphy, F. A., in *The Natural History of Rabies,* Vol 1, Baer, G. M., Ed., Academic Press, New York, 1975, 33. With permission.)

at 45S.[13] The virion RNA is not a messenger species and is thus minus stranded. Viral structural proteins consist of a glycopeptide of 80×10^3 daltons, called the G protein; a phosphorylated nucleoprotein of 62×10^3 daltons, called the N protein; a large peptide of 160×10^3 daltons called the L protein; and two membrane proteins of 40 and 25×10^3 daltons, called M_1 and M_2.[15,20] Virion densities are 1.20 and 1.22 in CsCl,[16] and the virion density is 1.17 g/ml in sucrose.[17] Infectivity is destroyed by lipid solvents and detergents, indicating the presence of lipids in the virion envelope.[18]

Viral Replication

The replication of rabies virus has not been studied in depth because it has not been possible, even with laboratory-adapted strains, to obtain replication to high titer in in vitro systems, and an ideal single-cycle growth curve has not been obtained.[21]

The replication of vesicular stomatitis virus[19,20] may be similar to that of rabies, although there is no assurance that this is true. Following adsorption and uncoating, the vesicular stomatitis minus-stranded helical genome undergoes primary transcription in the cytoplasm, with production of RNA plus strands sedimenting at 28 and 13 to 15S. Replication then proceeds with production of 40S plus strands and, secondarily, 40S negative strands which are either transcribed with the aid of a transcriptase or coated in the budding particles to form progeny virions.

A rabies transcriptase has been demonstrated only in very small amounts.[20] Transcriptase is readily demonstrated for vesicular stomatitis virus; however, rabies virus either requires very small quantities of transcriptase or the enzyme is labile and not measurable by current laboratory techniques.

Virus Strains and Antigenic Properties

Six members of the rabies serogroup are currently recognized.[4,5] Rabies virus is the only member recognized outside of Africa. The insect isolates Obodhiang and kotonkan are related very distantly to the other members of the group and do not cause disease in dogs or primates inoculated experimentally. Lagos bat, Mokola, and Duvenhage viruses are more closely related to rabies. In one instance rabies fluorescent conjugate did not diagnose Duvenhage infection,[22] although sufficient immunofluorescent cross-reactivity may generally be expected among rabies and these three closely related viruses so that a group diagnosis will be made. It is then necessary, especially in Africa, to perform neutralization tests or quantitative immunofluorescence studies to distinguish the specific infecting virus. In practice, this is not done outside of Africa, and it may not be necessary in other parts of the world, where rabies virus is the only known member of the serogroup.

The major reactive antigenic components associated with rabies infections are the G and N proteins. The G protein is localized primarily on the virion surface and plasma membranes, and the N protein is found in association with the virion nucleocapsid and in the cytoplasmic inclusion bodies.

The G protein is responsible for the formation of neutralizing antibody and is the reactive component in the hemagglutination inhibition test. The N protein is the primary reactant in the complement fixation and fluorescent antibody tests, although both antigens are capable of reacting in these tests. The role of the L and M proteins in serologic reactions is not definitively known. The neutralization test is generally more specific than immunofluorescence and complement fixation, and this difference has been proposed as the basis for a system of typing and grouping in the rabies serogroup.[23]

Clinical Features

Rabies is a neurotropic virus, and the clinical features reflect its pathogenesis. There is usually a history of exposure to bite of a dog, cat, or one of several types of wild animal including carnivores and bats. The incubation period ranges between 20 and 60 days; rarely, it is shorter or longer.[24] Postexposure treatment leads to a shorter incubation period, suggesting that either treatment is more effective in preventing long incubation-period disease[24] or induction of the immune response produces early death by an immunopathologic host reaction.

Hattwick and Gregg,[25] in an excellent and detailed account of clinical rabies, divide the disease into prodromal, acute neurological, coma, and death (or recovery) stages. During the prodrome, the site of the bite may also be the site of the initial symptoms of paresthesia, pain, or tingling. This is accompanied or followed shortly by fever, headache, malaise, and nausea. An acute neurological stage follows in 2 to 10 days. The encephalitis may be manifested by convulsions, paralyses, tics, anxiety, difficulty in swallowing, excess lacrimation and salivation, generalized excitation, hyperesthesias, and thrashing. The patient is often lucid, cooperative, and very much aware of his surroundings. There may or may not be an increase in cells in the spinal fluid during this phase. Coma ensues, and the patient dies of respiratory or cardiac arrest unless life-support systems are used. The total clinical course usually lasts 4 to 6 days, although it is emphasized that life support may prolong life[26] and may allow recovery.[2]

There have been two recognized cases of Mokola virus infection: a 3½-year-old Nigerian girl who developed fever, convulsions, and pharyngitis, but recovered without sequellae[27] and a 6-year-old Nigerian girl who had fever, confusion, and flaccid paralysis of limbs and became comatose and died.[6] Another rabies-related virus, Duvenhage, was isolated from a single case in South Africa.[22] A 31-year-old man was bitten on the lip by a bat and died after an illness characterized by fever, headache, muscle spasms, confusion, irritability, aggressiveness, hydrophobia, and convulsions. The differential diagnosis of rabies and other rabies serogroup virus infections includes lymphocytic choriomeningitis; arbovirus encephalitis; poliomyelitis; herpes encephalitis; tetanus; and rabies hysteria, in which the patient feigns epileptiform seizures. It may be very difficult to distinguish between rabies and a vaccine reaction in patients who have received rabies vaccine following exposure.[28]

Laboratory Diagnosis

Table 1 lists the preferred means of arriving at an etiologic diagnosis of rabies and its serorelatives. Diagnostic methods are described in detail by Johnson,[29] Kaplan and Koprowski,[30] Atanasiu,[31] Kissling,[32] and Thomas,[33] and these sources should be consulted. It is important to diagnose rabies early in order to eliminate from the diagnosis another illness for which specific chemotherapy is available. Moreover, although rabies is usually fatal, vigorous supportive treatment may result in survival of the patient without sequellae. However, it is questionable if life process support is warranted in patients with obviously severe neurologic deficit. This becomes an ethical issue to be decided on an individual basis. Antibody is rarely detectable before the eighth day of illness; therefore, the best chance of making an early ante-mortem diagnosis is by examination of biopsy material. Corneal impression smears, skin, and gingival biopsy material have been used successfully in a few instances and should be tried in all cases.[34] The material is fixed in acetone and stained by either the direct or indirect immunofluorescence method. After death, the diagnosis of rabies is best made by examination of brain and salivary gland tissue. The tissues most likely to yield positive results are the submaxillary glands, Ammon's horn, the cerebellum, the cerebral cortex, and the medulla. Figure 4 shows the isolation and identification procedure.

Histology

Inclusion bodies (Negri bodies) specific for rabies or rabies serogroup viruses are found in the cytoplasm of neurons in sectioned material. In the impression smear technique, the brain is cut and smears are made on clean microscope slides. While the

Table 1
RABIES SEROGROUP VIRUSES PATHOGENIC FOR MAN AND PREFERRED METHODS FOR THEIR DIAGNOSIS

			Laboratory diagnosis						
			Virus or antigen isolation[a]			Serologic techniques[b]			
Virus strain	Human disease	Diagnostic specimens	Cell culture	Animal	Histopathology	FFT	IFA	Mouse NT	Refs.
Rabies	Rabies	Brain, salivary gland, skin, mucosa, corneal cells	CER, neuro-blastoma (mouse)	Mouse	Negri body	1°	2°	3°	29—33, 38
Duvenhage (Africa only)	Rabies-like	Brain	Vero, CER	Mouse	Negri body	1°	2°	3°	5, 37
Mokola (Africa only)	Poliomyelitis, convulsions	Brain, cerebrospinal fluid	Vero, CER	Mouse	Negri body	1°	2°	3°	6, 37

[a] In Africa, rabies positive immunofluorescence or Negri body results do not exclude the possibility of Mokola or Duvenhage virus infection. Quantitative immunofluorescence or the neutralization test will identify the virus strain.

[b] FFT, fluorescent focus technique; IFA, indirect fluorescent antibody test; NT, neutralization test; 1°, primary technique (most valuable overall); 2°, secondary or alternate technique (less valuable); 3°, tertiary technique (least valuable).

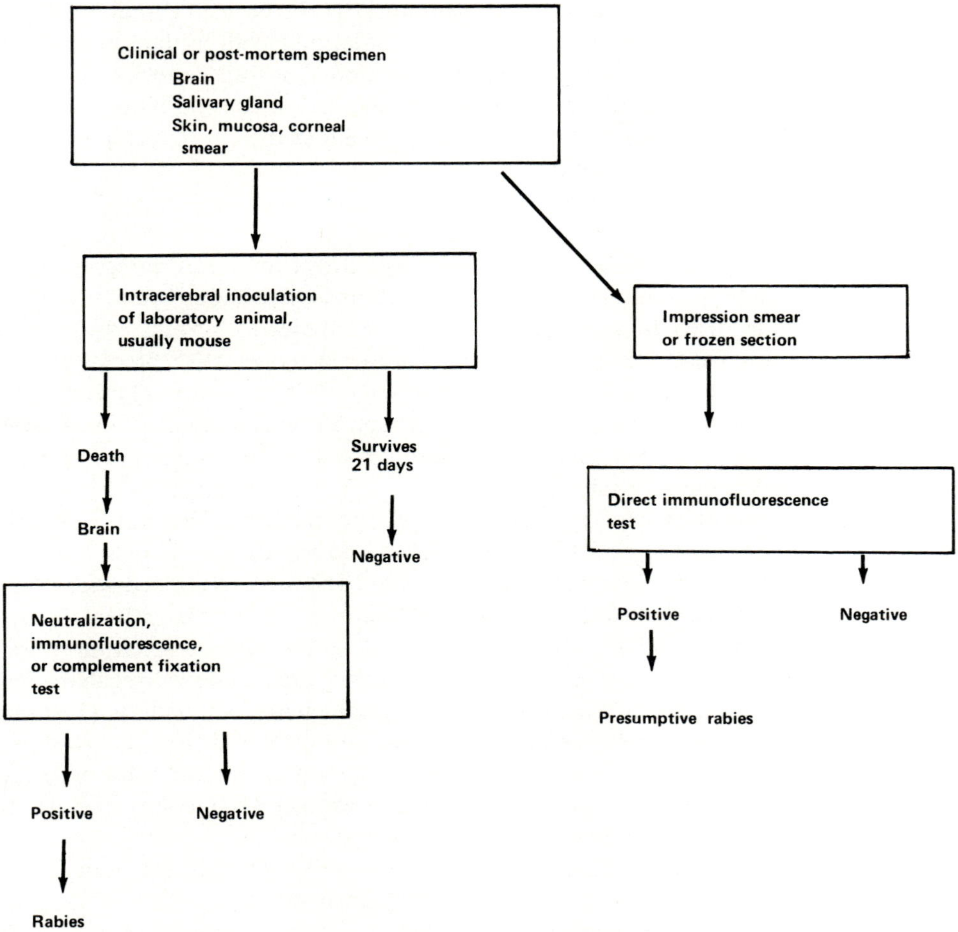

FIGURE 4. Flow chart for rabies virus and antigen isolation and identification. Note that a positive immunofluoresence test or Negri body should be confirmed by animal inoculation.

impressions are still moist, the slides are stained by Sellers' technique.[35] The stain consists of two parts of methylene blue and one part basic fuchsin, and the slides are immersed for 1 to 5 sec, rinsed in water, and air dried. The Negri body appears as a spherical or oval inclusion stained cherry red with inner blue granules and ranging from 0.5 to ≥ 25 μm in diameter.[36]

Virus Isolation

Virus isolation is routinely done in the 3- to 4-week-old Swiss mouse, although the infant mouse is slightly more sensitive. Specimens are triturated in phosphate-buffered diluent, pH 7.2 to 8.0, containing 2% serum or 0.75% bovine albumin, and the supernatant fluid is inoculated intracerebrally. The incubation period in mice is from 5 days to 3 weeks. Prior to death, mice become paralyzed, convulsive, or agitated and often demonstrate a serous conjunctivitis. Mouse brain should be tested by immunofluorescence, although to confirm the death as specific for rabies, complement fixation or the neutralization test may be used.

Most tissue cultures are less efficient than the mouse for primary isolation of street rabies; however, limited experience with the CER cell[37] and the N18 clone of mouse neuroblastoma cell[38] indicates sensitivity comparable to the weanling mouse. With most street viruses tested, a positive diagnosis was made within 48 hr by detecting specific rabies antigen in infected cells using immunofluorescence. It should be emphasized, however, that use of cell culture for rabies isolation is still in the research stage and requires comprehensive testing before its adoption as a routine technique. When processing specimens for animal or tissue culture inoculation, several tenfold dilutions of original material should be inoculated in order to avoid the autointerference sometimes seen with concentrated inoculum.

Immunofluorescence

The diagnostic test most widely used today is the direct immunofluorescence technique. Basically, a hyperimmune serum is prepared using infectious brain from a homologous source in order to avoid formation of antibodies to normal tissues of the animal being immunized. The resulting IgG is purified by selective precipitation or column chromatography and conjugated to fluorescein isothiocyanate. The conjugated IgG is used to locate rabies antigen in the cytoplasm of neurons by conventional immunofluorescence microscopy. Fluorescein isothiocyanate-conjugated antirabies IgG is readily obtainable through commercial sources.

Thin tissue impression smears or frozen sections from the brain (Ammon's horn and the cerebellum offer optimal sites for antigen detection) or salivary gland are air dried on slides and fixed in cold acetone for from 1 to 4 hr. The tissue is stained with diluted conjugate and incubated for 30 min at room temperature. The slide is then rinsed in buffered water, mounted in 90% glycerine at pH 8.6, and examined for fluorescence.

Each test is controlled by including a known positive and a known negative specimen. In addition, two identical smears of the test specimen are included; one is stained with conjugate mixed with 20% rabies-infected mouse brain, and the other is stained with conjugate mixed with normal mouse brain. The smear stained with conjugate mixed with rabies-infected brain must show no fluorescence (extinction) in order for positive fluorescence in the other smear to be considered specific.

The immunofluorescence technique correlates very well with results of mouse inoculation. In one study of 10,413 specimens with 321 positive cases, the correlation was 98%.[39] Unless the laboratory is experienced, however, it is recommended that confirmation of immunofluorescence results by mouse inoculation be obtained.

Measurement of Antibody

Because most patients and other animal hosts die as a result of rabies infection, the antibody tests are not usually employed for retrospective diagnosis of rabies as they are for other viral diseases. However, tests for antibody are useful late in the clinical course to confirm the diagnosis of rabies, in preexposure-vaccinated individuals to determine the immune status, and in postexposure-treated individuals to evaluate immune response to vaccine and titer of passively administered immune globulin.

The mouse neutralization test is carried out with the challenge virus standard (CVS) strain of rabies virus in infant mice inoculated intracerebrally. The inoculum consists of tenfold varying dilutions of virus mixed with undiluted test sera. The serum-virus mixture is incubated for 1 hr at 37°C. Virus dilutions mixed with a normal serum and incubated similarly and inoculated intracerebrally are used to determine the virus titer based on mouse mortality and calculated by the Reed and Muench[40] method. Results are expressed as neutralization indices. Variations of this test include use of different rabies strains, weanling mice instead of baby mice, serum dilutions with constant virus, and longer incubation periods for serum-virus mixtures.

The indirect immunofluorescence test[33] is much more rapid than the mouse neutralization test and gives comparable results. In this test, a conjugated antispecies serum is used as an indicator of the antigen-antibody reaction. The antigen consists of slide impressions of rabies-infected mouse brain, which is reacted with the test serum, washed, then reacted with the tagged antispecies globulin. Fluorescence indicates a positive test. As a control, the test serum is mixed with 20% rabies-infected mouse brain prior to reaction with the slide impression antigen, thus blocking the antigen-antibody reaction.

The fluorescent focus inhibition test has now become standard procedure in many laboratories. This neutralization test utilizes CVS rabies virus grown in BHK-21 13S cells pretreated with diethylaminoethyl dextran and plated in Lab-Tek® tissue culture slides containing the serum-virus mixture.[41,42] The infected cells are fixed in acetone 24 hr after infection and stained with a fluorescein isothiocyanate conjugate of antirabies IgG. Control chambers containing cells challenged with virus alone (approximately 16,000 PFU/0.1 ml) have fluorescent foci in at least 90% of 160 fields examined. Test sera are considered to have antibody if a 1:5 serum dilution reduces the number of positive fields by 50% or more. The fluorescent focus inhibition test shows a 95% correlation in results when compared to the mouse neutralization test with 512 human sera.[42]

Other serological techniques for determining rabies antibody include complement fixation,[43] lytic rabies antibody test,[44] hemagglutination inhibition,[45] agar gel diffusion precipitation,[46] and passive hemagglutination.[47] All have research application in the study of rabies, but none is currently used for routine diagnosis of rabies.

Pathogenicity in Experimental Animals

Rabies virus is pathogenic for mice, rats, guinea pigs, rabbits, hamsters, dogs, cats, monkeys, and cows, as well as almost all other mammalian and avian species tested in the laboratory. The virus multiplies first in the muscle[48] and proceeds via the axoplasm to the ganglia and CNS. After rapid spread through the CNS, antigen appears peripherally in nerve endings in the salivary glands, skin, mucosal surfaces, pancreas, gut, cornea, and most other organs. This pattern is typical of experimental animals studied. Atanasiu[31] has reviewed the clinical patterns in rabbits, mice, rats, guinea pigs, and hamsters. Generally, the incubation period of fixed rabies is approximately 6 to 7 days and that of street rabies is 10 days to 3 weeks. The suckling animal has a shorter incubation period than the adult, and some strains such as the high-egg passage Flury kill suckling mice by the intracerebral route, but do not kill adult mice.[49]

Mokola and Lagos bat viruses were studied in monkeys and dogs,[7] and the pathogenicity was similar to that of rabies. In mice, Mokola, Lagos bat, and Duvenhage viruses were also neurotropic and were similar to rabies virus. Obodhiang and kotonkan viruses killed baby mice by the intracerebral route of inoculation, but were not pathogenic to adult mice or to dogs or monkeys.[50]

Rabies in man and, under some conditions, in experimental animals can have a prolonged incubation period and can cause disease from which the animal recovers, after which time, under some conditions, the virus can become reactivated.[51] The mechanism of the long incubation period and reactivation is not known, although defective interfering particles may play a role.[52] In addition, animals can recover completely following disease, can become infected without disease, and can become chronically infected with or without disease.[2,51,53]In nature, the chronically infected animal may shed virus in the saliva and pose a potential public health hazard. However, the persistent infection patterns are clearly unusual and should not be emphasized to such an extent as to overshadow the truly great success in rabies control achieved through effective vaccination of dogs, cats, and other domestic animals; leash laws; and control of wildlife populations.

GENERAL REFERENCES

Baer, G. M., Ed., *The Natural History of Rabies,* Vols. 1 and 2, Academic Press, New York, 1975.

Bishop, D. H. L. and Smith, M. S., Rhabdoviruses, in *The Molecular Biology of Animal Viruses,* Vol 1, Nayak, D., Ed., Marcel Dekker, New York, 1977, 169.

Johnson, H. N., Rabiesvirus, in *Manual of Clinical Microbiology,* 2nd ed., Lennette, E. H., Spaulding, E. H., and Truant, J. P., Eds., American Society for Microbiology, Washington, D. C., 1974, 746.

Kaplan, M. and Koprowski, H., Eds., *Laboratory Techniques in Rabies,* 3rd ed., World Health Organization, Geneva, 1973.

REFERENCES

1. **Tierkel, E. S.,** Historical review of rabies in Asia, in*Rabies,* Nagano, Y. and Davenport, F. M., Eds., University of Tokyo Press, Tokyo, 1971, 309.
2. **Hattwick, M. A. W., Weis, T. T., Stechschulte, J., Baer, G. M., and Gregg, M. B.,** Recovery from rabies: a case report, *Ann. Intern. Med.,* 76, 931, 1972.
3. **Tillotson, J. R., Axelrod, D., Lyman, D. O.,** Follow-up on rabies — New York, *Morbid. Mortal. Weekl. Rep.,* 26, 249, 1977.
4. **Shope, R. E.,** Rabies virus antigenic relationships, in *The Natural History of Rabies,* Vol 1, Baer, G. M., Ed., Academic Press, New York, 1975, 141.
5. **Tignor, G. H., Murphy, F. A., Clark, H. F., Shope, R. E., Madore, P., Bauer, S. P., Buckley, S. M., and Meredith, C. D.,** Duvenhage virus: morphologic, biochemical, histopathologic and antigenic relationships to the rabies serogroup, *J. Gen. Virol.,* 37, 595, 1977.
6. **Familusi, J. B., Osunkoya, B. O., Moore, D. L., Kemp, G. E., and Fabiyi, A.,** A fatal human infection with Mokola virus, *Am. J. Trop. Med. Hyg.,* 21, 959, 1972.
7. **Tignor, G. H., Shope, R. E., Bhatt, P. N., and Percy, D. H.,** Experimental infection of dogs and monkeys with two rabies serogroup viruses, Lagos bat and Mokola (IbAn 27377): clinical, serologic, virologic and fluorescent-antibody studies, *J. Infect. Dis.,* 128, 471, 1973.
8. **Hummeler, K., Koprowski, H., and Wiktor, T. J.,** Structure and development of rabies in tissue culture, *J. Virol.,* 1, 152, 1967.
9. **Murphy, F. A.,** Morphology and morphogenesis, in *The Natural History of Rabies,* Vol. 1, Baer, G. M., Ed., Academic Press, New York, 1975, 33.
10. **Schneider, L. G., Horzinek, M., and Novický, R.,** Isolation of a hemagglutinating, immunizing, and non-infectious subunit of the rabies virion, *Arch. Gesamte Virusforsch.,* 34, 360, 1971.
11. **Crick, J. and Brown, F.,** An interfering component of rabies virus which contains RNA, *J. Gen. Virol.,* 22, 147, 1974.
12. **Kemp, G. E., Lee, V. H., Moore, D. L., Shope, R. E., Causey, O. R., and Murphy, F. M.,** Kotonkan, a new rhabdovirus related to Mokola virus of the rabies serogroup, *Am. J. Epidemiol.,* 98, 43, 1973.
13. **Sokol, F., Schlumberger, H. D., Wiltor, T. J., Koprowski, H., and Hummeler, K.,** Biochemical and biophysical studies on the nucleocapsid and on the RNA of rabies virus, *Virology,* 38, 651, 1969.
14. **Sedwick, W. D. and Wiktor, T. J.,** Reproducible plaquing system for rabies, lymphocytic choriomeningitis, and other ribonucleic acid viruses in BHK-21/13S agarose suspensions, *J. Virol.,* 1, 1224, 1967.
15. **Sokol, F., Stancěk, D., and Koprowski, H.,** Structural proteins of rabies virus, *J. Virol.,* 7, 241, 1971.
16. **Schneider, L. G., Horzinek, M., and Matheka, H. D.,** Purification of rabies virus from tissue culture, *Arch. Gesamte Virusforsch.,* 34, 351, 1971.
17. **Sokol, F., Kuwert, E., Wiktor, T. J., Hummeler, K., and Koprowski, H.,** Purifification of rabies virus grown in tissue culture, *J. Virol.,* 2, 836, 1968.
18. **Kissling, R. E. and Reese, D. R.,** Anti-rabies vaccine of tissue culture origin, *J. Immunol.,* 91, 362, 1963.
19. **Wagner, R. R.,** Reproduction of rhabdorviruses, in *Comprehensive Virology,* Vol. 4, Fraenkel-Conrat, H. and Wagner, R. R., Eds., Plenum Press, New York, 1975, 1.
20. **Bishop, D. H. L. and Smith, M. S.,** Rhabdoviruses, in *The Molecular Biology of Animal Viruses,* Vol. 1, Nayak, D., Ed., Marcel Dekker, New York, 1977, 169.
21. **Wiktor, T. J. and Clark, H. F.,** Growth of rabies virus in cell culture, in *The Natural History of Rabies,* Vol. 1, Baer, G. M., Ed., Academic Press, New York, 1975, 155.

22. **Meredith, C. D., Rossouw, A. P., and Van Praag Koch, H.,** An unusual case of human rabies thought to be of Chiropteran origin, *S. Afr. Med. J.,* 45, 767, 1971.
23. **Schneider, L. G., Dietzschold, B., Dierks, R. E., Matthaeous, W., Enzmann, P. J., and Strohmaier, K.,** Rabies group-specific ribonucleoprotein antigen and a test system for grouping and typing of rhabdoviruses, *J. Virol.,* 11, 748, 1973.
24. **Held, J. R., Tierkel, E. S., and Steele, J. H.,** Rabies in man and animals in the United States, 1946—65, *Public Health Rep.,* 82, 1009, 1967.
25. **Hattwick, M. A. W. and Gregg, M. B.,** The disease in man, in *The Natural History of Rabies,* Vol. 2, Baer, G. M., Ed., Academic Press, New York, 1975, 281.
26. **Rubin, R. H., Sullivan, L., Summers, R., Gregg, M. B., and Sikes, K. R.,** A case of human rabies in Kansas: epidemiologic, clinical, and laboratory consideration, *J. Infect. Dis.,* 122, 318, 1970.
27. **Familusi, J. B. and Moore, D. L.,** Isolation of a rabies-related virus from the cerebrospinal fluid of a child with "aseptic meningitis", *Afr. J. Med. Sci.,* 3, 93, 1972.
28. **Clark, H. F., Wiktor, T. J., and Koprowski, H.,** Human vaccination against rabies, in *The Natural History of Rabies,* Vol. 2, Baer, G. M., Ed., Academic Press, New York, 1975, 341.
29. **Johnson, H. N.,** Rabiesvirus, in *Manual of Clinical Microbiology,* 2nd ed., Lennette, E. H., Spaulding, E. H., and Truant, J. P., Eds., American Society for Microbiology, Washington, D. C., 1974, 746.
30. **Kaplan, M. and Koprowski, H., Eds.,** *Laboratory Techniques in Rabies,* 3rd ed., World Health Organization, Geneva, 1973.
31. **Atanasiu, P.,** Animal inoculation and the Negri body, in *The Natural History of Rabies,* Vol. 2, Baer, G. M., Ed., Academic Press, New York, 1975, 373.
32. **Kissling, R. E.,** The fluorescent antibody test in rabies, in *The Natural History of Rabies,* Vol. 2, Baer, G. M., Ed., Academic Press, New York, 1975, 401.
33. **Thomas, J. B.,** The serum neutralization, indirect fluorescent antibody, and rapid fluorescent focus inhibition tests, in *The Natural History of Rabies,* Vol. 2, Baer, G. M., Ed., Academic Press, New York, 1975, 417.
34. **Bryceson, A. D. M., Greenwood, B. M., Warrell, D. A., Davidson, N. McD., Pope, H. M., Laurie, J. H., Barnes, H. J., Bailie, W. E., and Wilcox, G. E.,** Demonstration during life of rabies antigen in humans, *J. Infect. Dis.,* 131, 71, 1975.
35. **Tierkel, E. S.,** Rapid microscopic examination for Negri bodies and preparation of specimens for biological test, in *Laboratory Techniques in Rabies,* 2nd ed., Monogr. No. 23, World Health Organization, Geneva, 1966, 35.
36. **Mann, G.,** *Physiological Histology,* Garandon Press, Oxford, 1902, 216.
37. **Smith, A. L., Tignor, G. H., Mifune, K., and Motohashi, T.,** Isolation and assay of rabies serogroup viruses in CER cells, *Intervirology,* 8, 92, 1977.
38. **Smith, A. L., Tignor, G. H., Emmons, R. W., and Woodie, J. D.,** Isolation of field rabies virus strains in CER and murine neuroblastoma cell cultures *Intervirology,* 9, 359, 1978.
39. **Lee, T. K. and Becker, M. E.,** *Public Health Lab.,* 31, 149—164, 1973.
40. **Reed, L. J. and Muench, H. A.,** A simple method of estimating fifty per cent endpoints, *Am. J. Hyg.,* 27, 493, 1938.
41. **Lennette, E. H. and Emmons, R. W.,** The laboratory diagnosis of rabies: review and perspective, in *Rabies,* Nagano, Y. and Davenport, F. M., Eds., University of Tokyo Press, Tokyo, 1970, 77.
42. **Smith, J. S. Yager, P. A., and Baer, G. M.,** A rapid reproducible test for determining rabies neutralizing antibody, *Bull. WHO,* 48, 535, 1973.
43. **Casals, J.,** Immunological techniques for animal viruses, in *Methods in Virology,* Vol. 3, Maramorosch, K. and Koprowski, H., Eds., Academic Press, New York, 1967, 113.
44. **Wiktor, T. J., Kuwert, E., and Koprowski, H.,** Immune lysis of rabies virus-infected cells, *J. Immunol.,* 101, 1271, 1968.
45. **Halonen, P. E., Murphy, F. A., Fields, B. N., and Reese, D. R.,** Hemagglutinin of rabies and some other bullet-shaped viruses, *Proc. Soc. Exp. Biol. Med.,* 127, 1037, 1968.
46. **Grasset, N. and Atanasiu, P.,** Etude de la diffusion en gelose d'antigenes de la rage fixe obtenus sur culture de tissus, *Ann. Inst. Pasteur Paris,* 101, 639, 1961.
47. **Gough, P. M and Dierks, R. E.,** Passive haemagglutination test for antibodies against rabies virus, *Bull. WHO,* 45, 741, 1971.
48. **Murphy, F. A., Bauer, S. P., Harrison, A. K., and Winn, W. C., Jr.,** Comparative pathogenesis of rabies and rabies-like viruses. Viral infection and transit from inoculation site to central nervous system. *Lab. Invest.,* 28, 361, 1973.
49. **Koprowski, H.,** Biological modification of rabies virus as a result of its adaptation to chicks and developing chick embryos, *Bull. WHO,* 10, 709, 1954.
50. **Tignor, G. H.,** personal communication, 1977.
51. **Bell, J. F.,** Latency and abortive rabies, in *The Natural History of Rabies,* Vol. 2, Baer, G. M., Ed., Academic Press, New York, 1975, 331.

52. **Holland, J. J., Villareal, L. P., Welsh, R. M., Oldstone, M. B. A., Kohne, D., Lazzarini, R., and Skolnick, E.,** Long-term persistent vesicular stomatitis virus and rabies virus infection of cells in vitro, *J. Gen. Virol.,* 33, 193, 1976.
53. **Fekadu, M.,** Atypical rabies in dogs in Ethiopia, *Ethiop. Med. J.,* 10, 79, 1972.

ARENAVIRIDAE: LYMPHOCYTIC CHORIOMENINGITIS, LASSA, AND RELATED DISEASES

Fredrick A. Murphy

Introduction

The arenavirus grouping — or, more precisely, the Arenaviridae family — was established in order to bring the viruses with common virion structure and physiochemical properties into one group. Most of the member viruses have also been shown to be serologically related.[1-3] The taxon contains several viruses which are significant human pathogens: lymphocytic choriomeningitis (LCM) virus, Lassa virus (the etiologic agent of Lassa fever), Machupo virus (the etiologic agent of Bolivian hemorrhagic fever), and Junin virus (the etiologic agent of Argentinian hemorrhagic fever). Six other member viruses have not been associated with human disease (Amapari, Latino, Parana, Pichinde, Tacaribe, and Tamiami viruses). With the exception of Tacaribe virus, the arenaviruses have similar ecologic characters; each is perpetuated in nature via transmission in a single or limited number of rodent species and very ofter in a limited geographic area.[4] In their natural reservoir hosts, these viruses are usually not acutely pathogenic, but infections tend toward persistence. Virus is shed in body excretions and secretions for long periods of time, and this shedding pattern is the source of human exposure.

The four pathogenic arenaviruses can become an extreme biohazard in conventional viral diagnostic and serology laboratories. Similarly, risk is involved in clinical patient care and in clinical chemistry and hematology laboratories. The risks encountered in dealing with the hemorrhagic fever agents Lassa, Machupo, and Junin viruses are well known, but it should also be remembered that LCM virus strains from nature may cause serious illness. Procedures in which Lassa, Machupo, or Junin viruses are used must be limited to laboratories with "class 3" containment, such as the Maximum Security Laboratory at the Center for Disease Control (CDC) in Atlanta. Procedures that require the use of field isolates of LCM virus should be limited to laboratories meeting standards defined by the U.S. Public Health Service.[5]

Morphology

LCM virus particles were first demonstrated by Dalton et al.[6] and were immediately considered to be different from those of any previously known virus or virus group. This identification was confirmed in other laboratories, and precise similarity was soon noted between LCM and Machupo viruses and later among these and all other viruses of the Tacaribe serocomplex.[7] All arenaviruses appear identical in ultrathin sections; they are round, oval, or pleomorphic, with a mean diameter of 110 to 130 nm and individual variations from 50 to 300 nm[8,9] (Figure 1). Particles have a surface projection layer (10 nm thick), a dense unit membrane envelope derived from host cell membrane, and a lucent interior containing a variable number of 20- to 25-nm granules which are indistinguishable from ribosomes. The distribution of ribosomes within virus particles is considered to be random, but in some cases, ribosomes connected by linear structures have been observed [9]. (Figure 2). Negative contrast microscopy of arenaviruses has proven difficult; nevertheless, the plastic envelope with club-shaped surface projections may be resolved even if internal structure is obscure (Figure 3).

Arenavirus maturation involves budding, primarily from the plasma membrane of the host cell[7-9](Figure 4). Ribosomes are contained in nascent buds, and increased membrane density and surface projections are limited to bud sites (except late in infection) (Figure 5). Intracytoplasmic inclusion bodies are prominent in many cell types

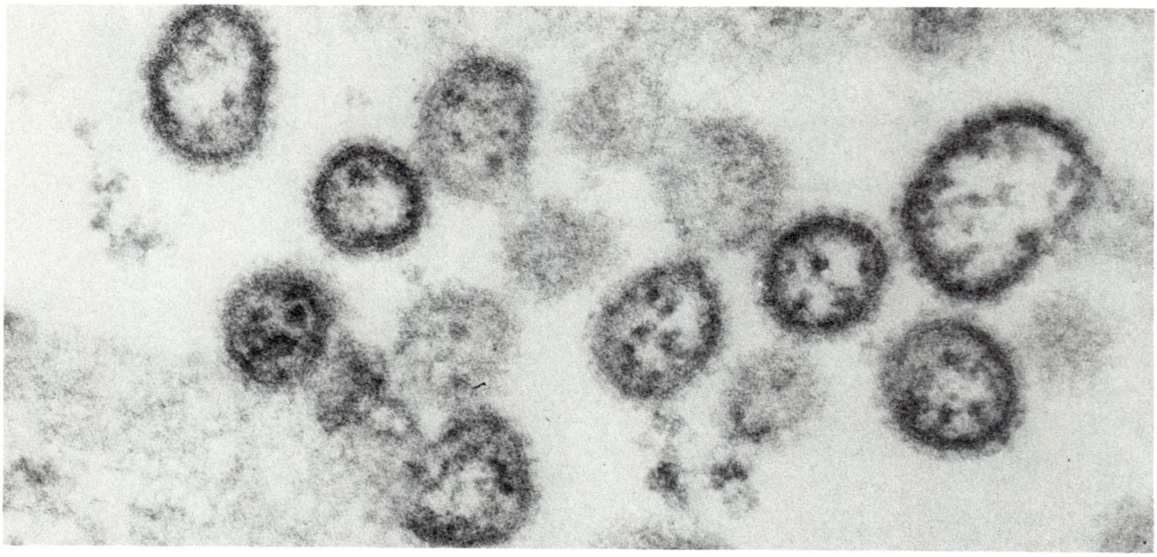

FIGURE 1. Lassa virus in first Vero cell passage after isolation from the pharynx of a patient in Sierra Leone. Arenavirus identification is certain when particles contain surface projections, a dense unit membrane envelope, and ribosomes in their interior. Magnification × 121,000.

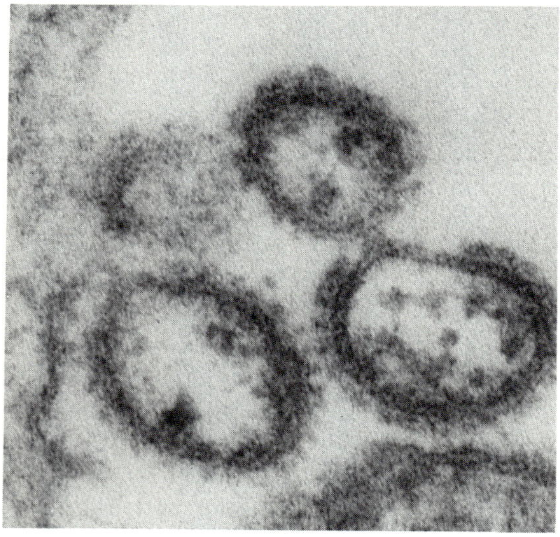

FIGURE 2. Lassa virus in human liver (biopsy). Although the position of ribosomes within arenavirus particles is usually random, a filamentous structure may link ribosomes in some cases. Magnification × 230,000.

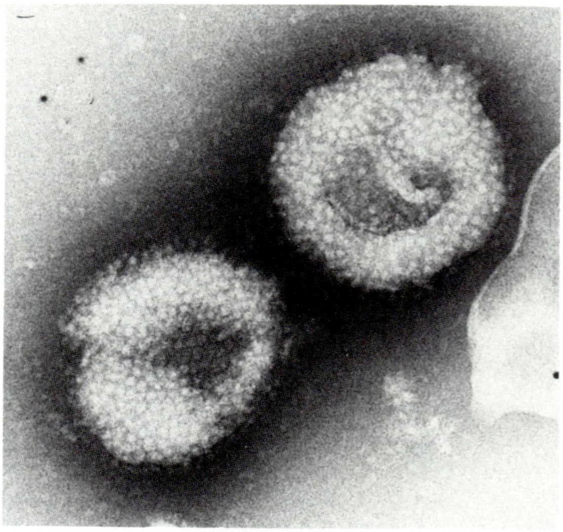

FIGURE 3. Parana virus stained with sodium silicotungstate to show the club-shaped surface projections on the plastic envelope. The fragile nature of arenaviruses limits the value of negative contrast electron microscopy in arenavirus diagnosis. Magnification × 260,000.

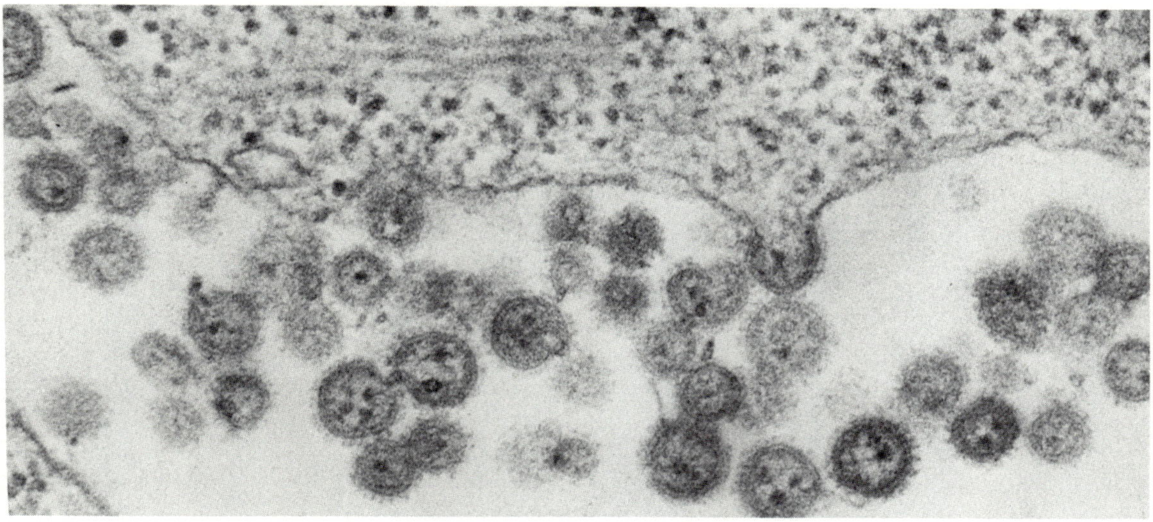

FIGURE 4. Machupo virus in human lymphoblastoid cell culture 33H. Virus particle budding from the plasma membrane of host cells in vivo and in culture delivers virus to extracellular spaces. Magnification × 114,000.

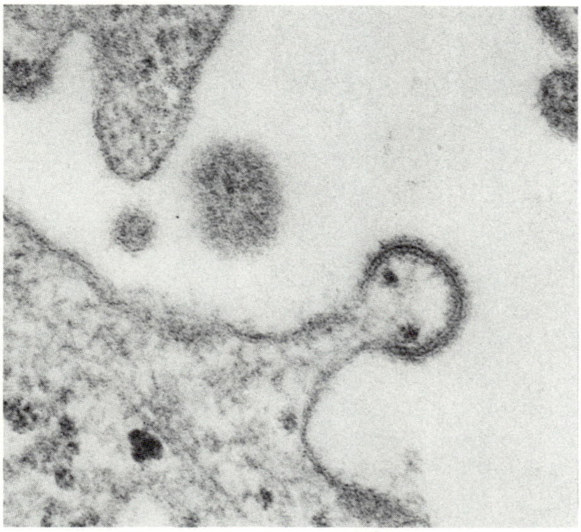

FIGURE 5. Junin virus budding from the plasma membrane of a Vero cell. The viral envelope is denser and more precise than the host cell membrane from which it is derived. Ribosomes are present in the nascent bud. Magnification × 173,000.

(in vivo and in vitro) infected with arenaviruses. These inclusions are made up of amassed ribosomes in a moderately dense matrix (Figure 6).

Physicochemical properties

Arenaviruses contain segmented single-stranded RNA — two virus specific species have molecular weights of 2.1 and 3.2×10^6 daltons; three host cell species have sedimentation constants of 28S and 18S (ribosomal RNAs) and 4 to 6S. Virion RNA does

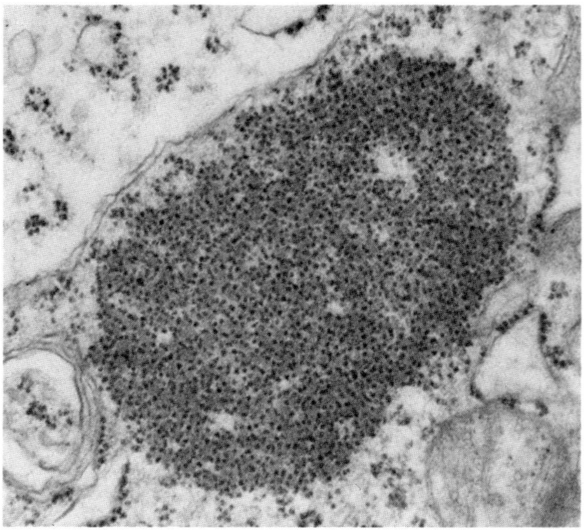

FIGURE 6. Tamiami virus inclusion body in the brain of a cotton rat (reservoir host of the virus). Inclusions have uniform, unbounded margins and consist of amassed ribosomes embedded in a matrix. Magnification × 40,000.

not possess messenger function; it is "minus stranded."[12] Viral proteins include two glycopeptides (72,000 and 34,000 to 44,000 daltons) and two polypeptides (72,000 and 12,000 daltons).[10] Density is 1.17 to 1.18 g/ml in sucrose. Infectivity is inactivated by lipid solvent, detergent, acid ($pH < 5.5$), heat, and ultraviolet (UV) or gamma radiation.[13]

Viral Replication

Arenaviruses probably invade cells by fusion of the viral envelope with the cell membrane followed by direct introduction of RNA into the cytoplasm.[7] This RNA is complementary to viral mRNA, but the viruses contain an endogenous RNA-dependent RNA polymerase for primary transcription.[12] Little is known of the details of primary or secondary transcription, translation, or genome replication. Likewise, comprehensive viral growth studies have not been undertaken; however, in one study of Pichinde virus in Vero cells, a single-cycle growth curve included an 8-hr eclipse, followed by rapid virus production which reached its peak at 24 hr.[14] Ultimately, extracellular virus titer exceeded that of cell-associated virus. Virus assembly and release occur via budding, primarily from the host cell plasma membrane. Budding may occur for a rather protracted length of time; in some cell culture systems a distinctive cytopathology ensues, but in many in vivo sites of arenavirus growth and with most LCM tissue culture systems there is little or no cell damage. Arenaviruses exhibit a propensity toward persistence in which carriage of viral genome and expression of viral antigens may be complex[15] and may be limited by defective interference.

Virus Strains and Antigenic Properties

At present, ten arenaviruses are recognized; the viruses are listed in Table 1 together with their geographic distribution and the vertebrate hosts which are involved in natural maintenance and epidemically important transmission to man. In the case of LCM virus, wild strains are carried by feral mice, but these strains cause important epizootics and persistent infections in laboratory mice and, more importantly, in laboratory ham-

Table 1
THE ARENAVIRUSES, THEIR GEOGRAPHIC DISTRIBUTION, AND THEIR VERTEBRATE HOSTS

Virus	Geographic distribution	Vertebrate hosts
LCM	Worldwide	*Mus musculus* (mouse) *Mesocricetus auratus* (hamster)
Lassa	West Africa	*Mastomys natalensis* (multimammate rat)
Machupo	Bolivia	*Calomys callosus*[a]
Junin	Argentina	*Calomys laucha*[a] *Calomys musculinus*[a] *Akodon azarae*[a]
Amapari	Brazil	*Oryzomys goeldi*[a] *Neacomys guianae*[a]
Latino	Bolivia	*Calomys callosus*[a]
Parana	Paraguay	*Oryzomys buccinatus*[a]
Pichinde	Colombia	*Oryzomys albigularis*[a]
Tacaribe	Trinidad	*Artibeus lituratus* (bat) *Artibeus jamaicensis* (bat)
Tamiami	Florida, U.S.A.	*Sigmodon hispidus*[a] (cotton rat)

[a] Cricetid rodent species.

sters exposed to feral mice. The complex nature of laboratory epizootics and the hazard of human exposure has been reviewed recently.[15] In those cases in which feral mice pose a hazard to man directly, control depends only on the implementation of proven mouse eradication methods. Human disease originating from infected laboratory or pet hamsters must be controlled by identifying contaminated breeding colonies (by virus isolation and serologic surveillance), eliminating contaminated animals and contacts, preventing reintroduction of virus via proper husbandry practices, upgrading facilities, and continuing monitoring for further evidences of infection. In the case of the other arenaviruses, geographic distribution is limited precisely; but with each of the human pathogens, control of human exposure involves an understanding of viral, vertebrate host, and local human ecology. This subject is best reviewed in Reference 15.

Arenaviruses express at least two major immunologic specificities. Antigens of the virion surface, which must be involved in neutralization, are virus specific unless extraordinary means are used to produce hyperimmune sera. There is usually no cross-neutralization or cross-protection. [4,16]Antigens which are found in infected cells, supernatant fluids, and the interior of virus particles do not induce neutralizing antibody, but are involved in the broadly cross-reactive complement fixation (CF) reaction. This antigenicity defines the Tacaribe serocomplex, the complex of arenaviruses of the Western hemisphere. The weak cross-reactivities of Lassa and LCM viruses with other arenaviruses is probably related to this same "internal" or "soluble" antigen. All of the same broad cross-reactivities seen in CF testing are demonstrated even more clearly by indirect immunoflourescence (IIF).[3]

Clinical Features

The incidence of arenaviral diseases in man has not been well documented, but the significance of each disease lies in its severity and not in the number of cases. In several surveys of acute (presumably viral) diseases of the central nervous system (CNS), it was found that LCM virus caused approximately 8% of cases and that the number of subclinical infections was small.[30] The annual incidence of Argentinean hemorrhagic fever varies, but there are usually hundreds of cases per year. At present, there are only a few cases of Bolivian hemorrhagic fever each year, but hundreds occurred in the epidemics during which the virus was first isolated.[4,22,23] Although more than one hundred of Lassa fever cases have been documented, the real incidence and geographic distribution of the infection in Africa is probably much greater.[16]

Hemorrhagic Fevers Caused by Lassa, Machupo, and Junin Viruses

These diseases initially exhibit symptomatology which is so nonspecific that clinical diagnosis may be uncertain unless similar cases occur together or cases occur in a geographic area in which disease has been seen previously.[31,32] Differential diagnosis usually requires consideration of malaria, typhoid fever, yellow fever, influenza, measles, meningococcal infection, typhus, leptospirosis, and Marburg/Ebola infections. The arenaviral hemorrhagic fevers are characterized by insidious onset of fever and myalgia beginning 3 to 16 days after exposure. There is malaise and headache. By the third to sixth day of illness, symptoms and signs may become intense; unremittant high fever is indicative of a poor prognosis. There are leukopenia and hemorrhagic manifestations such as petechiae and bleeding from the gingivae, stomach, and intestine. Ominous signs include a rising hematocrit and other signs of shock. The systemic toxemia and shock are difficult to treat; the judicious use of intravenous fluids and blood volume expanders is imperative in preventing irreversible shock. Plasma from convalescent patients has become the single specific therapeutic adjunct for patients severely ill with Lassa, Bolivian, or Argentinian hemorrhagic fevers.[4,32,33] Anecdotal clinical evidence indicates that efficacy is significant during the first 5 days of disease, but there have been no controlled studies. In practice, 1 or 2 units (250 mℓ each) of convalescent plasma (matched, if possible, to the patient's major blood group) are transfused over a period of 20 to 30 min. Convalescent plasma is in extremely short supply. In Machupo virus infections, there has been a negative correlation between survival rate and day of illness on which patients reached the hospital; i.e., the earlier patients receive comprehensive care, the higher the survival rate.[4] Convalescence is slow, and patients suffer fatigue for several weeks after the acute syndrome. The clinical manifestations of Lassa fever have been presented in detail by Monath[32] and Monath and Casals.[33] Similarly, detailed clinical manifestations of Bolivian and Argentinian hemorrhagic fevers have been described by Johnson and his colleagues.[31]

The clinical manifestations of hemorrhagic fever are consequences of infection and dysfunction of several organ systems including the liver, kidneys, heart, lungs, pharynx, pleura, intestines, and lymphoreticular tissues. Which are the primary viral target organs or the precise nature of dysfunctions are unknown. The relationships among viral proliferation and immunopathologic effects are uncertain in human hemorrhagic diseases. Pathologic descriptions have been published,[34] but, again, these have not clarified the underlying pathophysiology of these diseases.

The somewhat nonspecific early symptomatology has contributed to nosocomial spread of Lassa fever. Physicians practicing in parts of Africa in which this disease is known to occur and medical authorities elsewhere who receive febrile patients from possible endemic areas (including areas in which Machupo and Junin viruses are present) should remain alert to the possibility of these infections. Isolation and containment are essential;[32] consultative and diagnostic support is available from the CDC.

Lymphocytic Choriomeningitis

This infection of man may appear in three clinical forms: (1) "grippe," (2) meningitis, or (3) meningoencephalomyelitis.[30] In the extraneural syndrome there is fever; malaise; myalgia; and, in some cases, coryza, bronchitis, arthralgia, orchitis, and thrombocytopenia. The most common syndrome is meningitis; this may be preceded by "grippe" and is characterized by stiff neck and headache as well as fever, malaise, and myalgia. These signs may be mild or quite severe. The demarcation between this form of disease and meningoencephalomyelitis is vague, and the signs of CNS dysfunction are extremely variable. The pathophysiology of human LCM infection is obscure, as almost all patients recover without sequellae. Symptomatic treatment of a supportive nature is generally recommended. Clinical manifestations of LCM infection have been presented in detail by Lehmann-Grube.[30] Diagnostic services are usually provided by public health agency laboratories (e.g., state or provincial health department laboratories).

Laboratory Diagnosis

Diagnosis of arenaviral infections may be considered confirmed when virus is isolated and identified from blood, cerebrospinal fluid, throat washings, urine, or pleural fluid of patients or from visceral organs at autopsy. Diagnosis may also be confirmed by demonstration of an antibody titer rise between acute and convalescent serum specimens.[35,36] Because silent infections with the pathogenic arenaviruses are rare, it is possible to make a presumptive diagnosis by serologic testing on an unpaired convalescent serum in association with a compatible clinical diagnosis.

Virus Isolation Methods

Virus isolation methods vary somewhat for the different pathogenic arenaviruses (Table 2). Lassa virus is isolated by inoculation of Vero cells. The virus causes a characteristic cytopathology (progressive focal necrosis) within 4 to 5 days, but cells may be held longer and may be "blind passaged" if necessary (Figure 7). Several dilutions of original specimens should be used in order to avoid autointerference effects. Confirmatory identification of cytopathic agents may be made by using terminal dilution cell cultures as antigen in a CF test with reference antiserum. Similarly, direct immunofluorescence (IF) or indirect immunofluorescence (IIF) with reference antisera may be used on cell cultures.[37]

Primmary isolation of Machupo virus is made by intracerebral inoculation of newborn hamsters[4,31] and of Junin virus by inoculation of newborn mice or hamsters.[22,23] When animals become sick (7 to 12 days postinoculation), brain tissue is harvested and used as antigen for preliminary identification. Further passages and confirmatory identification (neutralization testing {NT}) are carried out in Vero cells.[38]

Many LCM virus isolation methods have been tried over many years, but present indications are that no method is as sensitive as intracerebral inoculations of weanling (3-week-old) mice. Several dilutions of clinical specimens should be inoculated, as the same autointerference phenomenon seen in infected cell cultures occures at low dilutions in vivo. LCM virus isolates may be identified with IF by using a touch impression of the brain and reference direct or indirect systems. Identity of isolates may also be confirmed by NT or CF. Two methods have been advocated for increasing the sensitivity of the mouse inoculation technique for LCM virus isolation. First, infection rather than lethality may be used as an indication of the presence of the virus; assay requires challenge of the immune status of surviving mice.[30] Second, viral lethality may be increased by an endotoxin challenge; mice infected with LCM virus are very sensitive to endotoxin.[39] A possible third method is to administer a larger dose per mouse (1 ml i.p. and 0.03 ml i.c.); the i.p. virus alone will induce meningitis if the blood-brain barrier is broken by the i.c. injection. Direct IF testing of touch impressions of tissues for the presence of LCM antigen may be of value in epidemic circumstances in which

Table 2
ARENAVIRUSES AND PREFERRED METHODS FOR THEIR DIAGNOSIS

Virus strain (prototype strain)	Human disease	Diagnostic specimen	Virus isolation techniques[a]		Serologic techniques[b]			Refs.
			Cell cultures	Animals	CF	NT	IF	
Lymphocytic choriomeningitis (Armstrong)	Lymphocytic choriomeningitis	Acute blood		Weanling mice 1° Guinea pigs 2°	3°	2°	1°	17—19
		Convalescent serum			3°	2°	1°	
Lassa (LP)	Lassa fever	Acute blood, fluid, or tissue	Vero cells 1°		2°		1°	20
		Convalescent serum			2°		1°	
Machupo (Carvallo)	Bolivian hemorrhagic fever	Acute blood, fluid, or tissue	Vero cells 2°	Newborn hamsters 1°	3°	1°	2°	21
		Convalescent serum			3°	1°	2°	
Junin (XJ)	Argentinian hemorrhagic fever	Acute blood, fluid, or tissue	Vero cells 2°	Newborn hamsters 1° Newborn mice 1°	3°	1°	2°	22, 23
		Convalescent serum			3°	1°	2°	
Amapari (BeAn 70563)	Nonpathogens	Acute specimen	Vero cells 3°	Newborn hamsters 1° Newborn mice 2°	3°	1°	2°	24
Latino (10924)		Convalescent serum			3°	1°	2°	25
Parana (12056)								26
Pichinde (An 3739)								27
Tacaribe (TRVL 11573)								28
Tamiami (W-10777)								29

[a] 1°, primary technique/most valuable overall; 2°, secondary or alternate technique/ less valuable; 3°, tertiary technique/least valuable.

[b] CF, complement fixation; NT, neutralization testing, IF, immunofluorescence. Brackets indicate paired serum testing.

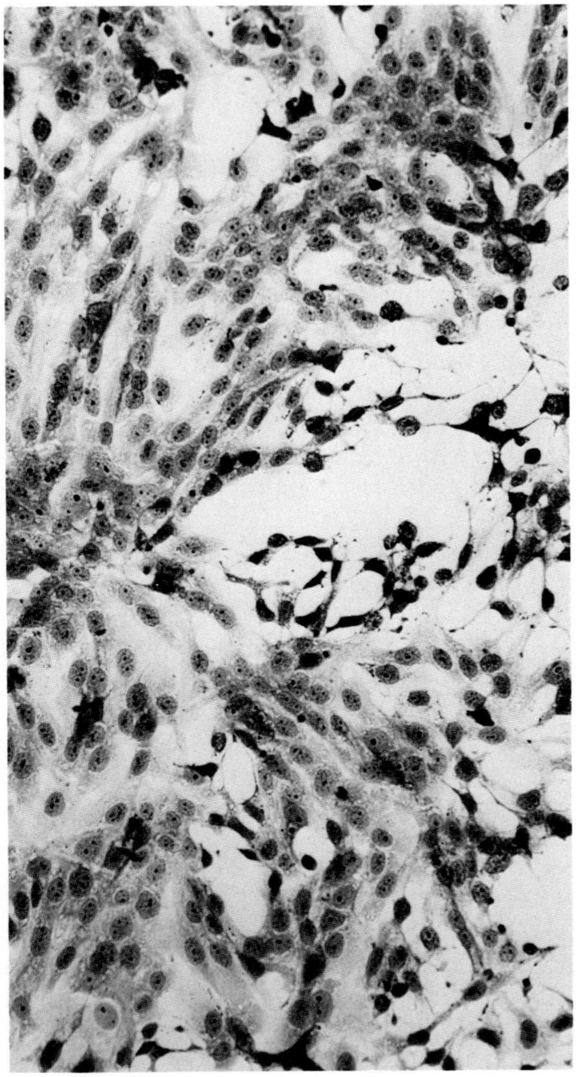

FIGURE 7. Lassa virus in Vero cells. These cells were inoculated with a human blood specimen 2 days before methanol fixation and Giemsa staining. The focal necrotic changes are characteristic and are easily seen in unstained cultures. Magnification × 110. (Courtesy of Dr. H. Wulff, Center for Disease Control, Atlanta.)

virus isolation per se is not practical. As in the IF diagnosis of rabies, the presence of viral antigen may be considered a confirmatory diagnosis.

Serologic Methods

Serologic methods vary in their specificity, sensitivity, speed, and volume capabilities, but no single method serves all needs. CF, IIF, and NT have been used in clinical, epidemiologic, and reference assays of human experience with the pathogenic arenaviruses (Table 2). These viruses do not hemagglutinate.

Complement Fixation Testing

CF testing is carried out in microtiter systems according to standardized techniques.[40] Antigens may be prepared from infected Vero cell cultures harvested late in infection (3 to 4+ CPE); supernatant fluids are clarified and used without further treatment. Antigens (except for Lassa virus) may also be made from infected suckling mouse brain by extraction with sucrose-acetone[41] or a fluorocarbon such as Genetron or Freon 113.[42,43] Antigens may be inactivated with β-propiolactone or UV irradiation,[44] but testing for residual infectivity must be rigorous. CF testing for LCM antibody has advantages when integrated into an antigen battery for use on all aseptic meningitis specimens. In this case, inactivated antigen may be used in a clinical laboratory in which handling of the virus itself would be hazardous. The method may also have value when very large numbers of specimens must be tested. However, CF antibody if (1) relatively unspecific for distinguishing among the arenaviruses, (2) relatively low in titer and subject to insensitivity problems, (3) subject to anticomplementary problems, (4) slow to appear after infection (sometimes remaining undetectable for 3 to 4 weeks after onset of illness and, in a few cases, remaining negative thereafter), and (5) the first assay to return to negative after infection. These shortcomings significantly limit the usefulness of CF testing in clinical as well as epidemiologic settings.

Indirect Immunofluorescence Testing

IIF testing is done in various ways, but the variations really involve only the preparation of viral antigen substrates. For all of the arenaviruses except LCM viruses, Vero cells[37,45] are grown alternatively in (1) multiple chamber slides (Lab-Tek® four- or eight-chamber slides; Miles Laboratories, Westmont, Ill.),* (2) bottles as monolayers, or (3) suspension cultures. Cells are infected; when at least 30% are shown by IF to contain viral antigen, the cells may be processed as substrates (viral growth usually takes approximately 3 days, but occasionally takes as long as 7 days postinoculation). Multiple chamber slides are processed directly, and suspended cells from monolayer or spinner cultures are dried as antigen spots in the uncoated circles on epoxy-coated slides (Cel-Line Associates, Minotola, N.J.). For LCM virus the same methods may be used, but BHK-21 or L cells are substituted for the Vero cells.[46,47] Whether in chambers or spots, infected cell substrates are dried, acetone fixed (for 10 min at room temperature or 20 min to 24 hr at −20 or −70°C), dried, and stored at −70°C. At this temperature they are stable indefinitely. For use, substrates are thawed and flooded with human serum (1:2 to 1:10 for screening or serial twofold dilutions for titrations) and held at room temperature for 30 min in a humid chamber. They are washed (usually one dip and two 10-min immersions in phosphate-buffered saline (PBS) with agitation), and fluorescein isothiocyanate (FITC) conjugated anti-human Ig (or anti-human IgG and IgM for class-specific testing or anti-Ig of other species contributing the antiserum of the first layer) at a predetermined working dilution is added for 30 min. The same washing procedure is followed by dipping the slides in water (to remove salts), air drying them, and mounting them under coverslips in PBS-buffered glycerin (1:9). Conventional IF microscopy equipment is used, and dilution end points of the serum are chosen where fluorescence brightness just begins to diminish. The FITC conjugate chosen must be thoroughly evaluated for potency and specificity; commercially available conjugates may be excellent, but some yield nonspecific fluorescence with some sera (usually a pale mint-green, smooth fluorescent pattern as opposed to the rich apple-green color and coarsely granular pattern of the specific reaction) (Figure 8). The choice of a substrate with 30 to 50% of the cells infected provides a built-in control for nonspecific reactions in which all cells fluoresce equally. Likewise, this kind of substrate provides the built-in control related to antigen distribution. Arena-

* Use of trade names is for identification only and does not constitute endorsement by the Public Health Service or by the U.S. Department of Health, Education, and Welfare.

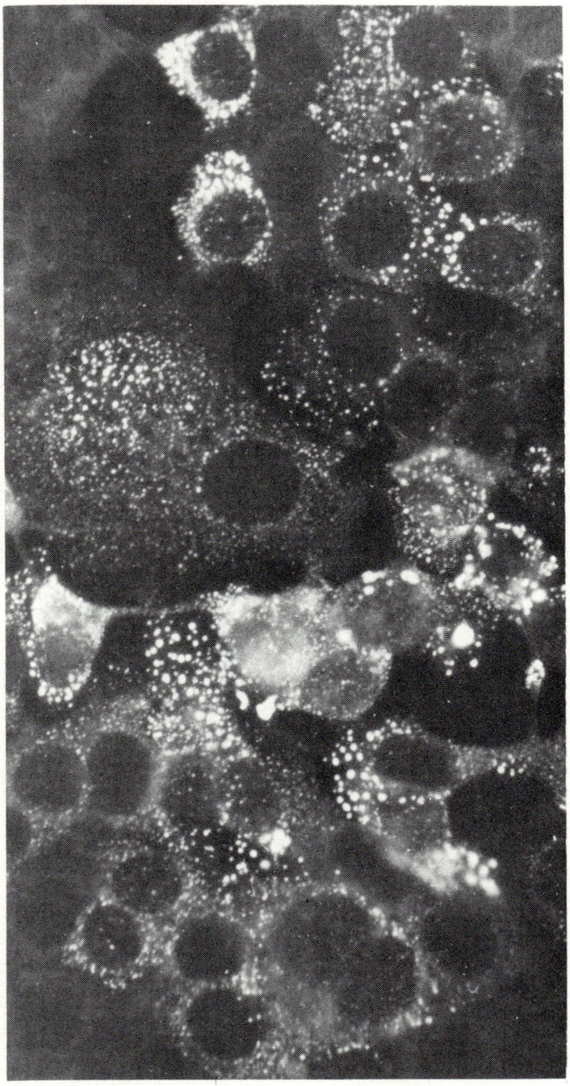

FIGURE 8. Indirect immunofluorescence (IIF) for detection of Lassa virus (antigen) or Lassa virus antibody. When infected Vero cell monolayers are used as the substrate for serial dilutions of human serum in the IIF test, this granular and aggregate form of the intracytoplasmic antigen should be seen. This serves as one control together with conventional controls. Magnification × 400. (Courtesy of Dr. H. Wulff, Center for Disease Control, Atlanta.)

viral antigens occur in infected cells in punctate, granular, or aggregate (inclusion body) form in cytoplasm; a smooth distribution or nuclear involvement of fluorescence should be discounted as artifact. All conventional controls should be included (positive and negative sera and specificity controls). The cell culture substrates for IIF testing are infectious, but Machupo virus-infected Vero cells have been shown to be inactivated during the initial drying and acetone fixation steps.[45] Lassa virus and LCM virus have been shown to survive acetone fixation procedures,[56] but UV irradiation totally inactivates substrates.[57] The IIF methods are becoming the most useful assays for arenavirus antibodies. These methods are extremely sensitive, are useful with improperly prepared or contaminated sera, and are safe when inactivated substrates are used. Moreover, they have the advantage of storability of substrates. The most significant advantage is that IIF antibodies are the first to become detectable after infection; they often become positive within a few days to a week after onset of illness. This has been shown in human LCM infections by comparative testing with CF and neutralization.[39,48] IIF antibody levels are maintained for very long periods after convalescence. The shortcoming lies in the lack of virus specificity among the arenaviruses. The geographic isolation of each arenavirus obviates this shortcoming, but LCM must be considered in all parts of the world. Earlier difficulties of using IIF methods in the field have been bypassed by newer interference filter systems equipped with a simple incandescent light source; if circumstances allow any microscopy to be performed in a given field situation, then IIF can also be done.

Neutralization Testing

NT for detecting arenaviral antibodies varies greatly, and no single method has been shown to have overall superiority. In vivo NTs in mice give nonreproducible results with most arenaviruses, but for many years LCM neutralization tests have been done by intracerebral inoculation of 3-week-old Swiss mice with mixtures of undiluted antiserum and tenfold dilutions of virus. Sensitivity may be increased by challenging animals which survive in the conventional lethal end point test with a known lethal dose of virus.[30] Animals which survive this challenge may be assumed to be immune by virtue of a sublethal infection in the initial test. Animals which die may be assumed to have been inoculated with a "completely" neutralized virus-serum mixture in the initial test. To convert the insensitive median lethal dose (LD_{50}) end points of the challenge test, numbers of animals surviving challenge are added to numbers of animals killed outright by the infection. From these LD_{50} values, neutralization indices may be calculated by usual means. More recently, LCM neutralizing antibody has been measured by its ability to inhibit the mouse footpad swelling evoked by inoculation of the virus;[49] this is the most sensitive in vivo method for determining the presence of LCM neutralizing antibodies.

NT methods for most of the arenaviruses are now most commonly performed in cell culture, but the precise techniques vary widely. An LCM NT may be carried out in tubes by using cells which support virus growth but do not exhibit cytopathology (e.g., L cells); end points are determined by CF tests on supernatant fluids.[30] Similarly, an NT with LCM or most of the other arenavirus may be done in Lab-Tek chambers with immunofluorescent focus inhibition (IFFI) used to indicate antiserum dilution end points.[50] For this test, virus is diluted to deliver 30 to 50 focus-forming units to each chamber aliquot. After serial dilutions of the serum being tested are added to the virus aliquots, mixtures are incubated for 1 hr at 37°C and absorbed to cells in chambers. At about 48 hr (when each focus of residual infectivity contains 4 to 10 cells), conventional IF staining (usually by the direct method) is done, and serum end points are determined from dilutions which inhibit 80% of the foci counted in normal serum controls. This method is sensitive and has distinct advantages when need is intermittent; it has not yet been used with Lassa virus.

Arenaviruses produce plaques in Vero cells; therefore, neutralization methods based upon conventional plaque reduction in monolayers are most useful. The Machupo virus neutralization test is done with Vero cells grown in disposable plastic trays containing 96 wells (Linbro Chemical Co., Inc., New Haven, Ct.).[51] After absorption of serum-virus mixtures (80 to 100 PFU virus preincubated at 30°C for 1 hr with serum) in a 4% CO_2 atmosphere, agar is overlayed and an indicator overlay is added at 5 to 8 days. Antisera titers are calculated from 80% plaque reduction end points. LCM virus plaquing in monolayers of BHK-21 cells has been described by Pulkkinen and Pfau.[52]

5. Public Health Service Committee on the Safe Shipment and Handling of Etiologic Agents, Classification of Etiologic Agents on the Basis of Hazard, 4th ed., Center for Disease Control, Atlanta, 1974.
6. **Dalton, A. J., Rowe, W. P., Smith, G. H., Wilsnack, R. E., and Pugh, W. E.**, Morphological and cytochemical studies on lymphocytic choriomeningitis virus, *J. Virol.*, 2, 1465, 1968.
7. **Murphy, F. A., Webb, P. A., Johnson, K. M., Whitfield, S. G., and Chappell, W. A.**, Arenaviruses in Vero cells: ultrastructural studies, *J. Virol.*, 6, 507, 1970.
8. **Murphy, F. A., Whitfield, S. G., Webb, P. A., and Johnson, K. M.**, Ultrastructural studies of arenaviruses, *in Lymphocytic Choriomeningitis Virus and Other Arenaviruses*, Lehmann-Grube, F., Ed., Springer-Verlag, Berlin, 1973, 273.
9. **Murphy, F. A. and Whitfield, S. G.**, Morphology and morphogenesis of the arenaviruses, *Bull. WHO*, 52, 409, 1976.
10. **Rawls, W. E., Ramos, B. A., and Carter, M. F.**, Biophysical and biochemical studies of Pichinde virus, *in Lymphocytic Choriomeningitis Virus and Other Arenaviruses*, Lehmann-Grube, F., Ed., Springer-Verlag, Berlin, 1973, 259.
11. **Pfau, C. J.**, Biochemical and biophysical properties of the arenaviruses, *Prog. Med. Virol.*, 18, 64, 1974.
12. **Rawls, W. E. and Buchmeier, M.**, Arenaviruses: purification and physiochemical nature, *Bull. WHO*, 52, 393, 1976.
13. **Pfau, C. J., Bergold, G. H., Casals, J., Johnson, K. M., Murphy, F. A., Pederson, I. R., Rawls, W. E., Rowe, W. P., Webb, P. A., and Weissenbacher, M. C.**, Arenaviruses, *Intervirology*, 4, 207, 1974.
14. **Mifune, K., Carter, M., and Rawls, W.**, Characterization studies of the Pichinde virus — a member of the arenavirus group, *Proc. Soc. Exp. Biol. Med.*, 136, 637, 1971.
15. **Monath, T. P., Ed.**, Int. Symp. Arenaviral Infections of Public Health Importance, *Bull. WHO*, 52, 381, 1975.
16. **Casals, J.**, Arenaviruses, *Yale J. Biol. Med.*, 48, 115, 1975.
17. **Armstrong, C. and Lillie, R. D.**, Experimental lymphocytic choriomeningitis of monkeys and mice produced by a virus encountered in studies of the 1933 St. Louis encephalitis epidemic, *Public Health Rep.*, 49, 1019, 1934.
18. **Rivers, T. M. and Scott, T. F. M.**, Meningitis in man caused by a filterable virus, *Science*, 81, 439, 1935.
19. **Traub, E.**, A filterable virus recovered from white mice, *Science*, 81, 298, 1935.
20. **Buckley, S. M. and Casals, J.**, Lassa fever, a new virus disease of man from West Africa. III. Isolation and characterization of the virus, *Am. J. Trop. Med. Hyg.*, 19, 680, 1970.
21. **Johnson, K. M., Wiebenga, N. H., Mackenzie, R. B., Kuns, M. L., Turaso, N. M., Shelokov, A., Webb, P. A., Justines, G., and Beye, H. K.**, Virus isolations from human cases of hemorrhagic fever in Bolivia, *Proc. Soc. Exp. Biol. Med.*, 118, 113, 1965.
22. **Parodi, A. S., Greenway, D. J., Rugiero, H. R., Rivero, S., Frigerio, M., de la Barrera, J. M., Mettler, N., Garzon, F., Boxaca, M., de Guerrero, L., and Nota, N.**, Sobre la etiologia del brote epidemico de Junin, *Dia Med.*, 30, 2300, 1958.
23. **Pirosky, I., Zuccarini, J., Molinelli, E. A., Di Pietro, A., Barrera Oro, J. G., Martini, P., and Copello, A. R.**, Virosis Hemorragica del Noroeste Bonaerense, Endemo-epidemica, Febril, Enantematica y Leucopenica, Comision Nacional Ad Hoc para Estudiar el Brote de 1958, Institue Nacional de Microbiologie, Buenos Aires, 1959.
24. **Pinheiro, F. P., Shope, R. E., de Andrade, A. H. P., Bensabeth, G., Cacios, G. V., and Casals, J.**, Amapari, a new virus of the Tacaribe group from rodents and mites of Amapa territory, Brazil, Proc. Soc. Exp. Biol. Med., 122, 531, 1966.
25. **Webb, P. A., Johnson, K. M., Peters, C. J., and Hibbs, J. B.**, Immunological relationships among Tacaribe complex viruses and description of a new serotype from Bolivia, Latino virus, manuscript in preparation.
26. **Webb, P. A., Johnson, K. M., Hibbs, J. B., and Kuns, M. L.**, Parana, a new Tacaribe complex virus from Paraguay, *Arch. Gesamte Virusforsch.*, 32, 379, 1970.
27. **Trapido, H. and Sanmartin, C.**, Pichinde virus, a new virus of the Tacaribe group from Colombia, *Am. J. Trop. Med. Hyg.*, 20, 632, 1971.
28. **Downs, W. G., Anderson, C. R., Spence, L., Aitken, T. H. G., and Greenhall, A. H.**, Tacaribe virus, a new agent isolated from Artibens bats and mosquitoes in Trinidad, West Indies, *Am. J. Trop. Med. Hyg.*, 12, 640, 1963.
29. **Calisher, C. H., Tzianabos, T., Lord, R. D., and Coleman, P. H.**, Tamiami virus, a new member of the Tacaribe group, *Am. J. Trop. Med. Hyg.*, 19, 520, 1970.
30. **Lehmann-Grube, F.**, Lymphocytic choriomeningitis, *Virol. Monogr.*, 10, 1, 1971.
31. **Johnson, K. M., Halstead, S. B., and Cohen, S. N.**, Hemorrhagic fevers of Southeast Asia and South America: a comparative appraisal, *Prog. Med. Virol.*, 9, 105, 1967.

32. **Monath, T. P.,** Lassa fever, *Trop. Doctor,* 4, 155, 1973.
33. **Monath, T. P. and Casals, J.,** Diagnosis of Lassa fever and the isolation and management of patients, *Bull. WHO,* 52, 707, 1976.
34. **Winn, W. C. and Walker, D. H.,** Pathology of human Lassa fever, *Bull. WHO,* 52, 535, 1975.
35. **Murphy, F. A.,** Arenaviruses: diagnosis of lymphocytic choriomeningitis, Lassa, and other arenaviral infections, in *Comparative Diagnosis of Viral Diseases,* Vol. 1, Kurstak, E. and Kurstak, C., Eds., Academic Press, New York, 1978, 759.
36. **Murphy, F. A.,** Immune response to arenaviruses, in *Manual of Clinical Immunology,* Rose, N. R. and Friedman, H., Eds., American Society for Microbiology, Washington, D. C., 1976, 463.
37. **Wulff, H. and Lange, J. V.,** Indirect immunofluorescence for study of Lassa virus infections, *Bull. WHO,* 52, 429, 1975.
38. **Webb, P. A., Johnson, K. M., Mackenzie, R. B., and Kuns, M. L.,** Some characteristics of Machupo virus, causative agent of Bolivian hemorrhagic fever, *Am. J. Trop. Med. Hyg.,* 16, 531, 1967.
39. **Hotchin, J. and Sikora, E.,** Laboratory diagnosis of lymphocytic choriomeningitis, *Bull. WHO,* 52, 555, 1975.
40. **Casey, H.,** Standardized Diagnostic Complement Fixation Method and Adaptation to Microtest, Public Health Monogr. No. 74, Public Health Service Publ. No. 1228, U. S. Government Printing Office, Washington, D. C., 1965.
41. **Clarke, D. H. and Casals, J.,** Techniques for hemagglutination and hemagglutination-inhibition with arthropod-borne viruses, *Am. J. Trop. Med. Hyg.,* 7, 561, 1958.
42. **Gresikova, M. and Casals, J.,** A simple method of preparing a complement-fixing antigen for lymphocytic choriomeningitis virus, *Acta Virol.,* (Engl. Ed.), 7, 380, 1963.
43. **Chastel, C., and LeNoc, P.,** Antigene fixateur du complement de titre eleve et de preparation simple pour le virus de la choriomeningite lymphocytaire, *Ann. Inst. Pasteur Paris,* 114, 698, 1968.
44. **Casals, J. and Buckley, S. M.,** Lassa fever virus, in *Lymphocytic Choriomeningitis Virus and Other Arenaviruses,* Lehmann-Grube, F., Ed., Springer-Verlag, Berlin, 1973, 325.
45. **Peters, C. J., Webb, P. A. and Johnson, K. M.,** Measurement of antibodies to Machupo virus by the indirect fluorescent technique, *Proc. Soc. Exp. Biol. Med.,* 142, 526, 1973.
46. **Cohen, S. M., Triandaphilli, I. A., Barlow, J. L., and Hotchin, J.,** Immunofluorescent detection of antibody to lymphocytic choriomeningitis virus in man, *J. Immunol.,* 96, 777, 1966.
47. **Lewis, V. and Clayton, D.,** Detection of lymphocytic choriomeningitis virus antibody in murine sera by immunofluorescence, *Appl. Microbiol.,* 18, 289, 1969.
48. **Lewis, V. J., Walter, P. D., Thacker, W. L., and Winkler, W. G.,** Comparison of three serological tests for diagnosis of lymphocytic choriomeningitis virus infection, *J. Clin. Microbiol.,* 2, 193, 1975.
49. **Hotchin, J., Benson, L., and Sikora, E.,** The detection of neutralizing antibody to lymphocytic choriomeningitis virus in mice, *J. Immunol.,* 102, 1128, 1969.
50. **Winn, W. C., Jr., Murphy, F. A., and Flemister, M. R.,** The pathogenesis of Tamiami virus meningoencephalitis in newborn mice, in *Lymphocytic Choriomeningitis Virus and Other Arenaviruses,* Lehmann-Grube, F., Ed., Springer-Verlag, Berlin, 1973, 299.
51. **Webb, P. A., Johnson, K. M., and Mackenzie, R. B.,** The measurement of specific antibodies in Bolivian hemorrhagic fever by neutralization of virus plaques, *Proc. Soc. Exp. Biol. Med.,* 130, 1013, 1969.
52. **Pulkkinen, A. J. and Pfau, C. J.,** Plaque size heterogeneity: a genetic trait of lymphocytic choriomeningitis virus, *Appl. Microbiol.,* 20 ,123, 1970.
53. **Hotchin, J. and Kinch, W.,** Microplaque reduction: a new assay for neutralizing antibody to lymphocytic choriomeningitis virus, *J. Infect. Dis.,* 131, 186, 1975.
54. **Walker, D. H., Wulff, H., and Murphy, F. A.,** Experimental Lassa virus infection in the squirrel monkey, *Am. J. Pathol.,* 80, 261, 1975.
55. **Terrell, T. G., Stookey, J. L., Eddy, G. A., and Kastello, M. D.,** Pathology of Bolivian hemorrhagic fever in the rhesus monkey, *Am. J. Pathol.,* 73, 477, 1973.
56. **Lewis, V. and Wulff, H.,** personal communications, 1975.
57. **Johnson, K. M.,** personal communication, 1976.

CORONAVIRIDAE

Kenneth McIntosh

Introduction

Coronaviruses are medium-sized, membrane-bound, RNA-containing viruses which have recently been set apart in a distinct genus, the *Coronaviridae*.[1] They are of principal importance as pathogens of animal species other than humans, although there is increasing recognition of their capacity to produce respiratory disease in man. Virus particles resembling coronaviruses have recently been observed in the feces of adults with nonbacterial gastroenteritis.[2] Some members of the genus which infect animals are avian infectious bronchitis virus (IBV), transmissible gastroenteritis virus of swine, mouse hepatitis virus (MHV), and neonatal calf diarrhea virus. This chapter will concentrate primarily on the human respiratory coronaviruses (HCV) and their characteristic features in the diagnostic and research laboratory.

Morphology

It was because of their distinctive morphology in negatively stained preparations (Figure 1) that the coronaviruses were first recognized as being distinct from myxo- and paramyxoviruses.[3] When they are properly prepared, coronaviruses appear as round, oval, or moderately pleomorphic particles with widely spaced club- or petal-shaped projections. The overall diameter of the particles, including projections, is 80 to 180 nm, and the projections themselves are quite uniformly 18 to 20 nm long. It is common to see the outline of projections viewed on end against a pool of phosphotungstic acid in or on the center of the particle. It is also common to see particles in

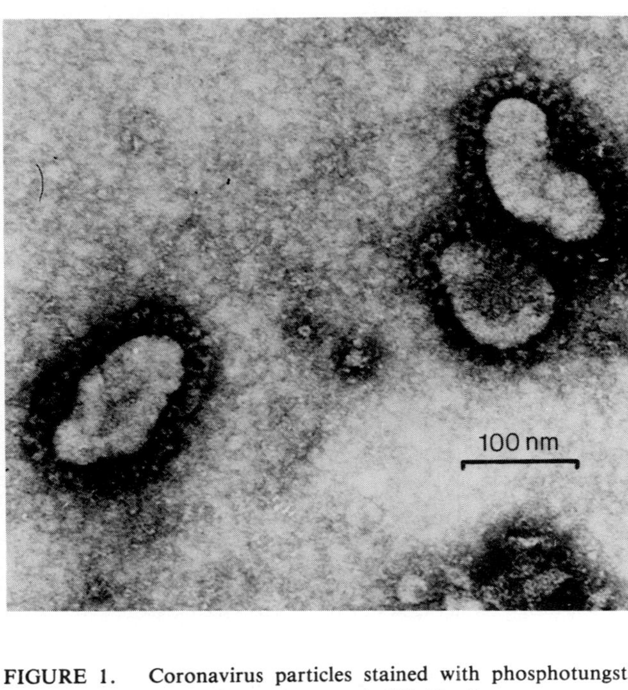

FIGURE 1. Coronavirus particles stained with phosphotungstic acid. Preparation of strain 229E grown in WI-38 cells.

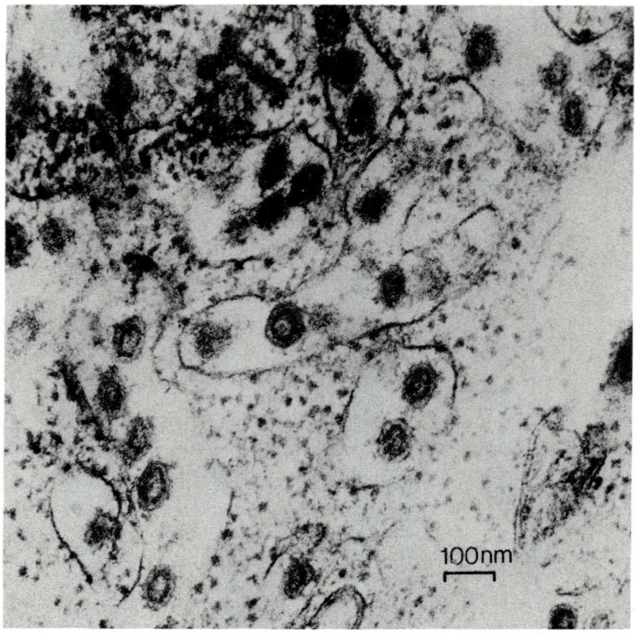

FIGURE 2. Thin section of a WI-38 cell infected with strain 229E, showing completed particles and some less distinct particles in various stages of budding. The outer double membrane and the dense inner shell or core are visible. Projections are not clearly seen in this type of preparation. (From Becker, W. B., McIntosh, K., Dees, J. H., and Chanock, R. M., *J. Virol.*, 1, 810, 1967. With permission.)

which projections appear to have been stripped away during the preparation of the specimen.

Attempts to elucidate the internal structure of coronaviruses with negative staining have not clearly been successful. Ether or detergents usually disrupt the particle in such a way that no internal component survives for clear characterization. Almeida has demonstrated inward-pointing "tongues" continuous with the outer membrane by preparing virus suspensions with antiserum plus complement,[4] but the function of these tongues is not clear, and they have not consistently been seen by other investigators.

A clearer picture of the internal structure is found in thin sections of infected tissue culture cells (Figure 2). The particles have an outer limiting unit membrane (the "outer shell" in Figure 2) and an "inner shell" or core which is separated from the outer membrane by a space. Between the two shells and inside the inner shell is poorly defined amorphous material. The projections are not well defined in these thin-section preparations.

The salient features of coronavirus morphology are summarized in Table 1 and compared with those of the myxo- and paramyxoviruses.

Physiochemical Properties

The coronaviruses are ether and chloroform labile.[4] The human strains are also acid labile, but this is not a consistent feature of all animal strains.[5] The nucleic acid is RNA, as demonstrated by both insensitivity to DNA inhibitors such as IUdR and BUdR[6-8] and direct examination of extracted core material from IBV.[9,10]There are six or seven polypeptides separable on acrylamide gels, four of which contain carbohydrate.[11] Neuraminidase is probably not a part of the virion. The density in sucrose is

Table 1
PROPERTIES OF CORONAVIRUSES IN RELATION TO INFLUENZA AND PARAINFLUENZA VIRUSES

	Particle		Projections		Internal component	
Virus	Shape	Diameter (nm)	Shape	Length (nm)	Morphology	Diameter (nm)
Coronavirus	Round or elliptical	80—180	Club	18-20	? shell	?
Influenza virus	Round or filamentous	80—140 if round	Stick	9—11	Helical	6
Parainfluenza virus	Pleomorphic	150—250	Stick	8—10	Helical	18

1.16 to 1.18 and in CsCl is 1.23.[12,13] One human serotype, strain OC43, hemagglutinates erythrocytes from various animal species including human, fowl, rat, and vervet monkey.[14]

Virus Replication

Little information is available concerning replication of the coronaviruses. The ultrastructural progress of infection by HCV in tissue cultures has been followed by several investigators.[15-18] Virus does not replicate in the nucleus, but in the cytoplasm. Most investigators agree that the virus forms by budding into cytoplasmic vesicles, probably taking altered smooth endoplasmic reticulum as its outer membrane. Budding from the cell surface rarely, if ever, occurs. In addition, several microscopists have observed crescentic shapes, apparently formed in the earliest stage of budding, made up of the apposition of a piece of the inner shell against the inwardly budding outer membrane. This process is shown schematically in Figure 3. It should be noted that there is no direct proof of this sequence of events.

Fluorescence studies confirm the absence of direct nuclear involvement in HCV replication.[16,18] However, there are no published studies of the biochemical changes in HCV-infected cells, and the information on animal strains is also scattered and preliminary. Single-step growth curves indicate that new virus appears after an eclipse phase of approximately 4 hr, and cell-associated virus reaches peak infectivity at approximately 22 hr.[19]

Hemadsorption can be demonstrated in OC43-infected cells, but probably occurs because whole virions adhere to the surface of infected tissue culture cells, and erythrocytes, in turn, adhere to the virus particles. This phenomenon has been called "pseudo-hemadsorption."[20]

Virus strains and Antigenic Properties

Many HCV strains have been isolated, but not all have undergone thorough antigenic characterization. This is because a number of strains have been grown only in organ cultures made from human embryonic trachea or nasal mucosa, and sophisticated immunologic studies have not been possible. Information on the best characterized strains is given in Table 2.

Those strains which have been recovered from the human respiratory tract using human diploid fetal cell monolayers appear to form a homogeneous group having prototype strain 229E.[21] Insofar as they have been studied, they are all antigenically identical by both neutralization and complement fixation (CF) tests. They have been recovered from adults, children, and infants with respiratory disease. They do not hemagglutinate, nor do they infect small laboratory animals or eggs. One strain, LP, was originally recovered in human embryonic tracheal organ culture (HETOC), but it

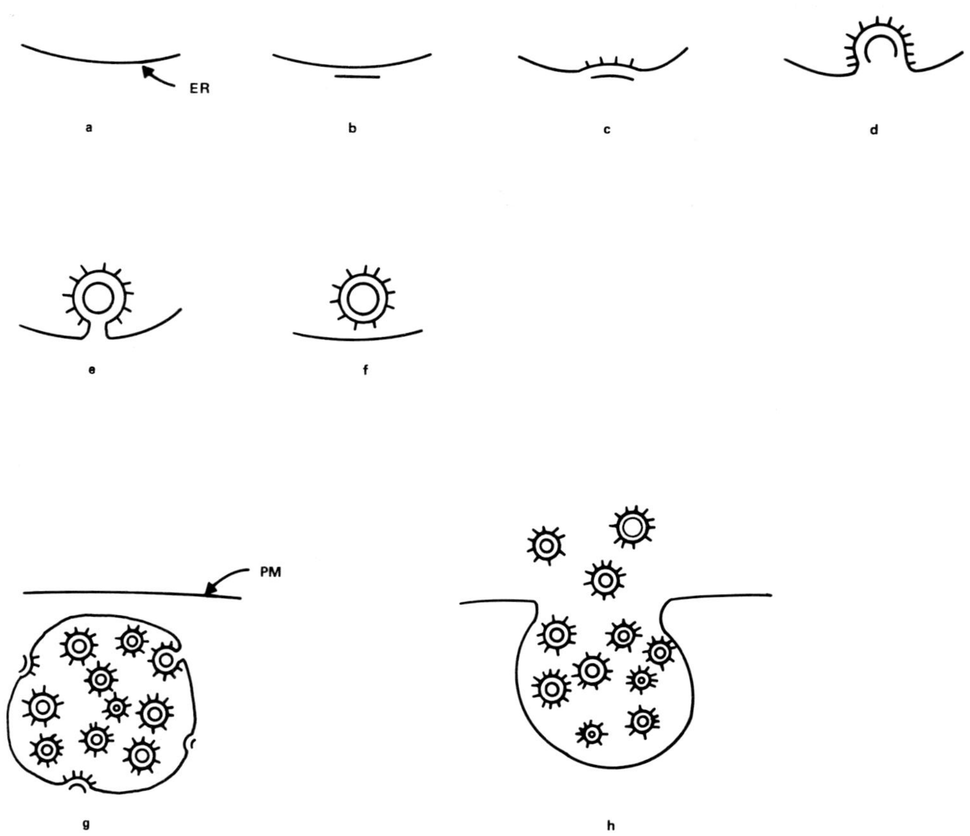

FIGURE 3. Schematic drawing of proposed stages of human coronavirus morphogenesis. ER, endoplasmic reticulum membrane; PM, plasma membrane. (a to f) Budding of an individual coronavirus particle. (g and h) Reverse pinocytotic release of coronavirus particles from the cell cytoplasm.

Table 2
HUMAN CORONAVIRUSES

			Laboratory diagnosis				
			Isolation technique				
Virus strain	Diseases associated	Source of specimen	Cell system	Animal	FA or EM[a]	Serologic diagnosis[b]	Ref.
B814	Upper respiratory tract infection	Throat washing	HETOC[c]	None	EM of organ culture harvests	None	33
229E	Upper respiratory tract infection, ? infant pneumonia	Throat swab, throat washing	WI-38, MA-177, other diploid	None	FA of nasal washes[d]	CF, neut., IHA	21
OC38, OC43	Upper respiratory tract infection	Throat washing	HETOC plus EM	Suckling mouse	FA of nasal washes,[d] EM of organ culture harvests	CF, HI	7

[a] EM, electron microscopy; FA, fluorescent antibody.
[b] CF, complement fixation; neut., neutralization; IHA, indirect hemagglutination; HI, hemagglutination inhibition.
[c] Human embryonic tracheal organ culture.
[d] Preliminary data from human volunteers (see text).

was subsequently adapted to tissue culture growth and shown to resemble 229E, but differ from it in neutralization kinetics.[22]

There are two (possibly three) other HCV strains which form a second homogeneous group. These viruses, designated OC38, OC43, and perhaps OC44, were recovered in HETOC from adults with colds.[7] Strains OC38 and OC43 were successfully adapted to growth in suckling mouse brain. Both strains cross-reacted in CF with several antigenic types of MHV. Both strains hemagglutinate, and both strains have subsequently been adapted to grow in primary monkey kidney cell monolayers and WI-38 cells.[23] Strains OC38 and OC43 are antigenically indistinguishable, and OC44 is similar, if not identical.

The remaining HCV strains are heterogeneous. One strain, B814, has been moderately well characterized and appears to be antigenically unique. This strain was originally recovered in HETOC, but was adapted by Bradburne and Tyrrell[8] to grow in the continuous L132 line. However, this adaptation has been difficult to reproduce in other laboratories.

Other strains (EVS, OC16, OC37, and OC48) grow only in HETOC and are poorly characterized. They are variably related in several tests of antigenicity to each other and to OC43, 229E, and MHV.[18,22] The HCV strain recovered from human feces[2] has not been examined for its possible antigenic relationships to other human or animal coronaviruses.

Clinical Features

The clinical features of HCV infection are those of respiratory disease. Infections, with consequent upper respiratory tract illness, apparently occur worldwide. The highest incidence of infection is in the winter and spring, but infections have been detected in most months of the year.[29]

The most common illness due to coronavirus infection is the common cold. The precise characterization of these colds is best made by analysis of studies in volunteers.[25] The incubation period under these conditions is approximately 3 days, and rhinorrhea, sore throat, and malaise are the most common symptoms. Fever occurs, but not commonly.

Other respiratory illnesses have been described associated with coronavirus infection. These include pneumonia, acute wheezing (in asthmatic children), and exacerbations of respiratory symptoms in adults with chronic pulmonary disease.

Laboratory Diagnosis

Because of the difficulty of cultivating HCV strains in vitro, most diagnoses will inevitably be based on serologic evidence of infection. The following section will describe in as much detail as possible the methods used for virus isolation and characterization, as well as the most useful serologic techniques. Only the respiratory HCV will be considered.

Specimens used for virus isolation are not different from those employed for cultivation of other respiratory viruses. They include nasal washes, oropharyngeal swabs immersed in suitable carrying media, suction specimens, etc. Coronaviruses have been successfully propagated from such specimens after freezing at −70°C and thawing, and there is no reason to believe that HCV are more labile in human secretions or specimens obtained from them than are, for example, paramyxoviruses. Usually, when coronaviruses are being cultured, other conventional viruses have already been sought and not found. Thus, most specimens have already been frozen and thawed once.

Two in vitro cell systems may be used for the cultivation of coronaviruses, organ culture, and tissue culture. Because the former technique is not widely used, a description of it will be included here.

The trachea from a human fetus of 15 to 22 weeks' gestational age is removed aseptically from the larynx to the first bifurcation. It is washed several times in antibiotic-supplemented Hanks' balanced salt solution, trimmed, and slit length-wise along the posterior aspect with pointed, straight-tipped scissors. At this stage ciliary beating is easily observed in adequate tissue specimens by observing the epithelial surface under incident light using a dissecting microscope. It is helpful also to wash excess mucus from the epithelium by squirting a stream of medium at it from a syringe fitted with a No. 25 needle.

The trachea is cut with a scalpel into 2- to 3-mm square fragments, one to four of which are placed ciliated side up on the scratched surface of a 6-cm Petri dish and 1.25 ml of medium (either Medium 199 or Leibovitz L-15 supplemented with glutamine, antibiotics, and 0.5% bovine albumin) is placed in the dish. It is incubated at 33°C in either humidified CO_2 (if Medium 199 is used) or in a sealed, humidified plastic box (if L-15 is used). It has been the custom of this author to use a rocker platform for incubation, but this may not be necessary. The fragments are observed daily for ciliary activity.

Inoculation of duplicate plates is performed by dropping the specimen directly on the fragments with a Pasteur pipette. The medium is changed daily or every other day, and the removed (and, presumably, virus-containing) medium is pooled and stored at −70°C. Cultures may be kept for several weeks, but useful harvests are obtained for only approximately 14 days.

Virus detection is perhaps the most difficult part of coronavirus recovery in organ culture. Several methods have been used, none of which is entirely satisfactory. First, harvested fluids pooled from days 3 to 9 of cultivation can be inoculated back into tissue culture. No coronaviruses have been detected by this technique, although it should theoretically be possible to recover some strains in which titer may have been enhanced by the HETOC passage and which then will have become cytopathic in tissue culture.

Second, the observer can measure ciliary activity in inoculated HETOC and compare it with that in control plates. Certain strains induce a premature cessation of ciliary motion, which is seen consistently on subpassage and is evidence of presumed microbial activity. Identification of the source of such ciliary immobilizing effect must then be made by other means.

A third method of detecting viruses is examination of pooled cell pellets or concentrated HETOC harvests by electron microscopy. One to five milliliters of medium harvested on days 3 to 10 after inoculation is placed in a swing-out rotor and centrifuged at 40,000 r/min for 2 hr. The pellet is taken up in a few drops of distilled water and stained on Formvar®-coated grids with phosphotungstic acid. These are then scanned at magnification × 20,000 for characteristic coronavirus particles. Final identification of coronaviruses may depend on neutralization with known antisera, immune electron microscopy,[26] or other methods.

As the entire field of coronavirus culture is still experimental, other methods of virus detection in infected cultures may prove to be superior to those mentioned above. British workers have used human volunteer inoculation as a screening procedure with some success.[27] Fluorescence microscopy of infected ciliary cells may prove to be a valuable method,[18] and interference with the growth of other viruses may also be useful. Immune electron microscopy may be helpful in increasing the sensitivity of recognition and identification of HCV strains.[26]

Several tissue culture lines and strains are susceptible to both growth and cytopathic effects (CPE) of 229E-like coronaviruses.[4] For recovery of virus from clinical specimens, this author has found that best results are obtained with either WI-38 (or some similar diploid human fetal lung fibroblast strain) or MA 177 cells (human embryonic intestinal cells, obtainable from Microbiological Associates, Bethesda, Maryland).

CPE may not be clear on first or even second passage. The cells may simply appear "dirty" after 7 to 10 days in culture. Eventually, however, a focal progressive CPE develops which is characteristic. Cells in foci develop granulation and fall off the glass without significant rounding, swelling, etc. In some cell systems syncytia may be produced.

Identification of the agent producing CPE depends on traditional methods (chloroform lability, absence of hemadsorption, resistance to IUdR) and, ultimately, both electron microscopy of cell supernates or lysates and neutralization with specific antiserum. The latter is not available commercially at this time.

Because of the difficulty of cultivating respiratory coronaviruses in vitro, it would appear that other methods directed at detecting virus in nasal or other respiratory secretions would be especially useful. Reports of direct electron microscopy or fluorescent antibody staining have not appeared. However, preliminary experiments using the methods of Gardner and McQuillin[28] have demonstrated brightly fluorescing cells in the secretions of volunteers infected with strains 229E and OC38.[29]

Serologic diagnosis depends on the demonstration of a fourfold or greater rise in antibody. CF (for strains 229E and OC43 or OC38) and hemagglutination inhibition (for strain OC43 or OC38) are readily available, and other tests (neutralization, indirect hemagglutination, indirect fluorescence) have been described but not widely applied.[16,30,31]

CF antigens for strain 229E are best made in either diploid fibroblast or L-132 cells, using either cell pack antigens or concentrated supernatant fluids. Infected suckling mouse brain suspensions make highly satisfactory antigens for strain OC43.[32] Likewise, infected suckling mouse brain suspensions may be used for measurement of hemagglutination inhibition antibody to strain OC43. Although human type O, vervet, mouse, rat, or fowl erythrocytes may be used, this author has found the last to be the most satisfactory. The test is best carried out at room temperature, using 2 to 4 hemagglutinating units.

Although two coronavirus strains, OC38 and OC43, were successfully adapted to growth in suckling mouse brain, the adaptation was slow, and attempts to repeat it with other organ culture grown strains or 229E were unsuccessful. The disease produced was an encephalitis, with death of 0- to 1-day old animals in 2 to 4 days after intracranial or intraperitoneal injection. Virus was not present in lungs or liver. A similar illness can be produced in suckling hamsters.

GENERAL REFERENCES

Bradburne, A. F. and Tyrrell, D. A. J., Coronaviruses of man, *Prog. Med. Virol.*, 13, 373, 1971.

McIntosh, K., Coronaviruses: a comparative review, *Curr. Top. Microbiol. Immunol.*, 63, 85, 1974.

Monto, A. S., Medical review; Coronaviruses, *Yale J. Biol. Med.*, 47, 234, 1974.

Kapikian, A. Z., The coronaviruses, *Dev. Biol. Stand.*, 28, 42, 64, 1975.

McIntosh, K., Coronaviruses as causes of disease: clinical observations and diagnosis, in *Comparative Diagnosis of Viral Diseases,* Vol 1 (Part A) Kurstak, E., Ed., Academic Press, New York, 1977.

REFERENCES

1. **Tyrrell, D. A. J., Almeida, J. D., Berry, D. M., Cunningham, C. H., Hamre, D., Hofstad, M. S., Mallucci, L., and McIntosh, K.**, Coronaviruses, *Nature*, 220, 650, 1968.
2. **Caul, E. O. and Clarke, S. K. R.**, Coronavirus propagated from patient with nonbacterial gastroenteritis *Lancet*, 2, 953, 1975.
3. **Berry, D. M., Cruickshank, J. G., Chu, H. P., and Wells, R. J. H.**, The structure of infectious bronchitis virus, *Virology*, 23, 403, 1964.
4. **McIntosh, K.**, Coronaviruses: a Comparative Review, *Curr. Top. Microbiol. and Immunol.*, 63, 85, 1974.
5. **Cowen, B. S. and Hitchner, S. B.**, pH stability studies with avian infectious bronchitis virus (coronavirus) strains, *J. Virol.*, 15, 430, 1975.
6. **Kapikian, A. Z., James, H. D., Jr., Kelly, S. J., Dees, J. H., Turner, H. C., McIntosh, K., Kim, H. W., Parrott, R. H., Vincent, M. M., and Chanock, R. M.**, Isolation from man of "avian infectious bronchitis virus-like" viruses (coronaviruses) similar to 229E virus, with some epidemiological observations, *J. Infect. Dis.*, 119, 282, 1969.
7. **McIntosh, K., Dees, J. H., Becker, W. B., Kapikian, A. Z., and Chanock, R. M.**, Recovery in tracheal organ cultures of novel viruses from patients with respiratory disease, *Proc. Natl. Acad. Sci.* U.S.A., 57, 933, 1967.
8. **Bradburne, A. F. and Tyrrell, D. A. J.**, The propagation of "coronaviruses" in tissue culture, *Arch. Gesamte Virusforsch.*, 28, 133, 1969.
9. **Tannock, G. A.**, The nucleic acid of infectious bronchitis virus, *Arch. Gesamte Virusforsch.*, 43, 259, 1973.
10. **Watkins, H., Reeve, P., and Alexander, D. J.**, The ribonucleic acid of infectious bronchitis virus, *Arch. Virol.*, 47, 279, 1975.
11. **Hierholzer, J. C., Palmer, E. L., Whitfield, S. G., Kaye, H. S., and Dowdle, W. R.**, Protein composition of coronavirus OC43, *Virology*, 48, 516, 1972.
12. **Pokorny, J., Bruckova, M., and Ryc, M.**, Biophysical properties of coronavirus strain OC43, *Acta Virol.*, 19, 137, 1975.
13. **Bradburne, A. F. and Tyrrell, D. A. J.**, Coronaviruses of man, *Prog. Med. Virol.*, 13, 373, 1971.
14. **Kaye, H. S. and Dowdle, W. R.**, Some characteristics of hemagglutination of certain strains of "IBV-like" virus, *J. Infect. Dis.*, 120, 576, 1969.
15. **Becker, W. B., McIntosh, K., Dees, J. H., and Chanock, R. M.**, Morphogenesis of avian infectious bronchitis virus and a related human virus (strain 229E), *J. Virol.*, 1, 1019, 1967.
16. **Hamre, D., Kindig, D. A., and Mann, J.**, Growth and intracellular development of a new respiratory virus, *J. Virol.*, 1, 810, 1967.
17. **Oshiro, L. S., Schieble, J. H., and Lennette, E. H.**, Electron microscopic studies of coronavirus, *J. Gen. Virol.*, 12, 161, 1971.
18. **McIntosh, K., Kapikian, A. Z., Hardison, K. A., Hartley, J. W., and Chanock, R. M.**, Antigenic relationships among the coronaviruses of man and between human and animal coronaviruses, *J. Immunol.*, 102, 1109, 1969.
19. **Witte, K. H., Tajima, M., and Easterday, B. C.**, Morphologic characteristics and nucleic acid type of transmissible gastroenteritis virus of pigs, *Arch. Gesamte Virusforsch.*, 23, 53, 1968.
20. **Bucknall, R. A., Kalica, A. R., and Chanock, R. M.**, Intracellular development and mechanism of hemadsorption of a human coronavirus, OC43, *Proc. Soc. Exp. Biol.*, 139, 811, 1972.
21. **Hamre, D. and Procknow, J. J.**, A new virus isolated from the human respiratory tract, *Proc. Soc. Exp. Biol. Med.*, 121, 190, 1966.
22. **Bradburne, A. F.**, Antigenic relationships amongst coronaviruses, *Arch. Gesamte Virusforsch.*, 31, 352, 1970.
23. **Bruckova, M., McIntosh, K., Kapikian, A. Z., and Chanock, R. M.**, The adaptation of two human coronavirus strains (OC38 and OC43) to growth in cell monolayers, *Proc. Soc. Exp. Biol. Med.*, 135, 431, 1970.
24. **Monto, A. S.**, Medical review; coronaviruses, *Yale J. Biol. Med.*, 47, 234, 1974.
25. **Bradburne, A. F., Bynoe, M. L., and Tyrrell, D. A. J.**, Effects of a "new" human respiratory virus in volunteers, *Br. Med. J.*, 3, 767, 1967.
26. **Kapikian, A. Z., James, H. D., Kelly, S. J., and Vaughn, A. L.**, Detection of coronavirus strain 692 by immune electron microscopy, *Infect. Immun.*, 7, 111, 1973.
27. **Tyrrell, D. A. J., Bynoe, M. L., and Hoorn, B.**, Cultivation of "difficult" viruses from patients with common colds, *Br. Med. J.*, 1, 606, 1968.
28. **Gardner, P. S. and McQuillin, J.**, *Rapid Virus Diagnosis: Application of Immunofluorescence*, Butterworths, London, 1974.
29. McIntosh, K., McQuillin, J., Reed, S., and Gardner, P. S., *J. Med. Virol.*, in press.

30. **Cavallaro, J. J. and Monto, A. S.,** Community-wide outbreak of infection with a 229E-like coronavirus in Tecumseh, Michigan, *J. Infect. Dis.,* 122, 272, 1970.
31. **Kaye, H. S. and Dowdle, W. R.,** Seroepidemiologic survey of coronavirus (strain 229E) infections in a population of children, *Am. J. Epidemiol.,* 101, 238, 1975.
32. **McIntosh, K., Becker, W. B., and Chanock, R. M.,** Growth in suckling-mouse brain of "IBV-like" viruses from patients with upper respiratory tract disease, *Proc. Natl. Acad. Sci.* U.S.A., 58, 2268, 1967.
33. **Tyrrell, D. A. J. and Bynoe, M. L.,** Cultivation of a novel type of common-cold virus in organ cultures, *Br. Med. J.,* 1, 1467, 1965.

RETRAVIRIDAE: RNA TUMOR VIRUSES

John E. Dahlberg

Introduction

RNA tumor viruses are found in many vertebrate species. Even in species from which infectious virus cannot be recovered, the genome of these viruses may be present in a latent form ("provirus") in the cellular DNA. This results from what appears to be a unique feature of the retraviruses: their ability to reverse the normal flow of genetic information from DNA to RNA by utilizing their own viral encoded RNA-dependent-DNA-polymerase. When an RNA tumor virus infects a cell, it is capable of producing a DNA copy of its RNA genome. This viral DNA may then integrate into the cellular DNA and become a permanent part of the cellular genome. Progeny cells continue to inherit the viral genome and may be induced to produce virus many generations later. Not only can this be demonstrated for infectious laboratory strains of virus, but additional evidence has proven that such events have occurred for millions of years, and that the integrated viral genome may eventually lose the ability to be activated. Thus, DNA sequences related to infectious viral RNA can often be observed in closely related species of animals, even where infectious virus cannot be obtained by any method.

This mechanism of transfer of virus is termed vertical transmission, in contrast to the more usual lateral spread of virus from cell to cell or animal to animal which is termed horizontal transmission. Thus, although a human RNA tumor virus has not been obtained at present, it is known that human cellular DNA contains several different sets of sequences related to viruses isolated from several species of monkeys. It is not yet known whether it will be possible to induce human cells to produce infectious virus from these or additional latent viral genomes. In primates carrying similar DNA sequences, it is sometimes possible to obtain virus while in other species efforts have been unsuccessful.

Over the past 60 years, and particularly during the past 20 years, RNA tumor viruses have been isolated from chickens, mice, rats, hamsters, cats, monkeys, and many other species. Many of these viruses cause leukemias and sarcomas in appropriate host animals, and some of these viruses, but not all, cause malignant transformation of normal cells grown in culture. These viruses tend to be highly infectious and grow to high titer in tissue cultures and/or animals. More recently, a number of chemical treatments have been successfully used to activate latent endogenous RNA tumor viruses from cells where they normally are not expressed. In contrast to the usual laboratory strains of virus, these endogenous viruses tend to be less infectious and to grow poorly.

To date, a human RNA tumor virus has not been isolated. This chapter will present a brief overview of the retraviruses occurring in animals and concentrate on those properties of general interest that should approximate those expected of a putative human virus.

A note on classification — The classification of RNA tumor viruses and other viruses related to them is still a controversial area because new isolates with differing morphological and serological properties are being obtained regularly. A recent symposium[1] summarized current findings and has led to the general adoption of a tentative grouping of the "retraviridae" (viruses with an RNA-dependent-DNA-polymerase) (Table 1). Only groups B, C, and D are known to include oncogenic viruses and, in fact, only a minority of type C viruses are known to be oncogenic.

Morphology

All retraviruses have certain morphological properties in common. They bud

Table 1
CLASSIFICATION OF RETRAVIRIDAE

Virus group	Oncogenic members[a]	Host species
Intracisternal A (+R)	—	Mouse, hamster, guinea pig, rat, cat, gerbil, others
Oncornavirus B[b]	+	Mouse, guinea pig, baboon
Oncornavirus C	+	Mouse, chicken, rat, cat, hamster, pig, baboon, viper, etc.
Oncornavirus D[b]	+	Monkeys of New World and Old World (e.g., squirrel monkey, langur, rhesus monkey), etc.
Oncornavirus E	—	Sheep (visna and maedi viruses)
Oncornavirus F[b]	—	Foamy viruses of monkeys, man, cat, and cow

[a] + = includes oncogenic viruses.
[b] Includes an intracytoplasmic A particle as potential virion core precursors.

through a suitably altered cellular membrane and are usually 90 to 110 nm in diameter. They all contain a single- or double-shelled nucleoid and surface spikes. Within this general description, however, characteristic differences are used to define the various groups.

The illustrations in this chapter depict the morphogenesis of the major groups of retraviruses. Included for each is its intracellular form where it can be detected, budding morphology, and appearance of extracellular virions. Except for intracisternal A particles, whose inclusion in the retraviruses can best be considered a historical accident, all members of these groups are released from the cell, contain a high molecular weight RNA genome, and a reverse transcriptase.

Figure 1 illustrates murine intracisternal A particles present in a rhabdomyosarcoma cell. Note how the particles bud either from the membrane of the endoplasmic reticulum or from other particles. Although these particles are not related to other retraviruses and are not released from the cell (and thus are not transmitted horizontally), they do occur at a high level predominantly in early embryonic cells and certain types of murine tumors. Their role in the neoplastic process is unknown.

Figures 2, 3, and 4 illustrate the intracisternal A particles seen in leukemic cells of the guinea pig. Although they also bud from the endoplasmic reticulum, in contrast to the murine A particle, they are released from necrobiotic tumor cells and undergo maturation of the nucleoid (Figure 4) in much the same way as type C oncornaviruses. Nevertheless, they are not immunologically related to any other group of oncornavirus and should be considered an intracisternal A particle with atypical morphogenesis.

Figures 5 to 8 depict the morphological development of the murine mammary tumor virus (MMTV), the prototype B virus. Figure 5 shows a cluster of intracytoplasmic A particles present in the cytoplasm of a cell from a murine mammary tumor. These particles are immunologically related to type B virions and may occasionally act as precursors of budding virions. Figure 6 shows a row of budding virions and illustrates how most of the double-shelled nucleoids are formed prior to the complete budding of the membrane. Preformed nucleoids also are characteristic of type D viruses (not illustrated). Figure 7 shows that many of the mature type B virions have eccentrically located condensed nucleoids. Figure 8 depicts negatively stained mammary tumor virus, showing the long (100 Å) well-defined surface spikes.

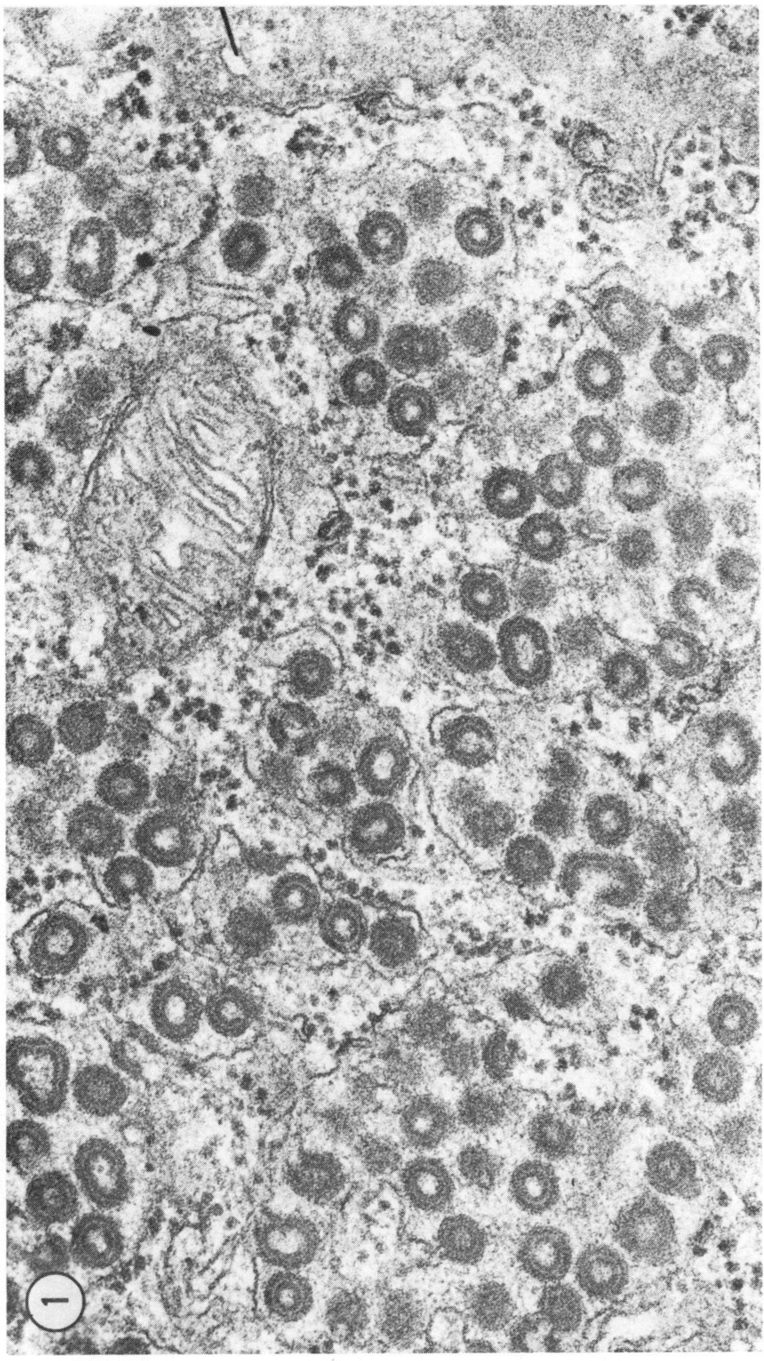

FIGURE 1. Intracisternal A particles (IAP) in a tissue culture derived from a murine rhabdomyosarcoma. These mouse tumor cells contain many thousand such virus-like particles. They can bud either from the membranes of the endoplasmic reticulum or from each other. The two electron dense layers are separated by an electron lucent layer, and the particles average about 80 nm in diameter. Extracellular murine intracisternal A particles are not observed, and detergents are required to separate them from cell membranes. (Magnification × 80,000.)

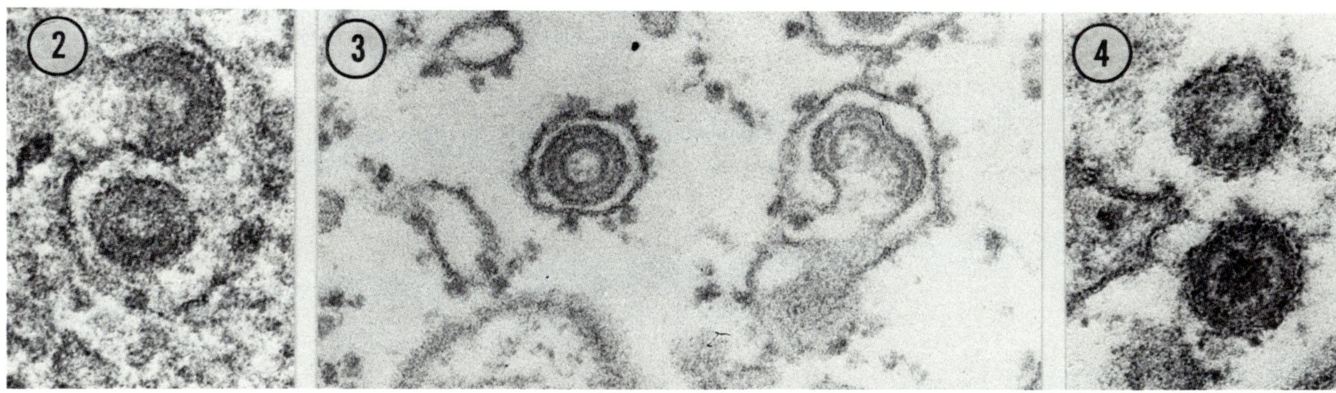

FIGURE 2. Intracisternal A particles present in the transplantable L_2C leukemia of guinea pigs. As in the mouse, these particles bud from membranes of the endoplasmic reticulum. They can be distinguished from murine IAP by their larger size (~ 100 nm) and an intermediate electron dense layer between the particle core and membrane. (Magnification x 160,000.) **FIGURE 3.** Guinea pig IAP located in cell debris and still associated with endoplasmic reticulum. The intermediate layer is particularly noticeable in these particles. (Magnification x 160,000.) **FIGURE 4.** Extracellular guinea pig IAP with condensed nucleoids are readily detected in cellular debris associated with late stages of the L_2C leukemia. This process of morphological maturation is not observed with murine IAP and is reminiscent of the maturation process characteristic of type B and type C oncornaviruses. (Magnification x 160,000.)

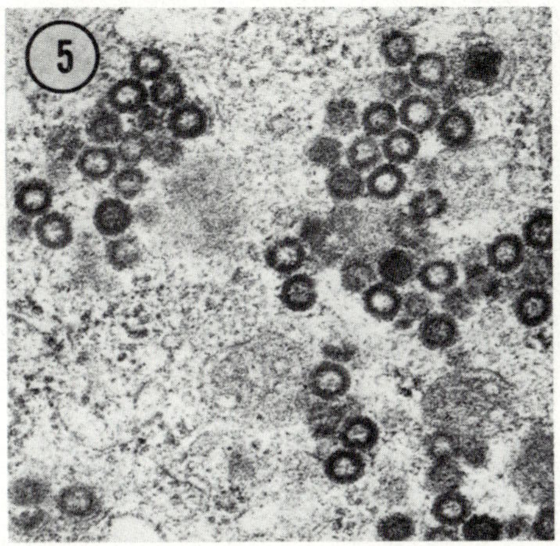

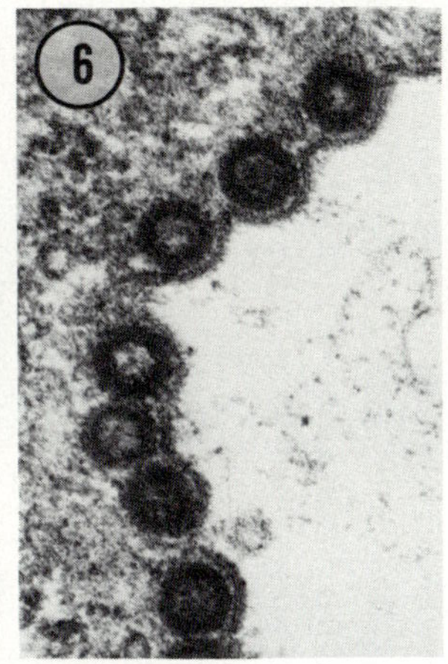

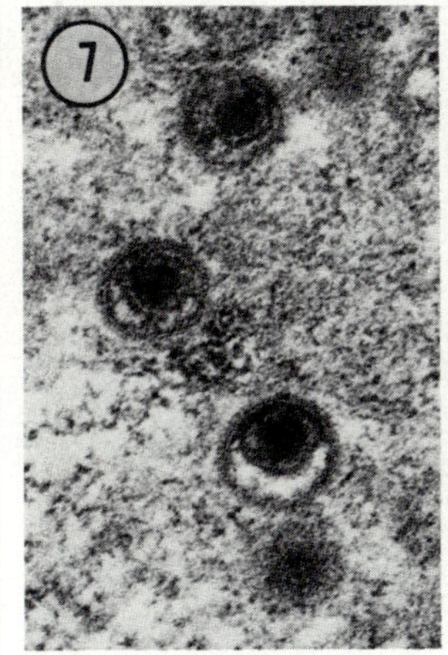

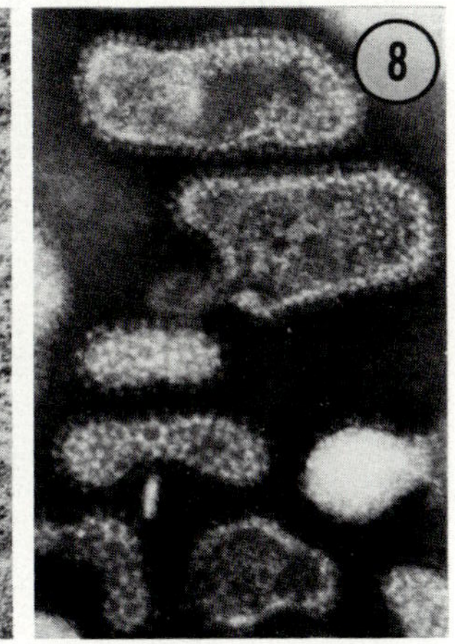

FIGURE 5. Murine intracytoplasmic A particles. This cluster of particles was detected in a cell from a spontaneous mammary carcinoma of a C_3H mouse. They can be distinguished from intracisternal A particle by their cytoplasmic location, rather than cisternal, and by their forming without budding through a membrane. (Magnification × 40,000.) **FIGURE 6.** Budding virions of mouse mammary tumor virus, MMTV, the prototype type B virus. In most cases, buds utilize preformed nucleoids morphologically indistinguishable from intracytoplasmic A particles. (Magnification × 80,000.) **FIGURE 7.** Extracellular MMTV virions present in C_3H mammary tumors have condensed nucleoids with a distinct intermediate layer. Many such virions have eccentrically located nucleoids. The characteristic fringe of 10-nm long spikes is apparent on the upper half of the lower virus particle. (Magnification × 80,000.) **FIGURE 8.** Negatively stained MMTV. These particles tend to be pleomorphic primarily due to the surface tension developed during specimen preparation, but the prominent surface spikes are readily apparent. (Magnification × 80,000.)

Figures 9 to 13 illustrate the morphology of type C oncornaviruses. This morphology is generally characteristic of type C viruses regardless of the species of origin. Figure 9 is a micrograph of a murine tumor cell showing that budding virions can often be observed both at the external plasma membrane and in internal vacuole membranes. It should be emphasized that these vacuole membranes are internalized plasma membranes, and the virus is, in reality, limited to using the plasma membrane to bud from. Many tumor cells do not contain such a network of vacuole membranes and internal budding is correspondingly rare. Figure 10 shows that budding virions are characterized by the synchronous formation of the nucleoid and budding membrane. Figures 11 and 12 are of immature and mature extracellular type C virions. Figure 12 illustrates how the mature condensed nucleoids of type C virions (in contrast to MMTV; Figure 7) tend to be centrally located. Mature type D virions are often distinct from both type B and type C particles, with the mature nucleoid of Mason-Pfizer monkey virus (MPMV) resembling a hollow cylinder (not shown).

As an example of a retravirus that is not a tumor virus (oncornavirus groups E and F), Figures 14 and 15 depict stages in the morphogenesis of a foamy virus observed in Rhesus monkey tissue cultures. Foamy viruses contain a RNA-dependent-DNA-polymerase and a high molecular weight RNA genome, but it is not known whether they can be vertically transmitted, and they do not appear to be oncogenic. Foamy viruses often cause a sudden and extensive cytopathic effect, characterized by extensive vacuolization (thus the foamy appearance), when cells are placed in culture. There are no known clinical symptoms or diseases associated with infection in vivo. By electron microscopy, intracytoplasmic particles can be found free in the cytoplasm and in association with the endoplasmic reticulum (Figure 14). Budding through the endoplasmic reticulum, as well as through the plasma and vacuole membranes, is accompanied by the insertion of long, distinctively shaped spikes into the membrane. After cell lysis, large quantities of extracellular virus, often associated with cellular debris, can be found. Note in Figure 15, the presence of long spikes on the periphery of the virions and the relatively small nucleoid.

Physicochemical Properties

Those retraviruses which have been extensively studied have 6 to 7 major structural proteins ranging in size from 10 to 70 × 10^3 daltons. There typically are two glycoproteins located in the virion envelope and a major internal protein of approximately 30,000 daltons. The single-stranded RNA genome is a dimer with a combined molecular weight of 6.5 to 7 × 10^6 daltons (with a sedimentation coefficient of 60 to 70 S). The RNA-dependent-DNA-polymerase comprises about 1% of the mass of the virion protein and has a molecular weight of approximately 70,000 daltons. Retraviruses have a density of 1.16 to 1.19 g/ml in sucrose and 1.17 to 1.21 in cesium chloride.

Virus Replication

Retraviruses, with foamy viruses being a possible exception, are unusual in their ability to generally avoid destroying a cell after productively infecting it. In most cases, even where enormous quantities of virus are being produced, the cell continues to metabolize and divide normally. Nevertheless, for many retraviruses, assay systems are available where infection can lead to the formation of plaques under suitable conditions.

It is important to distinguish between two types of replication for retraviruses based on whether exogenous virus infects a cell, or whether a latent virus is activated within a cell. In the first case, the virus must successfully penetrate the cell, uncoat, and with its transcriptase produce a DNA copy of its RNA genome. The viral DNA then can synthesize messenger RNA which in turn enables viral structural proteins to be made,

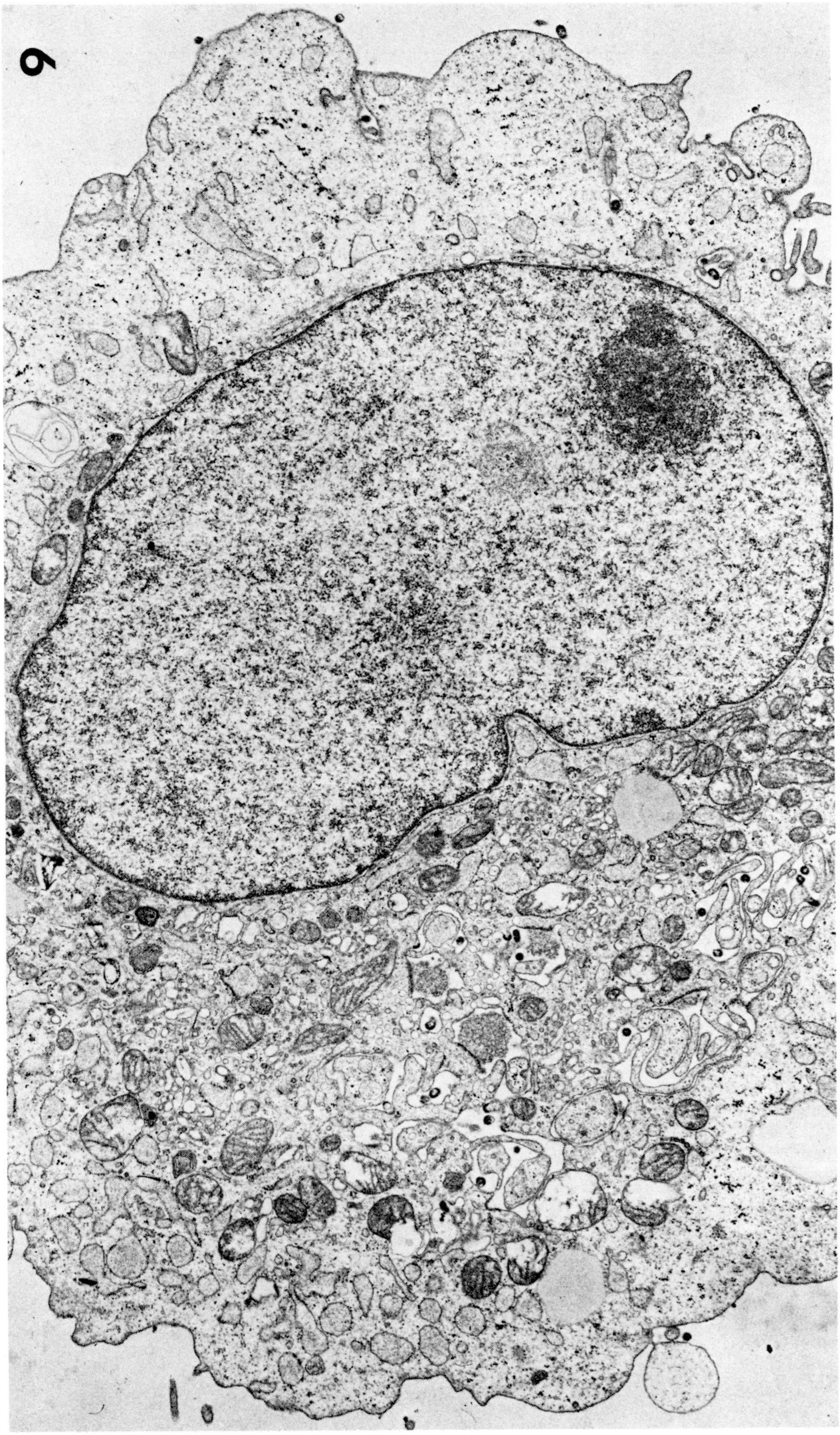

FIGURE 9. This figure is a low power electron micrograph of a mouse cell producing type C virions at both the cell periphery and from internal vacuole membranes. It should be emphasized that these vacuole membranes are internalized plasma membranes, and that type C viruses inevitably bud from the cell through the plasma membrane, regardless of its location. This micrograph should be contrasted with Figure 1 to underscore the distinction between vacuole membranes used by type C (and other) oncornaviruses, and the endoplasmic reticulum used by intracisternal A particles. Faulty fixation occasionally leads to confusion and misinterpretation of the identity of intracellular virus-like particles. (Magnification ×16,000.)

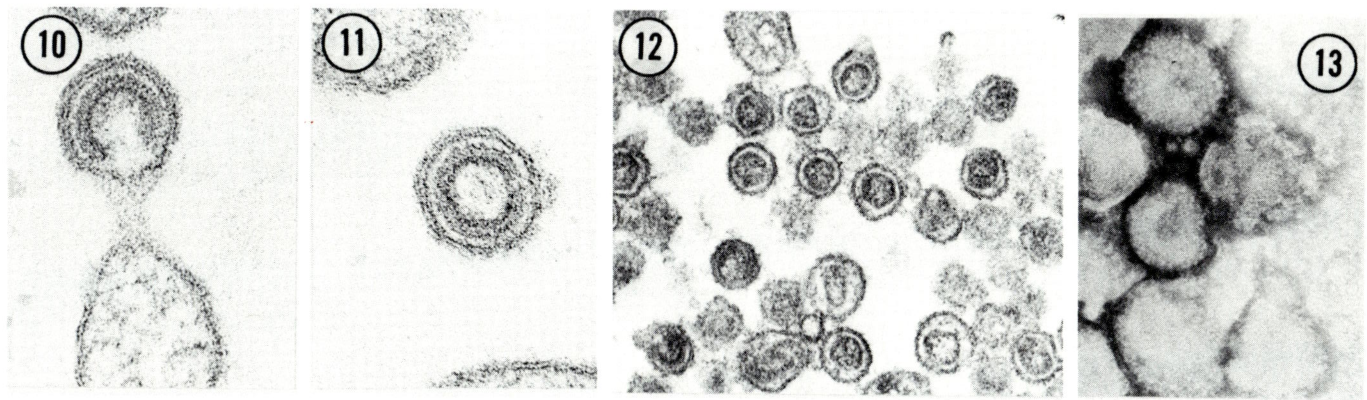

FIGURE 10. This figure is a typical budding type C virus on the surface of a mouse cell. Note how the nucleoid is formed during budding and its degree of completion parallels the degree to which the membrane and portion of the bud has been completed. Contrast this bud to those of MMTV (Figure 6). (Magnification x 160,000.) **FIGURE 11.** A murine type C virus particle recently released from a cell. (Magnification x 160,000.) **FIGURE 12.** A cluster of extracellular murine type C virus. These virions have matured and are characterized by condensed nucleoids without intermediate layers (mature avian type C viruses do have intermediate layers). Eccentrically located nucleoids are only rarely encountered. Contrast to Figure 7. (Magnification x 80,000.) **FIGURE 13.** This micrograph depicts negatively stained murine type C virions and shows that their surface spikes, while readily apparent, are much shorter than those on the type B virus, as shown in Figure 8. (Magnification x 160,000.)

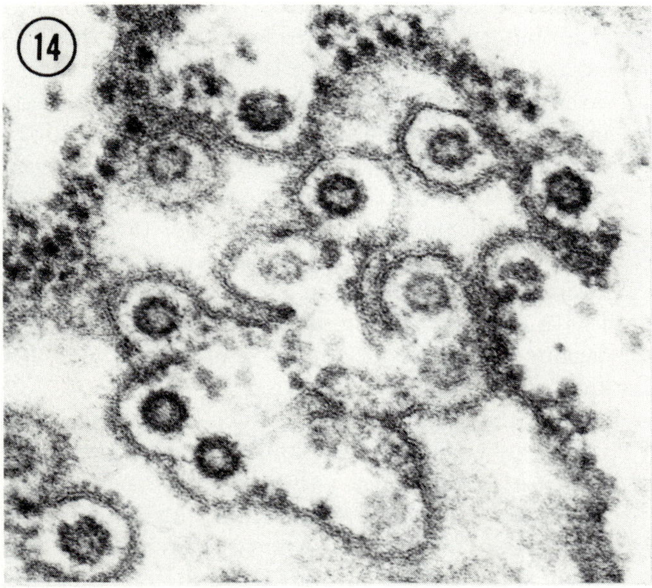

FIGURE 14. This figure illustrates the intracellular appearance of foamy virus growing in rhesus monkey kidney cells in tissue culture. Although foamy virus infection is inapparent in vivo, the culturing of cells from infected animals rapidly leads to the development of cytopathogenicity and, at the ultrastructural level, massive quantities of cell associated and released foamy virus particles. The virus buds from both the membrane of the endoplasmic reticulum and from plasma and vacuole membranes. (Magnification x 160,000.)

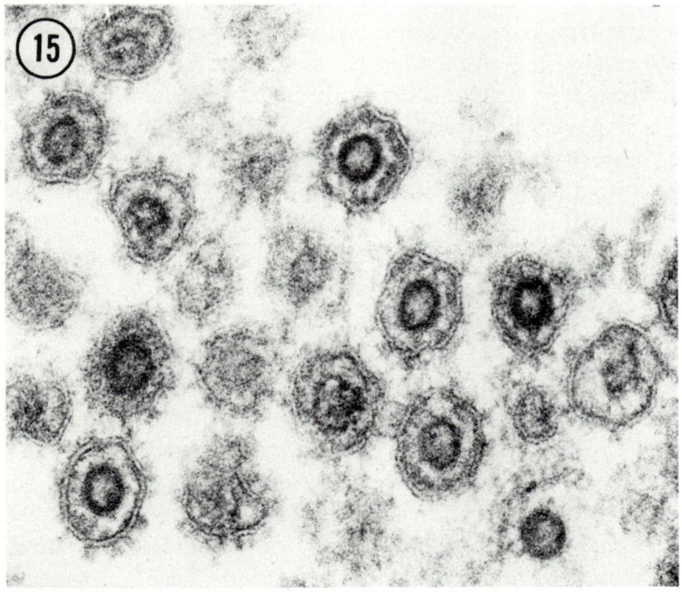

FIGURE 15. Extracellular foamy virus in cultures of rhesus monkey kidney cells tend to form clusters and are characterized by having a nucleoid with a single electron dense layer and relatively few but large and distinctive surface spikes. (Magnification x 160,000.)

and progeny virus to be assembled and released. The requirement for DNA synthesis is presumably universal, but it is not established whether or not the DNA must become stably integrated into the cellular DNA. Details concerning the mechanism of DNA synthesis, integration into cellular DNA, and whether this integration is essential for virus production remain to be worked out for even the carefully studied system.

In the second case, messenger and virion RNA are apparently produced directly from the proviral DNA present in the cellular genome. Activation of virus production by drugs is transitory and indicates that what happens is the temporary interruption of a control mechanism which normally prohibits virus production. Activation of virus during oncogenic transformation, or during long-term tissue culture, on the other hand, appears to involve a permanent alteration of the cellular genome which, inadvertently perhaps, permits continuous deregulation of control over the proviral DNA and resultant constitutive virus production.

Figure 16 is a schematic diagram of a cell in which several types of oncornaviruses are continuously replicating. This diagram is limited to what can be seen at the ultrastructural level. In all cases, there appears to be a requirement for messenger RNA being produced from nuclear viral DNA (either integrated or not), but viral protein synthesis is cytoplasmic, and in some cases is known to involve both membrane bound and free polysomes. Type C oncornaviruses are known to synthesize large polycistronic polypeptides, equivalent to their entire genome, that are subsequently proteolytically cleaved to produce the virion structural proteins. At this point, presumably, ultrastructural changes such as the formation of intracytoplasmic structures and budding particles become apparent.

Intracisternal A and R particles are formed by budding through the membranes of the endoplasmic reticulum. In all cases, except for the guinea pig intracisternal A particles, these particles remain restricted to the cisternal spaces. Although different stages of budding can be observed, in the mouse they do not become detached from the endoplasmic reticulum membrane.

Types B, D, and F oncornaviruses have an intracellular form called intracytoplasmic A particles. In the case of MMTV, these intracytoplasmic particles are immunologically related to the type B virion nucleoid, and they may frequently become the nucleoid of the budding virion. As seen in Figure 16, intracytoplasmic A particles can be observed as clusters often quite distant from the plasma membrane. Incomplete particles can often be observed, both within clusters and at the cell periphery within budding forms, but the majority of buds have preformed nucleoids. Following budding, the nucleoids of type B and type D virions undergo a distinctive morphological change to form mature virions. It is not known whether this maturation is essential for the production of infectious virus, or whether maturation produces noninfectious virions. Foamy viruses (type F) can bud with preformed nucleoids from endoplasmic reticulum, plasma, and vacuole membranes. Infection by these viruses causes extensive vacuolization and swelling of the intracisternal spaces, and virus appears to be produced so rapidly that the cell is eventually destroyed.

Type C viruses are uniquely different from the other retraviruses shown because an intracytoplasmic stage is not apparent at the ultrastructural level. The structural proteins self-assemble only at the plasma membrane, and the nucleoid forms synchronously with the bud itself. Released virions undergo maturation to a form generally distinct from the other groups of retraviruses, although individual virions of one group can often be confused with those of another group. Type C virions usually have a centrally located electron dense core, and may or may not contain a thin intermediate layer between the condensed core and the envelope membrane.

This section has emphasized ultrastructural features of retravirus replication, particularly for continuously produced viruses, because relatively little is known at the mo-

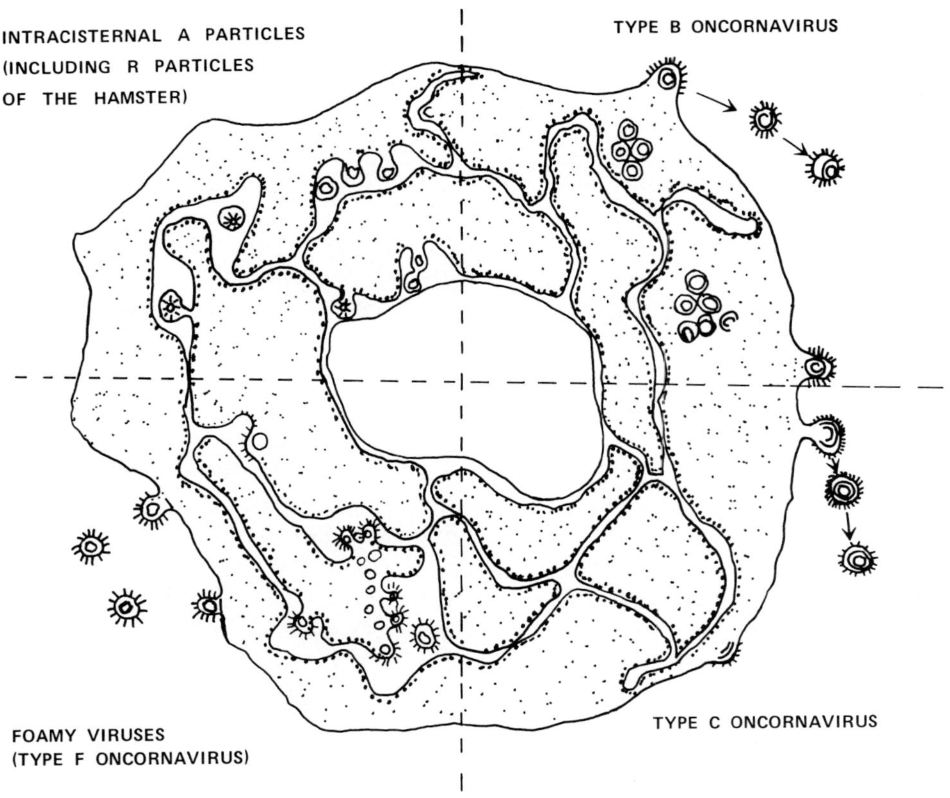

FIGURE 16. Schematic diagram of the replication of different types of oncornavirus particles. This depiction of a cell in which intracisternal A particles, type B, type C, and type F oncornaviruses are replicating, is limited to features observed at the ultrastructural level and is described in the text.

lecular level regarding the establishment of infection, the mechanism of host range control, or the mechanism of nucleic acid synthesis. Although much is now known regarding type C viral precursor polypeptide synthesis and subsequent cleavage to produce structural proteins, very little is known about how these proteins are packaged into virions and where this happens, except that which can be observed in the electron microscope.

Virus Strains and Antigenic Properties

There are no known human retraviruses, except for one or more foamy viruses (oncornavirus F), for which little information is available. This chapter will not list known animal viruses, because it would be a very long list and of doubtful value in this context. However, a few comments on overall antigenic properties of the retraviruses seem to be in order. Intracisternal A and R particles are extraordinarily difficult to work with, and with the exception of the murine intracisternal A particle, no immunological relationships have been demonstrated either between different A and/or R particles or to other viruses. Murine intracisternal A particles have been shown to be immunologically unrelated to murine type B and type C viruses.

Murine type B viruses are interrelated but distinct from the type B guinea pig virus and from any other known oncornaviruses. All type C viruses dicovered to date can be shown to be immunologically related. This cross-reactivity resides principally within the major internal structural protein (28 to 30 × 10^3 daltons) and suitable immunological assays should react with similar proteins in human type C viruses. Of course, depending upon the serum and the assay system used, one can define many serotypes and distinguish between not only type C viruses of different species, but also different isolates from a single species. Type D oncornaviruses include several viruses recently isolated from both New and Old World monkeys which have cross-reactive internal structural proteins, as well as several additional viruses which are unrelated to other oncornaviruses. This group includes the bovine leukemia virus, which clearly is oncogenic.

Type E and type F viruses are distinct from all other groups and contain no members which are oncogenic. Relatively little information is available at the molecular level for either group and immunological relationships are not yet worked out in any detail.

Clinical Features and Laboratory Diagnosis

Since no human isolates are available, no clinical disease is known. Isolation or diagnosis of a putative oncornavirus of human origin has been repetitively attempted in many hospitals and laboratories. Initially, efforts were concentrated on detecting type C (or type B) virions in human tissue, blood, and tumor samples. The techniques most often used rely on assays for properties unique to oncornaviruses. These include centrifuging cell or tissue extracts in sucrose gradients and looking for virus particles or reverse transcriptase activity at densities of 1.15 to 1.17 g/ml. A modification of this technique involved a search for reverse transcriptase activity associated with a 60 to 70 S RNA. This "simultaneous detection test" involves the incorporation of small amounts of radioactive precursors to DNA onto the 70 S genome of an RNA tumor virus and is dependent upon whether reverse transcriptase is present in virus-infected cells. Positive results (a peak of radioactivity at 70 S) would indicate that cell or tissue extracts contained reverse transcriptase activity associated with a 70 S RNA even when complete oncornavirus particles could not be detected. Control experiments in virus-infected animal cells indicated that the technique was sensitive and quite specific. When conducted with extracts of normal and malignant human tissues, positive results were occasionally obtained in malignant tissues but not in control tissues. Complete virus has not been recovered from positive tissues, and a current working hypothesis is that if human cells do contain oncornavirus genomes, only a portion of that genome may be expressed, i.e., the replication is defective.

This has led in recent years to concentration on a search for viral markers, rather than for complete virus. For the most part, this means a search for nucleic acid sequences related to viral probes and assaying for individual proteins in human cells which are related to primate oncornavirus structural proteins. To date, this search has inconclusive results.

Pathogenicity in Experimental Animals

Infectious RNA tumor viruses exhibit an enormous range of pathogenic potential in their normal animal hosts, and of course the leukemia and sarcoma viruses of one species can often cause malignant disease in a variety of other animal species. Pathogenic responses range from a nonapparent infection which may last for the lifetime of the individual to a rapidly developing and fatal neoplasia. The vertically transmitted latent endogenous retraviruses, on the other hand, are rarely activated under natural conditions until after the animal has passed its breeding age. When endogenous viruses are expressed, as in populations of wild mice or in inbred strains of mice selected for high tumor incidence, they are frequently capable of causing leukemia and a variety of neuropathological diseases.

A striking property of the various laboratory strains of murine leukemia and sarcoma viruses is the specificity of the choice of target cell. Certain strains of virus transform only spleen cells, while others always produce myelogenous leukemia. Furthermore, specificity may include the ability to infect only certain strains of animals. It has been shown in the mouse that this is due to a set of genetic markers that define host range for different sets of murine oncornaviruses.

Many type C retraviruses do not appear to transform any cell, either in vivo or in vitro. These viruses may still be oncogenic under conditions not yet tested, and they may still be infectious for a wide variety of cells and animals. Retraviruses, other than types B, C, and perhaps D, apparently do not transform cells under any condition, although intracisternal A and R particles are often associated with transformed cells. Foamy viruses are associated with an asymptomatic infection in animals, but are cytopathic in vitro. It is likely that the immune system of the animal is involved in the control of these viruses in vivo.

A variety of oncornaviruses can infect and sometimes transform human cells in vitro. Of course, in vivo inoculation of human subjects cannot be attempted, except for laboratory accidents. To date, there are no known cases of human disease being caused by animal retraviruses.

GENERAL REFERENCES

Gross, L., *Oncogenic Viruses,* 2nd ed., Pergamon Press, New York, 1970.

Dalton, A. J. and Haguenau, F., *Ultrastructure of Animal Viruses and Bacteriophages: An Atlas,* Academic Press, New York, 1973.

Tooze, J., *The Molecular Biology of Tumor Viruses,* Cold Spring Harbor Laboratory, Cold Spring Harbor, New York, 1973.

Dalton, A. J., Melnick, J. L., Bauer, H., Beaudreau, G., Bentvelzen, P., Bolognesi, D., Gallo, R., Graffi, A., Haguenau, F., Heston, W., Huebner, R., Todaro, G., and Heine, U. I., The case for a family of reverse transcriptase viruses: retraviridae, *Intervirology,* 4, 201, 1974.

Gilden, R. F., Interrelationships among RNA tumor viruses and host cells, in *Advances in Cancer Research,* Vol. 22, Klein, G. and Weinhouse, S., Eds., Academic Press, New York, 1975, 157.

Aaronson, S. A. and Stephenson, J. R., Endogenous type-C RNA viruses of mammalian cells, *Biochim. Biophys. Acta,* 458, 323, 1976.

MYXOVIRIDAE: ORTHOMYXOVIRUSES – INFLUENZA VIRUSES

Peter Palese and Mary B. Ritchey

A human influenza virus was first isolated in 1933.[1] Since that time, much information regarding the disease and its viruses has been accumulated. Despite extensive study, influenza is still a major threat to man, especially to infants and elderly people, and causes major economic losses each year. Estimates of the costs of influenza pandemics or epidemics range from 1.7 to 3.9 billion dollars per year in the United States, based on calculations by Kavet[2] for the years 1963, 1966, and 1969. There is constant public concern about influenza, especially because every decade new pandemic strains emerge against which the population at large has little immunity.*

Morphology and Physical-chemical Properties

Three types of influenza viruses have been identified (Table 1). These viruses are morphologically similar, but antigenically not cross-reactive. They differ in their epidemiology and in their virulence for man. Influenza A viruses share common internal antigens (nucleoprotein and membrane protein) which can be differentiated from those of influenza B and influenza C viruses. Influenza A virus particles are approximately spherical in shape (diameter 100 nm) and have characteristic spikes projecting from the viral envelope (Figure 1). However, it should be noted that many preparations contain pleomorphic, elongated particles and that sometimes long, filamentous forms are found. Influenza B viruses have a similar structure to that of influenza A viruses, while influenza C viruses possess a somewhat larger diameter of the internal component and appear to have a hexagonal lattice underlying the viral surface.[3]

Influenza A viruses consist of approximately 1% ribonucleic acid (RNA), 7% carbohydrate, 22% lipid, and 70% protein.[4] These values vary depending on the strain and the host system used. Different estimates have also been obtained for the molecular weight of influenza virions. The most recent data[5] suggest a molecular weight of 360×10^6 for influenza A virions.

A schematic diagram shows the structure of influenza A viruses (Figure 2).

Table 1
COMPARISON OF INFLUENZA A, B, AND C VIRUSES

	A	B	C
Morphology	Pleomorphic particles with spikes	Pleomorphic particles with spikes	Pleomorphic particles with spikes, slightly different from influenza A and B viruses
Genome	Eight single-stranded RNA segments	Eight single-stranded RNA segments	At least four single-stranded RNA segments
Antigenic variation	Frequent	Less frequent	Not well studied
Natural hosts	Man, birds, pigs, horses	Man only	Man only
Human disease	Common cause of respiratory disease, pandemics, epidemics	Less common in adults, symptoms possibly less severe, complications in children, epidemics	Mostly subclinical infections

* Literature references up to 1976 have been added

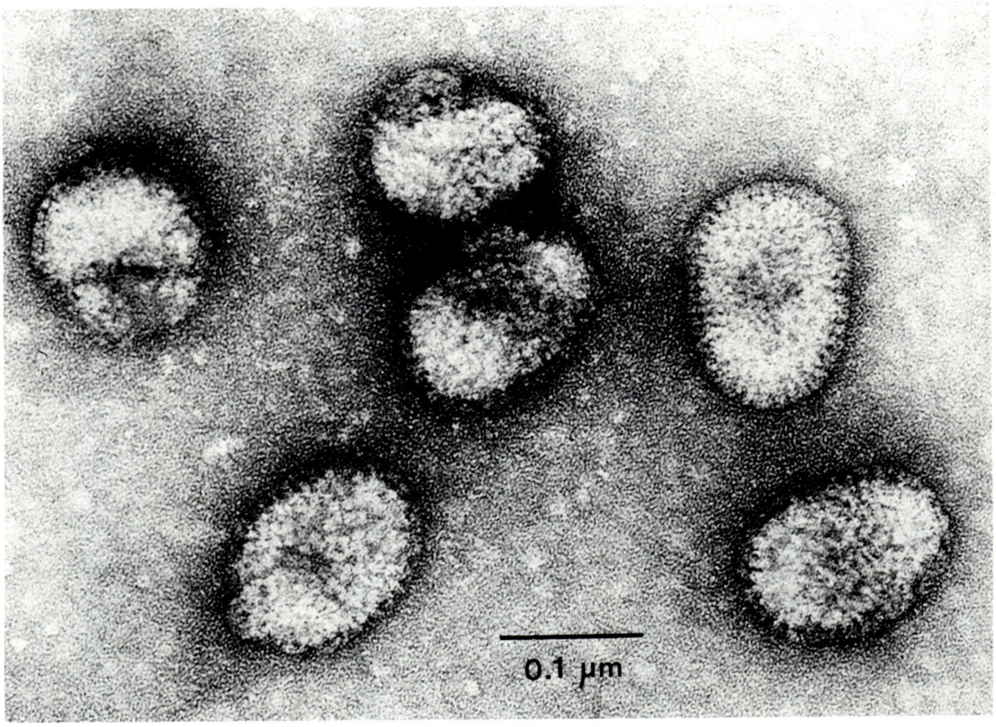

FIGURE 1. Influenza viruses (A/PR/8/34) negatively stained with uranyl oxalate. The particles have a diameter of approximately 100 nm and are characterized by distinct spikes (approximately 10 nm long) on the viral surface. Magnification × 390,000. (Electronmicrograph courtesy of Robert P. Aaronson, Mount Sinai School of Medicine of CUNY, New York.)

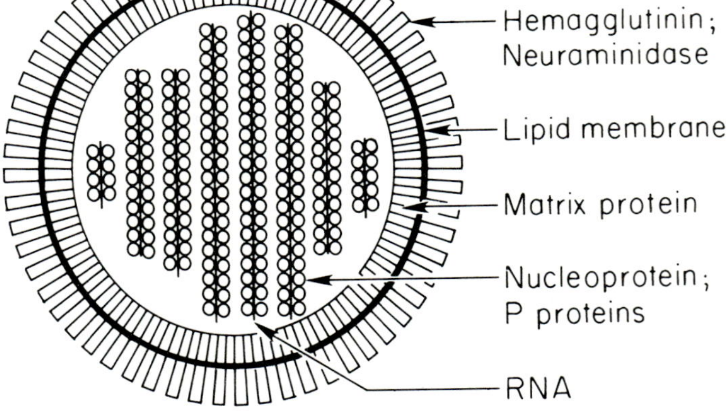

FIGURE 2. Schematic diagram of an influenza A virus particle. The virus contains seven virus-specific proteins, eight single-strand RNA segments, and a small percentage of lipids derived from the host membrane.

Characteristic for this virus are well-defined spikes which project from the unit membrane of the viral envelope. These spikes consist of the hemagglutinin, a glycoprotein which permits the virus to bind to neuraminic acid receptors on cell surfaces,[6,7] and a second glycoprotein, the neuraminidase, which has a complex function and prevents possible autoaggregation of the virus.[8,9] It is presently unclear how far the spikes "reach" into the lipid membrane of the virus and whether a direct contact between the spikes and the underlying layer of matrix or membrane (M) protein exists. Most studies show that the lipid content of the membrane is host determined and host derived.[4] It appears that the M protein layer provides structural support for the virus, but other functions of the M protein during the assembly or budding process cannot be excluded. The internal structure of influenza viruses consists of the ribonucleoprotein complex. There is general agreement that this complex is segmented and that each RNA is associated with multiple copies of nucleoprotein (NP) molecules and one (or possibly all three) P proteins. It is assumed that one (or all) of these proteins is (are) responsible for the virus-associated RNA-dependent RNA polymerase activity. The protein components of influenza B and C viruses have been studied less extensively. Glycoproteins have been identified for both influenza B and C viruses, but the influenza C viruses do not appear to possess a neuraminidase.[10,11] A summary of the proteins of influenza A viruses is given in Table 2.

Recent analysis[27-29] of the RNA structure of influenza viruses has revealed that influenza A and B viruses possess eight RNA segments each and influenza C viruses contain at least four RNA segments. Figure 3 shows the complete map of an influenza A virus. All eight RNA segments are identified with respect to the gene product for which they code.[30-32]

Virus Replication

A schematic representation of the multiplication of influenza A virus is shown in Figure 4. The kinetics and efficiency of replication vary widely with the strain and host employed. In one of the most efficient systems, influenza A/WSN (HONl) virus in MDBK (bovine kidney) cells, the latent period is 3 to 4 hr, with virus production reaching a plateau within 8 to 12 hr.[33] In addition to natural hosts and experimental animals (see

Table 2
VIRUS-SPECIFIC PROTEINS OF INFLUENZA A VIRUSES

Protein	Molecular weight	Percent of total protein in the virion	Function	Ref.
P1	90,000–110,000	<2.5	RNA synthesis?	4, 12–14
P2	80,000–94,000	<2.5	RNA synthesis?	4, 12–15
P3	80,000–94,000	<2.5	RNA synthesis?	15
HA[a,b]	75,000–80,000	20–30	Adsorption to cells, infectivity	7, 16–18
NA[b]	55,000–75,000	1–10	Neuraminidase activity, disaggregation of virus	6, 8, 9, 19, 20
NP	50,000–65,000	17–30	Component of ribonucleoprotein, RNA synthesis?	4, 21, 22
M	21,000–35,000	22–46	Structural component	4, 12, 23
NS	25,000–37,000	Only found in infected cells	Unknown	23–26

[a] The amount of cleavage of the hemagglutinin (HA) into HA1 (49,000 to 58,000 mol wt) and HA2 (25,000 to 30,000 mol wt) subunits depends on the cell system.
[b] Glycoproteins.

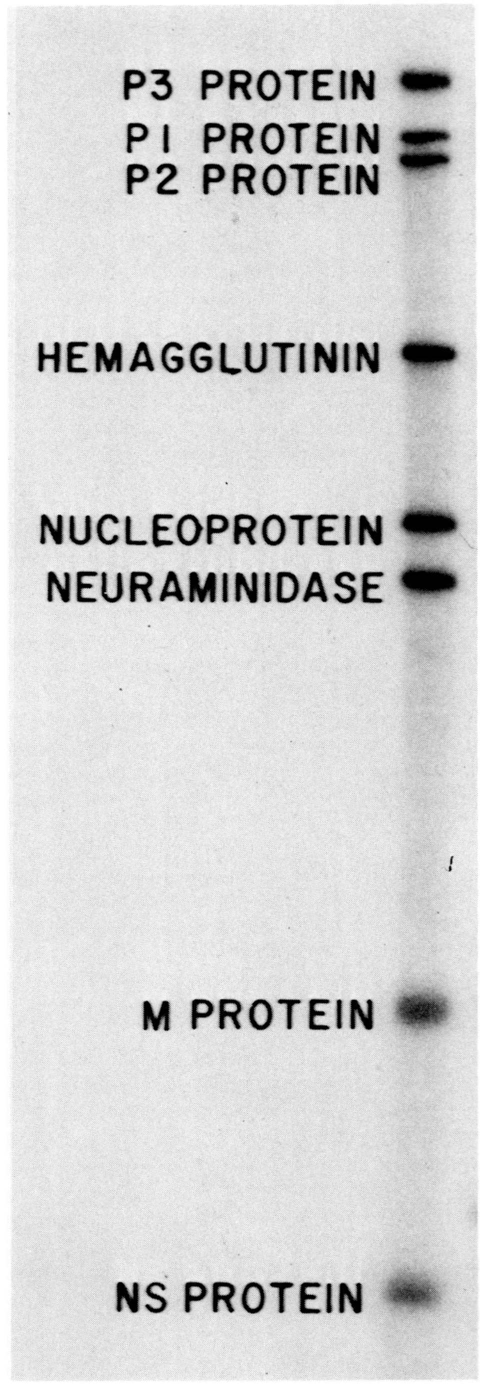

FIGURE 3. Genetic map of the influenza A virus genome. The RNAs of influenza A/PR/8/34 virus were separated by electrophoresis on 2.6% urea polyacrylamide gels. There are three slow-moving RNA segments, three RNAs in the center, and two fast-moving RMAs. The gene products for which each of these RNA segments codes are indicated to the left of the RNA band. For details see References 15 and 30 to 32.

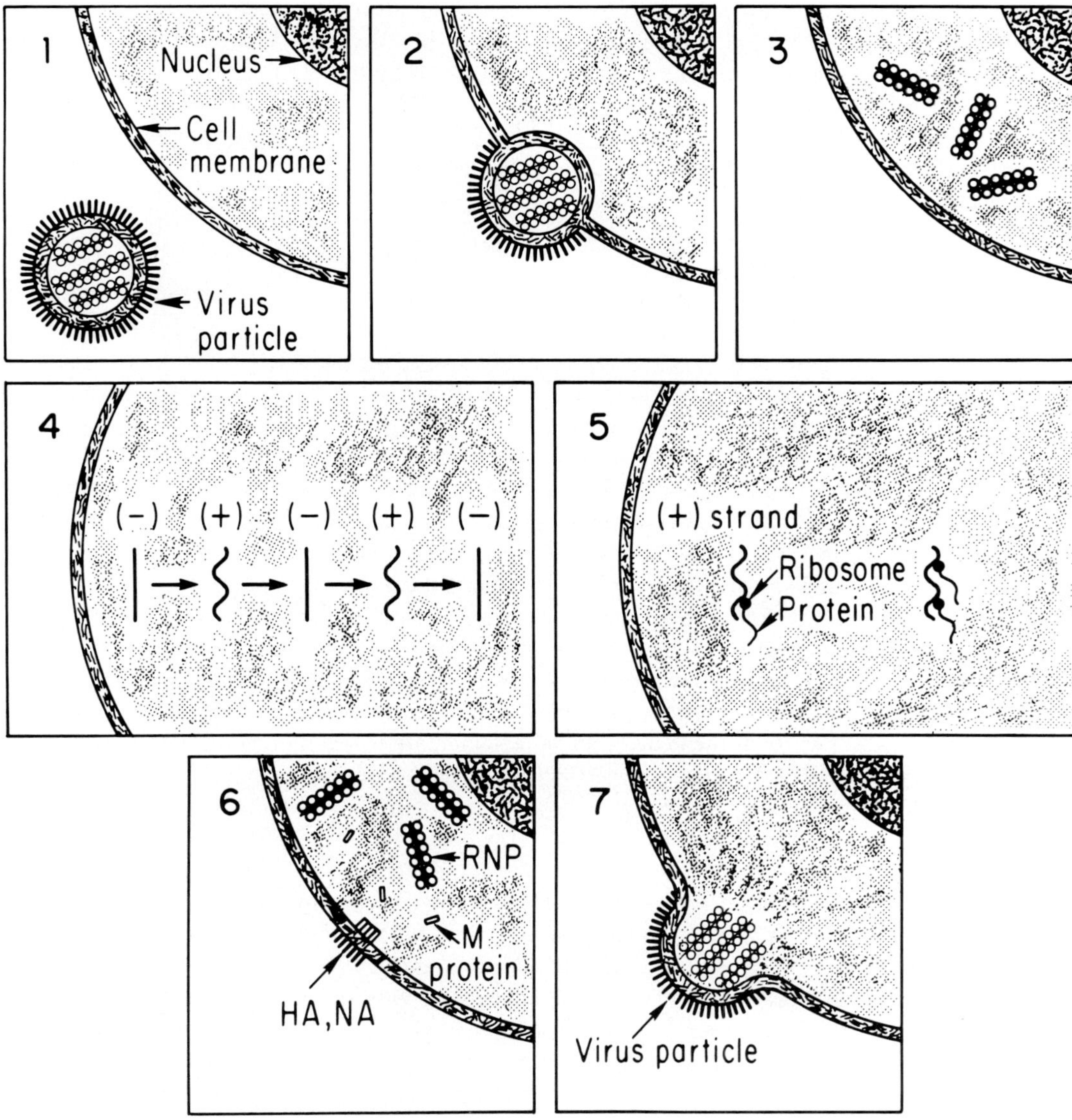

FIGURE 4. Scheme of influenza virus replication. (1) Adsorption: virus attaches to cell receptors via the hemagglutinin. (This diagram shows only three ribonucleoprotein complexes rather than eight.) (2) Penetration: engulfment of virus particle or fusion with cell membrane. (3) Uncoating: release of the ribonucleoprotein complex into the cytoplasm (and nucleus?). (4) RNA synthesis: Incoming viral RNA (−) is transcribed by virus-associated polymerase to make complementary RNA(+). This is used as message and as a template for further virus-specific RNA synthesis (+ and −). Details of these steps and sites of RNA synthesis are not well characterized. (5) Protein synthesis: takes place in cytoplasm, but P, NP, and NS proteins can be found in nucleus; glycosylation in rough and smooth membranes. (6) Assembly: RNA incorporated into ribonucleoprotein complex (RNP), hemagglutinin (HA), and neuraminidase (NA) are inserted into cell membrane; the matrix (M) protein probably associates with inner surface of the cell membrane. (7) Release: Components assemble at the cell membrane and virus buds out into the extracellular space; neuraminidase prevents autoaggregation.

below), influenza viruses can replicate in embryonated hen's eggs; some also grow in tissue culture cells, including those of bovine and monkey kidney.[34,35]

Adsorption, Penetration, and Uncoating (Steps 1, 2, and 3)

Adsorption of the virus to the susceptible host occurs by attachment of the hemagglutinin to a glycoprotein receptor containing neuraminic acid on the surface of the cell.[13,36] Penetration is thought to occur either by "viropexis," in which the whole virus particle is engulfed into the cell by pinocytosis, or by fusion of the viral membrane with the host cell membrane. Uncoating involves the release of the ribonucleoprotein complex into the cell and occurs in phagocytic vesicles or at the cell surface during fusion with the cell membrane. The penetration-uncoating step of many influenza A viruses is sensitive to amantadine.[37]

RNA Synthesis (Step 4)

After uncoating, virus-specific biosynthesis of RNA and proteins takes place. The RNA which enters the cell with the ribonucleoprotein complex is single stranded and of one-strand sense. This RNA is copied by an RNA-dependent RNA polymerase which is brought into the cell with the virus.[38-40] This newly synthesized complementary RNA (cRNA, plus strand) is used as a template to make more of the original, negative-strand RNA (vRNA), and the process is amplified. Kinetic studies indicate that synthesis of the cRNA peaks before that of the vRNA.[41,42] The exact cellular location of RNA synthesis is not known, but cell fractionation studies indicate that RNA polymerase activity is highest in the microsomal fraction, and some RNA-dependent RNA polymerase activity can be found in the nucleus.[43,44] Actinomycin D, an antibiotic which inhibits DNA-dependent RNA polymerase, has been shown to inhibit an early step in RNA synthesis, although this drug does not inhibit the in vitro polymerase reaction.[38,43] The significance of this observation is not clear. The proteins involved in RNA synthesis possibly include the nucleoprotein and the P proteins. Analysis of temperature-sensitive mutants shows that there are four complementation recombination groups which are defective in RNA synthesis, suggesting that at least four viral proteins[45,46] are necessary for virus-specific RNA synthesis.

Protein Synthesis (Step 5)

After the onset of virus-specific RNA synthesis, virus-specific proteins begin to be made. Studies of RNA on infected cell ribosomes[47,48] and in in vitro translation systems[24,49-52] indicate that the cRNA (plus strand) serves as the messenger. In addition to those proteins found in the purified virus, a virus-specific nonstructural protein is observed in infected cells.[23] Kinetic studies suggest that the P proteins, nucleoprotein, and nonstructural protein appear first in infected cells.[53,54] The relative amounts of each protein found in infected cells late in the cycle are similar to those found in purified virus, except for the matrix protein. This is the most abundant in the virus, but appears to be limiting in infected cells.[23,55] All of the proteins are found in the cytoplasm, but the P protein(s), nucleoprotein, and nonstructural protein can be found in the nucleus.[23,56,57]

Two of the virus proteins, the hemagglutinin and neuraminidase, are glycosylated. Studies of the hemagglutinin indicate that sugars are added in a stepwise fashion in distinct cellular compartments. It appears that glucosamine is added in the rough membranes and fucose in the smooth membranes.[58,59] The hemagglutinin also undergoes a posttranslational cleavage to yield two polypeptides designated HA1 and HA2. This cleavage appears to be required in order to produce an infectious particle.[17,18]

Assembly and Release (Steps 6 and 7)

The assembly of the virus first involves the incorporation of RNA into the ribonucleoprotein (RNP) complex. Both cRNA and vRNA are found in the RNP complexes throughout the infectious cycle and are detected in both the nucleus and cytoplasm.[42,60,61] However, only the RNPs containing the vRNA (negative strand) are incorporated into the virions. The hemagglutinin and neuraminidase are inserted into the cell membrane of the host and the matrix protein becomes attached to the inner surface of the cell membrane. The RNPs and P proteins gather at the cell membrane and, in the process of budding, whole virus particles are formed containing virus-specific proteins and a lipid layer derived from the host cell membrane. The mechanism by which the virus packages appropriate RNAs to give an infectious particle is unknown. Estimates indicate that only 1 out of 60 particles formed is infectious in a productive system.[62]

The function of the neuraminidase has been studied using neuraminidase inhibitors such as 2-deoxy-2,3-dehydro-*N*-trifluoroacetylneuraminic acid (FANA) and temperature-sensitive mutants defective in neuraminidase. If the enzyme is inactive, the virus particles contain neuraminic acid on their surface. The neuraminic acid molecules act as receptor for other virus particles, resulting in the formation of large virus aggregates and a decrease in infectivity. Thus, the neuraminidase appears to function in viral release to prevent autoaggregation of virus particles.[8,19,63]

Genetics

Burnet and Lind[64] were the first to describe recombination between influenza viruses by isolating recombinant viruses which derived biological properties from two different parent strains. The fact that influenza viruses undergo frequent antigenic variation has provided investigators with different strains possessing characteristic markers.[65,66] It was found that a high frequency of recombination among different strains occurs due to reassortment of the eight individual RNA genes. Recent studies with temperature-sensitive mutants of influenza WSN virus provide additional genetic evidence that the influenza virus genome is segmented. It was found that there are seven to eight complementation and recombination groups of ts mutants, each probably with a defect in one gene product.[45,67–70] Further support for the existence of eight genes and their ability to reassort comes from polyacrylamide gel analysis of RNAs and proteins of recombinant viruses. These techniques permitted for the first time the identification of the gene product of each RNA segment and the identification of the origin of a specific gene in a recombinant.[15,30–32,71,72] The genetics of influenza viruses have recently been reviewed in detail by Sugiura.[73]

Incomplete Replication

In addition to the normal replicative cycle, influenza virus also undergoes abnormal forms of multiplication in some systems. These include host-dependent abortive infections in which infectious progeny are not measurable although viral components[74–80] are made and the von Magnus phenomenon, which describes the increase in incomplete virus formed during serial passage of influenza viruses at multiplicities higher than one.[81] The precise mechanism by which von Magnus particles are formed remains to be elucidated.[27]

Virus Strains and Antigenic Properties

Influenza viruses are classified into A, B and C viruses according to the type of internal antigen.[82] It should be noted that the three groups of influenza viruses are unrelated, but members within one group share related internal antigens. The numbers and designations of the types and subtypes are listed in Tables 3 and 4, together with prototype strains. Strains are named by giving their type, place of origin, strain serial number, and year of isolation, in that order. For influenza A viruses, the designation of the hemagglutinin and neuraminidase subtype is also included in the description (e.g., influenza A/Hong Kong/1/68(H3N2) virus). For animal isolates, the animal origin is given between the designation of type and place of origin. It should be noted that serological differences can be detected among strains within one subtype.

Table 3
ANTIGENIC CHARACTERISTICS OF HUMAN INFLUENZA VIRUSES[82]

Influenza A viruses

Prototype	Hemagglutinin	Neuraminidase
A/PR/8/34	H0	N1
A/FM/1/47	H1	N1
A/Singapore/1/57	H2	N2
A/Hong Kong/1/68	H3	N2

Influenza B viruses

Strains with different hemagglutinins and neuraminidases have been isolated, but no subtypes have been established

Influenza C viruses

A limited number of strains have been isolated, but their antigenic differences have not been well characterized; viruses do not possess neuraminidase activity

Table 4
ANTIGENIC CHARACTERISTICS OF ANIMAL INFLUENZA A VIRUSES[82]

Prototype	Hemagglutinin	Neuraminidase
Avian viruses		
A/FPV/Dutch/27	Hav1	Neq1
A/chicken/Germany "N"/49	Hav2	Neq1
A/duck/England/56	Hav3	Nav1
A/duck/Czech/56	Hav4	Nav1
A/tern/S. Africa/61	Hav5	Nav2
A/turkey/Mass./65	Hav6	N2
A/duck/Ukraine/1/63	Hav7	Neq2
A/turkey/Ontario/6118/68	Hav8	Nav4
A/turkey/England/63	Hav1	Nav3
A/Shearwater/E. Australia/1/72	Hav6	Nav5
Equine viruses		
A/equine/Prague/1/56	Heq1	Neq1
A/equine/Miami/1/63	Heq2	Neq2
Swine virus		
A/swine/Iowa/15/30	Hsw1	N1

Clinical Features

The following briefly summarizes the clinical aspects of influenza virus infections. For details, an excellent review is available (see Reference 83). Specific procedures for treatment or differential analysis of influenza should be obtained from textbooks of medicine.

Transmission, Symptoms, and Signs

It is very likely that influenza virus is transmitted by the airborne route in the form of small droplet nuclei.[84] Virus in these droplet nuclei initially infects epithelial cells of the upper respiratory tract and may spread to other parts of the respiratory system. It should be noted that, in some instances, virus may directly attack cells in the alveoli. An excellent cytopathological description of specimens obtained from human biopsies or fatal cases has recently been published.[85]

Influenza caused by A viruses is characterized by a rapid rise of fever (up to 100 to 106°F) which usually persists for approximately 3 days; in a few cases a second episode of fever has been reported. Systemic symptoms (Table 5) usually diminish by the fourth day after the onset of illness, but respiratory symptoms may continue for several days. A long convalescence characterized by general malaise is relatively common as a result of influenza A virus infections. Symptoms of influenza A virus infections in children are generally milder than those of adults, but maximal temperatures are frequently higher among children.[83]

In contrast, although symptoms are similar, infections with influenza B virus are reported to be less severe,[88] and infections caused by influenza C viruses are usually afebrile and frequently asymptomatic.[89] Table 6 lists some host responses to influenza A and B virus infections.

Complications

In most adults influenza follows a predictable course. However, in certain epidemic years, up to 20% of all cases suffer from chest complications.[97] The most important complications, which have been recognized for a long time, are those resulting in pneumonia. Secondary bacterial pneumonias caused by pneumococci, staphylococci and *Hemophilus influenzae* are most frequently seen. Table 7 compares different aspects of secondary bacterial pneumonia with those of less frequently observed primary viral pneumonia. Both secondary bacterial and primary viral pneumonia are predominantly a threat to people over 60 years of age.[97] Young and middle-aged adults are less frequently affected by complications in the lower respiratory tract following an attack of influenza.

Table 5
SYMPTOMS OBSERVED IN PATIENTS WITH PROVEN INFLUENZA A

	Influenza caused by virus with serotype[a]	
Symptoms	H1N1 (1947)[86]	H3N2 (1957)[87]
Feverishness	99	88
Chilliness	87	43
Headache	86	79
Myalgia	60	40
Sneezing	–	44
Nasal discharge	70	71
Sore throat	49	62
Hoarseness	–	26
Cough	97	87
Anorexia	–	62
Nausea	29	19
Vomiting	9	22
Diarrhea	4	2
Abrupt onset	67	61

[a] Percent of patients having each symptom.

Table 6
HOST RESPONSES TO CLINICALLY PROVEN (UNCOMPLICATED) INFLUENZA

	Remarks	Refs.
Formation of antibodies	Development of hemagglutinin and neuraminidase antibodies in serum and respiratory secretions	90
Interferon	Has been demonstrated in nasal washings and serum	91, 92
Bacterial flora in respiratory tract	Positive association with *Meningococcus,* probably no correlation to *Streptococcus, Staphylococcus,* or *Hemophilus influenzae*	87, 93
Pulmonary function	Frequently impaired	94
Blood cell count	Leukocytosis in early stage of disease followed by leucopenia	83, 95
Viremia	Rare	96

Table 7
COMPARISON OF INFLUENZAL AND SECONDARY BACTERIAL PNEUMONIA

Primary influenza viral pneumonia
- Typical onset of influenza; frequent history of rheumatic heart disease
- Rapid progression without remission; increasing fever and dyspnea
- Cyanosis and extreme tachypnea
- Diffuse pulmonary findings; no localized areas of consolidation
- Few pathogens in sputum (smear and culture)
- No response to antimicrobial drugs
- High mortality rate

Secondary bacterial pneumonia following influenza
- Typical onset and symptoms of influenza
- Afebrile period of recovery
- Symptoms of pneumonia (fever, cough, chills, blood sputum)
- Localized areas of consolidation demonstrable by physical and X-ray examinations
- Pathogenic bacteria (*Pneumococci, Staphylococci, Streptococci, Hemophilus influenzae*) in sputum
- Response to antimicrobial drugs
- Low mortality rate

From Schulman, J. L., *Postgrad. Med.,* 50, 171, 1971. With permission.

Finally, it should be noted that the overall contribution of influenza to acute lower respiratory illness is slighter in children than in adults.[99]

Less frequently, influenza has been associated with exacerbations of chronic bronchitis and with obstructive laryngotracheitis. Reye's syndrome, which is an acute encephalopathy with fatty degeneration of the viscera, may be caused by influenza B virus infections,[100] although other etiological agents have been implicated. Evidence that influenza viruses are associated with other neurological disorders such as Guillain-Barré syndrome, transverse myelitis, encephalitis, or parkinsonism is inconclusive.

Many contradictory reports have appeared concerning the association of influenza with congenital malformations or subsequent neoplasms in childhood, but a careful analysis suggests that there is no direct association between influenza and these diseases.[101] A summary of all complications is shown in Table 8.

Epidemiology

Influenza viruses are unique among animal viruses in their capacity to undergo rapid antigenic changes. Every 10 to 15 years new pandemic A strains emerge which are unaffected by antibodies generated against earlier influenza virus strains. A list of the pandemics in this century is given in Table 9.

Attempts have been made to understand the reasons for the emergence of new strains. According to one hypothesis, the virus changes by mutation, "evading" the immunological pressure of antibodies to earlier strains. An alternative hypothesis suggests that new viruses may result from recombination among human and animal strains and that these pandemic strains derive novel antigenic characteristics from animal strains.[102,103]

It should be noted that influenza A viruses not only undergo dramatic antigenic changes when new pandemic strains emerge, but also that the antigenic characteristics of influenza viruses change to a lesser extent in interpandemic periods. The first is referred to as antigenic "shift," while the second is called the antigenic "drift" of influenza viruses.[102] It is puzzling that different serotypes of influenza A viruses are present at the same time in animals while only one serotype is present in the human population at any given time. Although the majority of the human influenza A virus isolates in 1976 were of the Hong Kong serotype, it should be noted that several viruses containing the surface proteins of swine influenza viruses were also isolated during this period. Even more intriguing is the fact that influenza B viruses do not undergo the dramatic antigenic changes characteristic of A viruses. We do not have a convincing explanation for this phenomenon.

For a detailed discussion of environmental determinants and seasonal and other factors which influence transmission and the course of influenza, see Reference 103.

Treatment and Prevention

No effective chemotherapeutic treatment of influenza is currently available. Amantadine and related compounds seem to be the only drugs developed to date for prevention of infection in animals and man,[104-106] but these compounds have never been widely used. Therefore, the alternative is the administration of vaccines to prevent infection. Table 10 summarizes the currently available approaches to influenza virus vaccination. It

Table 8
COMPLICATIONS OF INFLUENZA

Primary viral pneumonia
Secondary bacterial pneumonia
Exacerbation of chronic bronchitis
Obstructive laryngotracheitis (young children)
Reye's syndrome (possibly associated with influenza B virus)
Myositis
Guillain-Barré syndrome (unproven)
Encephalitis (few cases documented)
Transverse myelitis (few cases documented)
Congenital malformations and childhood malignancies (probably no direct association with influenza viruses)

Table 9
PANDEMICS IN THE TWENTIETH CENTURY CAUSED BY INFLUENZA A VIRUS

		Serotype	Pandemic
1918	Swine virus	(HswN1)	Severe
1929[a]	A0 virus	H0N1	
1946	A1 virus	H1N1	Mild
1957	Asian virus	H2N2	Severe
1968	Hong Kong virus	H3N2	Moderate

[a] In 1929 the virus changed its serotype, but the H0 hemagglutinin still crossreacted with the swine hemagglutinin. Because only retrospective serological studies are available, it is unclear whether or not a pandemic was associated with the H0N1 serotype.

Table 10
TYPES OF INFLUENZA VIRUS VACCINES

Inactivated
- Whole virus[a] (chemically inactivated)
- Disrupted or "split" virus[a] (detergent or ethyl ether disrupted)

Live virus
- Attenuated by passage[a] (occasionally under selective conditions), variants

[a] Vaccine strain may be a recombinant derived from a laboratory strain (high yielder) and a current isolate.[107]

should be noted that all vaccines (inactivated and live virus) become ineffective as major antigenic changes of the virus occur.

Commercial, inactivated influenza virus vaccines have been in use for many years and their effectiveness varies (50 to 90%), depending on the particular vaccine strain.[108] Despite recently improved methods of preparing inactivated vaccines, complete protection has not been achieved by vaccination. This may be explained by the fact that influenza viruses undergo constant antigenic changes (within a pandemic period) and that the most common route (parenteral) of administration of inactivated vaccines is an "unnatural" one.

Considerable efforts have been made to develop attenuated live virus vaccines which would be administered through the respiratory route and which may mimic a natural infection. At present, attempts are being made to use as live virus vaccine candidates temperature-sensitive mutants,[109] cold-adapted mutants,[110] and strains that have been attenuated by passage.[111] It should be noted that only inactivated influenza virus vaccines are currently in use and commercially available in the U.S. In addition, attempts are being made to develop neuraminidase-specific inactivated vaccines. Natural infection (or secondary challenge with current strains) of individuals vaccinated with neuraminidase-specific vaccine would result in limited replication of the virus without producing disease. This may lead to a more "natural" and, hence, more protective immune response than that elicited by standard vaccines.[112]

Laboratory Diagnosis

The laboratory diagnosis of influenza depends on isolation of the virus and/or demonstration of an increase in antibody titer. Detailed descriptions of serological assays and laboratory procedures can be found in References 113, 114, and 115. The excellent manual entitled "Advanced Laboratory Techniques for Influenza Diagnosis"[113,*] completely describes techniques for virus isolation, antigenic characterization, and serologic diagnosis.

It should also be noted that the Influenza Program established by the World Health Organization (WHO) provides a surveillance system to detect new antigenic variants as early as possible and to record the epidemiology and antigenic character of current virus strains.

Specimens

Throat and nasal specimens (either washings or swabs) are collected for virus isolation and should be taken within the first 3 days after the onset of disease. Specimens should be

* It is highly recommended and can be obtained by writing to the U.S. Department of Health, Education, and Welfare, Public Health Service, Center for Disease Control, Bureau of Laboratories, Atlanta, GA 30333, c/o Drs. W. R. Dowdle, D. F. Palmer, and M. T. Coleman.

kept cold or frozen at −70°C if they are not to be used within 48 hr. In fatal cases virus isolation should be attempted from lung tissue, tracheal mucosa, and blood.

For serological diagnosis, serum should be collected within the first 3 days of the disease and again 2 to 3 weeks later, during convalescence.[113]

Virus Isolation

Influenza viruses are most easily propagated in embryonated hen's eggs and rhesus monkey kidney cells. Both systems should be used for each specimen. A flow chart for the isolation of influenza viruses is shown in Figure 5.[113]

The embryonated eggs should be inoculated both intraamniotically and intraalantoically. Several blind passages may be necessary before enough virus is produced to enable detection of hemagglutinating activity. In order to isolate influenza C viruses, 7-day-old rather than 10-day-old eggs should be inoculated amniotically and incubated for 5 days at 33°C.

Monolayers of monkey kidney cells should be washed to prevent inhibition of virus growth by serum factors. Uninoculated controls should be used to detect the presence of adventitious agents. In general, a longer peiod of time is required to detect the presence of influenza A viruses than influenza B viruses in monkey kidney cells.

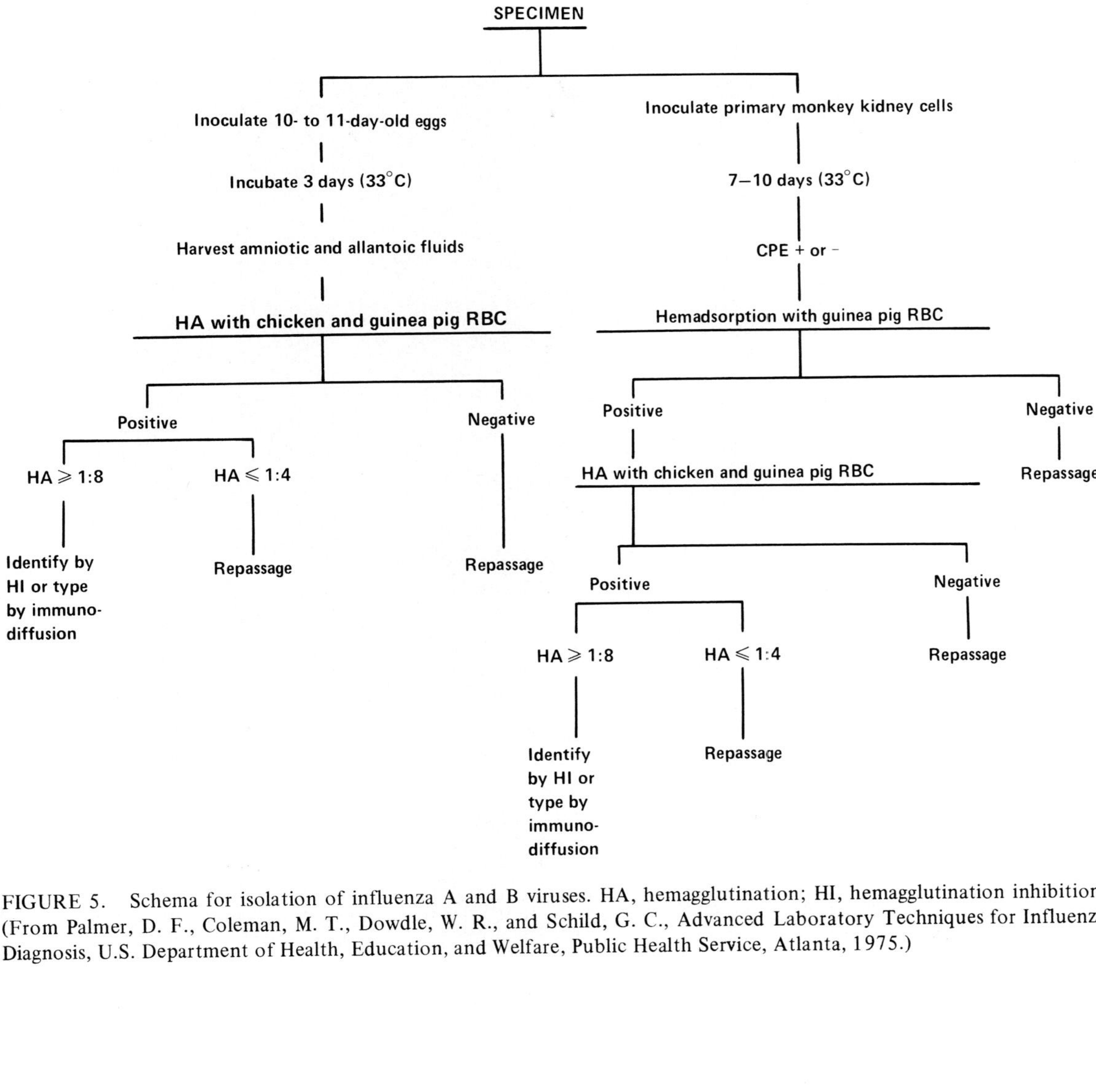

FIGURE 5. Schema for isolation of influenza A and B viruses. HA, hemagglutination; HI, hemagglutination inhibition. (From Palmer, D. F., Coleman, M. T., Dowdle, W. R., and Schild, G. C., Advanced Laboratory Techniques for Influenza Diagnosis, U.S. Department of Health, Education, and Welfare, Public Health Service, Atlanta, 1975.)

Virus Recognition

Hemagglutination of chicken or guinea pig red blood cells is used to detect the presence of virus in the allantoic or amniotic fluid of the embryonated eggs.

The growth of virus in monkey kidney cells can be demonstrated by the presence of cytopathic effects (CPE) or hemadsorption.[113] Hemadsorption can often be detected in the absence of extensive CPE or detectable hemagglutination of the supernatant medium. Essentially, this test involves demonstration of the specific adsorption of red blood cells to the monolayer of infected cells. A typical hemadsorption pattern is shown in Figure 6.

Serological Tests

The serological diagnosis of influenza depends on the demonstration of at least a fourfold increase in antibody titer between acute and convalescent sera. Although the collection of appropriate serum samples from one individual takes 2 to 3 weeks, an outbreak of influenza can be diagnosed in 1 day by comparing groups of sera from closely matched individuals in the acute and convalescent stages of illness.

The list below describes the serological tests currently available for analyzing isolated influenza viruses or detecting the presence of antibody in sera (a summary is given in Tables 11 and 12). In general, the laboratory diagnosis is performed with reagents specific for the strains currently causing disease. More detailed characterization and subtyping is done in specialized laboratories. The components of influenza viruses against which antibodies are formed are described in Table 2.

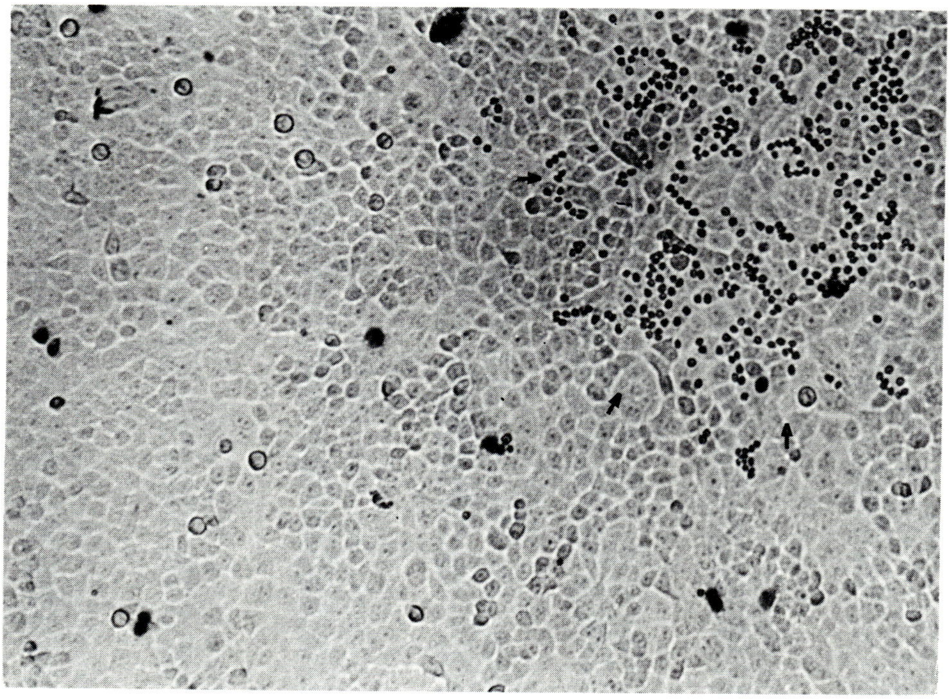

FIGURE 6. Hemadsorption of influenza virus infected cells. Human conjunctival (clone 1-5C-4) cells were infected with influenza A/PR/8/34 virus under an agar overlay. Two days after infection the agar overlay was removed and human red blood cells were added to the monolayer. After washing, red blood cells only adhere to monolayer cells which were infected by influenza virus. A plaque of infected conjunctival cells (identified by red blood cells adhering to them) is clearly seen in the upper right corner (arrows). (Micrograph courtesy of Jerome L. Schulman, Mount Sinai School of Medicine of CUNY, New York.)

Table 11
SUMMARY OF ANTIGENIC CHARACTERIZATION[a,113,115]

Antigen characterized	Test	Test antigen	Antisera	Recommended use[b] Type	Subtype	Strain
HA[c]	HI	Whole virus	Chicken (single injection)		++	+++
HA[c]	HI	Whole virus	Ferret (infection)		++	++++
HA[c]	HI	Whole virus	Rabbit (hyperimmune)		++++	+
HA	SRH	Whole virus	Ferret (infection)		+	++++
HA	SRH	Whole virus	Guinea pig (infection)		+	++++
HA	IDD	Disrupted virus	Anti-HA, chicken		++++	++
NA[c]	NI	Whole virus	Chicken (single injection)		++	+++
NA[c]	NI	Whole virus	Ferret (infection)		++++	++
NA[c]	NI	Whole virus	Rabbit (hyperimmune)		++++	++
NA	IDD	Disrupted virus	Anti-NA		++++	++
HA, NA	CF	Whole virus "V"	Guinea pig (injection)			+
NP	CF	CAM extract "S"	Anti-NP	+		
NP	IDD	Disrupted virus	Anti-NP	+		
M	IDD	Disrupted virus	Anti-M	+		

[a] Abbreviations: HA, hemagglutinin; NA, neuraminidase; NP, nucleoprotein; M, matrix protein; HI, hemagglutination inhibition; NI, neuraminidase inhibition; SRH, single radial hemolysis; IDD, immuno double diffusion; CF, complement fixation; CAM, chorioallantoic membrane.
[b] Usefulness indicated on a scale of + (least useful) to ++++ (most useful).
[c] Serum containing antibody to the other surface antigen may cause inhibition at low dilution.

Neutralization[116,117] – This assay demonstrates the presence of antibody to hemagglutinin by showing a reduction in virus titer. This test is complex and subject to variability because of differences in host susceptibility to virus.

Hemagglutination inhibition (HI)[113,118,119] – This test involves the inhibition of hemagglutination of a virus preparation in the presence of specific antisera. It is the test most commonly used for influenza diagnosis. The problems encountered in this test are as follows: Nonspecific inhibitors are often present in serum and, in order to eliminate them, sera are usually treated with neuraminidase (RDE) or periodate before use; failure of influenza virus antiserum to inhibit agglutination can indicate the presence of other hemagglutinating agents (viruses, bacteria) in the preparation; because influenza viruses contain host components (carbohydrate?), antibody to host antigens can often interfere with this test (these host-specific antibodies can be absorbed out using heterotypic influenza virus); and finally, antineuraminidase antibodies in the test serum may occasionally cause inhibition of the hemagglutination reaction. Use of the appropriate recombinants with genetically segregated surface antigens overcomes this problem and makes this test specific for hemagglutinin. Note that in addition to identifying virus or antisera, this test can be used to distinguish antigenic characteristics among viruses within one subtype.

Table 12
METHODS TO IDENTIFY INFLUENZA VIRUS-SPECIFIC ANTIBODIES[a,113,115]

Antibody	Method	Test antigen	Recommended use[b]	
			Serodiagnosis	Serosurvey
HA[c,d]	HI	Whole virus	++++	++++
NA[c,d]	NI	Whole virus		++++
NA	IHA	NA		++++
HA, NA[d]	IDD	Disrupted virus	++	+
HA, NA	CF	Whole virus "V"	+++	
HA, NA[d]	SRID	Whole virus	++++	+++
NP	CF	CAM extract "S"	++	
NP, M	IDD	Disrupted virus	++	+
NP, M	SRID	Disrupted virus	+++	

[a] Abbreviations: HA, hemagglutinin; NA, neuraminidase; NP, nucleoprotein; M, matrix protein; HI, hemagglutination inhibition; NI, neuraminidase inhibition; IHA, indirect hemagglutination; IDD, immuno double diffusion; CF, complement fixation; SRID, single radial immuno diffusion; CAM, chorioallantoic membrane.
[b] Usefulness on a scale of + (least useful) to ++++ (most useful).
[c] Serum containing antibody to the other surface antigen may cause inhibition at low dilution.
[d] Specificity achieved by selecting viruses or recombinants with the appropriate antigenic composition.

Single radial hemolysis (SRH)[115,120] – This test is performed by measuring the zone of hemolysis produced when antiserum is added to wells in an agarose gel containing complement and erythrocytes coated with virus. The assay appears to be specific for the hemagglutinin, and the zones of hemolysis produced are proportional to the amount of antiserum present. This test is strain specific, slightly more sensitive than the conventional HI, and is not sensitive to nonspecific inhibitors.

Neuraminidase inhibition (NI)[113,121] – This test permits the identification of the antigenic characteristics of neuraminidases and the determination of antineuraminidase antibody titers in sera. Essentially, the procedure consists of assay of the virus for neuraminidase activity, adjustment of the virus concentration to a standard unit of neuraminidase activity, and assay of antibody by testing serial dilutions of antisera for neuraminidase inhibition. A high-molecular weight neuraminidase substrate must be used in the assay because low-molecular weight substrates are hydrolyzed despite the presence of antiserum. The explanation for this appears to be that antibodies may not be formed against the active site, but rather against other parts of the enzyme. This test is also subject to interference from antibodies to host antigen and hemagglutinin.

Immuno double diffusion (IDD)[113,122-124] – This technique is used to detect the presence of viral antigens and antibodies by the formation of precipitin lines using a conventional Ouchterlony double diffusion test. It can be employed to identify the hemagglutinin, neuraminidase, nucleoprotein, or matrix protein and antibodies to these antigens. Disrupted virus or purified antigens are generally used. The test is strain specific for the hemagglutinin and neuraminidase and type specific for the nucleoprotein and matrix protein. It has the disadvantage of requiring larger quantities of antigen than other tests and is qualitative rather than quantitative. Because of its simplicity, it is recommended that this test replace the more conventional complement fixation (CF) test for typing influenza viruses.

Single radial immuno diffusion (SRID)[113,125] – This test is performed by incorporating the antigen into a solid matrix and adding antiserum into wells in the matrix. Reaction of the antibody and antigen is detected by a visible zone of

Table 13
EXPERIMENTAL ANIMAL MODELS

	Remarks	References
Mice	Excellent model for study of transmission of virus, new immunization procedures, effects of antiviral compounds, pathological changes, immune responses; used by many different groups	90, 130–133
Ferrets	Initially used to isolate virus; influenza in ferrets similar to that in man, preferential infection of respiratory tissue; convenient model for study of immunity	35, 134, 135
Swine	First isolation of influenza virus	136, 137
Horse	Experimental infections have been reported	138
Birds	Many different species have been used; in vivo recombination was demonstrated using turkeys	102, 137, 139
Hamsters	Recently used for the study of temperature sensitive vaccine strains	140
Nonhuman primates	Gibbons, baboons not frequently used	141, 142

precipitation around the well. It can be made quantitative and can be used to detect antibodies to hemagglutinin, neuraminidase, nucleoprotein, and matrix protein. It is slightly less sensitive than HI for detecting antibodies to hemagglutinin, but is more sensitive for detecting differences in antibody levels. In addition, it can be used to show antigenic relationships among different strains of one serotype. SRID is more sensitive for detecting antibody to nucleoproteins than the CF test.

Complement fixation (CF)[126] – This method is used to type influenza viruses and identify antibodies to their internal components. Initially, the test was performed using the "soluble" or S viral antigen (essentially ribonucleoprotein) from infected mouse lung tissue or chorioallantoic membranes. This test is not routinely used, as diagnosis is usually done by HI tests. The assay system is relatively complex and has the disadvantage that not all antibodies made against influenza virus fix complement.

Neuraminidase hemagglutination inhibition (N-HI)[115] – This test is used to detect antibodies to neuraminidase by exploiting the observation[127] that hemagglutination by some influenza type A strains possessing the N2 neuraminidase can be inhibited by antibodies against N2 neuramindase.

Indirect hemagglutination (IHA)[128] – This test detects neuraminidase antibodies by the agglutination of erythrocytes sensitized with neuraminidase. The test can be used to demonstrate the presence of low levels of antibodies against the N2 neuraminidase.

Fluorescent antibody[129] – It is possible to detect the presence of influenza virus-infected cells by direct or indirect immunofluorescence of infected cells.

Experimental Animal Models

Most important for the study of influenza is the use of mice and ferrets as experimental models (Table 13). Less frequently, other animals such as pigs, horses, nonhuman primates, and hamsters have been employed (Table 13). It should be noted that experimental infections in human volunteers have been extensively studied, and much of our current knowledge of influenza is derived from these studies.[83]

GENERAL REFERENCES

White, D. O., Influenza viral proteins: identification and synthesis, *Curr. Top. Microbiol. Immunol.*, 63, 1, 1974.

Kilbourne, E. D., Ed., *The Influenza Viruses and Influenza,* Academic Press, New York, 1975, 573.

Compans, R. W. and Choppin, P. W., Reproduction of myxoviruses, in *Comprehensive Virology,* Vol. 14, Fraenkel-Conrat, H. and Wagner, R. R., Eds., Plenum Press, New York, 1975, 179.

Palese, P., The genes of influenza virus, *Cell,* 10, 1, 1977.

REFERENCES

1. **Smith W., Andrewes, C. H., and Laidlaw, P. P.,** A virus obtained from influenza patients, *Lancet,* 2, 66, 1933.
2. **Kavet, J.,** as cited in Fox, J. P. and Kilbourne, E. D., Epidemiology of influenza – summary of Influenza Workshop IV, *J. Infect. Dis.,* 128, 361, 1973.
3. **Apostolov, K., Flewett, T. H., and Kendal, A. P.,** Morphology of influenza A, B, C and infectious bronchitis virus (IBV) virions and their replication, in *The Biology of Large RNA Viruses,* Barry, R. T. and Mahy, B. W., Eds., Academic Press, London, 1970, 3.
4. **Choppin, P. W. and Compans, R. W.,** The structure of influenza virus, in *The influenza Viruses and Influenza,* Kilbourne, E. D., Ed., Academic Press, New York, 1975, 15.
5. **Reimer, C. B., Baker, R. S., Newlin T. E., and Havens, M. L.,** Influenza virus purification with the zonal ultracentrifuge, *Science,* 152, 1379, 1966.
6. **Hirst, G. K.,** Adsorption of influenza hemagglutinins and virus by red blood cells, *J. Exp. Med.,* 76, 195, 1942.
7. **Schulze, I. T.,** The biologically active proteins of influenza virus: the hemagglutinin, in *The Influenza Viruses and Influenza,* Kilbourne, E. D., Ed., Academic Press, New York, 1975, 53.
8. **Palese, P., Tobita, K., Ueda, M., and Compans, R. W.,** Characterization of temperature sensitive influenza virus mutants defective in neuraminidase, *Virology,* 61, 397, 1974.
9. **Bucher, D. and Palese, P.,** The biologically active proteins of influenza virus: neuraminidase, in *The Influenza Viruses and Influenza,* Kilbourne, E. D., Ed., Academic Press, New York, 1975, 84.
10. **Kendal, A. P.,** A comparison of "influenza C" with prototype myxoviruses: receptor-destroying activity (neuraminidase) and structural polypeptides, *Virology,* 65, 87, 1975.
11. **Choppin, P. W., Lazarowitz, S. G., and Golberg, A. R.,** Studies on proteolytic cleavage and glycosylation of the hemagglutinin of influenza A and B viruses, in *Negative Strand Viruses,* Vol. 1, Mahy, B. W. J. and Barry, R. D., Eds., Academic Press, London, 1975, 105.
12. **Compans, R. W., Klenk, H. D., Caliguiri, L. A., and Choppin, P. W.,** Influenza virus proteins. I. Analysis of polypeptides of the virion and identification of spike glycoproteins, *Virology,* 42, 880, 1970.
13. **Schulze, I. T.,** The structure of influenza virus. I. The polypeptides of the virion, *Virology,* 42, 890, 1970.
14. **Simpson, R. W. and Bean, W. J., Jr.,** The biologically active proteins of influenza virus: influenza transcriptase activity of cells and virions, in *The Influenza Viruses and Influenza,* Kilbourne, E. D., Ed., Academic Press, New York, 1975, 125.
15. **Ritchey, M. B., Palese, P., and Schulman, J. L.,** Differences in protein patterns of influenza viruses, *Virology,* 76, 122, 1977.
16. **Lazarowitz, S. G., Compans, R. W., and Choppin, P. W.,** Proteolytic cleavage of the hemagglutinin polypeptide of influenza virus. Function of the uncleaved polypeptide HA, *Virology,* 52, 199, 1973.
17. **Klenk, H. D., Rott, R., Orlich, M., and Blödorn, J.,** Activation of influenza A viruses by trypsin treatment, *Virology,* 68, 426, 1975.
18. **Lazarowitz, S. G. and Choppin, P. W.,** Enhancement of the infectivity of influenza A and B viruses by proteolytic cleavage of the hemagglutinin polypeptide, *Virology,* 68, 440, 1975.

19. **Palese, P. and Schulman, J. L.**, Inhibitors of viral neuraminidase as potential antiviral drugs, in *Chemoprophylaxis and Viral Infections of the Respiratory Tract,* Oxford, J., Ed., CRC Press, Cleveland, 1977, 189.
20. **Drzeniek, R., Seto, J. T., and Rott, R.**, Characterization of neuraminidases from myxoviruses, *Biochim. Biophys. Acta,* 128, 547, 1966.
21. **Duesberg, P. H.**, Distinct subunits of the ribonucleoprotein of influenza virus, *J. Mol. Biol.*, 42, 485, 1969.
22. **Pons, M. W., Schulze, I. T., and Hirst, G. K.**, Isolation and characterization of ribonucleoprotein of influenza virus, *Virology,* 39, 250, 1969.
23. **Lazarowitz, S. G., Compans, R. W., and Choppin, P. W.**, Influenza virus structural and nonstructural proteins in infected cells and their plasma membranes, *Virology,* 46, 830, 1971.
24. **Ritchey, M. B. and Palese, P.**, In vitro translation of influenza virus messenger RNAs, *Virology,* 72, 410, 1976.
25. **Scholtissek, C. and Klenk, H.-D.**, Influenza virus replication, in *The Influenza Viruses and Influenza,* Kilbourne, E. D., Ed., Academic Press, New York, 1975, 215.
26. **Compans, R. W. and Choppin, P. W.**, Reproduction of Myxoviruses, in *Comprehensive Virology,* Vol. 4, Fraenkel-Conrat, H. and Wagner, R. R., Eds., Plenum Press, New York, 1975, 179.
27. **Bean, W. J., Jr. and Simpson, R. W.**, Transcriptase activity and genome composition of defective influenza virus, *J. Virol.,* 18, 365, 1976.
28. **Pons, M. W.**, A reexamination of influenza single- and double-stranded RNAs by gel electrophoresis, *Virology,* 69, 789, 1976.
29. **Ritchey, M. B., Palese, P., and Kilbourne, E. D.**, RNAs of influenza A, B and C viruses, *J. Virol.,* 18, 738, 1976.
30. **Palese, P. and Schulman, J. L.**, Mapping of the influenza virus genome: identification of the hemagglutinin and the neuraminidase genes, *Proc. Natl. Acad. Sci. U.S.A.,* 73, 2142, 1976.
31. **Palese, P., Ritchey, M. B., and Schulman, J. L.**, Mapping of the influenza virus genome. II. Identification of the P1, P2 and P3 genes, *Virology,* 76, 114, 1977.
32. **Ritchey, M. B., Palese, P., and Schulman, J. L.**, Mapping of the influenza virus genome. III. Identification of the genes coding for nucleoprotein, membrane protein and nonstructural protein, *J. Virol.,* 20, 307, 1976.
33. **Choppin, P. W.**, Replication of influenza virus in a continuous cell line: high yield of infective virus from cells inoculated at high multiplicity, *Virology,* 39, 130, 1969.
34. **Hoyle, L.**, The influenza Viruses, *Virol. Monogr.,* 4, 1968
35. **Dowdle, W. R. and Schild, G. C.**, Laboratory propagation of human influenza viruses, experimental host range, and isolation from clinical materials, in *The Influenza Viruses and Influenza,* Kilbourne, E. D., Ed., Academic Press, New York, 1975, 243.
36. **Hughes, R. C.**, Glycoproteins as components of cellular membranes, *Prog. Biophys. Mol. Biol.,* 26, 189, 1973.
37. **Davies, W. L., Grunert, R. R., Haff, R. F., McGahen, J. W., Neumayer, E. M., Paulshock, M., Watts, J. C., Wood, T. R., Hermann, E. C., and Hoffman, C. E.**, Antiviral activity of 1-adamantanamine (amantadine), *Science,* 144, 862, 1964.
38. **Chow, N. L. and Simpson, R. W.**, RNA dependent RNA polymerase activity associated with virions and subviral particles of myxoviruses, *Proc. Natl. Acad. Sci. U.S.A.,* 68, 752, 1971.
39. **Penhoet, E., Miller, H., Doyle, M., and Blatti, S.**, RNA dependent RNA polymerase activity in influenza virions, *Proc. Natl. Acad. Sci. U.S.A.,* 68, 1369, 1971.
40. **Skehel, J. J.**, RNA-dependent RNA polymerase activity of the influenza virus, *Virology,* 45, 793, 1971.
41. **Scholtissek, C. and Rott, R.**, Synthesis *in vivo* of influenza virus plus and minus strand RNA and its preferential inhibition by antibiotics, *Virology,* 40, 989, 1970.
42. **Krug, R. M.**, Cytoplasmic and nucleoplasmic viral RNPs in influenza virus-infected MDCK cells, *Virology,* 50, 103, 1972.
43. **Scholtissek, C. and Rott, R.**, Ribonucleic acid nucleotidyltransferase induced in chick fibroblasts after infection with an influenza virus, *J. Gen. Virol.,* 4, 125, 1969.
44. **Hastie, N. D. and Mahy, B. W. J.**, RNA dependent RNA polymerase in nuclei of cells infected with influenza virus, *J. Virol.,* 12, 951, 1973.
45. **Sugiura, A., Ueda, M., Tobita, K., and Enomoto, C.**, Further isolation and characterization of temperature sensitive mutants of influenza virus, *Virology,* 65, 363, 1975.
46. **Krug, R. M., Ueda, M., and Palese, P.**, Temperature sensitive mutants of influenza WSN virus defective in virus specific RNA synthesis, *J. Virol.,* 16, 790, 1975.
47. **Etkind, P. R. and Krug, R. M.**, Influenza viral messenger RNA, *Virology,* 62, 38, 1974.
48. **Pons, M. W.**, Studies on the replication of influenza virus RNA, *Virology,* 47, 823, 1972.

49. **Etkind, P. R. and Krug, R. M.**, Purification of influenza viral complementary RNA: its genetic content and activity in wheat germ cell-free extracts, *J. Virol.*, 16, 1464, 1975.
50. **Kingsbury, D. W. and Webster, R. G.**, Cell-free translation of influenza virus messenger RNA, *Virology*, 56, 654, 1973.
51. **Content, J.**, Cell-free translation of influenza virus m-RNA, *J. Virol.*, 18, 604, 1976.
52. **Tekamp, P. A. and Penhoet, E. E.**, Message activity of influenza viral RNA, *J. Virol.*, 18, 812, 1976.
53. **Skehel, J. J.**, Polypeptide synthesis in influenza virus-infected cells, *Virology*, 49, 23, 1972.
54. **Skehel, J. J.**, Early polypeptide synthesis in influenza virus-infected cells, *Virology*, 56, 394, 1973.
55. **Klenk, H. D. and Rott, R.**, Formation of influenza virus proteins, *J. Virol.*, 11, 823, 1973.
56. **Krug, R. M. and Etkind, P.**, Cytoplasmic and nuclear virus specific proteins in influenza virus-infected MDCK cells, *Virology*, 56, 334, 1973.
57. **Krug, R. M. and Soeiro, R.**, Studies on the intranuclear localization of influenza virus-specific proteins, *Virology*, 64, 378, 1975.
58. **Compans, R. W.**, Distinct carbohydrate components of influenza virus glycoproteins in smooth and rough cytoplasmic membranes, *Virology*, 55, 541, 1973.
59. **Schwarz, R. T. and Klenk, H. D.**, Inhibition of glycosylation of the influenza virus hemagglutinin, *J. Virol.*, 14, 1023, 1974.
60. **Pons, M. W.**, Isolation of influenza virus ribonucleoprotein from infected cells. Demonstration of the presence of negative stranded RNA in viral RNP, *Virology*, 46, 149, 1971.
61. **Krug, R. M.**, Influenza viral RNPs newly synthesized during the latent period of viral growth in MDCK cells, *Virology*, 44, 125, 1971.
62. **Schulze, I. T.**, The structure of influenza virus. II. A model based on the morphology and composition of subviral particles, *Virology*, 47, 181, 1972.
63. **Palese, P. and Compans, R. W.**, Inhibition of influenza virus replication in tissue culture by 2-deoxy-2,3-dehydro-N-trifluoro-acetylneuraminic acid (FANA): mechanism of action, *J. Gen. Virol.*, 33, 159, 1976.
64. **Burnet, F. M. and Lind, P. E.**, A genetic approach to variation in influenza viruses: recombination of characters between the influenza virus A strain NWS and strains of different serological subtypes, *J. Gen. Microbiol.*, 5, 67, 1951.
65. **Hirst, G. K.**, Genetic recombination with Newcastle disease virus, polio viruses and influenza, *Cold Spring Harbor Symp. Quant. Biol.*, 27, 303, 1962.
66. **Kilbourne, E. D., Lief, F. S., Schulman, J. L., Jahiel, R. I., and Laver, W. G.**, Antigenic hybrids of influenza viruses and their implications, *Perspect. Virol.*, 5, 87, 1967.
67. **Simpson, R. W. and Hirst, G. K.**, Temperature sensitive mutants of influenza A virus: isolation of mutants and preliminary observations on genetic recombination and complementation, *Virology*, 35, 41, 1968.
68. **MacKenzie, J. S.**, Isolation of temperature sensitive mutants and the construction of a preliminary genetic map for influenza virus, *J. Gen. Virol.*, 6, 63, 1970.
69. **Sugiura, A., Tobita, K., and Kilbourne, E. D.**, Isolation and preliminary characterization of temperature sensitive mutants of influenza virus, *J. Virol.*, 10, 639, 1972.
70. **Hirst, G. K.**, Mechanism of influenza virus recombination. I. Factors influencing recombination rates between temperature sentitive mutants of strain WSN and the classification of mutants into complementation recombination groups, *Virology*, 55, 81, 1972.
71. **Palese, P. and Schulman, J. L.**, Differences in RNA patterns of influenza A viruses, *J. Virol.*, 17, 876, 1976.
72. **Schulman, J. L. and Palese, P.**, Selection and identification of influenza virus recombinants of defined genetic composition, *J. Virol.*, in press.
73. **Sugiura, A.**, Influenza virus genetics, in *The Influenza Viruses and Influenza*, Kilbourne, E. D., Ed., Academic Press, New York, 1975, 171.
74. **Schlesinger, R. W.**, The relation of functionally deficient forms of influenza virus to viral development, *Cold Spring Harbor Symp. Quant. Biol.*, 18, 55, 1953.
75. **Henle, G., Girardi, A., and Henle, W.**, A nontransmissible cytopathogenic effect of influenza virus in tissue culture accompanied by formation of non-infectious hemagglutinins, *J. Exp. Med.*, 101, 25, 1955.
76. **Franklin, R. M. and Breitenfeld, P. M.**, The abortive infection of Earle's L-cells by fowl plague virus, *Virology*, 8, 293, 1959.
77. **ter Meulen, V. T. and Love, R.**, Virological, immunochemical, and cytochemical studies of four HeLa cell lines infected with two strains of influenza virus, *J. Virol.*, 1, 626, 1967.
78. **Rott, R. and Scholtissek, C.**, Investigations about the formation of incomplete forms of fowl plague virus, *J. Gen. Microbiol.*, 33, 303, 1963.

79. **Fraser, K. B.,** Immunofluorescence of abortive and complete infections by influenza A virus in hamster BHK 21 cells and mouse L-cells, *J. Gen. Virol.,* 1, 1, 1967.
80. **Avery, R. J.,** Abortive infection of L-cells by influenza virus: absence of virion RNA synthesis, *J. Virol.,* 16, 311, 1975.
81. **von Magnus, P.,** Propagation of the PR8 strain of influenza A virus in chick embryos: formation of "incomplete" virus following inoculation of large doses of seed virus, *Acta Pathol. Microbiol. Scand.,* 28, 278, 1951.
82. **Chanock, R. M., Cockburn, W. C., Davenport, F. M., Dowdle, W. R., Fazekas de St. Groth, S., Fukumi, H., Kilbourne, E. D., Schild, G. C., Schulman, J. L., Sohier, R., Soloviev, V. D., Tumova, B., Webster, R. G., Zakstelskaja, L. J. A., and Zdanov, V. M.,** A revised system of nomenclature for influenza viruses, *Bull. WHO,* 45, 119, 1971.
83. **Douglas, R. G., Jr.,** Influenza in man, in *The Influenza Viruses and Influenza,* Kilbourne, E. D., Ed., Academic Press, New York, 1975, 395.
84. **Schulman, J. L. and Kilbourne, E. D.,** Airborne transmission of influenza virus infection in mice, *Nature,* 195, 1129, 1962.
85. **Mulder, J. and Hers, J. F.,** *Influenza,* Wolters-Noordhoff, Groningen, The Netherlands, 1972.
86. **Kilbourne, E. D. and Loge, J. P.,** Influenza a prime: a clinical study of an epidemic caused by a new strain of virus, *Ann. Intern. Med.,* 33, 371, 1950.
87. **Jordan, W. S., Denny, F. W., Badger, G. F., Curtiss, C., Dingle, J. H., Oseasohn, R., and Stevens, D. A.,** A study of illness in a group of Cleveland families. XVII. The occurrence of Asian influenza, *Am. J. Hyg.,* 68, 190, 1958.
88. **Maynard, J. E., Feltz, E. T., Wulff, H., Fortuine, R., Poland, J. D., and Chin, T. D. Y.,** Surveillance of respiratory virus infections among Alaskan Eskimo children, *JAMA,* 200, 927, 1967.
89. **Mogabgab, W. J.,** Viruses associated with upper respiratory illnesses in adults, *Ann. Intern. Med.,* 59, 306, 1963.
90. **Schulman, J. L.,** Immunology of influenza, in *The Influenza Viruses and Influenza,* Kilbourne, E. D., Ed., Academic Press, New York, 1975, 373.
91. **Ray, C. G., Gravelle, C. R., and Chin, T. D. H.,** Circulating interferon in infants and children with acute respiratory illness, *J. Pediatr.,* 71, 27, 1967.
92. **Jao, R. L., Wheelock, E. F., and Jackson, G. G.,** Production of interferon in volunteers infected with Asian influenza, *J. Infect. Dis.,* 121, 419, 1970.
93. **Young, L. S. LaForce, F. M., Head, J. J., Feeley, J. C., and Bennett, J. V.,** A simultaneous outbreak of meningococcal and influenza infections, *N. Engl. J. Med.,* 287, 5, 1972.
94. **Camner, P., Jarstrand, C., and Philipson, K.,** Tracheobronchial clearance in patients with influenza, *Am. Rev. Respir. Dis.,* 108, 131, 1973.
95. **Douglas, R. G., Jr., Alford, R. H., Cate, T. R., and Couch, R. B.,** The leukocyte response during viral respiratory illness in man, *Ann. Intern. Med.,* 64, 521, 1966.
96. **Khakpour, M., Saidi, A., and Naficy, K.,** Proved viraemia in Asian influenza (Hong Kong variant) during incubation period, *Br. Med. J.,* 5677, 208, 1969.
97. **Fry, J.,** Influenza 1959: the story of an epidemic, *Br. Med. J.,* 2, 135, 1959.
98. **Schulman, J. L.,** Influenza, *Postgrad. Med.,* 50, 171, 1971.
99. **Glezen, W. P., Loda, F. A., Clyde, W. A., Jr., Senior, R. J., Sheaffer, C. I., Conley, W. G., and Denny, F. W.,** Epidemiologic patterns of acute lower respiratory disease of children in a pediatric group practice, *J. Pediatr.,* 78, 397, 1971.
100. **Glick, T. H., Likosky, W. H., Levitt, L. P., Mellin, H., and Reynolds, D. W.,** Reye's syndrome: an epidemiologic approach, *Pediatrics,* 46, 371, 1970.
101. **MacKenzie, J. S. and Houghton, M.,** Influenza infections during pregnancy: association with congenital malformations and with subsequent neoplasms in children, and potential hazards of live virus vaccines, *Bacteriol. Rev.,* 38, 356, 1974.
102. **Webster, R. G. and Laver, W. G.,** Antigenic variation of influenza viruses, in *The Influenza Viruses and Influenza,* Kilbourne, E. D., Ed., Academic Press, New York, 1975, 269.
103. **Kilbourne, E. D.,** Epidemiology of influenza, in *The Influenza Viruses and Influenza,* Kilbourne, E. D., Ed., Academic Press, New York, 1975, 483.
104. **Grunert, R. R., McGahen, J. W., and Davies, W. L.,** The *in vivo* antiviral activity of 1-adamantanamine (amantadine). I. Prophylactic and therapeutic activity against influenza viruses, *Virology,* 26, 262, 1965.
105. **Stanley, E. D., Muldoon, R. E., Akers, L. W., and Jackson, G. G.,** Evaluation of antiviral drugs: the effect of amantadine on influenza in volunteers, *Ann. N.Y. Acad. Sci.,* 130, 44, 1965.
106. **Togo, Y., Hornick, R. B., and Dawkins, A. T., Jr.,** Studies on induced influenza in man. I. Double-blind studies designed to assess prophylactic efficacy of amantadine hydrochloride against A2/Rockville/1/65/strain, *JAMA,* 203, 1089, 1968.

107. **Kilbourne, E. D.,** Future influenza vaccines and the use of genetic recombinants, *Bull. WHO,* 41, 643, 1969.
108. **Leibovitz, A., Coultrip, R. L., Kilbourne, E. D., Legters, L. J., Smith, C. D., Chin, J., and Schulman, J. L.,** Correlated studies of a recombinant influenza-virus vaccine. IV. Protection against naturally occurring influenza in military trainees, *J. Infect. Dis.,* 124, 481, 1971.
109. **Murphy, B. R., Chalhub, F. G., Nusinoff, S. R., and Chanock, R. M.,** Temperature-sensitive mutant of influenza virus. II. Attenuation of ts recombinants for man, *J. Infect. Dis.,* 126, 170, 1972.
110. **Maassab, H. F.,** Adaptation and growth characterization of influenza virus at 25 degrees C, *Nature,* 213, 612, 1967.
111. **Huygelen, C., Peetermans, J., Vascobionic, E., Berge, E., and Colinet, G.,** Live attenuated influenza virus vaccine: "in vitro" and "in vivo" properties, *Symp. Ser. Immunobiol. Stand.,* 20, 152, 1973.
112. **Kilbourne, E. D., Palese, P., and Schulman, J. L.,** Inhibition of viral neuraminidase as a new approach to the prevention and treatment of influenza, *Perspect. Virol.,* 9, 99, 1974.
113. **Palmer, D. F., Coleman, M. T., Dowdle, W. R., and Schild, G. C.,** Advanced Laboratory Techniques for Influenza Diagnosis, U.S. Department of Health, Education, and Welfare, Public Health Service, Atlanta, 1975.
114. **Dowdle, W. R. and Coleman, M. T.,** Influenza virus, in *Manual of Clinical Microbiology,* 2nd ed., Lennette, E. H., Spaulding, E. H., and Truant, J. P., Eds., American Society for Microbiology, Washington, D.C., 1974, 678.
115. **Schild, G. C. and Dowdle, W. R.,** Influenza virus characterization and diagnostic serology, in *The Influenza Viruses and Influenza,* Kilbourne, E. D., Ed., Academic Press, New York, 1975, 316.
116. **Lafferty, K. J.,** The interaction between virus and antibody. II. Mechanism of the reaction, *Virology,* 21, 76, 1963.
117. **Fazekas de St. Groth, S. and Webster, R. G.,** The antibody response, *Ciba Found. Symp.,* p. 246, 1964.
118. **Hirst, G. K.,** The quantitative determination of influenza virus and antibodies by means of red cell agglutination, *J. Exp. Med.,* 75, 49, 1942.
119. **Salk, J. E.,** Simplified procedure for titrating hemagglutinating capacity of influenza-virus and the corresponding antibody, *J. Immunol.,* 49, 87, 1944.
120. **Russell, S. M., McCahon, D., and Beare, A. S.,** A single radial haemolysis technique for the measurement of influenza antibody, *J. Gen. Virol.,* 27, 1, 1975.
121. **Aymard-Henry, M., Coleman, M. T., Dowdle, W. R., Laver, W. G., Schild, G. C., and Webster, R. G.,** Influenzavirus neuraminidase and neuraminidase-inhibition test procedures, *Bull. WHO,* 48, 199, 1973.
122. **Beard, C. W.,** Demonstration of type-specific influenza antibody in mammalian and avian sera by immunodiffusion, *Bull. WHO,* 42, 779, 1970.
123. **Schild, G. C., Winter, W. D., and Brand, C. M.,** Serological diagnosis of human influenza infections by immunoprecipitin techniques, *Bull. WHO,* 45, 465, 1971.
124. **Dowdle, W. R., Galphin, J. C., Coleman, M. T., and Schild, G. C.,** A simple double immunodiffusion test for typing influenza viruses, *Bull. WHO,* 51, 213, 1974.
125. **Schild, G. C., Henry-Aymard, M., and Pereira, H. G.,** A quantitative single-radial-diffusion test for immunological studies with influenza virus, *J. Gen. Virol.,* 16, 231, 1972.
126. **Lief, F. S. and Henle, W.,** Methods and procedures for use of complement-fixation technique in type- and strain-specific diagnosis of influenza, *Bull. WHO,* 20, 411, 1959.
127. **Kilbourne, E. D., Laver, W. G., Schulman, J. L., and Webster, R. G.,** Antiviral activity of antiserum specific for an influenza virus neuraminidase, *J. Virol.,* 2, 281, 1968.
128. **Holston, J. L., Jr. and Dowdle, W. R.,** Influenza A neuraminidase antibody assay with sensitized erythrocytes, *Appl. Microbiol.,* 25, 97, 1973.
129. **Liu, C.,** Rapid diagnosis of human influenza infection from nasal smears by fluorescein-labeled antibody, *Proc. Soc. Exp. Biol. Med.,* 92, 883, 1956.
130. **Schulman, J. L.,** Effects of immunity on transmission of influenza: experimental studies, *Prog. Med. Virol.,* 12, 128, 1970.
131. **Schulman, J. L., Khakpour, M., and Kilbourne, E. D.,** Protective effects of specific immunity to viral neuraminidase on influenza virus infection of mice, *J. Virol.,* 2, 778, 1968.
132. **Virelizier, J. L., Postlethwaite, R., Schild, G. C., and Allison, A. C.,** Antibody responses to antigenic determinants of influenza virus hemagglutinin. I. Thymus dependence of antibody formation and thymus independence of immunological memory, *J. Exp. Med.,* 140, 1559, 1974.
133. **Suzuki, F., Ohya, J., and Ishida, N.,** Effect of antilymphocyte serum on influenza virus infection in mice, *Proc. Soc. Exp. Biol. Med.,* 146, 78, 1974.

134. **Haff, R. F., Shriver, P. W., Engle, C. G., and Stewart, R. C.,** Pathogenesis of influenza in ferrets. I. Tissue and blood manifestations of disease, *J. Immunol.,* 96, 659, 1966.

135. **Barber, W. H. and Small, P. A., Jr.,** Dissemination of influenza virus between anatomically isolated sites in ferrets, *Infect. Immun.,* 9, 530, 1974.

136. **Shope, R. E.,** Swine influenza: filtration experiments and etiology, *J. Exp. Med.,* 54, 373, 1931.

137. **Easterday, B. C.,** Animal influenza, in *The Influenza Viruses and Influenza,* Kilbourne, E. D., Ed., Academic Press, New York, 1975, 449.

138. **Bláskovič, D., Szánto, I., Kapitánčik, B., Lesso, J., Lackovic, V., and Skarda, R.,** Experimental pathogenesis of A/equi/influenza virus infection in horses, *Acta Virol.* (Engl. Ed.), 10, 513, 1966.

139. **Slemons, R. D. and Easterday, B. C.,** Host response differences among 5 avian species to an influenzavirus-A/turkey/Ontario/7732/66 (Hav5N?), *Bull. WHO,* 47, 521, 1972.

140. **Mills, J. V. and Chanock, R. M.,** Temperature-sensitive mutants of influenza virus. I. Behavior in tissue culture and in experimental animals, *J. Infect. Dis.,* 123, 145, 1971.

141. **Johnsen, D. O., Woodling, W. L., Tanticharoenyos, P., and Karnjanaprakorn, C.,** An epizootic of A2/Hong Kong/68 influenza in gibbons, *J. Infect. Dis.,* 123, 365, 1971.

142. **Kalter, S. S., Heberling, R. L., Vice, T. E., Lief, F. S., and Rodriguez, A. R.,** Influenza (A2/Hong Kong/68) in the baboon (*Papio* sp.), *Proc. Soc. Exp. Biol. Med.,* 132, 357, 1969.

MYXOVIRIDAE: PARAMYXOVIRUSES—PARAINFLUENZA, MUMPS, AND NEWCASTLE DISEASE

Richard W. Compans and Kiyoto Nakamura

Introduction

The recognized members of this group include parainfluenza viruses types 1 to 5, Newcastle disease, and mumps viruses (Table 1). Newcastle disease virus (NDV) is not a human pathogen except in certain laboratory infections, but it is one of the best studied paramyxoviruses and has provided much important information; therefore, it is included in this chapter. In addition to morphological and biochemical similarities, the paramyxoviruses share important biological properties including adsorption to mucoprotein receptors, hemagglutinin and neuraminidase activities, and the ability to induce hemolysis and cell fusion. Furthermore, the available information indicates close similarities in the biochemical fine structure and molecular events in the replicative cycle of these viruses. Other recently identified viruses have been tentatively classified as paramyxovirus on the basis of morphological and biological properties. These include Yucaipa virus of chickens[1] and Nariva virus of rodents.[2]

Measles, canine distemper, and rinderpest viruses are similar in morphology and biochemistry to the paramyxoviruses described above; however, they lack neuraminidase and do not bind to neuraminic acid-containing receptors. They are frequently included in the paramyxovirus group, but their classification remains uncertain. Respiratory syncytial (RS) virus is sometimes also grouped with the paramyxoviruses, but it is structurally distinct, and insufficient information is available to determine if it is related in its replication process. Measles virus and RS virus are considered elsewhere in this volume. As will be discussed below, certain antigenic relationships exist among mumps, NDV, and parainfluenza viruses, but there is no single antigen shared by all members of the group.

Morphology

Virions are markedly pleomorphic in unfixed, negatively stained preparations and their size ranges from 120 to 500 nm or larger.[3-5] The envelope is frequently disrupted in such preparations, releasing the nucleocapsid component. In fixed preparations, less pleomorphism is seen and filamentous forms are frequently observed. These are approximately 120 nm in diameter, 1 μm or greater in length, and contain the nucleocap-

Table 1
MEMBERS OF THE PARAMYXOVIRUS GROUP

Virus type	Strain or subtype	Natural host
Mumps		Man
Newcastle disease	Many	Chicken, other birds
Parainfluenza 1	Sendai (hemagglutinating virus of Japan)	Mouse
	Hemadsorption type 2	Man
Parainfluenza 2		Man
Parainfluenza 3	Shipping fever	Cattle, sheep
	Hemadsorption type 1	Man
Parainfluenza 4	A	Man
	B	Man
Parainfluenza 5	SV5, DA, SA	Dog, monkey (?), man (?)

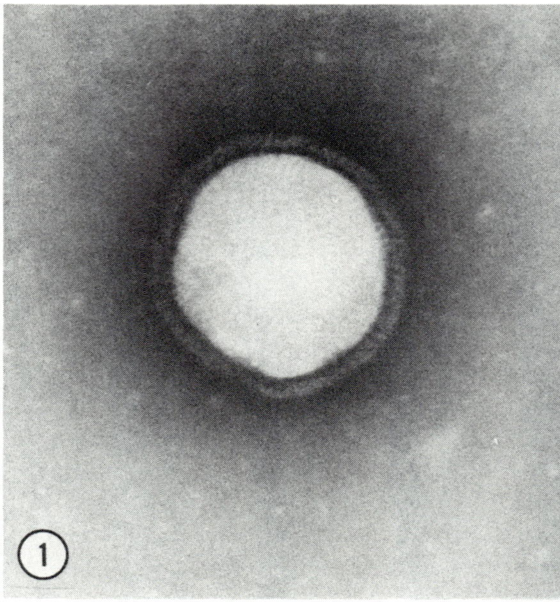

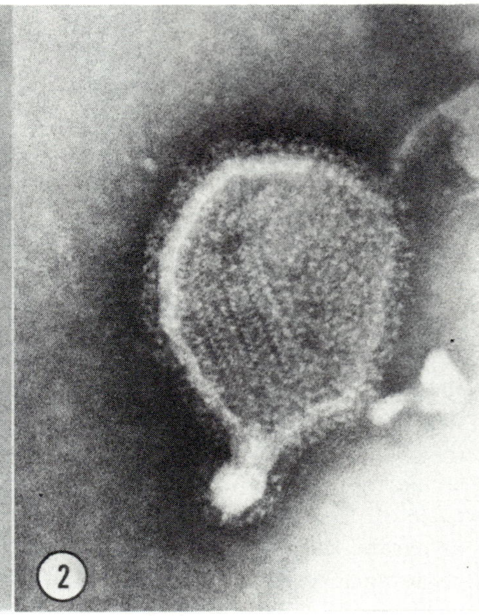

FIGURE 1. Newcastle disease virion negatively stained with sodium phosphotungstate. A layer of spikes covers the surface of the particle. Magnification × 175,000. **FIGURE 2.** Parainfluenza virion (SV5) negatively stained with sodium phosphotungstate. The envelope is penetrated by stain, revealing the helical nucleocapsid. Magnification × 175,000.

sid in a regular, coiled arrangement.[6] All paramyxoviruses possess a spike-covered, lipid-containing envelope which encloses a helical nucleocapsid (Figures 1 and 2). The nucleocapsids (Figure 3) are 18 nm in diameter, and have a length of approximately 1 μm, which corresponds to one viral RNA genome.[4,7-11] Most nucleocapsids are linear, but circular forms are occasionally found.[8] The nucleocapsid is a single helix,[12] and its sense has been determined to be left handed.[13] In thin section, virions exhibit a unit membrane structure similar in appearance to the plasma membrane of the host cell.

Physical Properties

The available information indicates close similarities among the paramyxoviruses in the physical properties which have been studied. Virions have a high sedimentation coefficient (approximately 1000 S). However, it is possible to separate particles of different sizes from the pleomorphic population,[14,15] and the value of 1000 S probably applies to the minimal infectious particle. A minimal particle weight to 6 to 7 × 10^8 daltons has been estimated, based on the estimated molecular weight of the genome and the RNA content of 0.91% or, alternatively, from the estimated size of the nucleocapsid and its contribution to the mass of virion.[16] The buoyant density of the virion is 1.22 to 1.24 g/cc in potassium tartrate or sucrose solutions. The isolated nucleocapsid sediments at 250 to 300 S and has an estimated mass of 1.65 × 10^8 daltons.[7,9,17] The nucleocapsid structure is stable to centrifugation in cesium chloride, in which it has a buoyant density of 1.297 g/cc.[7]

Many physical properties of paramyxoviruses are shared by other lipid-containing viruses. Thermal lability, sensitivity to inactivation by lipid solvents and detergents, and instability on freezing and thawing in the absence of added stabilizers are some examples of such properties.

Chemical Composition

The overall chemical composition and the molecular weights of the nucleic acid and

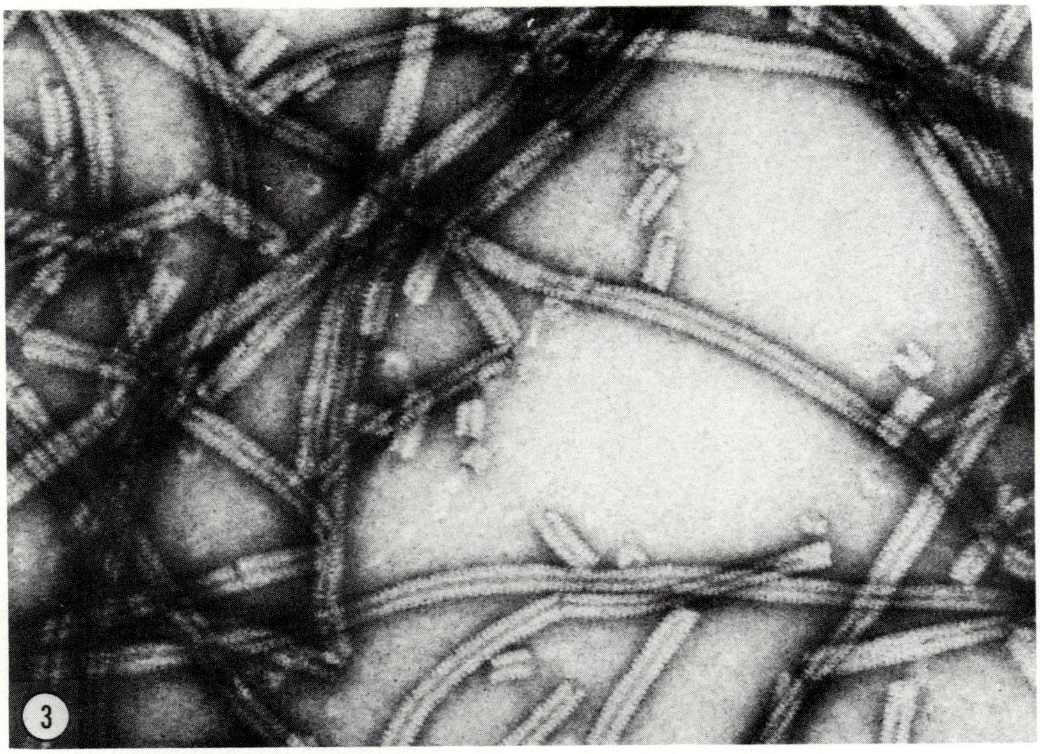

FIGURE 3. Purified helical nucleocapsids of the paramyxovirus SV5 negatively stained with uranyl acetate. Some of the nucleocapsids are fragmented, whereas others are typically 1 μm in length.[7] The central hollow core and periodicity of 50 A are clearly evident. Magnification × 200,000.

polypeptide components of paramyxoviruses are summarized in Table 2. The RNA genome is a single molecule of approximately 6×10^6 daltons mol wt. Genome RNA of this size has been identified in all paramyxoviruses studied to date.[17,19-24] Similar size estimates for the viral RNA have been made, based on sedimentation coefficient,[19-23] electron microscopic measurements of contour length,[24] and size and RNA content of the nucleocapsids.[7] A high uracil content is a common feature of paramyxovirus RNA. Further information on paramyxovirus RNA is available in detailed reviews on this subject (see References 25 to 27).

The virion polypeptides of several paramyxoviruses have been well characterized,[28-30] and similar results have been obtained for the different virus types (Table 2). Major polypeptides include a nucleocapsid subunit designated NP (approximately 60,000 daltons mol wt) and a membrane polypeptide designated M (approximately 40,000 daltons mol wt). Both of these are internal to the lipid of the viral envelope. Two glycoproteins form the spikes on the external surface of the envelope; these have molecular weights of 68,000 to 75,000 and 53,000 to 56,000 daltons, respectively. In contrast to the influenza viruses, in which the hemagglutinin and neuraminidase are distinct polypeptides, in the case of paramyxoviruses both biological activities are associated with the larger glycoprotein,[30] and this component is designated HN. The smaller glycoprotein, designated F, is involved in hemolysis and cell fusion.[31,32] With Sendai virus grown in certain cell types, this protein may be incorporated into virions as a larger precursor (designated Fo) which is inactive in cell fusion or hemolysis. Proteolytic cleavage converts the precursor to the active F glycoprotein, and it is of particular interest that this conversion is essential for viral infectivity.[31,32]

Table 2
STRUCTURAL COMPONENTS OF PARAMYXOVIRUSES[a]

Component	Percent in virion	Molecular weight (daltons)	Function
RNA	0. 9	6×10^6	Genome
Protein	73		
P	4	69,000	Polymerase (?)
HN	11	67,000—74,000	Hemagglutinin, neuraminidase
NP	21	60,000	Nucleocapsid structural protein
F	13	53,000—56,000	Cell fusion, hemolysis
M	15	38,000—41,000	Membrane structural protein
Minor species	9		
Lipid	20		Envelope structural component
Carbohydrate	6.1		Linked to glycoproteins and glycolipids

[a] Data on percent RNA, protein, lipid, and carbohydrate are for SV5.[16] The values for percent and molecular weights of individual viral polypeptides are those obtained for Sendai virions.[18]

One to four other polypeptides have been identified as minor components in paramyxovirus particles. An RNA transcriptase activity has been identified in purified virions.[33-35] One or more of the major polypeptides which are found in association with nucleocapsids are likely to possess this activity.[36,37]

Detailed studies of the lipids of parainfluenza viruses show close similarity with those of the plasma membrane of the host cell, and wide variation in lipid composition is observed with virions from different cell types.[38,39] The lipids of paramyxoviruses are arranged in a bilayer structure,[40] as has been observed with all of the other lipid-containing viruses.

Carbohydrates are covalently linked to both glycolipids and glycoproteins of paramyxoviruses.[41,42] Because carbohydrate compositions vary with the host cell, it is thought that cellular transferases are responsible for glycosylation.[29] However, presumably because of the action of viral neuraminidase, paramyxoviruses as well as myxoviruses lack neuraminic acid residues, in contrast to most other enveloped viruses.[43]

A schematic diagram of a paramyxovirus particle indicating the arrangement of structural components is shown in Figure 4. Several uncertainties still exist, including the location of minor polypeptides and the extent of protein penetration through the lipid bilayer. The exact morphology of each glycoprotein has not been established, although the isolated HN protein of SV5 has been found to be morphologically distinct from the F glycoprotein.[30] Further discussion of the structural components and their arrangement in the virions may be found in other recent reviews (see References 29 and 44).

Virus Replication

The kinetics of paramyxovirus replication are highly variable, depending on both the virus strain and host cell type. The most rapidly multiplying strains of NDV have a latent period of 2.5 hr, and the replication cycle may be completed in 8 hr.[45] Other viruses have a prolonged growth cycle. SV5 infection of primary monkey kidney cells

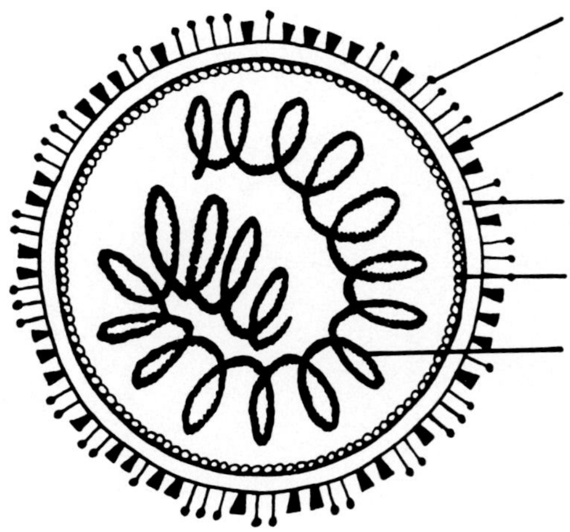

FIGURE 4. Schematic diagram of the probable arrangement of the structural components of a paramyxovirus. The shape of glycoprotein F is suggested from results of Scheid et al.[30]

causes virtually no cytopathic effects; the infected cells divide normally, and virus production continues indefinitely.[46]

A schematic diagram of the events in the replication cycle of a paramyxovirus is presented in Figure 5. Only a brief summary of the steps in replication is presented here, and further information is available in other reviews (see References 27 and 29).

Adsorption

The initial stage of infection by paramyxoviruses involves adsorption to neuraminic acid-containing receptors. Adsorption can be prevented by pretreatment of cells with neuraminidase or by antiviral antibody. Adsorption to host cells appears to be similar to adsorption to receptors on erythrocytes, which causes hemagglutination.

Penetration and Uncoating

Most evidence with paramyxoviruses now favors a mechanism of penetration and uncoating in which the viral envelope fuses with the plasma membrane of the host cell. Electron microscopic studies[47] and the simultaneous activation of infectivity and cell fusion and hemolysis activities observed on cleavage of the Sendai precursor Fo glycoprotein[31,32] support the idea that a fusion event is involved in infectivity. As a consequence of such fusion, the nucleocapsid would be discharged directly into the cytoplasm, where it begins to function in the replication cycle; thus, no further event in uncoating of viral RNA is required.

RNA and Protein Synthesis

As depicted in Figure 5, the biosynthetic events in paramyxovirus replication are complex, and many are still poorly understood. The following steps are thought to occur. RNA transcriptase activity is associated with paramyxovirus nucleocapsids,[36,37] and this enzyme directs the synthesis of complementary (+ strand) RNA species using the virion RNA (− strand) as template. This process is termed "primary transcription." These positive strands are of various sizes and function as messenger RNA species which direct the synthesis of the individual viral polypeptides.[48] In addition, some full-length + strands must be produced which serve as templates for replication, producing further genomic (− strand) RNA. These are assembled into nucleocapsids, which may either add to the pool of transcriptive complexes, producing more mRNA (secondary transcription), or associate with viral envelope proteins and become incorporated into progeny virions.

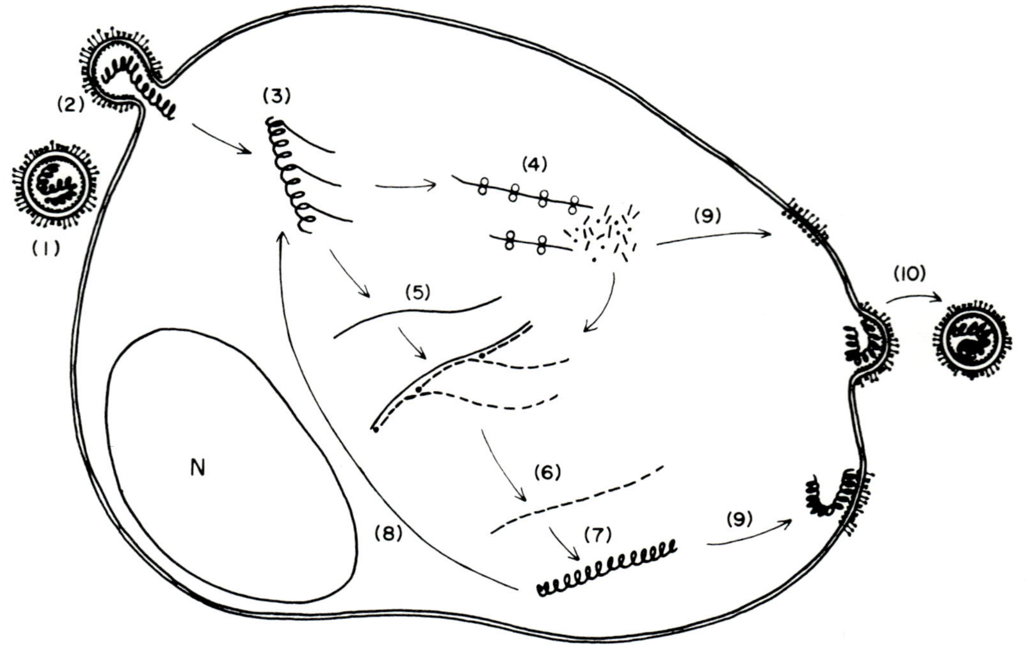

FIGURE 5. Diagram of the replication cycle of a paramyxovirus. All of the intracellular steps are thought to occur in the cytoplasm. (1) Adsorption to neuraminic acid-containing receptors occurs. (2) Fusion of the viral envelope with plasma membrane, with release of helical nucleocapsid into cytoplasm takes place. (3) Primary transcription-production of complementary strands (+ strands) of various sizes by the parental virion transcriptase occurs. (4) + strands serve as messenger RNA; they are incorporated into the polyribosomes and translated into viral polypeptides. (5) Some full length complementary RNA molecules are produced which serve as templates for RNA replication. (6) Progeny genomes (− strands) are produced. (7) Progeny genomes (− strands) are assembled into helical nucleocapsids. (8) Some of the newly formed nucleocapsids are involved in the production of more messenger RNA (secondary transcription). (9) Envelope polypeptides are incorporated into the plasma membrane and nucleocapsids associate specifically with such regions of membrane. (10) Virion formation occurs by budding and release at the cell surface.

The available information indicates that all replicative events are confined to the cytoplasm. This is in contrast to the orthomyxoviruses, which have an essential nuclear phase in replication. As with most other cytoplasmic RNA viruses, paramyxovirus replication is insensitive to actinomycin D, further distinguishing them from the orthomyxoviruses. The structures involved in transcription (transcriptive complexes) have been identified as nucleocapsid complexes, but those involved in production of progeny RNA species have not been characterized.

Assembly

As in most enveloped viruses, assembly of paramyxoviruses occurs by a process of budding at the cell surface. The first detectable event is incorporation of glycoproteins into the plasma membrane, which can be recognized by labeling with ferritin-conjugated antibody or by specific hemadsorption.[49,50] It is thought that a layer of M protein is incorporated into regions of the internal surface of the plasma membrane and that this serves as a recognition site for the alignment of nucleocapsid. Nucleocapsids are seen in a regular arrangement beneath regions of modified membrane and are incorporated into either spherical or filamentous virions by a budding process.[6,51] Filamentous particles contain multiple copies of the nucleocapsid. Examples of the stages in assembly are shown in Figures 6 to 8.

Dual infection of cells with the paramyxovirus SV5 and the rhabdovirus vesicular stomatitis virus (VSV) produces phenotypically mixed particles with the bullet shape

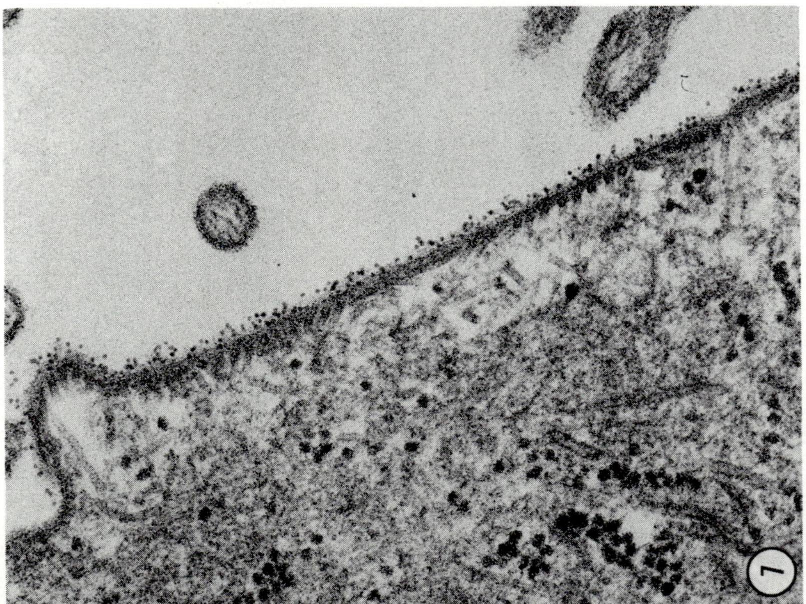

FIGURE 6. Stage in the assembly process of a paramyxovirus (SV5) in MDBK bovine kidney cells. Budding virions containing helical nucleocapsids, many in cross section. Ferritin-conjugated antibody to SV5 is shown tagging the surfaces of budding virions. Magnification × 120,000.
FIGURE 7. Stage in the assembly process of a paramyxovirus (SV5) in MDBK bovine kidney cells. Alignment of nucleocapsids under a region of the cell membrane. Ferritin is distributed on the external surface. Magnification × 75,000.

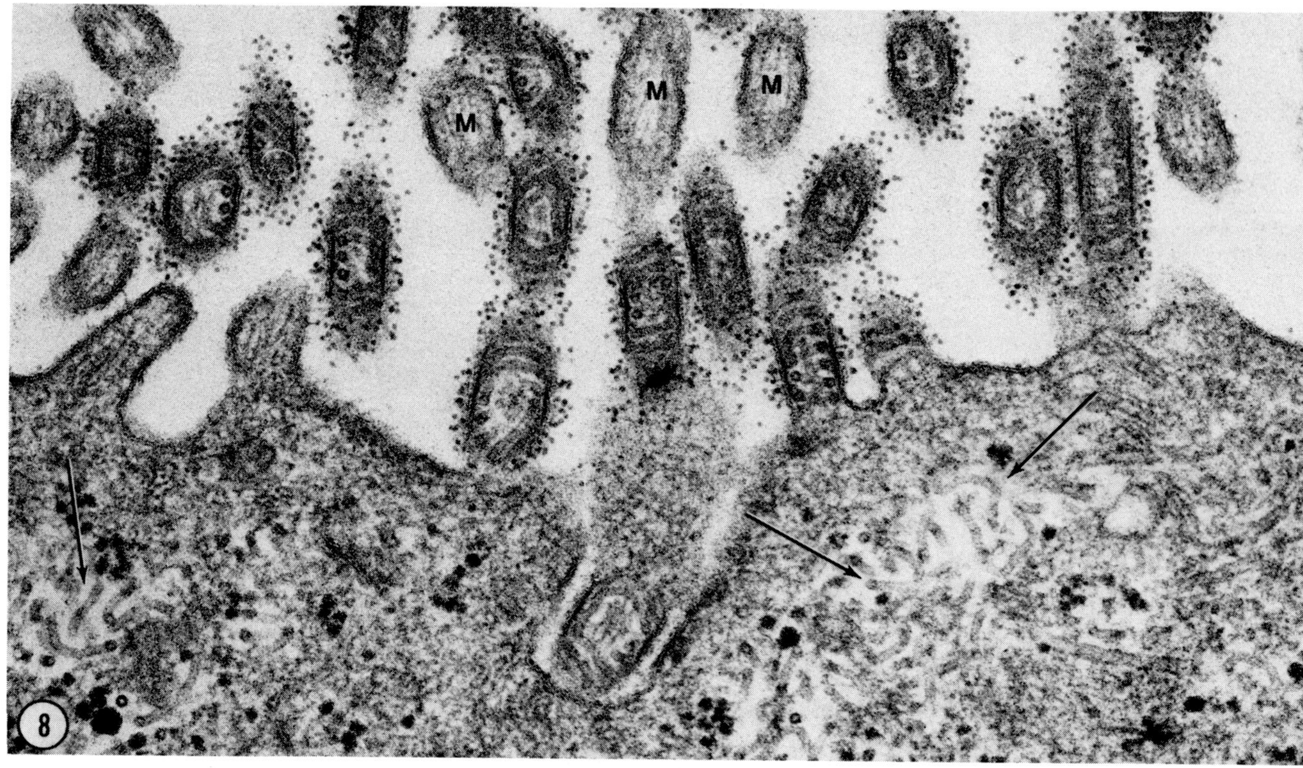

FIGURE 8. Stage in the assembly process of a paramyxovirus (SV5) in MDBK bovine kidney cells. Filamentous virions in the process of budding, sectioned at various angles. Ferritin is localized on the surfaces of virions, which contain nucleocapsid strands, but is absent on the adjacent cell surface as well as cellular microvilli (M). Nucleocapsid strands are also present in small groups within the cytoplasm (arrows). Magnification × 90,000.

of VSV, but containing envelope antigens derived from SV5.[52] Such particles contain both HN and F glycoproteins of SV5, but lack any detectable SV5 M protein, suggesting that each viral nucleocapsid specifically recognizes its M protein, whereas similar specificity does not extend to the glycoproteins.[53]

Culture Systems and Cytopathic Effects

Parainfluenza Viruses

Cytopathic changes induced by human parainfluenza viruses are not extensive. Indeed, persistently infected cells are readily obtained as described below. A common feature is syncytium formation.

Primary monkey kidney cells are widely used for isolation and titration of human parainfluenza viruses. Newly isolated strains of type 2 produce focal syncytia in these cells, whereas other types (types 1, 3, and 4) produce little or no cytopathic effect. However, growth of the latter viruses can be recognized by hemadsorption and immunofluorescence techniques. The formation of acidophilic inclusion bodies in the cytoplasm is also a marker of virus infection. These inclusions are known to consist of large accumulations of nucleocapsids.[6,51,54,55] When the viruses are serially propagated in primary monkey kidney cells, type 3 virus produces elongated satellite cells and syncytia, whereas types 1 and 4 still fail to produce a distinct cytopathic effect except for minimal rounding of cells.

In addition to primary monkey kidney cells, a continuous monkey kidney cell line (Vero) supports the growth of all types of human parainfluenza virus.[56] The viruses can be also adapted to human epithelial cell lines (e.g., HeLa, HEpII, etc.), and type 2 and 3 viruses produce marked syncytium formation in these cells. However, growth of type 1 virus in these cells as well as other continuous cell lines including Vero is limited to one cycle, and the virus fails to induce visible cytopathic change in infected cells. Studies with Sendai virus have revealed that the absence of multiple growth cycles of the virus in such cells is due to the lack of infectivity and cell-fusing activity of the virus obtained from the cells. Treatment of the inactive virus with trypsin restores such activities in parallel with the cleavage of Fo protein into F and other minor components.[3,32,57,58] The finding that co-cultivation of HeLa cells infected with type 1 virus of human origin (HA2) and treated with trypsin with normal HeLa cells produces marked syncytium formation[59] suggests that this is also the case with human type 1 virus. Furthermore, the observation that the growth of human parinfluenza viruses (at least types 1, 3 and 4) in Vero cells is accelerated by addition of trypsin into culture media[60] also suggests that the phenomenon observed for Sendai virus may be common to all human parainfluenza viruses, although it may be most critical with type 1. Therefore, we suggest including trypsin treatment in attempts to isolate parainfluenza viruses, and it is hoped that this may facilitate their isolation and assay.

Mumps Virus

Mumps virus grows well in the yolk sac or amniotic cavity of embryonated chick eggs without inducing death of the embryo. It is best to use 7- to 8-day-old eggs, and maximum titer is obtained after incubation for 4 to 5 days. Presence of the virus is detected by a positive hemagglutination reaction utilizing erythrocytes of chicken, human, guinea pig, or other species. Adaptation of the virus to allantoic membranes is also possible after several passages in the amniotic cavity. The virus can be propagated in a variety of cell cultures, especially in cells of human (HeLa, HEp II, human amnion cells, etc.) or simian origin. However, the susceptibility of such cells to mumps virus varies depending on the passage history of the virus in chick embryo. The virus obtained in primary isolation from patients or from early passages in chick embryo grows well in such tissue culture cells.[61,62] In contrast, virus which has been maintained for many passages in chick embryo fails to grow in these cells.[63] On the other hand, the chick embryo supports the growth of virus irrespective of the passage history of the virus in tissue culture cells. The growth of virus in cell cultures is usually recognized by the appearance of cytopathic effects. The typical cytopathic effect induced by mumps virus is syncytium formation. The virus yield from cell cultures is usually too low to be detected by a hemagglutination reaction, but positive hemadsorption will reveal the presence of virus even in such cases. Cytoplasmic inclusions are recognized in cells after various staining procedures.[55,64]

Newcastle Disease Virus

Embryonated chicken eggs are excellent hosts for propagation of NDV. NDV reaches maximum titer in the allantoic cavity of 10- to 11-day-old chick embryos within 24 to 36 hr after inoculation, and embryos inoculated usually die. However, the period required for induction of death of embryos is variable according to the virulence of strains used. Most virulent strains kill the embryos within 48 hr after inoculation. Recently, evidence has been presented that the difference in virulence among strains of NDV has a parallel relationship with the extent of cleavage of precursor Fo protein.[65] NDV also multiplies in a variety of cultured cells such as chick embryo fibroblast, bovine and monkey kidney, human embryonic tissues, and human heteroploid (HeLa, HEp II, etc.) cells. However, virus yield per infected cell is much lower in cell cultures than allantoic membranes.[66] The typical cytopathic change induced by NDV is syncytium formation. The extent of syncytium formation is affected by various factors including multiplicity of infection, virus strain, host cell type, etc.

Persistent Infection

One of the important properties of paramyxoviruses is their ability to cause persistent infection in tissue culture cells. Establishment of persistent infection in cultured cells has been reported with almost all members of this group (Table 3). Persistent infections caused by viruses have been classified into two types.[81] One is the so-called "carrier culture" type, in which only a fraction of cells in the population are infected and producing virus. Although such infected cells are destroyed by the cytopathic effect, the cultures can be serially passed containing some infected cells because the spread of virus is limited by the presence of genetically resistant cells or inhibitory factors such as interferon, defective interfering particles, or antibodies.[78,81,82] The other type is termed "steady-state infection" or "regulated infection," in which all cells in a culture are infected. However, destruction of the infected cells does not occur, and the cells continue to divide while containing viral components. In this situation, no inhibitory factors are required for establishment and maintenance of persistent infection, and the persistent state cannot be cured by the presence of antiviral antibody.

The latter type of persistent infection is observed in cells infected with several major groups of enveloped viruses, and paramyxoviruses are one of the most important groups which induce it. Elucidation of the control mechanism in regulated infection is important in understanding the pathogenesis of slow virus infections. It has recently been found in several systems that viruses recovered from cells persistently infected with paramyxoviruses are temperature sensitive.[73,79,80] Furthermore, it has been shown that a persistent state can be established more readily by such temperature-sensitive (ts) mutants than by wild strains originally used.[73] These findings suggest that selection of ts mutants may be important in establishment or maintenance of persistent infection. However, it appears unlikely that this idea can be generalized to include all persistent infections with paramyxoviruses because, as described above, infection of primary monkey kidney cells with primary isolates of human parinfluenza viruses produces little or no cytopathic change and initial infection of monkey kidney cells

Table 3
PERSISTENT INFECTION OF TISSUE CULTURE CELLS WITH PARAMYXOVIRUSES

Virus	Cell	Remarks	Ref.
Parainfluenza			
Type 1 (HA2)	HeLa	Regulated infection[a] Establishment of carrier state on initial infection Specific aberration of chromosome Decrease in alkaline phosphatase activity	69, 70 71
Type 1 (Sendai)	HeLa	Regulated infection Increased sensitivity to superinfection with NDV	72
	BHK	Regulated infection ts[b] mutants are isolated Infection of BHK or L cells with ts mutants establishes carrier state more readily than wild type	73
Type 3 (SF4)	Human conjunctiva cells	Regulated infection Establishment of carrier state on initial infection Maintenance of carrier state requires the media which supports active cell growth	74
Type 5 (SV5)	Rhesus monkey kidney	Regulated infection Continuous production of high yield of infectious virus No interference with superinfecting viruses	46, 75
Mumps	Human conjunctiva cells	Regulated infection	76
NDV	MCN and Lung-To	Production of interferon is essential for establishment and maintenance of carrier state	77
	L	Continuous production of interferon ts mutants are isolated	78, 79
	BHK	Regulated infection No interferon production ts mutants are isolated	80

[a] Criteria of "regulated infection" are as follows: (1) There is no requirement for inhibitory factors or antibody, (2) there is not a selection of resistant cells, (3) almost all cells are infected, and (4) infected cells are capable of repeated division.

[b] ts, temperature-sensitive.

with SV5 or HeLa cells with parainfluenza type 1 virus (HA2) induces persistent infection without any detectable cytopathic effect.[46,69] These facts suggest that some strains of paramyxoviruses have the ability to establish persistent infection without requirement for selection of a ts mutant. Thus, further studies are necessary in order to understand the mechanism of establishment and maintenance of persistent infection with these viruses.

Table 4
STRUCTURAL COMPONENTS OF PARAMYXOVIRUSES DETECTED BY SEROLOGICAL TESTS

Test	Structural components detected[a]
Hemadsorption inhibition	HN
Hemagglutination inhibition (HI)	HN
Complement fixation (CF)	HN, F, NP, and other internal proteins
Neutralization	HN, F

[a] Analysis of structural polypeptides of paramyxoviruses has been accomplished in greatest detail with Sendai virus, SV5, and NDV. However, it is likely that human parinfluenza viruses as well as mumps virus consist of the same components as those of the above viruses.

Serological Responses to Paramyxovirus Infections

Parainfluenza Viruses

The specificity of antibodies produced in man following infections with parainfluenza viruses for specific structural components of the virion is not established. Table 4 summarizes the possible participation of each structural component in the commonly used serological procedures on the basis of present data.

No immunological relationship between influenza and parainfluenza viruses is detected, and there is no antigen common to all parainfluenza viruses. However, a heterotypic antibody response to infection with one type of parainfluenza virus is frequently observed in patients who have had prior infections with other types. Furthermore, a rise in the level of antibodies against parainfluenza viruses is also observed in patients with mumps. Such heterotypic antibody response can be detected by hemagglutination inhibition(HI), complement fixation (CF), or neutralization tests. These facts suggest that there are antigenic relationships not only among types of parainfluenza viruses, but also between parainfluenza and mumps viruses.[83-86]

In the case of primary infection in man, the serological response is type specific, as seen also in experimental infection of guinea pigs and hamsters. Most adults have high levels of circulating antibodies to all virus types,[87] but lack IgA neutralizing antibodies in nasal secretions. Recurrent infections are observed despite the presence of neutralizing serum antibodies.[87] The levels of IgA antibodies in nasal secretions rise in response to each recurrent infection, but disappear within a few months, whereas circulating antibodies remain at high levels. These facts suggest the importance of IgA antibodies in protection from parainfluenza virus infections.[88] Little is known about the role of cellular immunity in parainfluenza virus infection.

Mumps Virus

Structural components of mumps virus are believed to be similar to those of parainfluenza viruses. Antigens of mumps virus have been differentiated into S and V antigens. S antigens appear to correspond to NP and other internal proteins, and V antigens probably contain two glycoproteins, HN and F. There are antigenic relationships between mumps and parainfluenza viruses as described above, and a relationship is also recognized between mumps and NDV.[89,90] No antigenic differences have been noted among various mumps virus strains.

Antibody to S antigen usually appears within a few days of onset of illness and reaches a peak within approximately 10 days. On the other hand, anti-V antibody becomes detectable after 2 to 3 weeks and attains maximum level within 3 to 4 weeks. Thus, during the first week of illness, a high titer of anti-S antibody with or without a low level of anti-V antibody may be detectable. Anti-S antibody disappears within 8 to 9 months, whereas antibody to V antigen persists as evidence of past infection.

Man acquires immunity after either clinical or subclinical mumps virus infection, and it persists through life in almost all cases. The protective mechanism in such life-long immunity to mumps remains to be elucidated.

Newcastle Disease Virus

A rise in the level of anti-NDV antibodies is often recognized in patients with mumps,[89,90] suggesting the presence of common antigens between mumps and NDV. As the CF test with S antigen fails to detect such heterotypic response,[91] common antigens may be associated with surface antigens HN or F.

Infection of Animals by Paramyxoviruses

Parainfluenza Viruses

Naturally occuring strains of human parainfluenza viruses grow poorly or not at all in embryonated chicken eggs. (Two animal parainfluenza viruses, Sendai virus and SV5, can grow very well in eggs). However, each of the four human types can be adapted to grow in the amniotic sac of 8- to 10-day-old eggs after serial passage,[83,85,92,93] and type 1 and type 2 virus can be adapted to grow in the allantoic sac after serial amniotic propagation.

Intranasal inoculation of parainfluenza viruses into guinea pigs results in infection without producing apparent disease.[86,94] Hamsters are also infected by type 1, 2, and 3 viruses.[95,96] Relatively high titers of the viruses are recovered from lung, but no obvious sign of disease is observed in this case either. Both guinea pigs and hamsters develop type-specific serum antibodies, enabling the preparation of antisera required for identification of primary isolates. However, it should be noted that some populations of these as well as other experimental animals must be spontaneously infected with parainfluenza viruses, as the presence of antibodies against the viruses are often observed in their sera.[86] Therefore, it is important to rule out spontaneous infection of the animals before immunization, or heterotypic antibody responses may occur.

Because of the possibility that parainfluenza viruses may cause slowly progressive neurological disorders, intracerebral inoculation of type 2 virus into newborn mice has been attempted.[97] Although no apparent symptoms are observed, viral antigens are detectable in the cerebral cortex as long after inoculation as the 40th day.

Natural infection of mice with Sendai virus has often been recognized in Japan, China, and the U.S.S.R.;[98] more recently, the wide distribution of this virus in mice has also been found in the United States.[99] Therefore, these results should be carefully considered when mice are used as experimental animals.

Monkeys are hosts of SV5 and SV41. These viruses are frequently recovered from primary monkey kidney cell cultures. Therefore, controls for contamination with the viruses are required when primary monkey kidney cells are used for isolation of human parainfluenza viruses. The available evidence indicates that monkeys acquire SV5 infection while in quarantine.[100] Monkeys bled at the time of capture do not have detectable antibodies to SV5,[101,102] suggesting that the virus is not of simian origin. SV5 has frequently been isolated from dogs with the respiratory disease "kennel cough," and dogs may be a natural host for the virus. Various animal species have been infected experimentally with SV5 and little evidence of disease has been observed.[103] SV5 has also been recovered from humans,[104,105] but the significance of the virus in human disease has not been established.

Mumps Virus

Man is the only natural host for mumps virus. However, the animals listed in Table 5 are known to be susceptible to the virus recoverd from patients with mumps. Typical parotitis develops in monkeys only by inoculating parotid gland with mumps virus via Stensen's duct.[106] The virus can be passed serially in monkeys by injection of homogenates of parotid gland. In addition to the animals listed, the virus grows well in the yolk sac or amniotic cavity of 7- to 8-day-old chick embryos. No destruction of cells is observed in these embryos.

Newcastle Disease Virus

Newcastle disease is an important epizootic infection of chickens. The respiratory, gastrointestinal, and central nervous system are usually involved. Many other species of birds also are natural and experimental hosts.[113]

Intranasal inoculation of mice with large amounts of NDV induces extensive consolidation of the lung, and the mice usually die within 3 to 4 days. However, this effect cannot be transmitted serially,[114] indicating that it is not a manifestation of virus multiplication. Encephalitis due to virus multiplication in the brain may develop in surviving mice 5 to 29 days after intranasal inoculation.[115] Meningoencephalitis develops after intracerebral inoculation of NDV into hamsters, pigs, cats, or monkeys.

Diseases Induced by Paramyxoviruses (Table 6)

Parainfluenza Virus

The parainfluenza viruses are very important pathogens causing acute respiratory

Table 5
PATHOGENICITY OF MUMPS VIRUS IN EXPERIMENTAL ANIMALS

Species	Route of inoculation	Response	Remarks	Ref.
Monkey	Via Stensen's duct	Typical parotitis with swelling	Serial passage of the virus is possible by inoculating with a suspension of the involved parotid gland	106
Mouse	Intracerebral	Encephalitis or death	Suckling mice are used; passage history of the virus is an important factor; strains adapted to mouse embryo cells fail to induce disease; strains passaged in suckling hamster brains are more virulent	107
Guinea pig	Intraocular	Corneal reaction with rise in CF antibodies		108
Hamster	Intranasal	Bronchitis and interstitial pneumonitis with rise in HI and CF antibodies	Serological response is similar to natural human infection	109
	Intracerebral	Hydrocephalus		110
Ferret	Intranasal or subcutaneous	Bronchitis and interstitial pneumonitis with antibody response	Antibody response occurs more effectively by intranasal than subcutaneous inoculation	111
Dog	Via Stensen's duct	Subclinical infection with development of CF and HI antibodies	CF antibodies are detectable in 18% of normal dogs	112

Note: CF, complement-fixing; HI, hemagglutination-inhibiting.

Table 6
DISEASES CAUSED BY PARAMYXOVIRUSES

	Disease	
Virus	ommon	Less common
Parainfluenza		
Type 1	Upper respiratory infection	
	Bronchitis	
	Bronchopneumonia	
	Croup	
Type 2	Croup	
Type 3	Upper respiratory infection	
	Bronchitis	
	Bronchopneumonia	
Type 4		Upper respiratory infection
Mumps	Parotitis	Pancreatitis
	Orchitis	Oophoritis
	Meningitis	Encephalitis
		Thyroiditis
Newcastle disease		Conjunctivitis

tract infection, primarily during infancy and childhood. The severity ranges from mild upper respiratory infections to croup or pneumonia and appears to be influenced by age. The most serious manifestations involving the lower respiratory tract are observed in the case of primary infection of infants or very young children. Reinfections which sometimes occur in older children produce only minor upper respiratory diseases. The viruses may also cause mild upper respiratory infections in adults. They have been isolated from acute respiratory diseases of adults, and inoculation of type 1, 2, and 3 into volunteers has produced common cold-like illness.[116]

Approximately one third of all cases of croup appear to be caused by types 1, 2, and 3,[117] with type 1 the most important agent. These three types are also responsible for other forms of acute respiratory syndromes. The majority of infections with type 4 virus are clinically inapparent.

The clinical features of acute respiratory tract illness induced by parainfluenza viruses are not distinguishable from those induced by bacterial agents or other viruses. Therefore, etiological diagnosis on the basis of clinical features is not possible.

As measles virus has been established as an etiologic agent for subacute sclerosing panencephalitis,[118-120] it is possible that other chronic or subacute diseases such as multiple sclerosis, systemic lupus erythematosus, polymyositis, etc. may also be caused by viruses. Parainfluenza viruses have been among the most likely candidates as etiologic agents for such diseases because infection of tissue culture cells with parainfluenza viruses readily establishes persistent noncytocidal infection. Indeed, a virus which has characteristics similar to those of parainfluenza type 1 virus has been isolated from the brain of a patient with multiple sclerosis.[121] However, there is presently insufficient evidence to establish type 1 virus as causative agent for the disease.

Mumps Virus

The clinical aspects of mumps are briefly summarized here. For further details, excellent textbooks are available (see References 122 and 123). In almost all cases, symptoms develop after an incubation period of 18 to 21 days. Bilateral or unilateral parotitis with marked swelling is the feature most commonly observed. In addition to parotid glands, submaxillary glands are frequently involved, whereas the involvement of sublingular glands is less frequent. Furthermore, inflammation of organs other than salivary glands sometimes occurs. In postpubertal males, infection with mumps orchitis may lead to atrophy of the affected testicle, but sterility is rare because the involvement is usually unilateral. Bilateral involvement is observed in 2 to 6% of the cases with orchitis. Aseptic meningitis is one of the most important complications and develops in 0.5 to 10% of the cases with parotitis. It should be noted that aseptic meningitis is often observed in cases without parotitis. Fortunately, the prognosis is usually good and sequelae are rarely found

In contrast to measles, there are many inapparent cases of mumps infection. The finding that specific antibody detected by CF can be recognized in the sera of 30 to 50% of the tested adults with no history of past disease[124,125] indicates the high frequency of inapparent infections.

Newcastle Disease Virus

Infection of man with NDV is confined to poultry workers and laboratory personnel. The illness is restricted to conjunctivitis without corneal involvement.

Laboratory Diagnosis

Parainfluenza Viruses

It is not possible to identify the specific type of virus responsible for an infection by serologic procedures alone because of the frequent occurrence of heterotypic antibody response, which is observed in all immunological tests including HI, CF, and neutralization tests. The titers of antibodies to heterotypic strains may exceed those to the infecting strains. Therefore, specific etiologic diagnosis can be achieved only after virus isolation. Virus isolation from patients suffering from acute respiratory infection is most frequently accomplished by inoculating nasal and throat washings or swabs into primary monkey kidney cell cultures. The majority of type 1, 3, and 4 viruses do not usually produce detectable cytopathic effect in these cells. However, the presence of such viruses can be detected most easily by the hemadsorption test with guinea pig erythrocytes. Type 2 virus usually produces typical focal syncytia, but the growth of the virus can also be detected more rapidly by hemadsorption. Type 1 and type 3 virus can usually be detected by hemadsorption by the 5th day after inoculation, whereas type 2 and type 4 require 10 days or more. A scheme for virus isolation utilizing primary monkey kidney cells is summarized in Figure 9.

Primary monkey kidney cells are often latently infected with simian parainfluenza virus SV5, which sometimes interferes with primary isolation from humans. This must be avoided by careful examination of controls. An alternative approach which avoids this problem is the use of continuous cell lines to isolate human parainfluenza viruses. The cultivation of Vero or MDCK cells inoculated with specimens in the presence of a low concentration of trypsin yields satisfactory results.[60] Comparison of this system with primary monkey kidney cells shows not only that a higher frequency of virus isolation is obtained in this system, but also that the presence of virus is detectable more rapidly.

Isolates can be identified as to specific type by serologic tests employing immune sera obtained from guinea pigs despite the presence of common antigens among parainfluenza viruses. Such sera give much higher titers against homotypic strains than heterotypic ones. Hemadsorption inhibition is most frequently used for typing of isolates because of feasibility, although other tests (HI, CF, neutralization, and immunofluorescence techniques) can be employed for this purpose.

Investigation of the possible involvement of parainfluenza viruses in chronic diseases such as multiple sclerosis and systemic lupus erythematosus requires attempts to isolate parainfluenza viruses from patients with such conditions. The recovery of measles virus from subacute sclerosing panencephalitis shows that isolation is successful only

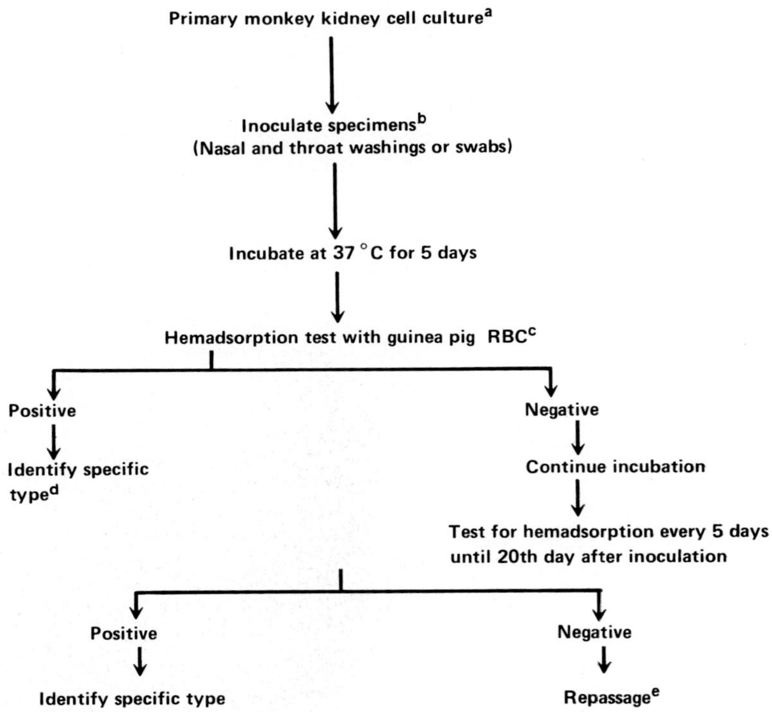

FIGURE 9. Scheme for isolation of human parainfluenza viruses from acute respiratory tract infection. [a]Check for latent simian virus infection by hemadsorption with guinea pig red blood cells. [b]The specimens should be stored at 4°C if inoculation is possible within 48 hr after preparation of specimens. If this is not possible, they should be frozen at −70°C with 0.5 to 1% bovine serum albumin. [c]Uninoculated cells must be tested in parallel at any time when a hemadsorption test is made. The test should be made under aseptic conditions, which permits the continued cultivation of cells after testing. [d]Specific virus type can be identified most easily by the hemadsorption inhibition test. Other serological tests including hemagglutination inhibition, complement fixation, neutralization, and immunofluorescence tests are also available. [e]If repassages give negative results, it is concluded that isolation is unsuccessful.

when lesions are digested with trypsin and co-cultivated with cells susceptible to measles virus.[118-120] Therefore, when isolation of parainfluenza viruses from such disease is attempted, co-cultivation of trypsin-treated specimens with susceptible cells (primary monkey kidney cells, VERO, MDCK, HeLa, etc.) should be included in the isolation procedure. Trypsin treatment may be required for activation of virus infectivity as discussed above.

Mumps Virus

In contrast to parainfluenza viruses, serological procedures are useful in the diagnosis of mumps virus infection. It should be noted that serological evidence can be obtained for any type of infection caused by mumps virus by demonstrating a significant rise in antibody titer following onset of illness.

Immunological Tests

The immunological tests commonly used for diagnosis of mumps virus are listed in Table 7 and are described below.

Complement fixation (CF) test — CF is the most practical and reliable test for diagnosis of mumps infection. The antibodies against both S and V antigens can be detected utilizing each antigen. As mentioned above, the appearance of anti-S antibody precedes that of anti-V antibody, and the latter persists longer. Thus, high levels of

Table 7
SEROLOGICAL TESTS FOR DIAGNOSIS OF MUMPS

Test	Antigens which react in test	Antigen source	Comments	Cross-reactions with parainfluenza viruses and Newcastle disease virus
Hemagglutination inhibition	V (HN)	Amniotic or allantoic fluids	Red blood cell; chicken; nonspecific inhibitors in sera should be removed by pretreatment with KIO_4 or receptor-destroying enzyme (from *Vibrio cholera*)	+
Complement fixation[a]	V (HN and F)	Amniotic or allantoic fluids		+
	S (NP and other internal proteins)	Preparation from allantoic membranes[b]		−
Neutralization	V (HN and F)	Stock virus adapted to cell system used	Cell system; monkey kidney, HeLa, HEpII, chick embryo fibroblast, etc.	+

[a] If isolated viral components are available, antibodies against specific structural components can be detected by this test.
[b] The procedures for preparation of S antigens are described by Henle et al.[126,127]

anti-S antibody with or without low titer of anti-V antibody are often observed in the sera taken early in illness. A presumptive diagnosis of current mumps infection is possible by this finding. However, it is necessary to examine another serum sample taken later to establish the diagnosis by demonstrating a significant increase (fourfold or greater) in antibody level. The rise in antibody against S antigens is usually detected within 7 to 10 days after onset. After 14 days, the CF test with V antigens or the HI test is recommended because of the rapid decrease in anti-S antibodies. It should be noted that cross-reaction with parainfluenza viruses and NDV occurs in both CF and V antigens and HI, but not in CF with S antigens.[91]

Hemagglutination inhibition test (HI) — Antibodies against V antigens can be detected by this test. As cross-reactions with parainfluenza viruses and NDV are observed,[84,89,90] it is necessary to test all related antigens simultaneously. The highest antibody rise occurs in response to antigens of the infecting agent.

Neutralization test (NT) — Because of laboriousness compared with CF and HI, this test is not used as commonly as the above two tests. Antibodies against V antigens are detectable by this test.

Skin test — this test is useful in surveys for estimating immunity, but should not be used for diagnosis. The process of skin testing itself induces a rise in antibody level, which confuses the interpretation of the results obtained from other immunological tests.

Virus Isolation

Direct evidence for mumps virus infection can be obtained by virus isolation, for which the following comments are given.

Specimens — Saliva, mouth swabs, cerebrospinal fluids, and urine are usually used. As virions are unstable at room temperature or 4°C, the specimens should be used immediately or stored at −70 to −80°C.

Hosts used for isolation — Inoculation of specimens into the amniotic cavity of 7- to 8-day-old chick embryos or tissue culture cells (HeLa, HEpII, monkey kidney cells,

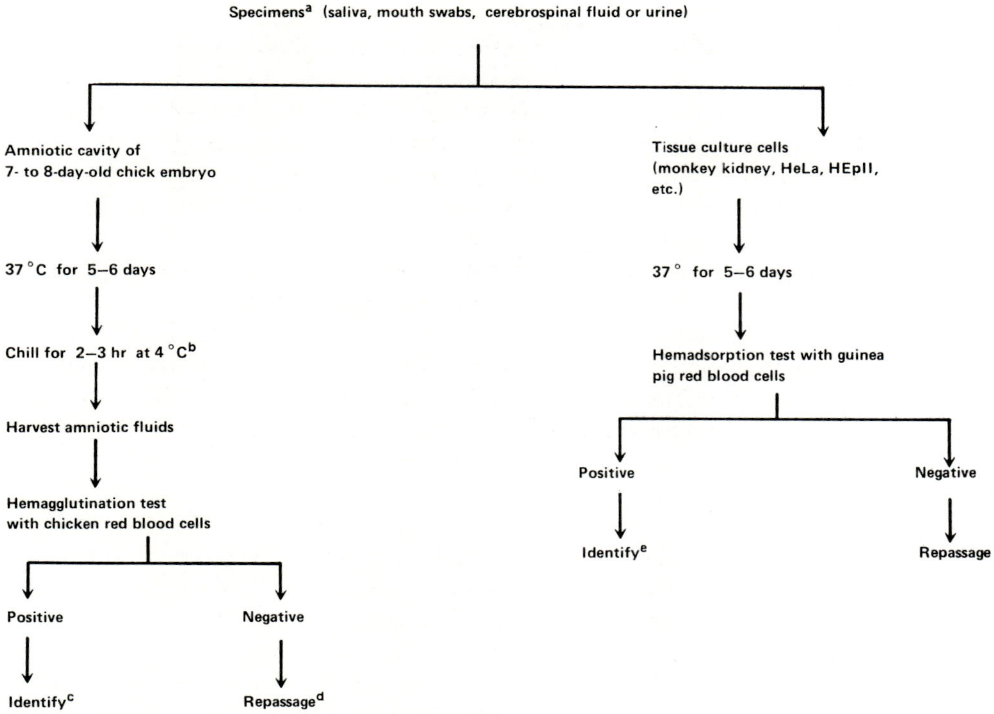

[a] Specimens should be used immediately or stored at – 70° C.
[b] This step is necessary to avoid damage to blood vessels.
[c] The isolate can be identified by complement fixation, hemagglutination inhibition, or neutralization tests.
[d] If the repassage gives a negative result, it is concluded that isolation is unsuccessful.
[e] If virus titer is too low to be identified by hemagglutination inhibition or complement fixation tests, either the neutralization or hemadsorption inhibition test may be employed. High titers of virus for hemagglutination inhibition or complement fixation tests can be obtained by inoculating the isolates into amniotic cavity of hen's eggs or by repassage in tissue culture cells.

FIGURE 10. Scheme for isolation of mumps virus.

etc.) is recommended. Comparative studies have shown that tissue culture cells are more sensitive than chick embryos for primary isolation.[61,62,128]

Procedure for isolation of mumps virus — See Figure 10.

Identification — Serological procedures including CF, HI, and neutralization tests are available for identification. Hemadsorption inhibition is also employed. Specific antisera against mumps virus can be obtained from guinea pigs immunized with the virus.

Newcastle Disease Virus

Serological diagnosis of human infection is usually not possible because of the low antibody response. The virus can be isolated by inoculating scrapings of conjunctiva into the allantoic cavity of chick embryos or the tissue culture cells listed above.

Prevention and Control

Parainfluenza Viruses

There are no specific treatment or control measures. Development of an effective vaccine which induces an antibody response in the respiratory tract would be an important advance in prevention of disease in young children, especially the outbreaks in hospitals and institutions.

Mumps Virus

Control by isolation of patients is not effective because there are many subclinical infections. However, children with mumps should be excluded from school for at least 2 weeks. Although the disease can be prevented by vaccination, many physicians doubt whether it is worthwhile to immunize all children, from the point of view that the disease is usually mild in children and they acquire a life-long immunity. Nevertheless, vaccination should be recommended, especially for susceptible adults, in whom the symptoms are usually more severe than children.

A live attenuated mumps vaccine obtained by passage of the Jeryl-Lynn strain in chick embryos and then in cultured chick fibroblast has been used in the U.S. since 1968. Subcutaneous inoculation of the vaccine induces the development of antibody in greater than 95% of the subjects without any side effects. Although the level of antibody remains low compared with natural infections, vaccinated individuals acquire resistance to subsequent mumps virus infections for at least 5 years.

Epidemiology

Parainfluenza Viruses

Parainfluenza viruses produce acute respiratory infections throughout the year, but a seasonal peak is observed from late fall to spring. Types 1 and 3 occasionally induce small epidemics, whereas type 2 virus infection usually occurs in a sporadic pattern. Type 3 sometimes produces explosive outbreaks in semiclosed populations such as hospitals and institutions. In such episodes, all children who are free of neutralizing antibodies become infected with the virus, whereas type 1 and type 2 produce disease in only 50% of children under similar conditions.[117] The viruses appear to spread by airborne routes.

Of the parainfluenza viruses, infection with type 3 virus generally occurs at the earliest age. Most infants acquire antibodies against type 3 by 2 years of age. They acquire antibodies against type 1 and 2 viruses later, and almost all children acquire such antibodies by 6 years of age. Infection with type 4 virus is rare and not of public health importance.

Mumps Virus

Man is the only reservoir of mumps virus, and the virus spreads primarily by droplets of saliva. Saliva usually contains infectious virus from 6 days before to 9 days after the appearance of glandular swelling. It should be noted that the presence of virus in saliva is detectable in subclinical cases as well as clinical cases lacking signs of involvement of salivary glands. Excretion of the virus in the urine has been reported.[128] However, it is not certain if urine plays any role in spread of the virus. In contrast to measles or varicella, there are many inapparent mumps infections. Furthermore, the fact that the disease often develops in adults suggests that many children escape mumps virus infection.

Although patients with mumps are found throughout the year, a seasonal peak is observed in winter and early spring. Mumps usually occur in endemic patterns, but large epidemics have been observed every 7 or 8 years.

GENERAL REFERENCES

Henle, W. and Enders, J. F., Mumps virus, in *Viral and Rickettsial Infections of Man, 4th ed.,* Horsfall, F. L., Jr. and Tamm, I., Eds., J. B. Lippincott, New York, 1965, 755.

Chanock, R. M., Bell, J. A., and Parrott, R. H., Natural history of parainfluenza infection, in *Perspectives in Virology,* Vol. 2, Pollard, M., Ed., Burgess, Minneapolis, 1961, 126.

Choppin, P. W. and Compans, R. W., Reproduction of paramyxoviruses, in *Comprehensive Virology*, Vol. 4, Fraenkel-Conrat, H. and Wagner, R. R., Eds., Plenum Press, New York, 1975, 95.
Compans, R. W. and Choppin, P. W., The structure and assembly of influenza and parainfluenza viruses, in *Comparative Virology*, Maramorosch, K. and Kurstak, E., Eds., Academic Press, New York, 1971, 407.
Hilleman, M. R., Mumps vaccination in *Modern Trends in Medical Virology*, Vol. 2, Heath, R. B. and Waterson, A. P., Eds., Appleton, New York, 1970, 241.
Hsiung, G. D., Parainfluenza 5 virus: infections in man and animals, *Prog. Med. Virol.*, 14, 241, 1972.
Kingsbury, D. W., Paramyxovirus replication, *Curr. Top. Microbiol. Immunol.*, 59, 1, 1973.
Walker, D. L., Persistent viral infection in cell cultures, in *Medical and Applied Virology*, Sanders, M. and Lennette, E. H., Eds., Warren H. Green, St. Louis, 1968, 99.

REFERENCES

1. **Dinter, Z., Hermodsson, S., and Hermodsson, L.**, Studies on Myxovirus yucaipa: its classification as a member of the paramyxovirus group, *Virology*, 22, 297, 1964.
2. **Walder, R.**, Electron microscopic evidence of Nariva virus structure, *J. Gen. Virol.*, 11, 123, 1971.
3. **Horne, R. W., Waterson, A. P., Wildy, P., and Farnham, A. E.**, The structure and composition of the myxoviruses. I. Electron microscope studies of the structure of myxovirus particles by negative staining techniques, *Virology*, 11, 79, 1960.
4. **Hosaka, Y., Kitano, H., and Ikeguchi, S.**, Studies on the pleomorphism of HVJ virions, *Virology*, 29, 205, 1966.
5. **Simon, E. H.**, The distribution and significance of multiploid virus particles, *Prog. Med. Virol.*, 14, 36, 1972.
6. **Compans, R. W., Holmes, K. V., Dales, S., and Choppin, P. W.**, An electron microscopic study of moderate and virulent virus-cell interactions of the parinfluenza virus SV5, *Virology*, 30, 411, 1966.
7. **Compans, R. W. and Choppin, P. W.**, Isolation and properties of the helical nucleocapsid of the parainfluenza virus SV5, *Proc. Natl. Acad. Sci. U.S.A.*, 57, 949, 1967.
8. **Compans, R. W. and Choppin, P. W.**, The length of the helical nucleocapsid of Newcastle disease virus, *Virology*, 33, 344, 1967.
9. **Kingsbury, D. W. and Darlington, R. W.**, Isolation and properties of Newcastle disease virus nucleocapsid, *J. Virol.*, 2, 248, 1968.
10. **Hosaka, Y. and Shimizu, K.**, Lengths of the nucleocapsids of Newcastle disease and mumps viruses, *J. Mol. Biol.*, 35, 369, 1968.
11. **Finch, J. T. and Gibbs, A. J.**, Observations on the structure of the nucleocapsids of some paramyxoviruses, *J. Gen. Virol.*, 6, 141, 1970.
12. **Choppin, P. W. and Stoeckenius, W.**, The morphology of SV5 virus, *Virology*, 23, 195, 1964.
13. **Compans, R. W., Mountcastle, W. E., and Choppin, P. W.**, The sense of the helix of paramyxovirus nucleocapsids, *J. Mol. Biol.*, 65, 167, 1972.
14. **Roman, J. M. and Simon, E. H.** Morphologic heterogeneity in egg- and monolayer-propagated Newcastle disease virus, *Virology*, 69, 287, 1976.
15. **Kingsbury, D. W., Portner, A., and Darlington, R. W.**, Properties of incomplete Sendai virions and subgenomic viral RNAs, *Virology*, 42, 857, 1970.
16. **Klenk, H.-D. and Choppin, P. W.**, Chemical composition of the parainfluenza virus SV5, *Virology*, 37, 155, 1969.
17. **Compans, R. W. and Choppin, P. W.**, The nucleic acid of the parinfluenza virus SV5, *Virology*, 35, 289, 1968.
18. **Lamb, R. A. and Mahy, B. W. J.**, The polypeptides and RNA of Sendai virus, in *Negative Strand Viruses*, Vol. 1 Barry, R. D. and Mahy, B. W. J., Eds., Academic Press, London, 1975, 65.
19. **Duesberg, P. H. and Robinson, W. S.**, Isolation of the nucleic acid of Newcastle disease virus (NDV), *Proc. Natl. Acad. Sci. U.S.A.*, 54, 794, 1965.
20. **Kingsbury, D. W.**, Newcastle disease virus RNA. I. Isolation and preliminary characterization of RNA from virus particles, *J. Mol. Biol.*, 18, 195, 1966.
21. **Blair, C. D. and Robinson, W. S.**, Replication of Sendai virus. I. Comparison of the viral RNA and virus-specific RNA synthesis with Newcastle disease virus, *Virology*, 35, 537, 1968.
22. **Barry, R. D. and Bukrinskaya, A. G.**, The nucleic acid of Sendai virus and ribonucleic acid synthesis in cells infected by Sendai virus, *J. Gen. Virol.*, 2, 71, 1968.

23. **East, J. L. and Kingsbury, D. W.,** Mumps virus replication in chick embryo lung cells: properties of ribonucleic acids in virions and infected cells, *J. Virol.,* 8, 161, 1971.
24. **Kolakofsky, D., Boy de la Tour, E., and Delius, H.,** Molecular weight determination of Sendai and Newcastle disease virus RNA, *J. Virol.,* 13, 261, 1974.
25. **Robinson, W. S. and Duesberg, P. H.,** The large RNA viruses. The myxoviruses, in *Molecular Basis of Virology,* Fraenkel-Conrat, H., Ed., American Chemical Society, Washington D.C., 1968, 255.
26. **Blair, C. D. and Duesberg, P. H.,** Myxovirus ribonucleic acids, *Annu. Rev. Microbiol.,* 24, 539, 1970.
27. **Kingsbury, D. W.,** Paramyxovirus replication, *Curr. Top. Microbiol.,* 59, 1, 1972.
28. **Mountcastle, W. E., Compans, R. W., Caliguiri, L. A., and Choppin, P. W.,** Nucleocapsid protein subunits of simian virus 5, Newcastle disease virus, and Sendai virus, *J. Virol.,* 6, 677, 1970.
29. **Choppin, P. W. and Compans, R. W.,** Reproduction of paramyxoviruses, in *Comprehensive Virology,* Vol. 4, Fraenkel-Conrat, H. and Wagner, R. R., Eds., Plenum Press, New York, 1975, 95.
30. **Scheid, A., Caliguiri, L. A., Compans, R. W., and Choppin, P. W.,** Isolation of paramyxovirus glycoproteins. Association of both hemagglutinating and neuraminidase activities with the larger SV5 glycoprotein, *Virology,* 50, 640, 1972.
31. **Homma, M. and Ohuchi, M.,** Trypsin action on the growth of Sendai virus in tissue culture cells. III. Structural differences of Sendai viruses grown in eggs and tissue culture cells, *J. Virol.,* 12, 1457, 1973.
32. **Scheid, A. and Choppin, P. W.,** Identification of biological activities of paramyxovirus glycoproteins. Activation of cell fusion, hemolysis, and infectivity by proteolytic cleavage of an inactive precursor protein of Sendai virus, *Virology,* 56, 475, 1974.
33. **Huang, A. S., Baltimore, D., and Bratt, M. A.,** Ribonucleic acid polymerase in virions of Newcastle disease virus: comparison with the vesicular stomatitis virus polymerase, *J. Virol.,* 7, 389, 1971.
34. **Robinson, W. S.,** Ribonuclei acid polymerase activity in Sendai virions and nucleocapsid, *J. Virol.,* 8, 81, 1971.
35. **Stone, H. O., Portner, A., and Kingsbury, D. W.,** Ribonucleic acid transcriptases in Sendai virions and infected cells, *J. Virol.,* 8, 174, 1971.
36. **Stone, H. O., Kingsbury, D. W., and Darlington, R. W.,** Sendai virus-induced transcriptase from infected cells: polypeptides in the transcriptive complex, *J. Virol.,* 10, 1037, 1972.
37. **Marx, P. A., Portner, A., and Kingsbury, D. W.,** Sendai virion transcriptase complex: polypeptide composition and inhibition by virion envelope proteins, *J. Virol.,* 13, 107, 1974.
38. **Klenk, H.-D. and Choppin, P. W.,** Lipids of plasma membranes of monkey and hamster kidney cells and of parainfluenza virions grown in these cells, *Virology,* 38, 255, 1969.
39. **Klenk, H.-D. and Choppin, P. W.,** Plasma membrane lipids and parainfluenza virus assembly, *Virology,* 40, 939, 1970.
40. **Landsberger, F. R., Compans, R. W., Choppin, P. W., and Lenard, J.,** Organization of the lipid phase in viral membranes. Effects of independent variation of the lipid and the protein composition, *Biochemistry,* 12, 4498, 1973.
41. **Klenk, H.-D. and Choppin, P. W.,** Glycosphingolipids of plasma membranes of cultured cells and an enveloped virus (SV5) grown in these cells, *Proc. Natl. Acad. Sci. U.S.A.,* 66, 57, 1970.
42. **Klenk, H.-D., Caliguiri, L. A., and Choppin, P. W.,** The proteins of the parainfluenza virus SV5. II. The carbohydrate content and glycoproteins of the virion, *Virology,* 42, 473, 1970.
43. **Klenk, H.-D., Compans, R. W., and Choppin, P. W.,** An electron microscopic study of the presence or absence of neuraminic acid in enveloped viruses, *Virology,* 42, 1158, 1970.
44. **Lenard, J. and Compans, R. W.,** The membrane structure of lipid-containing viruses, *Biochim. Biophys. Acta,* 344, 51, 1974.
45. **Wheelock, E. F. and Tamm, I.,** Biochemical basis for alterations in structure and function of HeLa cells infected with Newcastle disease virus, *J. Exp. Med.,* 114, 617, 1961.
46. **Choppin, P. W.,** Multiplication of a myxovirus (SV5) with minimal cytopathic effects and without interference, *Virology,* 23, 224, 1964.
47. **Morgan, C. and Howe, C.,** Structure and development of viruses as observed in the electron microscope. IX. Entry of parainfluenza I (Sendai) virus, *J. Virol.,* 2, 1122, 1968.
48. **Morrison, T. G., Weiss, S., Hightower, L., Spanier-Collins, B., and Bratt, M. A.,** Newcastle disease virus protein synthesis, in *In Vitro Transcription and Translation of Viral Genomes,* Haenni, A. L. and Beaud, G., Eds. Inserm, Paris, 1975, 281.
49. **Compans, R. W. and Choppin, P. W.,** The structure and assembly of influenza and parainfluenza viruses, in *Comparative Virology,* Maramorosch, K. and Kurstak, E., Eds., Academic Press, New York, 1971, 407.
50. **Compans, R. W. and Choppin, P. W.,** Orthomyxoviruses and paramyxoviruses, in *Ultrastructure of Animal Viruses and Bacteriophages, An Atlas,* Dalton, A. J. and Haguenau, R., Eds., Academic Press, New York, 1973, 213.

51. **Howe, C., Morgan, C., deVaux St. Cyr, C., Hsu, K. C., and Rose, H. M.,** Morphogenesis of type 2 parainfluenza virus examined by light and electron microscopy, *J. Virol.*, 1, 215, 1967.
52. **Choppin, P. W. and Compans, R. W.,** Phenotypic mixing of envelope proteins of the parainfluenza virus SV5 and vesicular stomatitis virus, *J. Virol.*, 5, 609, 1970.
53. **McSharry, J. J., Compans, R. W., and Choppin, P. W.,** Proteins of vesicular stomatitis virus and of phenotypically-mixed vesicular stomatitis virus-simian virus 5, *J. Virol.*, 8, 722, 1971.
54. **Kuhn, N. O. and Harford, C. G.,** Electron microscopic examination of cytoplasmic inclusion bodies in cells infected with parainfluenza virus, type 2, *Virology*, 21, 527, 1963.
55. **Brandt, C. D.,** Inclusion body formation with Newcastle disease and mumps viruses in cultures of chick embryo cells, *Virology*, 5, 177, 1958.
56. **Okawa, S., Nakajima, M., Katoh, T., Doi, Y., Morimoto, Y., and Itoh, H.,** Attempts at adaptation of parainfluenza type 1 (HA2) virus to Vero cell cultures cercopithecus kidney line, *Uirusu*, 20, 15, 1970 (in Japanese with English abstract).
57. **Homma, M.,** Trypsin action on the growth of Sendai virus in tissue culture cells. I. Restoration of the infectivity for L cells by direct action of trypsin on L cell-borne Sendai virus, *J. Virol.*, 8, 619, 1971.
58. **Homma, M. and Tamagawa, S.,** Restoration of the fusion activity of L cell-borne Sendai virus by trypsin, *J. Gen. Virol.*, 19, 423, 1973.
59. **Nakamura, K. and Homma, M.,** Cell fusion by HeLa cells persistently infected with haemadsorption type 2 virus, *J. Gen. Virol.*, 25, 117, 1974.
60. **Kaji, J., Onishi, H., Kimura, S., and Hatano, K.,** Isolation of myxo and paramyxoviruses from children with upper respiratory tract diseases by utilizing trypsin-added MDCK and Vero cells, in Abstr. Annu. Meet. Society of Japanese Virologists, Nagoya, Japan, 1976.
61. **Henle, G. and Deinhardt, F.,** Propagation and primary isolation of mumps virus in tissue culture, *Proc. Soc. Exp. Biol. Med.*, 89, 556, 1955.
62. **Utz, J. P., Kasel, J. A., Cramblett, H. G., Szwed, C. F., and Parrott, R. H.,** Clinical and laboratory studies of mumps. I. Laboratory diagnosis by tissue-culture technics, *N. Engl. J. Med.*, 257, 497, 1957.
63. **Henle, G., Deinhardt, F., and Girardi, A.,** Cytolytic effects of mumps virus in tissue cultures of epithelial cells, *Proc. Soc. Exp. Biol. Med.*, 87, 386, 1954.
64. **Gresser, I. and Enders, J. F.,** Cytopathogenicity of mumps virus in cultures of chick embryo and human amnion cells, *Proc. Soc. Exp. Biol. Med.*, 107, 804, 1961.
65. **Nagai, Y., Klenk, H.-D., and Rott, R.,** Proteolytic cleavage of the viral glycoproteins and its significance for the virulence of Newcastle disease virus, *Virology*, 72, 494, 1976.
66. **Wheelock, E. F. and Tamm, I.,** Effect of multiplicity of infection on Newcastle disease virus-HeLa cell interaction, *J. Exp. Med.*, 113, 317, 1961.
67. **Kohn, A.,** Polykaryocytosis induced by Newcastle disease virus in monolayers of animal cells, *Virology*, 26, 228, 1965.
68. **Bratt, M. A. and Gallaher, W. R.,** Preliminary analysis of the requirements for fusion from within and fusion from without by Newcastle disease virus, *Proc. Natl. Acad. Sci. U.S.A.*, 64, 536, 1969.
69. **Ishida, N., Homma, M., Osato, T., Hinuma, Y., and Miyamoto, T.,** Persistent infection in HeLa cells with hemadsorption virus type 2, *Virology*, 24, 670, 1964.
70. **Homma, M., Ohira, M., and Ishida, N.,** Specific chromosome aberrations in cells persistently infected with type 2 hemadsorption virus, *Virology*, 34, 60, 1968.
71. **Ebina, T., Takahashi, K., Homma, M., and Ishida, N.,** Alkaline phosphatase activity of HeLa-S3 cells persistently infected with hemadsorption type 2 virus, *Virology*, 39, 597, 1969.
72. **Maeno, K., Yoshii, S., Nagata, I., and Matsumoto, T.,** Growth of Newcastle disease virus in a HVJ carrier culture of HeLa cells, *Virology*, 29, 255, 1966.
73. **Kimura, Y., Ito, Y., Shimokata, K., Nishiyama, Y., Nagata, I., and Kitoh, J.,** Temperature-sensitive virus derived from BHK cells persistently infected with HVJ (Sendai virus), *J. Virol.*, 15, 55, 1975.
74. **Cole, F. E. and Hetrick, F. M.,** Persistent infection of human conjunctiva cell cultures by myxovirus parainfluenza 3, *Can. J. Microbiol.*, 11, 513, 1965.
75. **Choppin, P. W. and Holmes, K. V.,** Replication of SV5 RNA and the effects of superinfection with poliovirus, *Virology*, 33, 442, 1967.
76. **Walker, D. L. and Hinze, H. C.,** A carrier state of mumps virus in human conjuctiva cells. I. General characteristics, *J. Exp. Med.*, 116, 739, 1962.
77. **Henle, W., Henle, G., Deinhardt, F., and Berges, V. V.,** Studies on persistent infection of tissue cultures. IV. Evidence for the production of interferon in MCN cells by myxoviruses, *J. Exp. Med.*, 110, 525, 1959.
78. **Thacore, H. and Youngner, J. S.,** Cells persistently infected with Newcastle disease virus. I. Properties of mutants isolated from persistently infected L cells, *Virology*, 4, 244, 1969.
79. **Preble, O. T. and Youngner, J. S.,** Temperature-sensitive mutants isolated from L cells persistently infected with Newcastle disease virus, *J. Virol.*, 9, 200, 1972.

80. **Youngner, J. S. and Quagliana, D. O.**, Temperature-sensitive mutants isolated from hamster and canine cell lines persistently infected with Newcastle disease virus, *J. Virol.*, 16, 1332, 1975.
81. **Walker, D. L.**, Persistent viral infection in cell cultures, in *Medical and Applied Virology*, Sanders, M. and Lennette, E. H., Eds., Warren H. Green, St. Louis, 1968, 99.
82. **Henle, W.**, Some aspects of persistent viral infections in cell cultures, in *Medical and Applied Virology*, Sanders, M. and Lennette, E. H., Eds., Warren H. Green, St. Louis, 1968, 8.
83. **Chanock, R. M.**, Association of a new type of cytopathogenic myxovirus with infantile croup, *J. Exp. Med.*, 104, 555, 1956.
84. **DeMeio, J. L. and Walker, D. L.**, Demonstration of antigenic relationship between mumps virus and hemagglutinating virus of Japan, *J. Immunol.*, 78, 465, 1957.
85. **Chanock, R. M., Parrott, R. H., Cook, K., Andrews, B. E., Bell, J. A., Reichelderfer, J., Kapikian, A. T., Mastrota, F. M., and Huebner, R. J.**, Newly recognized myxorviruses from children with respiratory disease, *N. Engl. J. Med.*, 258, 207, 1958.
86. **Cook, M. K., Andrews, B. E., Fox, H. H., Turner, H. C., James W. D., and Chanock, R. M.**, Antigenic relationships among the "newer" myxoviruses (parainfluenza), *Am. J. Hyg.*, 69, 250, 1959.
87. **Chanock, R. M., Bell, J. A., and Parrott, R. H.**, Natural history of parainfluenza infection, in *Perspectives in Virology*, Vol. 2, Pollard, M., Ed., Burgess, Minneapolis, 1961, 126.
88. **Smith, C. B., Purcell, R. H., Bellanti, J. A., and Chanock, R. M.**, Protective effect of antibody to parainfluenza type 1 virus, *N. Engl. J. Med.*, 275, 1145, 1966.
89. **Kilham, L., Jungherr, E., and Luginbuhl, R. E.**, Antihemagglutinating and neutralizing factors against Newcastle disease virus (NDV) occurring in sera of patients convalescent from mumps, *J. Immunol.*, 63, 37, 1949.
90. **Jordan, W. S., Jr. and Feller, A. E.**, The relationship of complement-fixing and antihemagglutinating factors against the viruses of mumps and Newcastle disease, *J. Lab. Clin. Med.*, 36, 369, 1950.
91. **Numazaki, Y.**, Common antigens between hemagglutinating virus of Japan (HVJ) and mumps virus, *Uirusu*, 10, 359, 1962 (in Japanese).
92. **Jensen, K. E., Peeler, B. E., and Dulworth, W. G.**, Immunization against parainfluenza infections. Antigenicity of egg adapted type 1 and 3, *J. Immunol.*, 89, 216, 1962.
93. **DeMeio, J. L.**, Adaptation of parainfluenza 2 (group-associated) virus to the embryonated hen's egg, *J. Bacteriol.*, 85, 943, 1963.
94. **Johnson, K. M., Chanock, R. M., Cook, M. K., and Huebner, R. J.**, Studies of a new human hemadsorption virus. I. Isolation, properties and characterization, *Am. J. Hyg.*, 71, 81, 1960.
95. **Craighead, J. E., Cook, M. K., and Chanock, R. M.**, Infection of hamsters with parainfluenza 3 virus, *Proc. Soc. Exp. Biol. Med.*, 104, 301, 1960.
96. **Cook, M. K. and Chanock, R. M.**, In vivo antigenic studies of parainfluenza viruses, *Am. J. Hyg.*, 77, 150, 1963.
97. **Shimokata, K., Nishiyama, Y., and Nagata, I.**, Pathogenicity of parainfluenza viruses in central nervous system, in Abstr. Annu. Meet. Society of Japanese Virologists, Nagoya, Japan, 1976.
98. **Fukumi, H., Nishikawa, F., and Kitayama, T.**, A pneumotropic virus from mice causing hemagglutination, *Jpn. J. Med. Sci. Biol.*, 7, 345, 1954.
99. **Parker, J. C. and Reynolds, R. K.**, Natural history of Sendai virus infection in mice, *Am. J. Epidemiol.*, 88, 112, 1968.
100. **Atoynatan, T. and Hsiung, G. D.**, Epidemiologic studies of latent virus infections in captive monkeys and baboons. II. Serologic evidence of myxovirus infections with special reference to SV5, *Am. J. Epidemiol.*, 89, 472, 1969.
101. **Kalter, S. S., Ratner, J., Kalter, G. V., Rodriguez, A. R., and Kim, C. S.**, A survey of primate sera for antibodies to viruses of human and simian origin, *Am. J. Epidemiol.*, 86, 552, 1967.
102. **Shah, K. V. and Southwick, C. H.**, Prevalence of antibodies to certain viruses in sera of free-living rhesus and of captive monkeys, *Indian J. Med. Res.*, 53, 488, 1965.
103. **Chang, P. W. and Hsiung, G. D.**, Experimental infection of parainfluenza virus type 5 in mice, hamsters and monkeys, *J. Immunol.*, 95, 591, 1965.
104. **Hsiung, G. D., Isacson, P., and McCollum, R. W.**, Studies of a myxovirus isolated from human blood. I. Isolation and properties, *J. Immunol.*, 88, 284, 1962.
105. **Schultz, E. W. and Habel, K.**, SA virus. A new member of the myxovirus group, *J. Immunol.*, 82, 274, 1959.
106. **Swan, C. and Mawson, J.**, Experimental mumps: transmission of the disease to monkeys. Attempts to propagate the virus in developing hen's eggs, *Med. J. Aust.*, 1, 411, 1943.
107. **Kilham, L. and Murphy, H. W.**, Propagation of mumps virus in suckling mice and in mouse embryo tissue cultures, *Proc. Soc. Exp. Biol. Med.*, 80, 495, 1952.
108. **Bolin, V. S., Anderson, J. A., and Leymaster, G. R.**, Experimental intraocular infection of guinea pigs with mumps virus, *Proc. Soc. Exp. Biol. Med.*, 75, 166, 1950.

109. **Burr, M. M. and Nagler, F. P.**, Mumps infectivity studies in hamsters, *Proc. Soc. Exp. Biol. Med.*, 83, 714, 1953.
110. **Johnson, R. T., Johnson, K. P., and Edmonds, C. J.**, Virus-induced hydrocephalus: development of aqueductal stenosis in hamsters after mumps infections, *Science*, 157, 1066, 1967.
111. **Gordon, I., Parvi, K., and Cohen, S. M.**, Response of ferrets to mumps virus, *J. Immunol.*, 76, 328, 1956.
112. **Morris, J. A., Blount, R. E., and McCown, J. M.**, Natural occurrence in dog serum of antibodies against mumps virus, *Cornell Vet.*, 46, 525, 1956.
113. **Brandly, C. A.**, Recognition of Newcastle disease as a new disease, in Newcastle Disease Virus: an evolving Pathogen: Proceedings of an International Symposium, Hanson, R. P., Ed., University of Wisconsin Press, Madison, 1964, 53.
114. **Ginsberg, H. S.**, Mechanism of production of pulmonary lesions in mice by Newcastle disease virus (NDV), *J. Exp. Med.*, 94, 191, 1951.
115. **Liu, C. and Bang, F. B.**, Encephalitis and pneumonia following the intranasal inoculation of Newcastle disease virus in different strains of mice, *Am. J. Hyg.*, 55, 182, 1952.
116. **Reichelderfer, T. E., Chanock, R. M., Craighead, J. E., Nuebner, R. J., Ward, T. G., Turner, H. C., and James, W. D.**, Infection of human volunteers with type 2 hemadsorption virus, *Science*, 128, 779, 1958.
117. **Chanock, R. M., Parrott, R. H., Johnson, K. M., Kapikian, A. Z., and Bell, J. A.**, Myxoviruses: parainfluenza, *Am. Rev. Respir. Dis.*, 88, Suppl. 152, 1963.
118. **Horta-Barbosa, L., Fuccillo, D. A., Sever, J. L., and Zeman, W.**, Subacute sclerosing panencephalitis: isolation of measles virus from a brain biopsy, *Nature (London)*, 221, 974, 1969.
119. **Payne, F. E., Baublis, J. V., and Itabashi, H. H.**, Isolation of measles virus from cell cultures of brain from a patient with subacute sclerosing panencephalitis, *N. Engl. J. Med.*, 281, 585, 1969.
120. **Barbanti-Brodano, G., Oyanagi, S., Katz, M., and Koprowski, H.**, Presence of two different viral agents in brain cells of patients with subacute sclerosing panencephalitis, *Proc. Soc. Exp. Biol. Med.*, 134, 230, 1970.
121. **Lewandowski, L. J., Lief, F. S., Verini, M. M., Penkowski, M. M., Ter-Meulen, V., and Koprowski, H.**, Analysis of a viral agent isolated from multiple sclerosis brain tissue. Characterization as a parainfluenza virus type 1, *J. Virol.*, 13, 1037, 1974.
122. **McNair Scott, T. E.**, Mumps, in *Textbook of Pediatrics*, 8th ed., Nelson, W. E., Ed., W. B. Saunders, Philadelphia, 1964, 556.
123. **Henle, W. and Enders, J. F.**, Mumps virus, in *Viral and Rickettsial Infections of Man*, 4th ed., Horsfall, F. L., Jr. and Tamm, I., Eds., J. B. Lippincott, Philadelphia, 1965, 755.
124. **Penttinen, K., Somer, P., Klemola, E., and Kivalo, E.**, Studies of vaccination with formalized mumps virus vaccine in the Finnish defense focus, *Ann. Med. Exp. Biol. Fenn.*, 32, 248, 1954.
125. **Henle, W., Crawford, M. N., Henle, G., Faz-Tabio, H., Deinhardt, F., Guerra Chabeau, A., and Olshin, I. J.**, Studies on the prevention of mumps. VII. Evaluation of dosage schedules for inactivated mumps vaccine, *J. Immunol.*, 83, 17, 1959.
126. **Henle, G., Henle, W., and Harris, S.**, The serological differentiation of mumps complement-fixation antigens, *Proc. Soc. Exp. Biol. Med.*, 64, 290, 1947.
127. **Henle, G., Harris, S., and Henle, W.**, The reactivity of various human sera with mumps complement fixation antigens, *J. Exp. Med.*, 88, 133, 1948.
128. **Utz, J. P., Szwed, C. F., and Kasel, J. A.**, Clinical and laboratory studies of mumps. II. Detection and duration of excretion of virus in urine, *Proc. Soc. Exp. Biol. Med.*, 99, 259, 1958.

MYXOVIRIDAE: PSEUDOMYXOVIRUS — MEASLES VIRUS

Erling Norrby

Introduction: Measles Virus

Measles has been recognized as an important infectious disease for several centuries. In 1911 the disease was transmitted to experimental animals — macaque monkeys.[1] A useful tissue culture system for propagation of the virus was established in 1954 by Enders and Peebles.[2] This system allowed accumulation of data on structural-functional characteristics of virions and on features of virus-cell interactions. It also allowed the development of an efficient live measles vaccine. The use of this vaccine has markedly reduced the occurrence of measles in some industrialized countries. However, measles still remains a major health problem in many parts of the world, particularly in developing countries. During the last decade it has been realized that measles virus may cause not only acute disease, but also persistent infections. These infections include the uncommon disease subacute sclerosing panencephalitis (SSPE) and possibly other persistant and progressive diseases.

Morphology and Classification

Virions are spherical particles with a diameter ranging from 100 to 250 nm (Figure 1). The particles are comprised of an envelope, covered with projections with a length of 90 to 150 Å, which surrounds a helical ribonucleoprotein. This nucleocapsid (Figure 2) has a diameter of 17 to 18 nm and a length of about 1 μm.

This morphology corresponds to that of other members of the family *Paramyxoviridae* (with the exception of the genus *Pneumovirus*). Measles virus and the serologically related distemper and rinderpest viruses form a separate genus, *Morbillivirus*.[3,4]

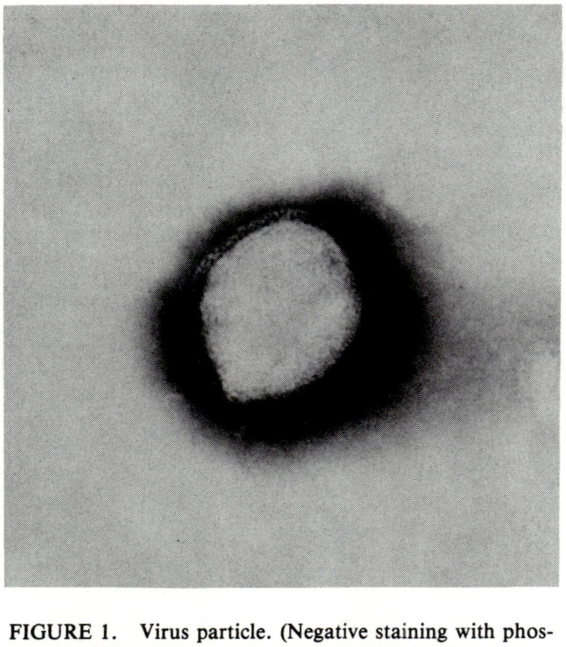

FIGURE 1. Virus particle. (Negative staining with phospho-tungstic acid; Magnification × 270,000.) (Electron micrograph by J. D. Almeida.)

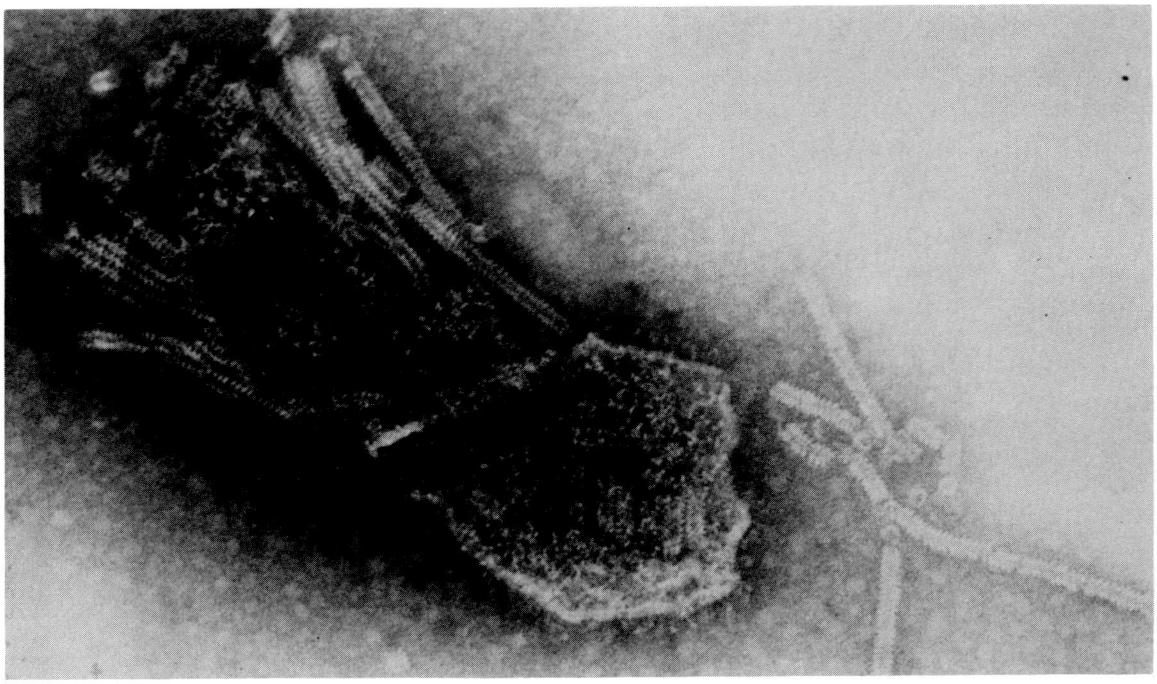

FIGURE 2. Free-lysing nucleocapsid. (Negative staining with phospho-lungstic acid; Magnification × 315,000.) (Electron micrograph by J. D. Almeida.)

Physicochemical properties

Virions contain one piece of single-stranded RNA with a sedimentation rate of 52 S.[5,6] The complex of this RNA with the nucleocapsid protein has a density of 1.305 g/ml in CsCl.[7,8] CsCl density of virions is 1.24 to 1.25 g/ml and for hemagglutinin rosettes prepared by extraction with Tween 80 and ether 1.29 g/ml.[7]

The polypeptide composition of measles virus shows similarities to other paramyxoviruses, but also has certain distinguishing features. The different polypeptides have been summarized from recent publications[9,10] in Table 1.

Infectivity of virions is a labile property and is readily inactivated by physical and chemical agents.[12,13] No marked loss occurs within a range of pH 5 to 10.5[14] At 37°C the half-life of infectivity in the presence of a protein stabilizer is only 2 hr. Exposure to visible light results in inactivation.[15]

Elimination of infectivity can be obtained by heating of freeze-dried material at 95°C for 30 min.[16,16a] Treatments with formalin or Tween 80 and ether, which were used for production of inactivated vaccines, destroy infectivity and also antigenicity of one envelope component, the hemolysin.[17,18] The general sensitivity of different biological activities to inactivating agents is:

Infectivity >hemolytic activity >hemagglutinating activity > complement fixation antigen, immunodiffusion activities etc. > immunogenicity

Virus replication

Measles virus replicates in the cytoplasm of cells. The replication seems not to be dependent on functioning cellular DNA as inferred from the effect of Actinomycin D.[19] The following are specific features of different steps of virus multiplication:

Absorption — Receptors on erythrocytes (and most likely on cytoplasmic membrane

Table 1
POLYPEPTIDES OF MEASLES VIRUS

Designation	Molecular weight (in thousands)	Characteristics
L	180	Unknown function
H	79	Glycosylated, hemagglutinin
P (?)	72	Associated with nucleocapsids; polymerase (?)
NP	60	Nucleoprotein
Ac	43	Cellular actin
F1[a]	40	Not glycosylated; fusion factor (hemolysin)
M	36	Matrix protein
F2[a]	20	Glycosylated

[a] A precursor of F1 and F2, denoted F0 has been assumed to occur in infected cells.[11] After cleavage of F0 all carbohydrates remain in F2.

of nucleated cells) are sensitive to trypsin, but not to neuraminidase treatment.[21] Thus they are different from those of other paramyxoviruses.

Penetration — Fusion between virus envelope and cytoplasmic membrane allows nucleocapsid penetration. Certain tripeptides block virus penetration and other biologic expressions of the same virus component, cell fusion, and hemolysis.[22]

Replication of virus RNA — In infected cells progeny 50S RNA and different species of messenger RNA can be distinguished. Blocking of protein synthesis by cycloheximide interferes with the synthesis of 50S RNA, but causes a threefold increase in synthesis of messenger RNA.[22] The messenger RNA represents a transcript of viron RNA. It contains polyadenylic acid sequences and shows sedimentation characteristics of 12 to 16S and 25 to 36S.[23]

Maturation — Structural antigens accumulate in the cytoplasm and are mainly discernible in the form of "fuzzy" nucleocapsids with a diameter of 30 to 40 nm.[24,25] The fuzzy material most likely represents matrix protein. At later stages certain virus strains give an intranuclear accumulation of nucleocapsids.[26] The morphology of intranuclear and intracytoplasmic nucleocapsids differs (Figure 3). Intranuclear nucleocapsids are smooth and have a diameter of 15 to 17 nm corresponding to that of nucleocapsids released from virions. Other paramyxoviruses, with the exception of the closely related distemper and rinderpest viruses and also of bovine parainfluenza virus type 3, do not accumulate smooth nucleocapsids in the nuclei of the infected cells and further display the occurrence of smooth nucleocapsids instead of fuzzy structures in the cytoplasm. Maturation and release of measles virus occurs by budding through the cytoplasmic membrane in a way corresponding to that of other paramyxoviruses.

Cytopathic changes caused by measles virus in tissue cultures are influenced by properties of the virus strain, the cells used for virus propagation and milieu conditions. Both syncytia formation and changes of polygonal cells into spindle-shaped or stellate forms may occur. In general, rapidly growing nonautointerfering strains show a greater tendency to induce cell fusion.[27] Both intracytoplasmic and, as a special feature, intranuclear inclusions can be identified in stained preparations.

Measles virus may cause both lytic and persistent infections in tissue cultures. Different forms of measles virus-cell interactions have been summarized in Table 2.

Virus Strains and Antigenic Properties

The first isolation of measles virus was made by Enders and Peebles.[2] Examples of

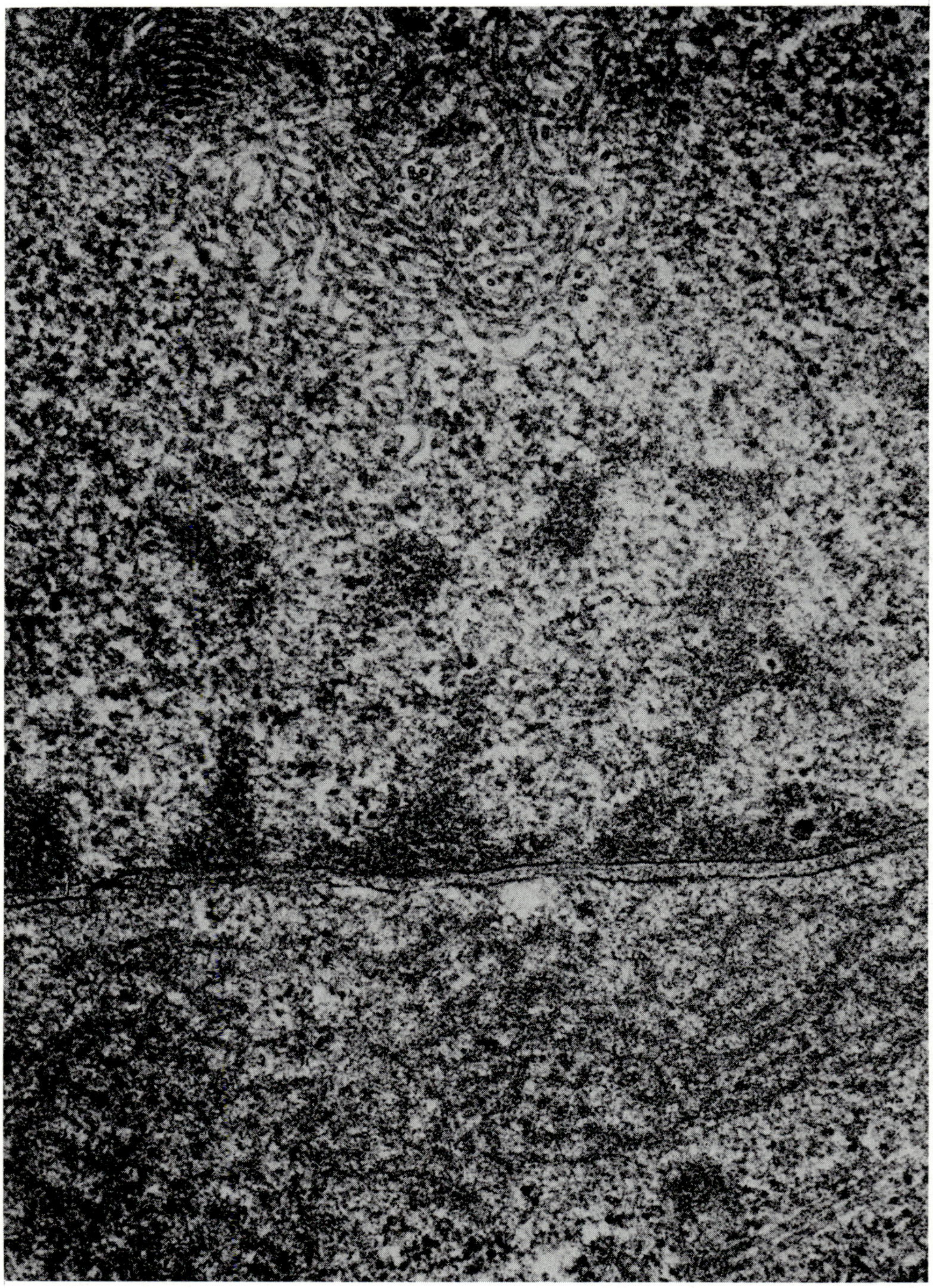

FIGURE 3. Morphology of intranuclear and intracytoplasmic nucleocapsid structures. (Magnification × 70,000.)

Table 2
MEASLES VIRUS-CELL INTERACTION

Form of interaction	Model system	Ref.
Lytic — no autointerference	Edmonston DP (diluted passage) virus in HeLa cells	27
Lytic — certain, autointerference	Edmonston UP (undiluted passage) virus in HeLa cells	27
Carrier infection — production of infectious virus	Edmonston virus adapted to persistent growth in HeLa cells	28, 29
Carrier infection — no production of infectious virus	Virus-producing carrier cells passaged in the presence of antibodies; explants of brain tissue from patients with SSPE	31

virus strains isolated from different kinds of measles virus infections are listed in Table 3.

There is no evidence for an antigenic heterogeneity among measles virus strains. Certain differences in neutralizability of different virus isolates have been described.[38,39]However, these differences do not represent stable biological characteristics and disappear upon passaging in tissue culture systems.

Antigenic properties of virus strains can be characterized by a range of different serological tests (of Table 6). Three structural components, the nucleocapsid, the hemagglutinin and the hemolysin are serologically identifiable. Recent findings[16a] have shown that preparations denoted "nucleocapsids" contain both nucleocapsid and matrix antigen. A comparative characterization of antigens of the three main structural components in different virus isolates including strains from cases of subacute sclerosing panencephalitis (SSPE) have not shown any significant differences. Comparative analysis with distemper and rinderpest viruses show that the main differences concern the hemagglutinin antigen (with detectable biological activity only in measles virus). Cross-reactions can be found between the hemolysin component of measles virus and the analogous components in the other two viruses and the nucleocapsid components of the three viruses appear indistinguishable.[40]

Clinical features

Four different forms of measles infections can be distinguished. These are acute measles and its immediate sequelae pneumonia or encephalitis and finally a late sequelae (SSPE). Some features of these diseases are summarized in Table 4.

The intensity of acute measles varies considerably between individuals. Severe forms occur under conditions of undernourishment which partly relates to an enteric involvement of the disease.[46] The infection heavily engages the lymphatic system as reflected in an abrogation of cell-mediated immune reactions.[47] Both cell-mediated and humoral immunity may prevent reinfection, but presence of the former is obligatory for recovery from acute infection. Individuals devoid of capacity to mount a cell-bound immunity may develop giant cell pneumonia. These patients do not show any rash, which has an immune-pathological origin, and may die from their disease.[41,42]

The symptomatology of acute measles encephalitis is variable both from the quantitative and qualitative point of view.[43,44]In a certain frequency of cases the disease leads to persistent brain damage, but the majority of cases appear to recover without detectable sequelae. SSPE is an uncommon late sequelae of the virus infection.[31,45]It afflicts primarily individuals who have contracted their regular measles infection before the age of 2 years and is three times more common in boys than in girls.[48] The form in which the virus persists in the organism after the primary infection is not known, and the mechanisms for activation of the virus infection has not been defined.

Table 3
MEASLES VIRUS STRAINS AND ORIGINAL ISOLATES

Virus strain	Cell culture; technique	Disease association (for details see section on virus infections)	Ref.
Edmonston	Primary human kidney	Acute measles	2
Various	Various	Acute measles	32
Unnamed	Vero cells	Acute measles encephalitis	33
Unnamed	CV-1 cells; cocultivation	Acute measles encephalitis	34
Unnamed	BS-C-1 cells; cocultivation	SSPE	35
Niigata-1	Explant of human brain (cell-associated virus)	SSPE	36
Unnamed	Explants of human and ferret brain (cell-associated virus)	SSPE	37

Table 4
CLINICAL FEATURES OF MEASLES VIRUS INFECTION

Clinical disease	Characterizing features	Ref.
Acute measles	Droplet infection; 10 days incubation; cough, coryza, conjunctivitis, fever for 4 to 5 days followed by rash	12, 13
Measles pneumonia (giant cell)	Uncommon; progressive pneumonia developing after prodromal symptoms; no rash	41, 42
Measles encephalitis	Frequency about 1/1000; occurs about 5 days after rash, demyelination	43, 44
Subacute sclerosing panencephalitis (SSPE)	Frequency about 1/300,000, occurs an average of 5 to 7 years after regular measles; progressive destruction of white and grey matter; relentlessly progressive	31, 45

Laboratory diagnosis

Samples to be collected for establishment of a laboratory diagnosis vary depending upon the form of the disease to be examined. A summarized presentation is given in Table 5. The acute form of the disease usually is clinically characteristic to the extent that there is no demand for laboratory diagnosis. When needed, attempts can be made to rapidly establish the diagnosis. Cytological examination can be made of cells in nasal secretions.[55] Characteristic intranuclear inclusions may aid in the diagnosis. However, immune fluorescence analysis is of greater value. This can be performed with cells from mucous membranes[56] and in urine[57] and also on skin biopsies.[58] Virus isolation is generally performed from throat washings. Primary human or monkey cells should be used for isolation.[13] The virus may be detected by the appearance of characteristic cytopathic changes, but at an earlier stage, presence of virus antigen can be shown by immune fluorescence or by hemadsorption. Virus isolation usually is not successful later than 1 to 2 days after rash. However, isolations from cells in urine have been made 5 days after rash.[59] Virus isolation in cases of suspected measles pneumonia are performed as in cases of acute measles.

Table 5
LABORATORY DIAGNOSIS OF MEASLES VIRUS INFACTION

Virus disease	Samples for detection of virus antigen or for virus isolation	Samples for serology	Remarks	Ref.
Acute measles	Blood (leukocytes), throat secretions, conjunctival secretions, urine	Acute and convalescent serum	Period of infectivity: prodromal stage until 1 to 2 days after rash; antibody rises occurs at appearance of rash	13
Measles pneumonia	Blood (leukocytes), throat secretions, conjunctival secretions, urine	Acute and convalescent serum		41, 42
Acute measles encephalitis	Brain specimen (biopsy or autopsy specimen), cells in cerebrospinal fluid	Serum and cerebrospinal fluid	In most cases no infectious virus is detectable; occasional local production of antibodies in the central nervous system	49, 50
SSPE	Brain specimen (biopsy or autopsy specimen), cells in cerebrospinal fluid, lymph node biopsy (?)	Serum and cerebrospinal fluid	Virus antigen detected in cerebrospinal fluid cells; virus isolation requires propagation of explant cultures and cocultivation with susceptible cells; hyperimmune antibody response; local production in antibodies in the central nervous system	31, 45, 51-54

Acute measles encephalitis generally is of a postinfectious type and therefore, by definition, virus isolations are negative. However, in exceptional cases virus isolations have been made from brain material or from cells in the cerebrospinal fluid.[33] In one case isolation of virus was made only after propagation of tissue from the central nervous system together with susceptible cells.[34] Cases in which virus isolations have been made, may represent forms of the disease different from the postinfectious encephalitis.

SSPE is caused by a virus which shows a defective interaction with cells in the central nervous system.[31,45] Virus antigen can be identified in cells in cerebrospinal fluid[51] and in brain tissue, but no direct isolation of virus can be made. Explant cultures of brain tissue may carry a persistent virus infection demonstrable by immune fluorescence or by hemadsorption. Release of cell-free virus from explant cultures has been described but represents exceptional circumstances. Generally explanted cells have to be cocultivated with susceptible cells to yield infectious extra-cellular virus particles. However, virus remains in a cell-associated form in a large fraction of the cases examined.

Serological analysis can be performed with a large number of tests which are summarized in Table 6. For practical purposes the hemagglutination-inhibition (HI) test preferably should be used both for detection of antibody rises and for screening of immunity. It is simple to perform and nonspecific inhibitors usually do not cause problems in the test even when this is performed with low serum dilutions. Nonspecific agglutinins can be removed from sera by absorption with green monkey erythrocytes. Corresponding results normally are obtained in HI and in neutralization tests (Nt).

Table 6
SEROLOGICAL TESTS FOR DETECTION OF MEASLES ANTIBODIES

Type of test	Type of antigen with which detectable antibodies react	Optimal conditions for test; remarks	Ref.
Neutralization (Nt)	Hemagglutinin (primarily) and hemolysin	Virus antigen from short time harvest of nonautointerfering strain (diluted passage)	60
Hemolysis-inhibition (HLI)	Hemagglutinin and hemolysin	Green monkey erythrocytes; 37°C; after removal of HI antibodies by absorption with Tween-ether treated antigen, a selective detection of hemolysin-reacting antibodies can be made	17, 60, 61
Hemagglutination-inhibition (HI)	Hemagglutinin	Green monkey erythrocytes; 37°C; Tween-ether treated antigen; special test with salt-dependent hemagglutinin	62, 63
Mixed hemadsorption	Hemagglutinin and hemolysin	Highly sensitive; careful calibration of reagents needed	16a
Immunofluorescence (IF)	Both envelope and internal components	Moderately sensitive; semiquantitative; intranuclear flourescence selectively indicates antibodies against nucleocapsid antigen	
Immune diffusion (ID)	Both envelope and internal components	Moderately sensitive; semiquantitative; treatment of antigen with nonionic detergents allows preferential identification of envelope-reacting antibodies; antibodies against nucleocapsid antigens are demonstrable by use of antigen treated with 0.25% sodium dodecyl sulphate	60
Complement-fixation (CF)	Both envelope and internal components	Moderately sensitive; isolated antigens, e.g., nucleocapsids, may allow detection of antibodies against specific structural components	60

The sensitivity of the two tests frequently is of about the same order of magnitude, but may vary depending upon the set-up in different laboratories. In a small fraction of human sera (less than 10%) antibodies against the second major envelope component, the hemolysin (HLI), dominate quantitatively over HI antibodies.[64] The Nt titers of these sera are comparatively low. It should be noted that HI antibodies, by indirect effects, cause HLI. In order to determine the absolute titers of antibodies against the hemolysin in sera also containing HI antibodies these have to be removed. This can be attained by absorption with Tween 80-ether treated antigen which removes HI antibodies without any concomitant reduction in the titer of antibodies reacting with the hemolysin.[17] Like the HLI test the mixed hemadsorption test can be used for simultaneous detection of antibodies against both hemagglutinin and HLI. However, the usefulness of this test needs to be further evaluated. Immunoflorescence (IF) immunodiffusion (ID), and complement fixation (CF) tests are relatively less sensitive than the HI test and only in the case of CF tests readily quantified. Antibodies of a certain specificity may be determined by use of these tests. Antibodies against the nucleocapsid can be measured by analysis of intranuclear IF or by CF tests employing purified nucleocapsids.

Pathogenicity in experimental animals

Several different species of animals have been used for experimental analysis of mea-

Table 7
PATHOGENICITY OF MEASLES VIRUS IN EXPERIMENTAL ANIMALS

Species	Strain of virus	Type of disease	Ref.
Monkeys	Wild strains or laboratory strains	Generally mild; cynomolgus monkeys more susceptible than rhesus monkeys	65
Monkeys	Cell-associated SSPE virus	SSPE-like condition	66
Calves, lambs	Cell-associated SSPE virus	Diffuse encephalitis	67
Dogs	Cell-associated SSPE virus	Diffuse encephalitis	68
Ferrets	SSPE measles virus; passage only with cell-associated material	Diffuse, nonfatal encephalitis	37, 69
Rats	Neuroadapted strains from hamsters	Nonfatal encephalitis in adult animals; certain persistence of infection	70, 71
Hamsters—newborn	Laboratory strains adapted to growth in hamster brain tissue	Acute necrotizising encephalitis; fatal in 4 to 10 days	72—74
Hamsters—weanling and adult	Laboratory strains adapted to growth in hamster brain tissue	Subclinical to mild infection; persistance of virus shown in some cases	75, 76
Mice	Laboratory strain adapted to growth in cerebral tissue of mice	Acute necrotizising encephalitis; fatal in 4 to 10 days	77, 78

sles infections. These are summarized in Table 7. The only system which allows an analysis of pathogenic events of acute measles infections corresponding to those in humans are infections in seronegative monkeys. However, these infections generally are mild. Several animal systems have been employed to analyze neurotropic characteristics of measles virus strains by adaptation to intracerebral passage of the virus in different types of rodents. Under certain conditions latent measles virus infections in the central nervous system can be established. These latent infections can be activated by treatment with immune-suppressive drugs.[79,80] In addition to rodents certain other animals have been used for characterization of SSPE measles virus strains. Cell-associated SSPE virus may give a diffuse nonfatal encephalitis in ferrets, calves, lambs, dogs, and an SSPE-like condition in monkeys.

GENERAL REFERENCES

Kingsbury, D. W., Bratt, M. A., Choppin, P. W., Hansen, E. P., Hosaka, Y., ter Meulen, V., Norrby, E., Plowright, W., Rott, R., and Wunner, W. H., Paramyxoviridae, *Intervirology,* 10, 137,1978.

Katz, S. L. and Enders, J. F., Measles virus in *Viral and Rickettsial Infections of Man,* 4th ed., Horsfall, F. L., Jr. and Tamm, I., Eds., Lippincott, Philadelphia, 1965, 784.

Katz, S. L. and Enders, J. F., Measles virus, in *Diagnostic Procedures for Viral and Rickettsial Infections,* 4th ed., Lennette, E. H., and Schmidt, N. J., Eds., American Public Health Association, New York, 1969, 504.

Payne, F. E. and Baublis, J. V., Measles virus and subacute sclerosing panencephalitis, *Perspect. Virol.,* 7, 179, 1971.

Miller, H. G., Slanton, J. B., and Gibbons, J., Para-infectious encephalomyelitis and related syndromes. A critical review of the neurological complications of certain specific fevers, *Q. J. Med.,* 25, 427, 1956.

ter Meulen, V., Katz, M., and Muller, D., Subacute sclerosing panencephalitis, *Curr. Top. Microbiol. Immunol.,* 57, 1, 1972.

Norrby, E., Measles virus infections in the central nervous system, in *Dynamic Aspects of Host-parasite Relationships,* Vol. 2, Zuckerman, A., Ed., Keter Press, Jerusalem, 1975.

REFERENCES

1. **Anderson, J. F. and Goldberger, J.**, Experimental measles in the monkey: a preliminary note, *Publ. Health Rep.*, 26, 847, 1911.
2. **Enders, J. F. and Peebles, T. C.**, Propagation in tissue culture of cytopathogenic agents from patients with measles, *Proc. Soc. Exp. Biol. Med.*, 86, 277, 1954.
3. **Fenner, F.**, The classification and nomenclature of viruses, Summary of the results of meeting of the International Committee on Taxonomy, in Madrid, September 5, 1975, *Intervirology,* 6, 1, 1976.
4. **Kingsbury, D. W., Bratt, M. A., Choppin, P. W., Hansen, R. P., Hosaka, Y., ter Meulen, V., Norrby, E., Plowright, W., Rott, R., and Wunner, W. H.**, Paramyxoviridae, *Intervirology,* **10, 137, 1978.**
5. **Schluederberg, A.**, Measles virus RNA, *Biochem. Biophys. Res. Commun.*, 42, 1012, 1971.
6. **Hammarskjöld, B. and Norrby, E.**, A comparison between virion RNA of measles virus and some other paramyxoviruses, *Med. Microbiol. Immunol.*, 160, 99, 1974.
7. **Norrby, E.**, Separation of measles virus components by equilibrium centrifugation in CsCl gradients. I. Crude and Tween and ether treated concentrated tissue culture material, *Arch. Gesamte Virusforsch.*, 14, 306, 1964.
8. **Norrby, E. and Hammarskjöld, B.**, Structural components of measles virus, *Microbios,* 5, 17, 1972.
9. **Mountcastle, W. E. and Choppin, P. W., A comparison of the polypeptides of four measles virus strains,**[78.463.1977.] ***Virology,*86, 254, 1978.**
10. **Tyrrell, D. L. J. and Norrby, E.**, Structural polypeptides of measles virus, *J. Gen. Virology.*, 39, 219, 1978.
11. **Graves, M. C., Silver, S. M., and Choppin, P. W., Measles virus polypeptide synthesis in infected cells, *Virology,* 86, 254, 1978.**
12. **Katz, S. L. and Enders, J. F.**, Measles virus, in *Viral and Rickettsial Infections of Man,* 4th ed., Horsfall, F. L., Jr. and Tamm, I., Eds., Lippincott, Philadelphia, 1965, 784.
13. **Katz, S. L. and Enders, J. F.**, Measles virus, in *Diagnostic Procedures for Viral and Rickettsial Infections,* 4th ed., Lennette, E. H. and Schmidt, N. J., Eds., American Public Health Association, New York, 1969, 504.
14. **Musser, S. J. and Underwood, G. E.**, Studies on measles virus. II. Physical properties and inactivation studies of measles virus, *J. Immunol.*, 85, 292, 1960.
15. **Cutchins, E. C. and Dayhuff, T. R.**, Photoinactivation of measles virus, *Virology,* 17, 420, 1962.
16. **Apostolov, K. and Damjanovic, V.**, Selective inactivation of the infectivity of freeze-dried Sendai virus by heat, *Cryobiology,* 10, 255, 1973.

16a. **Norrby, E.**, unpublished.

17. **Norrby, E. and Gollmar, Y.**, Identification of measles virus-specific hemolysis-inhibiting antibodies separate from hemagglutination-inhibiting antibodies, *Infect. Immunol.*, 11, 231, 1975.
18. **Norrby, E., Enders-Ruckle, G., and ter Meulen, V.**, Differences in the appearance of antibodies to structural components of measles virus after immunization with inactivated and live virus, *J. Infect. Dis.*, 132, 262, 1975.
19. **Hall, W. W. and ter Meulen, V.**, The effects of Actinomycin D on RNA synthesis in measles virus-infected cells, *J. Gen. Virol.*, 34, 391, 1977.
20. **Norrby, E.**, Hemagglutination by measles virus. 2. Properties of the hemagglutinin and of the receptors on the erythrocytes, *Arch. Gesamte Virusforsch.*, 12, 164, 1962.
21. **Norrby, E.**, The effect of a carbobenzoxy tripeptide on the biological activities of measles virus, *Virology,* 44, 599, 1971.
22. **Hall, W. W., Genius, D., and ter Meulen, V.**, The effect of cycloheximide on the replication of measles virus, *J. Gen. Virol.*, 35, 579, 1977.

23. **Hall, W. W. and ter Meulen, V.**, Polyadenylic acid (poly (a)) sequences associated with measles virus intracellular ribonucleic acid (RNA) species, *J. Gen. Virol.*, 35, 497, 1977.
24. **Nakai, M. and Imagawa, D. T.**, Electron microscopy of measles virus replication, *J. Virol.*, 3, 187, 1969.
25. **Nakai, T., Shand, F. L., and Howatson, A. F.**, Development of measles virus in vitro, *Virology*, 38, 50, 1969.
26. **Norrby, E.**, Intracellular accumulation of measles virus nucleocapsid and envelope antigens, *Microbios*, 5, 31, 1972.
27. **Oddo, F. G., Flaccomio, R., and Sinatra, A.**, "Giant-cell" and "Strand-forming" cytopathic effect of measles virus lines conditioned by serial propagation with diluted or concentrated inoculum, *Virology*, 13, 550, 1961.
28. **Rustigian, R.**, Persistent infection of cells in culture by measles virus. 1. Development and characteristics of HeLa sublines persistently infected with complete virus, *J. Bacteriol.*, 92, 1792, 1966.
29. **Norrby, E.**, A carrier cell line of measles virus in Lu106 cells, *Arch. Gesamte. Virusforsch.*, 20, 215, 1967.
30. **Rustigian, R.**, Persistent infection of cells in culture by measles virus. II. Effect of measles antibody on persistently infected HeLa sublines and recovery of a HeLa clonal line persistently infected with incomplete virus, *J. Bacteriol.*, 92, 1805, 1966.
31. **Payne, F. E. and Baublis, J. V.**, Measles virus and subacute sclerosing panencephalitis, *Perspect. Virol.*, 7, 179, 1971.
32. **Matumoto, M.**, Multiplication of measles virus in cell cultures, *Bacteriol. Rev.*, 30, 152, 1966.
33. **Purdham, D. R. and Batty, B. F.**, A case of acute measles meningoencephalitis with virus isolation, *J. Clin. Pathol.*, 27, 994, 1975.
34. **ter Meulen, V., Käckell, Y., Müller, D., Katz, M., and Meyermann, R.**, Isolation of infectious measles virus in measles encephalitis, *Lancet*, 2, 1172, 1972.
35. **Payne, F. E., Baublis, J. V., and Itabashi, H. H.**, Isolation of measles virus from cell cultures of brain from a patient with subacute sclerosing panencephalitis, *N. Engl. J. Med.*, 281, 585, 1969.
36. **Doi, Y., Sampe, T., Nakajima, M., Okawa, S., Katoh, T., Itoh, H., Sato, T., Ohuchi, K., Kumanishi, T., and Tsubaki, T.**, Properties of a cytopathic agent isolated from a patient with subacute sclerosing panencephalitis in Japan, *Jpn. J. Med. Sci. Biol.*, 25, 321, 1972.
37. **Thormar, H., Jervis, G. A., Karl, S. C., and Brown, H. R.**, Passage in ferrets of encephalitogenic cell-associated measles virus isolated from brain of a patient with subacute sclerosing panencephalitis, *J. Infect. Dis.*, 127, 678, 1973.
38. **Horta-Barbosa, L., Fuccillo, D. A., Hamilton, R., Traub, Ley, A., and Sever, J. L.**, Some characteristics of SSPE measles virus, *Proc. Soc. Exp. Biol. Med.*, 134, 17, 1970.
39. **Payne, F. E. and Baublis, J. V.**, Decreased reactivity of SSPE strains of measles virus with antibody, *J. Infect. Dis.*, 127, 505, 1973.
40. **Örvell, C. and Norrby, E.**, Further studies on the immunologic relationships among measles, distemper and rinderpest viruses, *J. Immunol.*, 113, 1850, 1974.
41. **Enders, J. F., McCarthy, K., Mitus, A., and Cheatham, W. J.**, Isolation of measles virus at autopsy in cases of giant cell pneumonia without rash, *N. Engl. J. Med.*, 261, 875, 1959.
42. **Mitus, A., Enders, J. F., Craig, J. M., and Holloway, A.**, Persistence of measles virus and depression of antibody formation in patients with giant-cell pneumonia after measles, *N. Engl. J. Med.*, 261, 882, 1959.
43. **Miller, H. G., Slanton, J. B., and Gibbons, J. L.**, Para-infectious encephalomyelitis and related syndromes. A critical review of the neurological complications of certain specific fevers, *Q. J. Med.*, 25, 427, 1956.
44. **McNair Scott, T. F.**, Postinfectious and vaccinal encephalitis, *Med. Clin. North Am.*, 51, 701, 1967.
45. **ter Meulen, V., Katz, M., and Müller, D.**, Subacute sclerosing panencephalitis. A review, *Curr. Topics Microbiol. Immunol.*, 57, 1, 1972.
46. **Morley, D.**, Severe measles in the tropics, I. *Br. Med. J.*, 1, 297, 1969.
47. **Starr, S. and Berkovich, S.**, Effects of measles, gamma globulin-modified measles and vaccine measles on the tuberculin test, *N. Engl. J. Med.*, 270, 386, 1964.
48. **Jabbour, J. T., Duenas, D. A., Sever, J. L., Krebs, H. M., and Horta-Barbosa, L.**, Epidemiology of subacute sclerosing panencephalitis (SSPE). A report of the SSPE registry, *JAMA*, 220, 959, 1972.
49. **Norrby, E.**, Measles virus infections in the central nervous system, in *Dynamic aspects of host-parasite relationships*, Vol. 2 Zuckerman, A., Ed., Keter Press Jerusalem, 1975.
50. **Sköldenberg, B., Carlström, A., Forsgren, M., and Norrby, E.**, Transient appearance of oligoclonal immunoglobulins and measles virus antibodies in the cerebrospinal fluid in a case of acute measles encephalitis, *Clin. Exp. Immunol.*, 23, 451, 1976.
51. **Dayan, A. D. and Stokes, M. I.**, Immunofluorescent detection of measles-virus antigens in cerebrospinal-fluid cells in subacute sclerosing panencephalitis, *Lancet*, 1, 891, 1971.

52. **Salmi, A. A., Norrby, E., and Panelius, M.,** Identification of different measles virus-specific antibodies in the serum and cerebrospinal fluid from patients with subacute sclerosing panencephalitis and multiple sclerosis, *Infect. Immunol.*, 6, 248, 1972.
53. **Vandvik, B. and Norrby, E.,** Oligoclonal IgG antibody response in the central nervous system to different measles virus antigens in subacute sclerosing panencephalitis, *Proc. Natl. Acad. Sci. U.S.A.*, 70, 1060, 1973.
54. **Vandvik, B., Norrby, E., Nordal, H., and Degré, M.,** Oligoclonal measles virus-specific IgG antibodies isolated by virus immunoabsorbtion of cerebrospinal fluids, brain extracts, and sera from patients with subacute sclerosing panencephalitis and multiple sclerosis, *Scand. J. Immunol.*, 5, 979, 1976.
55. **Mottet, N. K. and Szanton, V.,** Exfoliated measles giant cells in nasal secretions, *Arch. Pathol.*, 72, 434, 1961.
56. **Fulton, R. E. and Middleton, P. J.,** Immunofluorescence in diagnosis of measles infections in children, *J. Pediatr.*, 86, 17, 1975.
57. **Llanes-Rodes, R. and Liu, C.,** Rapid diagnosis of measles from urinary sediments stained with fluorescent antibody, *N. Engl. J. Med.*, 275, 516, 1966.
58. **Olding-Stenkvist, E. and Bjorvatn, B.,** Rapid detection of measles virus in skin rashes by immunofluorescence, *J. Infect. Dis.*, 134, 463, 1976.
59. **Gresser, I. and Katz, S. L.,** Isolation of measles virus from urine, *N. Engl. J. Med.*, 263, 452, 1960.
60. **Norrby, E. and Gollmar, Y.,** Appearance and persistence of antibodies against different virus components after regular measles infections, *Infect. Immunol.*, 6, 240, 1972.
61. **Saburi, Y. and Matomuto, M.,** Assay of measles virus hemolysin and its antibody, *Arch. Gesamte Virusforsch.*, 17, 29, 1965.
62. **Norrby, E.,** Hemagglutination by measles virus. 4. A simple procedure for production of high potency antigen for hemagglutination-inhibition (HI) tests, *Proc. Soc. Exp. Biol. Med.*, 111, 814, 1962.
63. **Schluederberg, A. and Nakamura, M.,** A salt-dependent hemagglutinating particle from measles-infected cells, *Virology*, 33, 297, 1967.
64. **Salmi, A., Gollmar, Y., Norrby, E., and Panelius, M.,** Antibodies against three different structural components of measles virus in patients with multiple sclerosis, their siblings and matched controls, *Acta Pathol. Microbiol. Scand. Sect. B*, 81, 627, 1973.
65. **Peebles, T. C., McCarthy, K., Enders, J. F., and Holloway, A.,** Behavior of monkeys after inoculation of virus derived from patients with measles and propagated in tissue culture together with observations on spontaneous infections of these animals by an agent exhibiting similar antigenic properties, *J. Immunol.*, 78, 63, 1957.
66. **Ueda, S., Otsuka, T., and Okuno, Y.,** Experimental subacute sclerosing panencephalitis (SSPE) in a monkey by subcutaneous inoculation with a defective SSPE virus, *Biken. J.*, 18, 179, 1975.
67. **Thein, P., Mayr, A., ter Meulen, V., Koprowski, H., Käckell, Y. M., Müller, D., and Meyermann, R.,** Subacute sclerosing panencephalitis: transmission of the virus to calves and lambs, *Arch. Neurol.*, 27, 540, 1972.
68. **Katz, M., Käckell, Y. M., Müller, D., ter Meulen, V., and Koprowski, H.,** Immunohistological, microscopical and neurochemical studies on encephalitides. VII. Subacute sclerosing panencephalitis susceptibility of dogs to the virus and characterization of host response, *Acta Neuropathol.*, 25, 81, 1973.
69. **Katz, M., Rorke, L. B., Masland, W. S., Koprowski, H., and Tucker, S. H.,** Transmission of an encephalitogenic agent from brains of patients with subacute sclerosing panencephalitis to ferrets; Preliminary report, *N. Engl. J. Med.*, 279, 793, 1968.
70. **Schumacher, H. P., Albrecht, P., and Tauroso, N. M.,** The effect of altered immune reactivity on experimental measles encephalitis in rats, *Arch. Gesamte Virusforsch.*, 37, 218, 1972.
71. **Byington, D. P. and Burnstein, T.,** Measles encephalitis produced in suckling rats, *Exp. Mol. Pathol.*, 19, 36, 1973.
72. **Waksman, B. H., Burnstein, T., and Adams, R. D.,** Histological study of the encephalomyelitis produced in hamsters by a neurotropic strain of measles, *J. Neuropathol. Exp. Neurol.*, 21, 25, 1962.
73. **Burnstein, T., Jensen, J. H., and Waksman, B. H.,** The development of a neurotropic strain of measles virus in hamsters and mice, *J. Infect. Dis.*, 114, 265, 1964.
74. **Baringer, J. R. and Griffith, J. F.,** Experimental measles virus encephalitis, A light, phase fluorescence and electron microscopic study, *Lab. Invest.*, 23, 335, 1970.
75. **Byington, D. P. and Johnson, K. P.,** Experimental subacute sclerosing panencephalitis in the hamster: correlation of age with chronic inclusion-cell encephalitis, *J. Infect. Dis.*, 126, 18, 1972.
76. **Byington, T. and Johnson, K., Subacute sclerosing panencephalitis (SSPE) agent in hamsters. II. The neuropathology of acute and chronic infections, *Exp. Mol. Pathol.*, 18, 345, 1973.**
77. **Imagawa, D. T. and Adams, J. M.,** Propagation of measles virus in suckling mice, *Proc. Soc. Exp. Biol. Med.*, 98, 567, 1958.

78. **Carlström, G.,** Comparative studies on measles and distemper viruses in suckling mice, *Arch. Gesamte Virusforsch.,* 8, 527, 1959.
79. **Kibler, R. and Cremer, N. E.,** Antithymocyte serum treatment of hamsters inoculated with a measles virus isolated from a patient with subacute sclerosing panencephalitis, *Immunol. Commun.,* 2, 303, 1973.
80. **Johnson, K. P., Feldman, E. G., and Byington, D. P.,** Effect of neonatal thymectomy on experimental subacute sclerosing panencephalitis in adult hamsters, *Infect. Immunol.,* 12, 1464, 1975.

MYXOVIRIDAE: PSEUDOMYXOVIRUS — RESPIRATORY SYNCYTIAL (RS) VIRUS

Erling Norrby

Respiratory Syncytial Virus

Human respiratory syncytial (RS) virus was first isolated in 1957 from young children with respiratory disease.[1] The virus has since been shown to be an important etiological agent of lower respiratory tract disease of varying severity during infancy. Morphologically and antigenically related viruses have been isolated from animals suffering from respiratory disease.

Morphology and Classification

RS virus is a pleomorphic enveloped virus. Spherical particles with a diameter of 100 to 300 μm and filamentous forms with a diameter of 100 nm have been observed by analysis of both stained preparations of infected cells (Figure 1) and free-lying particles using the negative contrast technique.[2-4] The virus envelope surrounds a helical nucleocapsid with a diameter of 13 to 14 nm. RS viruses belong to the family *Paramyxoviridae,* but due to their distinctive morphological features and biological characteristics they form a separate genus named *Pneumovirus*.[5]

Physicochemical Characteristics

RS virions contain RNA of a size corresponding to that of other paramyxoviruses.[6] About seven structural polypeptides have been identified in different studies,[7,8] but there are certain disagreements concerning molecular weights and occurrence of glycosidation. An analysis[9] of 11 different strains of RS virus showed proteins with the molecular weights and functions illustrated in Table 1.

Intact virions have a density in CsCl gradients of 1.23 g/ml[10,11] and in sucrose gradients of 1.18 to 1.19 g/ml.[12] Treatment with Tween 80® and ether gives envelope fragments with a density of 1.28 to 1.30 g/ml in CsCl. Nucleocapsid antigen has been purified[6,13] and has a density of 1.29 to 1.30 g/ml in CsCl.

RS virus infectivity is thermolabile.[14] Attempts have been made to stabilize virus infectivity by addition of sucrose,[15] salts in hypertonic concentrations,[16] and bentonite.[17]

Virus Replication

The molecular events of RS virus replication have been poorly analysed. Virus-specific intracellular RNA species have been examined.[18] The metabolic inhibitors actinomycin and proflavin do not restrict virus replication and may even enhance growth of virus.[18-20] A limited replication of RS virus can take place in enucleated cells. The virus matures by budding through the cytoplasmic membrane. (Figure 1).[2,3] Prevention of glycosidation of viral proteins by addition of 2-deoxy-D-glucose blocks the final stage of morphogenesis.[21] A complete growth cycle of lytic RS virus comprises 20 to 30 hr.[19,20] Presence of antibodies does not restrict the development of syncytia in infected cultures.[24] Temperature-sensitive mutants of RS virus have been described.[25-27] Persistent infections with RS virus have been proposed to be associated with the occurrence of the virus in a DNA provirus form.[28] This finding remains to be confirmed.

Virus Strains and Antigenic Properties

Biological activities of the virus include infectivity, fusion activity, and complement-fixation (CF) antigen activity. No hemolytic, hemagglutinating or neuraminidase activities have been identified.[29] Comparative analysis of strains has been based on CF and

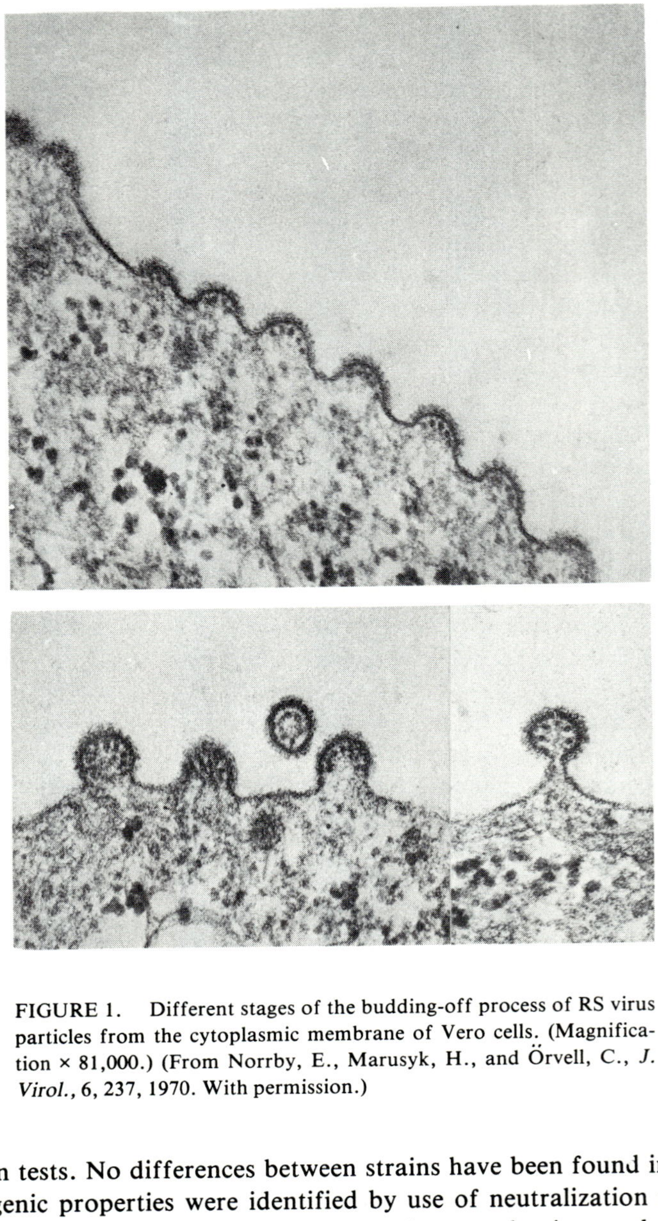

FIGURE 1. Different stages of the budding-off process of RS virus particles from the cytoplasmic membrane of Vero cells. (Magnification × 81,000.) (From Norrby, E., Marusyk, H., and Örvell, C., *J. Virol.*, 6, 237, 1970. With permission.)

neutralization tests. No differences between strains have been found in CF assays but distinct antigenic properties were identified by use of neutralization tests performed either as an end point assay in tubes or as a plaque reduction analysis. Differences between strains are most readily demonstrated by use of hyperimmune sera from rabbits or guinea pigs. Differences can also be found in tests with ferret postinfection sera. Repeated immunization of the animals, however, gives a broadening of the antibody response. Human convalescent sera generally are broad-reacting and do not detect antigenic differences. Different strains have been identified as summarized in Table 2.

Clinical features

RS virus is an important etiological agent in respiratory illnesses. Among hospitalized patients in the younger age groups about 50% of cases of bronchiolitis and 25% of cases of pneumonia are caused by RS virus.[34-39] The peak incidence of severe RS virus infections occur at 2 months of age.[40] The diseases are more common among boys.[40,41] Nosocomial spreading of infections may occur.[42] Occasional cases with rash

Table 1
STRUCTURAL POLYPEPTIDES OF RS VIRUS[9]

Designation[a]	Possible function
200	Possibly of viral origin
VGP 48	Major viral glycoprotein
VP 41	Nucleocapsid polypeptide
VP 38	Unknown function
VP 32	Unknown function. Certain variations in molecular size between strains
VP 27	Matrix polypeptide
VP 25	Unknown function
10	Possibly of viral origin

[a] Different figures denote molecular weights in thousands.

Table 2
ANTIGENIC CHARACTERISTICS OF STRAINS OF RS VIRUS

Virus strain	Year of isolation	Antigen characteristics; sera; neutralization assay	Ref.
Long	1956		1
Randall	1958	Similar to the Long strain	30
CH 18537	1962	Different from the Long strain; postinfection ferret sera; tube neutralization	31
87	1962	Different from the Long strain; rabbit and guinea pig hyperimmune sera; tube neutralization	32
8/60	1960	Different from the Long strain; postinfection ferret sera; syncytia foci reduction	33

have been described.[43] RS virus infections in adults have not been fully characterized. Mild illness of the common cold type has been described.[44-47] The incubation period is 3 to 7 days. Virus shedding from the respiratory tract of children occurs not only during the period of development of symptoms but frequently several days after this.[48] RS virus infections occur world-wide and with characteristic epidemiological patterns. In the northern hemisphere RS virus epidemics primarily occur in the late fall, winter, and spring. Overall, February and March are peak months for RS virus disease.

Reinfections occur with a high frequency.[48,49] Secondary attack rates are in the range of 20 to 30%.[39,46] They are not dependent on the appearance of new antigenic variants. Different variants may occur simultaneously during an epidemic.[50]

The occurrence of RS virus infections during the early years of life is reflected in the accumulation of a high percentage of seropositive individuals in young age groups.[51-53] About half of the children are infected during the first 2 years of life and at the age of 7 to 10 years the frequency of antibody carriers approaches 100%.

Laboratory diagnosis

Samples for virus identification or isolation are collected from the nose or orophar-

Table 3
CELL LINES USED FOR ISOLATION AND PROPAGATION OF RS VIRUS

Type of cell line	Ref.
HeLa, HEp-2	19, 22, 23, 60, 61
Human diploid lung	62
HL (heteroploid human)	63
U (heteroploid human amnion)	64
BS-C-1 (heteroploid monkey kidney)	65
RK 13 (heteroploid rabbit kidney)	66
Diploid bovine testis	58

ynx by a swab or by a washing procedure. A rapid diagnosis can be obtained by direct identification of specific viral antigens in cells by immunofluorescence.[54,55] A potential use of electron microscopy for direct identification of virus has been proposed,[56] and this technique was applied in the study of one epidemic.[57]

Materials to be used for virus isolation (Table 3) must be carefully handled to prevent inactivation of virus infectivity. Preferably samples should not be frozen,[30,58] but snap frozen materials have been used in certain studies.[14,15,59]Optimal recovery of virus is obtained when nasal washings are used for bed side inoculation of the proper cell cultures.[48]

RS virus grows relatively slowly in cells but before development of cytopathic effects infection can be detected by use of immune fluorescence[67,68] and by immunoperoxidase analysis.[69] Methods for preparation of reagents to be used for direct fluorescence antibody diagnosis were recently described.[70]

Developing cytopathic effects in most cases are readily distinguishable. However, in spite of the name of the virus, syncytial formation does not always dominate the cytopathic effects. In certain cell types a round cell degeneration is prevalent.[22] If the inoculum contains a high titer of infectious virus, distinct cytopathic changes can be seen in 4 days.

Serological tests

The double immunodiffusion test has been used only to a limited extent, and for practical purposes neutralization or CF tests are used in routine diagnostic work (Table 4). The neutralization test is more discriminative than the CF test and has therefore been used for separate identification of different strains of RS virus (see above). The neutralizing activity of certain sera can be increased by addition of complement or by use of filtered virus in the test.[74] Maximum sensitivity of the neutralization test is obtained when a plaque reduction assay is used.[50]

Serological analysis can be carried out with paired sera[40,75] and also with secretions from the respiratory tract.[76,77] The fact that the major part of clinically significant RS virus infections occur in children below the age of 6 months poses a certain problem since sera almost always contain maternal antibodies. The neutralization test is more useful than the CF test in detecting antibody rises in children of this age group. However, the sensitivity of the CF test can be improved by increasing the amount of CF antigen from 4 to 8 units.[34,75] Evidence has been presented that it is possible to distinguish between actively and passively acquired antibodies in infant sera on the basis of their fixation pattern in CF tests.[78] In infants of 1 to 3 months of age the level of acute-phase serum neutralizaing antibody was inversely related to the height of serum CF antibody response during convalescence.[40] Thus maternally derived antibody may

Table 4
SEROLOGICAL TESTS FOR DETECTION OF RS VIRUS ANTIBODIES

Type of test	Antigen	Ref.
Neutralization-cytopathic effects in tubes	extra-cellular virus	60
Neutralization-plaque reduction	extra-cellular virus	50, 71
Neutralization-syncytia foci reduction	extra-cellular virus	24
Neutralization-fluorescent foci reduction	extra-cellular virus	72
CF	cell-associated antigen	60
Double immunodiffusion	cell-associated antigen	73

Table 5
PATHOGENICITY OF RS VIRUS IN EXPERIMENTAL ANIMALS

Species	Symptoms	Ref.
Chimpanzee	Coryza	80
Calf	Biphasic pyrexia	81
Ferrets, mink, chinchilla	None	82
Guinea pig	None	82—84
Hamster, mouse adult	None	82
Mouse, suckling	Lethal acute encephalitis	85

suppress the immunological response of infants, a phenomenon of possible relevance for pathogenic events.

The use of lymphocyte transformation as an indicator of cell-mediated immunity to RS virus has been described.[79] However, the usefulness of this test for evaluation of significant immunological parameters needs further evaluation.

Pathogenicity in experimental animals

Studies of different animals used for RS infections are summarized in Table 5. RS virus was originally named chimpanzee coryza agent (CCA) because the first isolation was made from a chimpanzee suffering from a cold-like illness. Chimpanzees still remain the animal of choice for studies of pathogenesis. Ferrets and guinea pigs have been most widely used for preparation of convalescent sera. Both types of animals contract symptomless infections and develop good antibody responses.

GENERAL REFERENCES

Kingsbury, D. W., Bratt, M. A., Choppin, P. W., Hansen, E. P., Hosaka, Y., ter Meulen, V., Norrby, E., Plowright, W., Rott, R., and Wunner, W. H., Paramyxoviridae, *Intervirology,* 10,, 137, 1978.
Chanock, R. M., Kim, H. W., and Brandt, C., Respiratory syncytial virus, in *Viral Infections of Humans: Epidemiology and Control,* Evans, A. S., Ed. Plenum Press, New York, 1976, 365.

Beem, M. and Hamre, D., Respiratory syncytial virus, in *Diagnostic Procedures for Viral and Rickettsial Infections,* 4th ed. Lennette, E. H. and Schmidt, N. J., Eds., American Public Health Association, New York, 1969, Chap. 15.

REFERENCES

1. **Chanock, R. M., Roizman, B., and Myers, R.**, Recovery from infants with respiratory illness of virus related to chimpanzee coryza agent (CCA). I. Isolation, properties and characterization, *Am. J. Hyg.*, 66, 281, 1957.
2. **Norrby, E., Marusyk, H., and Orvell, C.**, Morphogenesis of respiratory syncytial virus in a green monkey kidney cell line (Vero), *J. Virol.*, 6, 237, 1970.
3. **Bächi, T. and Howe, C.**, Morphogenesis and ultrastructure of respiratory syncytial virus, *J. Virol.*, 12, 1173, 1973.
4. **Berthiaume, L., Joncas, J., and Pavilanis, V.**, Comparative structure, morphogenesis and biological characteristics of the respiratory syncytial (RS) virus and the pneumonia virus of mice (PVM), *Arch. Gesamte Virusforsch.*, 45, 39, 1974.
5. **Fenner, F.**, The classification and nomenclature of viruses, *Intervirology*, 6, 1, 1976.

5a. **Kingsbury, D. W., Bratt, M. A., Choppin, P. W., Hansen, E. P., Hosaka, Y., ter Meulen, V., Norrby, E., Plowright, W., Rott, R., and Wunner, W. H.**, Paramyxoviridae, *Intervirology*, 10, 137, 1978.

6. **Bussell, R. H., Waters, D. J., and Seals, M. K.**, Measles, canine distemper and respiratory syncytial virions and nucleocapsids. A comparative study of their structure polypeptide and nucleic acid composition, *Med. Microbiol. Immunol.*, 160, 105, 1974.
7. **Wunner, W. H. and Pringle, C.**, Respiratory syncytial virus proteins, *Virology*, 73, 228, 1976.
8. **Levine, S.**, Polypeptides of respiratory syncytial virus, *J. Virol.*, 21, 427, 1977.
9. **Cash, P., Wunner, W. H., and Pringle, C. R.**, A comparison of the polypeptides of human and bovine respiratory syncytial viruses and murine pneumonia virus, *Virology*, 82, 369, 1977.
10. **Bloth, B. and Norrby, E.**, Fractionation of respiratory syncytial (RS) virus components by centrifugation techniques, *Arch. Gesamte Virusforsch.*, 19, 385, 1966.
11. **Coates, H. V., Forsyth, B. R., and Chanock, R. M.**, Biophysical studies of respiratory syncytial virus. I. Density of respiratory syncytial virus and associated complement-fixing antigens in a cesium chloride density gradient, *J. Bacteriol.*, 91, 1263, 1966.
12. **Cline, G. B., Coates, H., Anderson, N. G., Chanock, R. M., and Harris, W. W.**, Respiratory syncytial virus isolation by combined continuous flow-isopycnic banding centrifugation, *J. Virol.*, 1, 659, 1967.
13. **Senterfit, L. B. and Baldridge, P. B.**, Separation and concentration of respiratory syncytial virus (RSV) antigens, *J. Immunol. Meth.*, 4, 349, 1974.
14. **Hambling, M. H.**, Survival of the respiratory syncytial virus during storage under various conditions, *Br. J. Exp. Pathol.*, 45, 647, 1964.
15. **Law, T. J. and Hull, R. N.**, The stabilizing effect of sucrose upon respiratory syncytial virus infectivity, *Proc. Soc. Exp. Biol. Med.*, 128, 515, 1968.
16. **Rechsteiner, J.**, Thermal inactivation of respiratory syncytial virus in water and hypertonic solutions, *J. Gen. Virol.*, 5, 397, 1969.
17. **Klein, J. D., Strangert, K., and Collier, A. M.**, Stability of respiratory syncytial virus in a new bentonite holding medium, *J. Clin. Microbiol.*, 1, 534, 1975.
18. **Hodes, D. S., Schauf, V., and Chanock, R. M.**, Isolation of RNA species from HeLa cells infected with respiratory syncytial virus, *Proc. Soc. Exp. Biol. Med.*, 146, 287, 1974.

19. **Hornsleth, A.,** Growth of respiratory syncytial (RS) virus in HEp-2 cell cultures, *Acta Pathol. Microbiol. Scand.,* 76, 637, 1969.
20. **Jong, J. C. de and Harmsen, M.,** Stimulated growth of respiratory syncytial virus in the presence of actinomycin D, *Antonie van Leeuwenhoek; J. Microbiol. Ser.,* 39, 409, 1973.
21. **Schnitzler, T. J., Hodes, D. S., Gerin, J., Camargo, E., and Chanock, R. M.,** Effect of 2-deoxy-D-glucose and glucosamin on the growth and function of respiratory syncytial and parainfluenza 3 viruses, *Virology,* 67, 306, 1975.
22. **Jordan, W. S., Jr.,** Growth characteristics of respiratory syncytial virus, *J. Immunol.,* 88, 581, 1962.
23. **Levine, S. and Hamilton, R.,** Kinetics of the respiratory syncytial virus growth cycle in HeLa cells, *Arch. Gesamte Virusforsch.,* 28, 122, 1969.
24. **Taylor-Robinson, D. and Doggett, J. E.,** An assay method for RS virus. *Br. J. Exp. Pathol.,* 44, 473, 1963.
25. **Gharpure, M., Wright, P., and Chanock, R.,** Temperature-sensitive mutants of respiratory syncytial virus (RSV), *J. Virol.,* 3, 414, 1969.
26. **Wright, P. F., Gharpure, M. A., Hodes, D. S., and Chanock, R. M.,** Genetic studies of respiratory syncytial virus temperature-sensitive mutants, *Arch. Gesamte Virusforsch.,* 41, 238, 1973.
27. **Faulkner, G. P., Shirodaria, P. V., Follett, E. A. C., and Pringle, C. R.,** Respiratory syncytial virus ts mutants and nuclear immunofluorescence, *J. Virol.,* 20, 487, 1976.
28. **Simpson, R. W. and Iinuma, M.,** Recovery of infectious proviral DNA from mammalian cells infected with respiratory syncytial virus, *Proc. Natl. Acad. Sci. U.S.A.,* 72, 3230, 1975.
29. **Richman, A. V., Pereira, F. A., and Tauriso, N.,** Attempts to demonstrate haemagglutination and haemadsorption by respiratory syncytial virus, *Appl. Microbiol.,* 21, 1099, 1971.
30. **Beem, M. O., Wright, F. H., Hamre, P., Egerer, R., and Oehme, M.,** Association of the chimpanzee coryza agent with acute respiratory disease in children, *N. Engl. J. Med.,* 263, 523, 1960.
31. **Coates, H. V., Kendrick, L., and Chanock, R. M.,** Antigenic differences between two strains of RS virus, *Proc. Soc. Exp. Biol. Med.,* 112, 958, 1963.
32. **Wulff, H., Kidd, P., and Wenner, H. A.,** Respiratory syncytial virus: observations on antigenic heterogeneity, *Proc. Soc. Exp. Biol. Med.,* 115, 240, 1964.
33. **Doggett, J. E. and Robinson, D. T.,** Serological studies with respiratory syncytial virus, *Arch. Gesamte Virusforsch.,* 15, 601, 1965.
34. **Chanock, R. M., Kim, H. W., Vargosko, A. J., Delem, A. J., Johnson, K. M., Cumming, C., and Parrott, R. H.,** Respiratory syncytial virus. I. Virus recovery and other observations during 1960 outbreak of bronchiolitis, pneumonia and minor respiratory diseases in children, *JAMA,* 176, 647, 1961.
35. **McClelland, L., Hilleman, M. R., Hamparian, V. V., Ketler, A., Reilly, C. M., Cornfeld, D., and Stokes, J.,** Studies of acute respiratory illness caused by respiratory syncytial virus. 2. Epidemiology and assessment of importance, *N. Engl. J. Med.,* 264, 1169, 1961.
36. **Parrott, R. H., Vargosko, A. J., Kim, H. W., Cumming, C., Turner, H., Huebner, R. J., and Chanock, R. M.,** Respiratory syncytial virus. II. Serologic studies over a 34-month period of children with bronchiolitis, penumonia and minor respiratory diseases, *JAMA,* 176, 653, 1961.
37. **Jacobs, J. W., Peacock, D. B., Corner, B. D., Ceul, E. O., and Clarke, S. K. K.,** Respiratory syncytial and other viruses associated with respiratory disease in infants, *Lancet,* 1, 871, 1971.
38. **Kim, H. W., Arrobio, J. O., Brandt, C. D., Jeffries, B. C., Pyles, G., Reid, J. L., Chanock, R. M., and Parrott, R. H.,** Epidemiology of respiratory syncytial virus infection in Washington, D. C. I. Importance of the virus in different respiratory tract disease syndromes and temporal distribution of infection, *Am. J. Epidemiol.,* 98, 216, 1973.
39. **Chanock, R. M., Kim, H. W., and Brandt, C.,** Respiratory syncytial virus, in *Viral Infections of Humans: Epidemiology and Control,* Evans, A. S., Ed. Plenum Press, New York, 1976, 365.
40. **Parrott, R. H., Kim, H. W., Arrobio, J. A., Hodes, D. S., Murphy, B. R., Brandt, C. D., Camergo, E., and Chanock, R. M.,** Epidemiology of respiratory syncytial virus infection in Washington, D. C. II. Infection and disease with respect to age, immunologic status, race and sex. *Am. J. Epidemiol.,* 98, 289, 1973.
41. **Rose, C. A. C., Pinkerton, J. W., and Assaad, F. A.,** Pathogenesis of respiratory syncytial virus diseases in infancy, *Arch. Dis. Child.,* 46, 702, 1971.
42. **Breese-Hall, C., Douglas, R. G., Jr., Geiman, J. M., and Messner, M. K.,** Nosocomial respiratory syncytial virus infections, *N. Engl. J. Med.,* 293, 1343, 1975.
43. **Berkovich, S. and Kibrick, S.,** Exanthem associated with respiratory syncytial virus infection, *J. Pediatr.,* 65, 368, 1964.
44. **Kravetz, H. M., Knight, V., Chanock, R. M., Morris, J. A., Johnson, K. M., Rifkind, D., and Utz, J. P.,** Respiratory syncytial virus. III. Production of illness and clinical observations in adult volunteers, *JAMA,* 176, 657, 1961.

45. **Johnson, K. M., Bloom, H. H., Mufson, M. A., and Chanock, R. M.,** Natural reinfection of adults by respiratory syncytial virus. Possible relation to mild upper respiratory disease, *N. Engl. J. Med.,* 267, 68, 1962.
46. **Hall, C. B., Geiman, J. M., Biggar, R., Kotch, D. I., Hogan, P. M., and Douglas, R. G., Jr.,** Respiratory syncytial virus infections within families, *N. Engl. J. Med.,* 294, 414, 1976.
47. **Cooney, M. K., Fox, J. P., and Hall, C. E.,** The Seattle virus watch. VI. Observations of infections with an illness due to parainfluenza, mumps, and respiratory syncytial viruses, and Mucoplasma pneumoniae, *Am. J. Epidemiol.,* 101, 532, 1975.
48. **Hall, C. B., Douglas, R. G., Jr., and Geiman, J. M.,** Quantitative shedding patterns of respiratory syncytial virus in infants, *J. Infect. Dis.,* 132, 151, 1975.
49. **Beem, M.,** Repeated infections with respiratory syncytial virus, *J. Immunol.,* 98, 1115, 1967.
50. **Coates, H. V., Alling, D. W., and Chanock, R. M.,** An antigenic analysis of respiratory syncytial virus isolates by a plaque reduction neutralization test, *Am. J. Epidemiol.,* 83, 299, 1964.
51. **Moss, P. D., Adams M. O., and Tobin, J. O. H.,** Serological studies with respiratory syncytial virus, *Lancet,* 1, 298, 1963.
52. **Beem, M., Egerer, R., and Anderson, J.,** Respiratory syncytial virus neutralizing antibodies in persons residing in Chicago, *Pediatrics,* 34, 761, 1964.
53. **Hambling, M. H.,** A survey of antibodies to respiratory syncytial virus in the population, *Br. Med. J.,* 1, 1223, 1964.
54. **McQuillin, J. and Gardner, P. S.,** Rapid diagnosis of respiratory syncytial virus infection by immunofluorescent antibody techniques, *Br. Med. J.,* 1, 602, 1968.
55. **Gray, K. G., MacFarlane, D. E., and Sommerville, R. G.,** Direct immunofluorescent identification of respiratory syncytial virus in throat swabs from children with respiratory illness, *Lancet,* 1, 446, 1968.
56. **Joncas, J. H., Berthiaume, L., Williams, R., Beaudry, P., and Pavilanis, V.,** Diagnosis of viral respiratory infections by electron microscopy, *Lancet,* 1, 956, 1969.
57. **Morrell, R. E., Marks, M. J., Champlin, R., and Spence, L.,** An outbreak of severe pneumonia due to respiratory syncytial virus in isolated arctic population, *Am. J. Epidemiol.,* 101, 231, 1975.
58. **Caul, E. O., Jacobs, J. W., and Clarke, S. K. R.,** Bovine testis cells for routine isolation of respiratory syncytial virus from infants, *J. Med. Microbiol.,* 7, 301, 1974.
59. **Suto, T., Yano, N., Ikeda, M., Miyamoto, M., Takai, S., Shigeta, S., Hinuma, Y., and Ishida, N.,** Respiratory syncytial virus infection and its serologic epidemiology, *Am. J. Epidemiol.,* 82, 211, 1966.
60. **Beem, M. and Hamre, D.,** Respiratory syncytial virus, in *Diagnostic Procedures for Viral and Rickettsial Infections,* 4th ed. Lennette, E. H. and Schmidt, N. J., Eds., American Public Health Association, New York, 1969, Chap. 15.
61. **Peacock, D. B. and Clarke, S. K. R.,** Respiratory syncytial virus in Britain, *Lancet,* 2, 466, 1961.
62. **Anderson, J. M. and Beem, M. O.,** Use of human diploid cell cultures for primary isolation of respiratory syncytial virus, *Proc. Soc. Exp. Biol. Med.,* 121, 205, 1966.
63. **Cavallero, J. J. and Monto, A. S.,** HL cells, a sensitive line for the isolation and propagation of respiratory syncytial virus, *Proc. Soc. Exp. Biol. Med.,* 140, 507, 1972.
64. **Berglund, B. and Strahlmann, C. H.,** Respiratory syncytial virus infections in hospitalised children: evaluation of virus isolation and complement fixation techniques in the virological diagnosis: clinical and epidemiological characteristics, *Acta Paediatr. Scand.,* 56, 269, 1967.
65. **Berglund, B., Vihma, C., and Wickstrom, J.,** Respiratory syncytial virus studies on children hospitalized during an outbreak of respiratory illness in Finland, *Am. J. Epidemiol.,* 81, 271, 1965.
66. **MacWilliam, K.,** Growth of respiratory syncytial virus in RK 13 cells, *J. Clin. Pathol.,* 27, 432, 1974.
67. **Schieble, J. H., Lennette, E. H., and Kase, A.,** An immunofluorescent staining method for rapid identification of respiratory syncytial virus, *Proc. Soc. Exp. Biol. Med.,* 120, 203, 1965.
68. **Gardner, P. S. and McQuillin, J.,** The application of the immunofluorescent antibody technique in the rapid diagnosis of respiratory syncytial virus infection, *Br. Med. J.,* 2, 340, 1968.
69. **Benjamin, D. R. and Ray, C. G.,** Use of immunoperoxidase for the rapid identification of human myxoviruses and paramyxoviruses in tissue culture, *Appl. Microbiol.,* 28, 47, 1971.
70. **Lyerla, H. C. and Hierholzer, J. C.,** Physicochemical and serological characteristics of respiratory virus fluorescein-isothiocyanate conjugates for fluorescent-antibody diagnosis, *J. Clin. Microbiol.,* 1, 451, 1975.
71. **Kisch, A. L. and Johnson, K. M.,** A plaque assay for respiratory syncytial virus, *Proc. Soc. Exp. Biol. Med.,* 112, 583, 1963.
72. **Schieble, J. H., Kase, A., and Lennette, E. H.,** Fluorescent cell counting as an assay method for respiratory syncytial virus, *J. Virol.,* 1, 494, 1967.
73. **Mäntyjarvi, R., Berglund, B., and Salmi, A.,** Precipitating-antibody response of guinea-pigs to respiratory syncytial virus, *Acta Virol.,* 11, 467, 1967.

74. **Baughman, R. H., Fenters, J. D., Marquis, G. S., Jr., and Holper, J. C.,** Effect of complement and viral filtration on the neutralization of respiratory syncytial virus, *Appl. Microbiol.,* 16, 1076, 1968.
75. **Ross, C. A. C., Stott, E. J., McMichael, S., and Crowther, I. A.,** Problems of laboratory diagnosis of respiratory syncytial virus infection in childhood, *Arch. Gesamte Virusforsch.,* 14, 553, 1964.
76. **Kim, H. W., Bellanti, J. A., Arrobio, J. O., Mills, J., Brandt, C. D., Chanock, R. M., and Parrott, R. H.,** Respiratory syncytial virus neutralizing activity in nasal secretions following natural infections, *Proc. Soc. Exp. Biol. Med.,* 131, 658, 1969.
77. **Scott, R. and Gardner, P. S.,** The local antibody response to RS virus infection in the respiratory tract, *J. Hyg.,* 72, 111, 1974.
78. **Jacobs, J. W. and Peacock, D. B.,** Differentiation of actively and passively acquired complement-fixing antibodies in infants with respiratory syncytial virus infection, *J. Med. Microbiol.,* 3, 313, 1970.
79. **Kim, H. W., Leikin, S. L., Arrobio, J., Brandt, C. D., Chanock, R. M., and Parrott, R. H.,** Cell-mediated immunity to respiratory syncytial virus induced by inactivated vaccine or by infection, *Pediatr. Res.,* 10, 75, 1976.
80. **Morris, J. A., Blount, R. E., Jr., and Savage, R. E.,** Recovery of cytopathogenic agent from chimpanzees with coryza, *Proc. Soc. Exp. Biol. Med.,* 92, 544, 1956.
81. **Jacobs, J. W. and Edington, N.,** Experimental infection of calves with respiratory syncytial virus, *Res. Vet. Sci.,* 18, 299, 1975.
82. **Coates, H. V. and Chanock, R. M.,** Experimental infection with respiratory syncytial virus in several species of animals, *Am. J. Hyg.,* 76, 302, 1962.
83. **Hambling, M. H.,** Antibody response in guinea pigs following intranasal inoculation of respiratory syncytial virus, *J. Pathol. Bacteriol.,* 91, 625, 1966.
84. **Berglund, B., Mantyjarvi, R., and Salmi, A.,** Neutralizing, complement-fixing and precipitating antibody response of guinea pigs to intranasal and intracardiac respiratory syncytial virus inoculations, *Ann. Med. Exp. Fenn.,* 45, 193, 1967.
85. **Cavallaro, J. J. and Maassab, H. F.,** Adaptation of respiratory syncytial (RS) virus to brain of suckling mice, *Proc. Soc. Exp. Biol. Med.,* 121, 37, 1966.

Index

INDEX*

A

* (1) indicates Volume I, Part 1. (2) indicates Volume I, Part 2.

D

E

F

G

H

I

J

K

L

M

N

O

P

Q

R

S

T

U

V

W

Y